AF682234

Fundamentals of Basin and Petroleum Systems Modeling

Thomas Hantschel · Armin I. Kauerauf

Fundamentals of Basin and Petroleum Systems Modeling

RWTH AACHEN
HOCHSCHULBIBLIOTHEK
Ausgesondert

Springer

Dr. Thomas Hantschel
Integrated Exploration
Systems GmbH
A Schlumberger Company
Ritterstr. 23
52072 Aachen
Germany
thantschel@slb.com

Dr. Armin I. Kauerauf
Integrated Exploration
Systems GmbH
A Schlumberger Company
Ritterstr. 23
52072 Aachen
Germany
akauerauf@slb.com

ISBN 978-3-540-72317-2 e-ISBN 978-3-540-72318-9
DOI 10.1007/978-3-540-72318-9
Springer Dordrecht Heidelberg London New York

Library of Congress Control Number: Applied for

© Springer-Verlag Berlin Heidelberg 2009
This work is subject to copyright. All rights are reserved, whether the whole or part of the material is concerned, specifically the rights of translation, reprinting, reuse of illustrations, recitation, broadcasting, reproduction on microfilm or in any other way, and storage in data banks. Duplication of this publication or parts thereof is permitted only under the provisions of the German Copyright Law of September 9, 1965, in its current version, and permission for use must always be obtained from Springer. Violations are liable to prosecution under the German Copyright Law.
The use of general descriptive names, registered names, trademarks, etc. in this publication does not imply, even in the absence of a specific statement, that such names are exempt from the relevant protective laws and regulations and therefore free for general use.

Cover design: Bauer, Thomas

Printed on acid-free paper

Springer is part of Springer Science+Business Media (www.springer.com)

Preface

It is with great satisfaction and personal delight that I can write the foreword for this book *Fundamentals of Basin and Petroleum Systems Modeling* by Thomas Hantschel and Armin Ingo Kauerauf. It is a privilege for us geoscientists that two outstanding physicists, with scientific backgrounds in numerical methods of continuum-mechanics and in statistical physics respectively could be won to deeply dive into the numerical simulation of complex geoprocesses. The keen interest in the geosciences of Thomas Hantschel and Armin I. Kauerauf and their patience with more descriptive oriented geologists, geochemists, sedimentologists and structural geologists made it possible to write this book, a profound and quantitative treatment of the mathematical and physical aspects of very complex geoprocesses. In addition to their investigative interest during their patient dialogue with afore mentioned geological specialists Thomas Hantschel and Armin I. Kauerauf gained a great wealth of practical experience by cooperating closely with the international upstream petroleum industry during their years with the service company IES, Integrated Exploration Systems. Their book will be a milestone in the advancement of modern geosciences.

The scientific and the practical value of modern geosciences rests to a large degree upon the recognition of the complex interrelationship of individual processes, such as compaction, heat-, fluid- and mass-flow, reaction kinetics etc. and upon the sequential quantification of the entire process chain. The intelligent usage of modern high speed computers made all this possible.

Basin modeling was for many years considered as "a niche discipline", mainly propagated and used by geochemists. What a fundamental error and misunderstanding! The absolute contrary is the truth. Basin modeling integrates practically all geoscientific disciplines, it allows an unprecedented quantitative understanding of entire process chains and it detects quickly inconsistencies or uncertainties in our knowledge base. In short, the basin modeling–approach is a big step forward in modern geosciences. This book is a challenge for academic teachers in the geosciences and likewise for scientists and engineers in the petroleum and mining industry. The challenge is to educate much more than in the past the younger ones among us to be able to walk along the borderline between the exact sciences with a physcial and mathematical background and the classical geosciences and vice versa.

In 1984 Prof. Bernard Tissot and I wrote in the Preface of the second edition of our book *Petroleum Formation and Occurrence*: "It is evident that computer modeling is here to stay, and may very well revolutionize the field. The computer can be used as an experimental tool to test geological ideas and hypotheses whenever it is possible to provide adequate software for normally very complicated geological processes. The enormous advantages offered by computer simulation of geological processes are that no physical or physico-chemical principles are violated and that for the first time the geological time factor, always measured in millions of years rather than in decades, can be handled with high speed computers with large memories. Thus, the age of true quantification in the geosciences has arrived. We believe that this computer-aided, quantitative approach will have an economic and intellectual impact on the petroleum industry, mainly on exploration." All this indeed is the case now. And even more so, basin modeling enhances and deepens the intelligent interpretation of geological data acquired by geophysical, geological and geo-chemical methods and thus converts static information into dynamic process understanding.

I congratulate the two authors for their excellent textbook. I urge the geoscientific community to dig into the wealth of scientific information offered in this book. It will help us to understand and quantify the dynamics occurring in the subsurface.

Dietrich Welte

In the late 1970s "Basin Modeling" was introduced as the term describing the quantitative modeling of geological processes in sedimentary basins on geological timescales. At that time basin models found their main application in heat and pore water flow modeling with regard to sediment compaction and temperature controlled chemistry of hydrocarbon generation. Since then geological, chemical, and transport related models have much improved. Basin modeling turned into a complex and integrated framework of many processes, such as multiphase fluid flow for hydrocarbon migration and accumulation, advanced reaction schemes for organic and mineral transformations or compressional and extensional tectonics.

The term "Basin Modeling" is not only used for the modeling of processes in sediments, but also for the modeling of crustal and mantle heat and mass flow processes to predict the sedimentary basin type and the related tectonic subsidence. We prefer the naming "Crustal Models"' for this type of analysis. Obviously, processes in the crust are tightly linked to the sedimentary basin and hence integrated basin and crustal models have also been developed.

In addition to pure scientific research there has always been a commercial motivation for basin modeling as a means to understand, quantify and predict petroleum repositories. From the start, the petroleum industry has been the main sponsor for the development of basin modeling tools for exploration and resource assessment. Over time, a number of specialized tools and different types of basin modeling simulators have been developed and with them new terminologies have been introduced, such as "Petroleum Systems Modeling", "Exploration Risk Assessment" or "Prospect and Play Analysis".

We, the authors of this book, are both physicists with a focus on numerical modeling and software design. Since 1990 and 1997 respectively, we have developed major parts of various generations of the commercial basin simulation software PetroMod®. Furthermore, we have offered many training courses on the subject of the theory and fundamental principles behind basin modeling. The training courses contain a fair amount of mathematics, physics and chemistry – the basic building blocks of the software tools. A complete simulation of an actual geological basin often displays complex fluid flow and accumulation patterns which are difficult to interpret. We believe that a basic understanding of the theory behind the tools is essential to master the models in detail.

Most basin modelers, in scientific research institutions or the petroleum industry, are expert geologists, coming from an entirely different academic domain. They may therefore be unfamiliar with the mathematics and quantitative science related to the software. This results in an abundance of excellent literature about basin modeling from the geological point of view but no comprehensive study regarding mathematics, physics and computer science.

The book is intended above all as an introduction to the mathematical and physical backgrounds of basin modeling for geologists and petroleum explorationists. Simultaneously, it should also provide (geo)physicists, mathematicians and computer scientists with a more in–depth view of the theory behind the models. It is a challenge when writing for an interdisciplinary audience to find the balance between the depth and detail of information on the one hand and the various educational backgrounds of the readers on the other. It is not mandatory to understand all of the details to comprehend the basic principles. We hope this book will be useful for all parties.

With this work we also wanted to create a handbook offering a broad picture of the topic, including comprehensive lists of default values for most parameters, such as rock and fluid properties and geochemical kinetics. We hope that our compilation will ease the work of many modelers. The book is not intended as an introduction to the geological principles of basin formation nor as a tutorial to practical basin modeling. Case studies have not been included. A second volume focusing on case studies and the practical aspects of the application is planned for the future.

Experts in sedimentology, petrology, diagenesis, fault seal analysis, fracturing, rock mechanics, numerics, and statistics may find the approach to some topics in this book too simplistic, but we deliberately came to the decision to open the book to a broader interdisciplinary understanding. At the same time we also feel that we present in many instances ideas which could inspire further studies.

The main focus has been on numerical models and features. Naturally, there is a tendency to focus on features which we ourselves developed for PetroMod®, but most of the basic models are also applicable for other academic and commercial software programs. Since there are not many publications by other development groups about the fundamentals, theory and parameters of their work, we were often unable to include appropriate references in our discussion.

Basin modeling is a multi–disciplinary science. We hope that students, researchers and petroleum explorers with very different experiences will benefit from the presented work.

Acknowledgments

We wish to acknowledge our families who tolerated many long hours on the computer, in the evening and at the weekend.

IES, which recently became part of Schlumberger, was very generous in supporting this work. It is hardly conceivable for us to write such a book without the infrastructure, support and friendly atmosphere at IES.

We appreciate the IES geologists, who tried to teach us basics about geology and supplied us continuously with test models and data. Many examples in this volume can be attributed to their work.

Special thanks go to the IES software development team, which provided us among others with excellent software for building and analyzing basin models.

We are indebted to Thomas Fuchs, Michael de Lind van Wijngaarden and Michael Fücker for many interesting discussions. Our understanding was again and again improved, in general and in detail.

It must be pointed out, that many rock data values and correlations which were not published up to now are originally from Doug Waples. He did a great job of collecting data over many years.

Special acknowledgments are due to the patient referees Michael Hertle, Bernd Krooss, Tim Matava, Ken Peters, Øyvind Sylta, René Thomson, Doug Waples, Dietrich Welte, Michael de Lind van Wijngaarden, Bjorn Wygrala and Gareth Yardley. Hopefully, we did not stress them too much.

Chris Bendall and Katrin Fraenzel checked spelling and grammar. Thanks for the hard job.

Finally we should gratefully mention all colleagues, customers and friends who accompanied us during the last years. We wanted to avoid missing a person, so we abdicated a huge list.

Further special acknowledgments from Armin are due Dr. Gerich–Düsseldorf, Dr. Kasparek and Dr. Schäfer who diagnosed a serious disease in April 2008. Prof. Dr. Autschbach and his team at the RWTH–Klinikum were able to save Armin's life in an emergency surgery. Many thanks. Unfortunately, this delayed the appearance of the book at least for two additional months and we have a suitable justification for some typos.

December, 2008 Thomas Hantschel and Armin I. Kauerauf

Contents

1

Introduction to Basin Modeling

1.1 History

Geology and geochemistry in sedimentary basins have been established sciences for centuries. Important textbooks, such as Tissot and Welte (1984); Hunt (1996); Gluyas and Swarbrick (2004); Peters et al. (2005); Allen and Allen (2005), summarize the knowledge especially related to petroleum geosciences.

The first basin modeling computer programs were developed around 1980 (Yükler et al., 1979). The main concept encompassed multi–1D heat flow simulation and subsequent geochemical models to construct petroleum generation and expulsion maps for the evaluation of source rock maturity. One of the key tasks was to calculate and calibrate the temperature history during the evolution of a geological basin. Heat flow calculation is one of the best investigated problems in applied engineering. A formulation and solution of the corresponding differential equations can be easily achieved. Once the paleo–temperatures were known, equations for chemical kinetics could be used to evaluate the cracking rates of petroleum generation. Another important part of the analysis was the prediction of pore fluid pressures. Transport equations for one fluid phase with a special term for the overburden sedimentation rate were used to calculate the compaction of the sediments. The compaction state and related porosity facilitated the determination of bulk thermal conductivities for heat flow calculations. At that time, practical studies were mainly performed as 1D simulations along wells, because the computer capabilities were still limited and multiphase fluid flow for migration and accumulation of petroleum had not been well implemented. Temperature profiles from multi–well analysis were used to calculate petroleum generation with source rock maturity maps over time and the determination of the peak phases of oil and gas expulsion. This concept is still used when data are scarce in early exploration or when the project requires some quick output.

From 1990 to 1998 a new generation of basin modeling programs became the standard in the petroleum industry. The most important new feature was

T. Hantschel, A.I. Kauerauf, *Fundamentals of Basin and Petroleum Systems Modeling*, DOI 10.1007/978-3-540-72318-9_1,
© Springer-Verlag Berlin Heidelberg 2009

the implementation of refined fluid flow models with three phases: water, liquid petroleum, and gas. In commercial packages, 2D Darcy flow models and map based flowpath analysis were realized (Ungerer et al., 1990; Hermanrud, 1993). Darcy flow models are able to model all relevant processes of flow, accumulation, and seal break through. They are based on differential equation systems for the competing fluid phases. However, they are restricted to 2D simulators, since they require a high computing and development effort. The map based flowpath technique redistributes pre-calculated expulsion amounts of petroleum along reservoir–seal interfaces within the reservoirs. Accumulation bodies are calculated under correct conservation of the petroleum mass and volume. The approach is based on some crude approximations concerning flow. However, it considers horizontal spilling from one drainage area to the next and simple break through when the column pressure exceeds the seal capability. Most models under study were first performed in 2D along cross sections because pre-interpreted horizons and faults along 2D seismic lines were readily available. Calculated generation and expulsion amounts were again used for the flowpath analysis afterwards. Although 2D Darcy flow models work very well, they were rarely used in practical exploration studies as horizontal petroleum migration in the third dimension can not be neglected. Another important innovation was the implementation of special geological processes such as salt dome tectonics, refined fault behavior, diffusion, cementation, fracturing, and igneous intrusions.

In 1998, a new generation of modeling programs were released changing the workflow of most basin modeling studies once again. Many new features were related to petroleum migration and the characteristics of reservoirs. Most programs and tools focused on 3D functions with improved features for model building and increased simulator performance. From that time on, most of the heat and pore pressure calculations were performed in full 3D. This required the interpretation and mapping of a relatively complete set of horizons instead of just the horizons of the reservoirs. Three–phase–Darcy flow models were also made available in 3D. However, high computation efforts were necessary while simplifying the model's premises to a large degree. Consequently the model's resolution was restricted which often led to unrealistic or oversimplified geometries. Pure Darcy flow models were not applicable in practice. Three alternatives for modeling migration were developed. One was the use of the well established flowpath models, the other two are new developments: hybrid flow simulators and the invasion percolation method. Hybrid fluid flow models use domain decomposition to solve the Darcy flow equations only in areas with low permeabilities and flowpath methods in areas with high permeabilities, resulting in a significant decrease of computing time. Invasion percolation is another rule based transport technique which focuses on capillary pressure and buoyancy without any permeability controlled flow timing. Another new feature was the implementation of multicomponent resolved petroleum phases and the development of fast thermodynamic PVT (Pressure Volume Temperature) controlled fluid analysis based on flash calculation for these components. Between four and fourteen fluid components (chemical species) are usually taken into consideration, replacing the traditional two component (oil–gas) black oil models. Reservoir composition and petroleum quality prediction were significantly improved. Simultaneously, better computer hardware especially PC clusters combined with parallelized simulators, reduced computing times significantly. Furthermore, statistics for calibration, risk analysis for quantification of probability for success or failure and the consideration of extensional and compressional tectonics significantly increased the applicability of basin modeling. Integrated exploration workflows, which incorporate basin modeling, became a standard in the industry.

1.2 Geological Processes

Basin modeling is dynamic modeling of geological processes in sedimentary basins over geological time spans. A basin model is simulated forward through geological time starting with the sedimentation of the oldest layer until the entire sequence of layers has been deposited and present day is reached. Several geological processes are calculated and updated at each time step (Fig. 1.1). Most important are deposition, compaction, heat flow analysis, petroleum generation, expulsion, phase dissolution, migration, and accumulation.

Deposition

Layers are created on the upper surface during sedimentation or removed during erosion. It is assumed that the geological events of deposition and hiatus are known. Therefore, paleo times of deposition can be assigned to the layers.

The depositional thickness of a new layer is calculated via porosity controlled backstripping from present day thickness or imported from structural restoration programs. The overall geometry may also change due to salt movement or magmatic intrusions. Estimated backstripping amounts yield calculated present day thicknesses which are not identical with the given present day geometry. The differences facilitate a better estimation of the depositional thicknesses in the next simulation run. This method of organizing multiple forward simulations to calibrate against the present day geometry is referred to as optimization procedure.

Pressure Calculation and Compaction

Pressure calculation is mainly a one–phase water flow problem which is driven by changes of the overburden weight due to sedimentation. Additionally, internal pressure building processes such as gas generation, quartz cementation and mineral conversions can be taken into account.

Pore pressure reduction entails compaction and leads to corresponding changes in the geometry of the basin. That is why pressure calculation and compaction have to be performed before heat flow analysis in each time step.

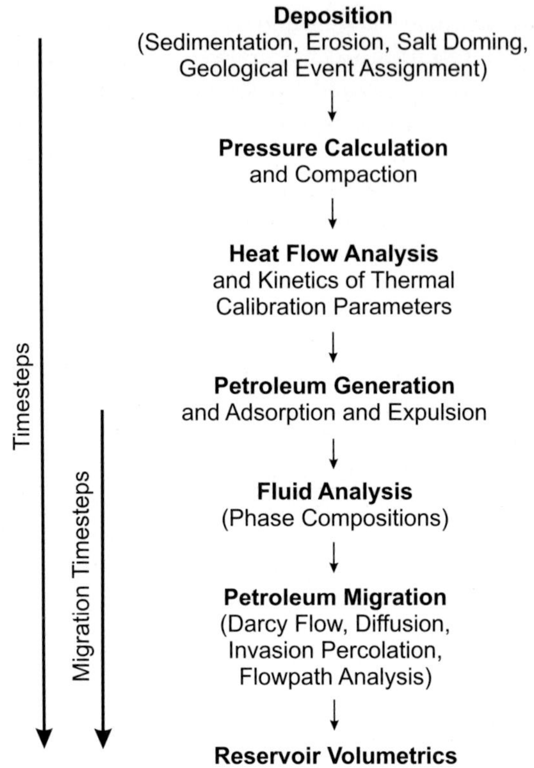

Fig. 1.1. Major geological processes in basin modeling

Heat Flow Analysis

Temperature calculation is the target of the heat flow analysis. It is a necessary prerequisite for the determination of geochemical reaction rates. Heat conduction and convection as well as heat generation by radioactive decay must be taken into consideration. Igneous intrusions require the inclusion of thermal phase transitions in sediments. Thermal boundary conditions with inflow of heat at the base of the sediments must be formulated. These basal heat flow values are often predicted with crustal models in separate preprocessing programs or are interactively calculated for each geological event.

Kinetics of Calibration Parameters

It is possible to predict vitrinite reflectance values, the concentration of molecular biomarkers and apatite fission tracks with suitable models which are based on Arrhenius type reaction rates and simple conversion equations. These predictions are temperature sensitive and can therefore be compared to measured data so that uncertain thermal input data, such as paleo–heat flow values, can be restricted or even calibrated.

Petroleum Generation

The generation of petroleum components from kerogen (primary cracking) and the secondary cracking of the petroleum is usually described with sets of parallel reactions of decomposition kinetics. The number of chemical components vary between two (oil, gas) and twenty. The cracking schemes can be quite complex when many components and secondary cracking are taken into account. Adsorption models describe the release of hydrocarbons into free pore space of the source rock.

Fluid Analysis

The generated hydrocarbon amounts are mixtures of chemical components. Fluid flow models deal with fluid phases which are typically liquid, vapor and supercritical or undersaturated phases. Therefore temperature and pressure dependent dissolution of components into the fluid phases is studied during fluid analysis. The two most important fluid models are the rather simple black oil model and the thermodynamically founded multicomponent flash calculations. Fluid phase properties, such as densities and viscosities, are also derived from fluid models. They are essential for accurate migration modeling and reservoir volumetrics.

Darcy Flow and Diffusion

Darcy flow describes multicomponent three phase flow based on the relative permeability and capillary pressure concept. It can be applied for migration. Migration velocities and accumulation saturations are calculated in one procedure. Special algorithms are used to describe break through and migration across or in faults. Diffusion effects can be evaluated for the transport of light hydrocarbons in the water phase.

Flowpath Analysis

In carriers lateral petroleum flow occurs instantaneously on geological timescales. It can be modeled with geometrically constructed flowpaths. Information about drainage areas and accumulations with compositional information can easily be obtained. Spilling between and merging of drainage areas must be taken into account. Flowpath analysis in combination with Darcy flow in low permeability regions is called the hybrid method. Migration modeling without sophisticated Darcy flow, instead using simplified vertical transport of generated hydrocarbons into carriers, is commonly called flowpath modeling.

Invasion Percolation

Migration and accumulation can alternatively be modeled with invasion percolation. This assumes that on geological timescales petroleum moves instantaneously through the basin driven by buoyancy and capillary pressure. Any time control is neglected and the petroleum volume is subdivided into very small finite amounts. Invasion percolation is very convenient to model in–fault flow. The method is especially efficient for one phase flow with the phase consisting of only a few hydrocarbon components.

Reservoir Volumetrics

The column height of an accumulation is balanced by the capillary entry pressure of the corresponding seal. Leakage and break through are therefore important processes reducing the trapped volume. Other processes such as secondary cracking or biodegradation also have a serious impact on the quality and quantity of the accumulated volume.

In principle all processes depend on each other. Therefore, at a given time, all these coupled processes must be solved together with the solution of the last time step as the initial condition. For numerical reasons such an approach can be performed implicitly in time and is thus called an implicit scheme. In practice it is found, that the processes can be decoupled, very often to some high order of accuracy. Finally it is possible to solve for all the processes which are shown in Fig. 1.1 in the given order. Extra loops with iterative updates for higher accuracy can easily be performed. Decoupled schemes are often called explicit schemes, especially if the processes itself are treated explicitly in time.

For example, migration and accumulation seldom has an important effect on basin wide compaction. Thus migration can often be treated independently. However, a coupling of migration with compaction might arise with pressure updates due to gas generation and subsequent local modification of the geometry. By re-running the entire simulation with consideration of the gas pressure of the previous run, the modified geometry can in principle be iteratively improved until convergence is reached. In practice, it is often found, that only very few iterative runs are necessary.

For the implicit scheme, the temporal evolution of the basin must obviously be calculated on the smallest timescale of all involved geological processes. A big advantage of an explicit scheme is the fact, that each explicitly treated process can be solved on its own timescale. On the other hand, time steps of implicitly treated processes can often, for numerical reasons, be longer than time steps of explicitly treated processes. This increases the performance of the implicit scheme, especially when iterative feedback loops have to be taken into account in explicit schemes. In practice, a combination of both schemes is found to be most advantageous. This yields three types of time steps, which are often called events, basic and migration time steps.

The outer time loops are identical with geological events. They characterize the period in which one layer has been uniformly deposited or eroded or when a geological hiatus occurred. Thus, the total number of events is almost equal to the number of geological layers and usually ranges between 20 and 50. Events are subdivided into basic time steps with one solution for pressure or compaction and the heat equations. The length of the basic time step depends on deposition or erosion amounts and on the total duration of the event. The total number of time steps usually lies between 200 and 500. The basic time steps are further subdivided into migration steps for an explicitly treated Darcy flow analysis. In one migration time step the transported fluid amount per cell is usually restricted to the pore volume of that cell. Therefore the total number ranges from 1000 up to 50000 and more and depends on the flow activity and the selected migration modeling method. All time loops for events, basic time steps and migration time steps are commonly managed automatically in most simulators. Mathematical convergence is often ensured by empirical rules for step length calculation.

Transport Processes

Heat flow, pore pressure and compaction, Darcy flow migration processes, and diffusion are transport processes. They follow a similar scheme of description, derivation, and formulation of the basic equations. The core problem is the interaction of two basic quantities, the state and the flow variable (Table 1.1). The influence of a flow variable acting from any location on any other neighboring location is the main part of the mathematical formulation. Modeling of transport problems requires a major computing effort.

For example, temperature and heat flow are the corresponding basic variables for heat conduction. Temperature is the state variable and heat flow is the corresponding flow variable. A temperature difference (or gradient) causes a heat flow, and the heat flow decreases the temperature difference. The heat flow is controlled by the thermal conductivity and the temperature response by the heat capacity.

State variable	Flow variable	Flow equation	Material property
Temperature T	Heat flow $\mathbf{q}$	$\mathbf{q} = -\boldsymbol{\lambda} \cdot \mathrm{grad}\, T$	Thermal conductivity $\boldsymbol{\lambda}$
Pressure p	Water flow $\mathbf{v}_w$	$\mathbf{v}_w = -\frac{\mathbf{k}}{\nu} \cdot \mathrm{grad}(p - \rho g z)$	Permeability $\mathbf{k}$ and viscosity ν
Fluid potential u_p	Fluid flow $\mathbf{v}_p$	$\mathbf{v}_p = -\frac{\mathbf{k}k_{rp}}{\nu_p} \cdot \mathrm{grad}\, u_p$	Relative perm. $\mathbf{k}k_{rp}$ and viscosities ν_p
Concentration c	Diffusion flux $\mathbf{J}$	$\mathbf{J} = -D\, \mathrm{grad}\, c$	Diffusion coeff. D

Table 1.1. Fundamental physical transport laws and variables

In general, an energy or mass balance can be used to formulate a boundary value problem with appropriate boundary conditions and to calculate the development of both the state and the flow variables through geological time. A solution to the boundary value problem requires in practice a discretization of the basin into cells and the construction and inversion of a large matrix. The matrix elements represent the change of the state variable caused by the flow between two neighboring cells. The number of cells is the number of unknowns. Finally, an inversion of the matrix results in the solution vector, e.g. containing a temperature inside of each cell.

The inversion of transport processes is often the major computing effort in basin modeling (Chap. 8). It depends strongly, almost exponentially, on the number of cells and therefore the resolution.

Examples of non-transport processes are fluid analysis, chemical kinetics and accumulation analysis, which depend only linearly on the number of cells if they are separated and explicitly treated. These processes can then be modeled very efficiently.

1.3 Structure of a Model

The general analysis of the basin type and the main phases of basin evolution precede the construction of the model input data. This encompasses information about plate tectonics, rifting events, location of the basin, and depositional environments through geological time, global climates, paleo–bathymetries, and tectonic events. The model input is summarized in Fig. 1.2, and includes: present day model data with depth horizons, facies maps, fault planes, the age assignment table for the geological event definition, additional data for the description of paleo–geometries, thermal and mechanical boundary conditions through geologic time, the property values for lithologies, fluids, and chemical kinetics.

1 **Present Day Model**
- Horizons (Depth/Structure Maps)
- Facies Maps
- Fault Surfaces

2 **Age Assignment**

3 **Paleo Geometry**
- Water Depth Maps
- Erosion Maps
- Salt Thickness Maps
- Paleo Thickness Maps

4 **Boundary Conditions**
- SWI-Temperature Maps
- Basal Heat Flow Maps

5 **Facies**
- Facies Definitions
- TOC & HI Maps
- Rock Composition Maps

6 **Seismic** (optional)
- Attributes (Cubes, Maps)
- Reference Horizons for Depth Conversion

Fig. 1.2. Basic elements of model input

Present Day Model Data

A sedimentary basin is a sequence of geological layers. Each of the layers contains all the particles which have been deposited during a stratigraphic event. A horizon is the interface between two layers (Fig. 1.3) and usually interpreted from a seismic reflection surface. Seismic interpretation maps and lines (in 2D) are usually not extended over the entire model area and have to be inter– and extrapolated and calibrated with well data. The construction of the horizon stacks often requires most of the time for the model building.

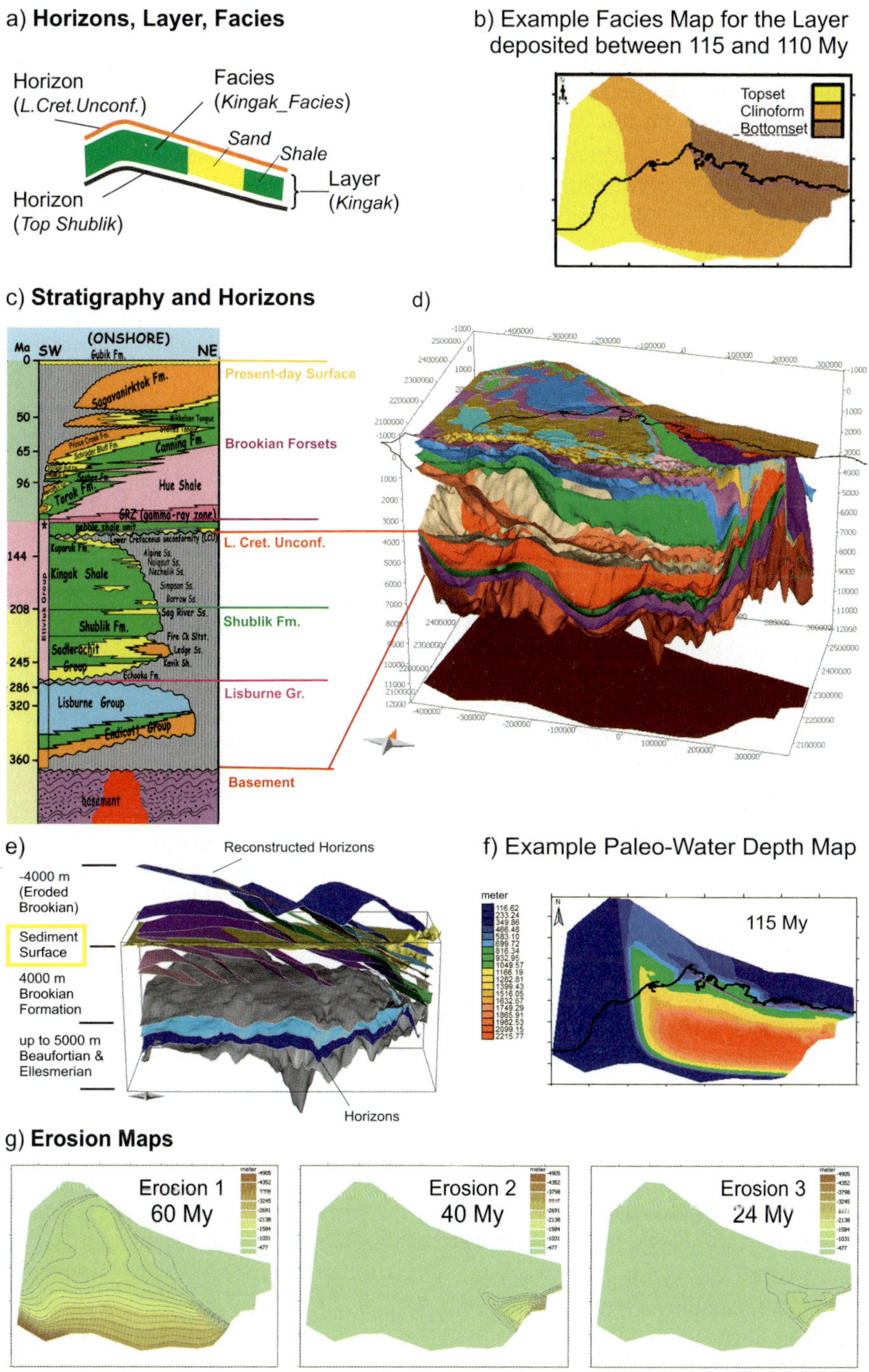

Fig. 1.3. Present day and paleo–geometry data: example from Alaska North Slope

A complete stack of horizon maps subdivides the space for volumetric property assignments. Parts of layers with similar sedimentation environments are called geological facies (Fig. 1.8). Facies are related to common property values of geological bodies. They are the main "material types" of the model. Layers can consist of several different facies and the same facies can appear in different layers. The distribution of facies is usually described with one facies map in each layer, based on well data information and sedimentological principles, e.g. clastic rocks are distributed corresponding to relationships between grain size and transport distances, particularly the distance from the coast (Fig. 1.3). In simple cases a layer can be characterized only by one unique facies type, whereas high resolution seismic facies maps allow the construction of very detailed facies maps (Fig. 1.10).

Fault planes are constructed from seismic interpretations, well data, and dips, which can also require a lot of effort. Depth horizons, facies maps, and fault planes constitute the present day model.

Age Assignment

The age assignment or stratigraphic table relates the present day horizons and layers with the geologic age of their deposition and erosion. In layer sequences without erosions, horizons represent all sedimentary particles, which are deposited during the same geological events (Fig. 1.3). If valid for the model, erosion and hiatus events also have to be included in the stratigraphic table. Erosion events require additional maps for the amounts of erosion and have to be combined with the corresponding water-depth for the description of the related uplift of the basin.

Stratigraphic diagrams with facies variations (Fig. 1.3) have to be simplified in order to get a relatively low number of model horizons in the range of $10-50$. Migrating patterns of facies through time generally require a Wheeler diagram instead of one single simplified age table. However, this feature is rather difficult to implement into a computer program.

Paleo-Geometry Data

The present day model can be built from measured data, such as seismic and well data. The paleo–model is mainly based on knowledge and principles from historical and regional geology, sedimentology and tectonics, which results in higher degrees of uncertainty. Water depth maps are derived from isostasy considerations of crustal stretching models together with assumptions on global sea level changes. They describe the burial and uplift of the basin. Water depth maps can also be derived from known distributions of sediment facies and vice versa (see e.g. the equivalence of the water–depth and facies map at 115 My in Fig. 1.3.b and f).

The construction of the erosion maps is usually more difficult. In the simplest case, one layer is partially eroded during one erosional event. The erosion thickness can be re–calculated by decompaction of the present day thickness and subtraction from an assumed relatively uniform depositional map. The

Age [My]	Horizon	Layer	Facies Maps	Erosion Maps	Paleo-Water Depth Maps
0	Present Day				PWD_0
		Brookian D	BrFac D		
24					PWD_24
		EROSION	*none*	**Erosion3**	
25	Top Oligocene				PWD_25
		Brookian C	BrFac C		
40					PWD_40
		EROSION	*none*	**Erosion2**	
41	Top Lutetian				PWD_41
		Brookian B	BrFac B		
60					PWD_60
		EROSION	*none*	**Erosion1**	
65	Top Upper Cretaceous				PWD_65
		Brookian A	BrFac A		
115	Top Lower Cretaceous				PWD_115
...	...	...	...		...
126	L.Cret.Unconf.				...
		Kingak	Kingak_Facies		...
208	Top Shublik				
		Shublik	Shublik_Facies		
260	Top Lisburne				
		Lisburne	Lisburne_Facies		
360	Top Basement				
		Basement	Basement_Facies		
400	Base Basement				

Fig. 1.4. Excerpt from the age assignment table of the Alaska North Slope model

sediment surface of the example model in Fig. 1.3.d acts as a unconformity and cuts many layers. A simple approach is to construct the missing erosion amount for each layer separately and to assume uniform erosion during the time period of erosion. This is illustrated in Fig. 1.3.e with the virtual horizons of the Brookian formation above the sediment surface. However, in the considered model it is further known that there were three main erosion periods and thus the corresponding erosion maps could be constructed (Fig. 1.3.g.). These maps together with the virtual Brookian horizons yield the erosion amounts for each of the layers in the three erosion events.

The above model description would have been sufficient, if the Brookian formation were eroded after complete deposition. In reality, compressional deformation in the Tertiary produced a fold–and–trust belt resulting in uplift and erosion and in a broad shift of the basin depocenters from WSW to ENE, which lead to mixed erosion and deposition events. A schematic description is illustrated in Fig. 1.5 which is finally realized in the age assignment table of Fig. 1.4. Note, that each erosion mentioned in the age assignment table consists of several layer specific maps with the erosion amounts related to the respective event. Unfortunately, such a complicated behavior is rather typical than exceptional. Input building tools often provide sophisticated map calculators with special features to make the construction of erosion maps easier. A preliminary simulation result of an ongoing Alaska North Slope study is shown in Fig. 1.6.

The occurrence of salt diapirs requires paleo-thickness maps for the main phases of salt doming. The reconstruction of the salt layers is usually based on geometrical principles, in the simplest case the present day thickness map

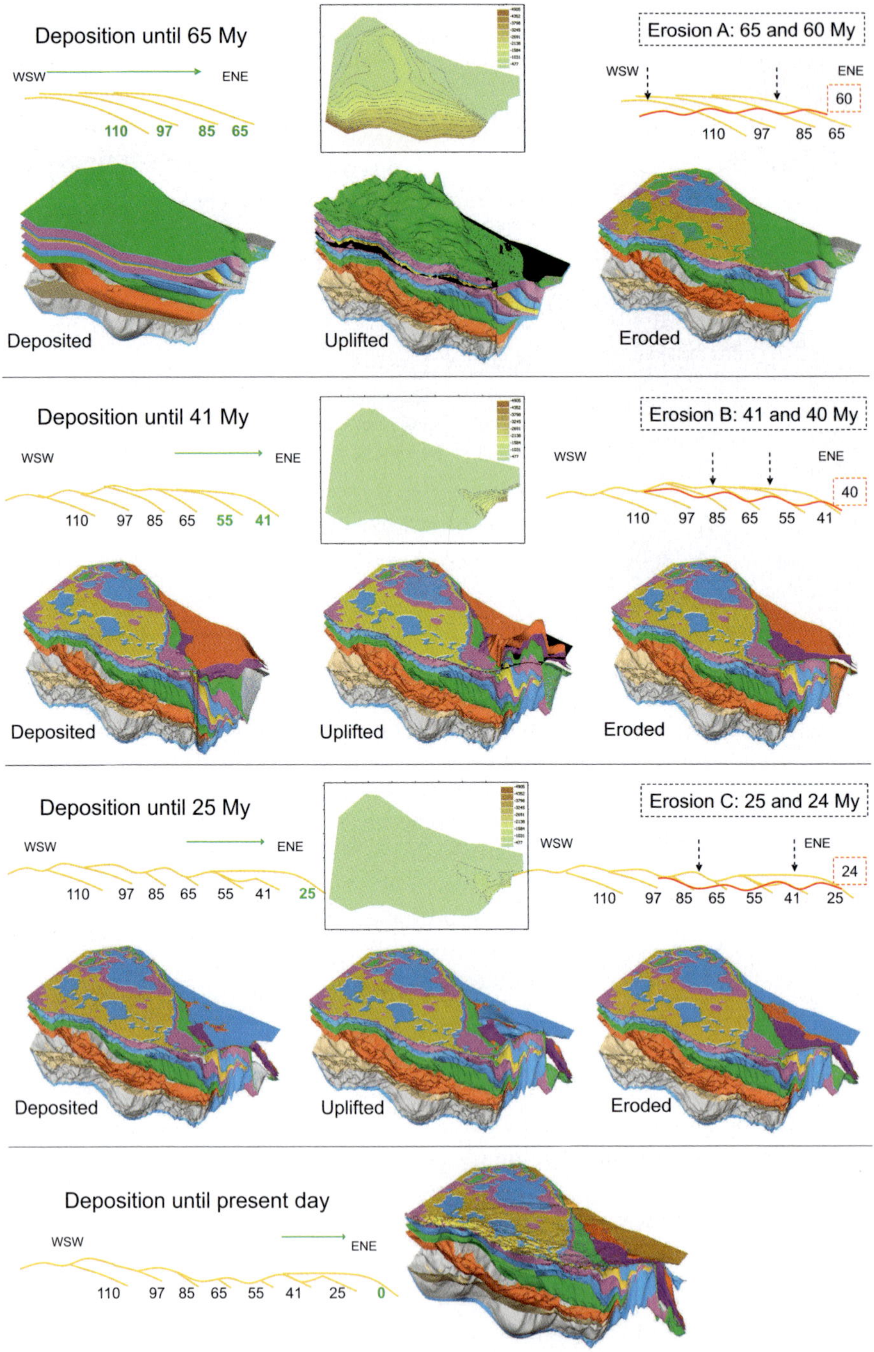

Fig. 1.5. Paleo–geometry data: example from the Alaska North Slope

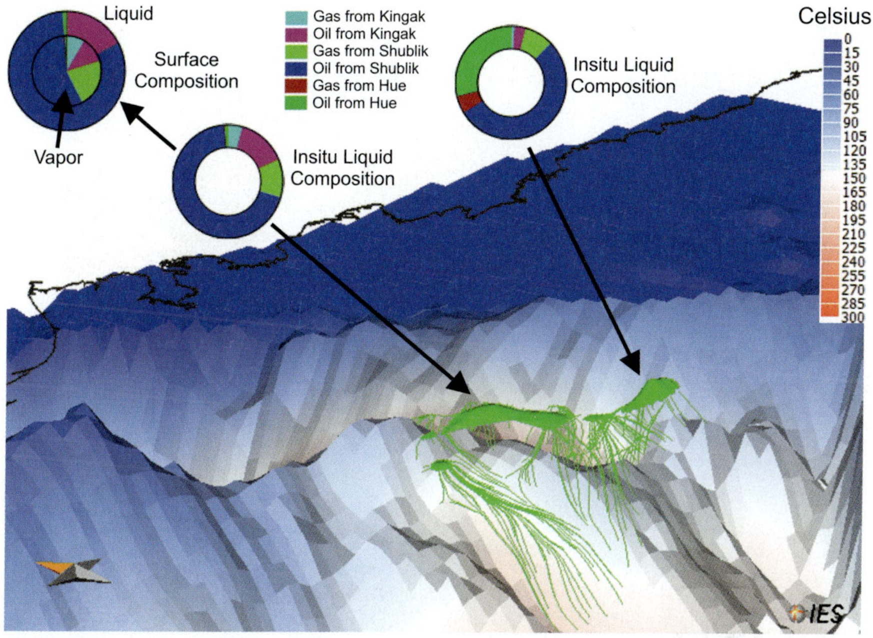

Fig. 1.6. Source rock tracking in Alaska North Slope. The two big visible accumulations are the Kuparuk (center) and Prudhoe Bay (right) fields

is linearly interpolated to an uniform deposition map. Corrections are made, if the resulting paleo-geometries show unrealistic kinks in the reconstructed base–salt maps. Salt layers can also be reconstructed based on calculated lithostatic pressures or total stresses at the salt boundaries because salt moves along the gradient of the lowest mechanical resistivity. The reconstructed salt thickness maps can be implemented in the input model by two methods: paleo–thicknesses for autochthonic salt layers and penetration maps for allochthonous salt bodies as illustrated in Fig. 1.7 for the Jurassic salt layer of the Northern Campos Model. Autochthonous salt maps through geologic times can be simply realized by adjusting the layer thickness in each gridpoint. The occurrence and timing of the salt windows is often very important for petroleum migration and pressure development as subsalt fluids and pressures are released afterwards.

The penetration of shallower sediments by salt and the formation of single allochthonous salt bodies is usually implemented with the replacement of the original sediment facies by the salt facies. Both methods have to be combined with adjustments of the other sediment thicknesses to maintain the mass balance. These correction maps can be added to the input data as paleo-thickness maps during the corresponding events.

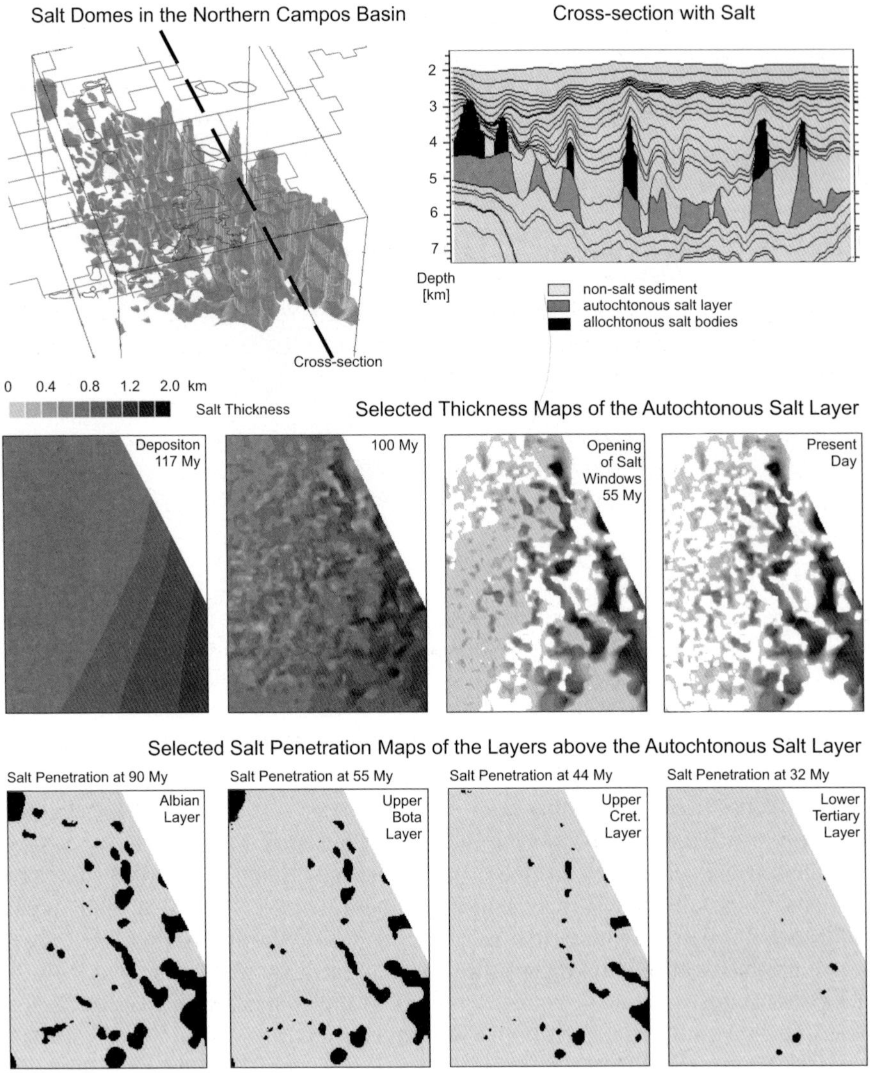

Fig. 1.7. Paleo–salt maps: example from the Northern Campos Basin in Brazil

The interplay of paleo-water depth, erosion, salt thickness, and other paleo-thickness maps finally determines the paleo-geometries and often requires some experience of the basin modeler to build geological reasonable scenarios.

Boundary Conditions

Boundary conditions need to be defined for the heat, pressure, and fluid flow analysis through the entire simulated geologic history. The usual boundary condition data for the heat flow analysis are temperature maps on the sediment surface or the sediment–water interface and basal heat flow maps for the respective events. The surface temperature maps are collected from general paleo–climate databases. The basal heat-flow maps can be estimated from crustal models and calibrated with thermal calibration parameters, which is explained in more detail in Chap. 3. Specific inner and upper igneous intrusion temperature maps should be added for magmatic intrusion and extrusion events, respectively.

The boundary conditions for the pore pressure and fluid flow analysis are often defined as ideal open (e.g. at sediment surface) and ideal closed (e.g. at base sediment). Exceptions are onshore basins or erosion events, which require the definition of groundwater maps to calculate the groundwater potential as the upper boundary condition for the pore pressure analysis. Herein, the sediment surface could be a good approximation.

It is a common method to determine the boundary values through geologic history as trend curves at single locations (gridpoints) first and calculate boundary value maps for the geological events by inter– and extrapolation afterwards.

Facies Properties

Facies are sediment bodies with common properties. The name facies is widely used in geoscience for all types of properties. Here, the facies is characterized by two sub–group facies types: the rock facies (or lithology) and the organic facies (or organofacies, Fig. 1.8).

A classification of lithologies is also shown in Fig. 1.8. It is used for the rock property tables in the appendix. The main rock properties are thermal conductivities, heat capacities, radiogenic heat production, permeabilities, compressibilities, and capillary entry pressures. Most of them depend on temperature and porosity. Functions for fracturing and cementation are also rock specific properties.

A classification of the organic facies is discussed in Chap. 4. The organic facies encompass all kinetic parameters for the generation and cracking of petroleum and the parameters to specify the quantity and quality of organic matter. The kinetic parameters are mainly Arrhenius–type activation energy and frequency data for primary and secondary cracking of hydrocarbon components. The total organic content (TOC) and the hydrogen index (HI) are usually defined by distribution maps. Furthermore, adsorption parameters are also related to the organic facies type. Fluid properties are either given directly for the different fluid phases or calculated from compositional information. Fluid phase properties are e.g. densities or viscosities. Typical fluid component properties are critical temperatures, pressures, and specific volumes.

Seismic

Seismic attribute cubes or maps can be used to refine the facies distribution maps in some layers, e.g. the ratio of shear to compressional velocity is correlated to the average grain size of clastic rock. The conversion of seismic

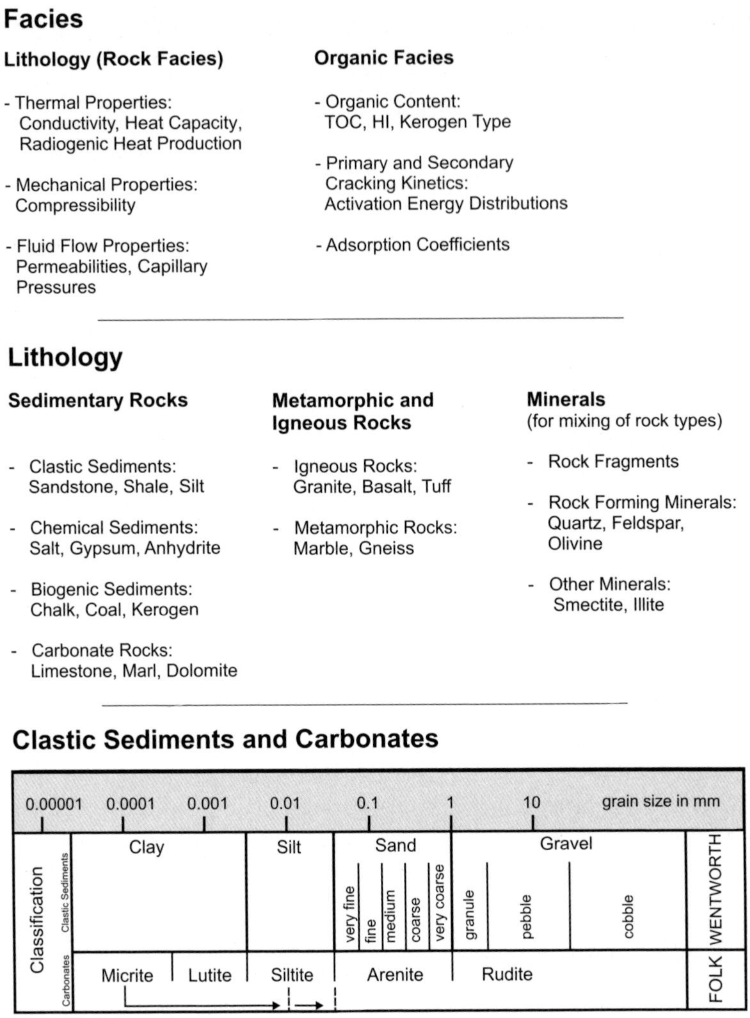

Fig. 1.8. Classification of facies, lithologies with the most important examples and terminology of clastic sediments and carbonates according to grain sizes. The picture is from Bahlburg and Breitkreuz (2004)

attributes to a "lithocube" requires a lot of effort and is only available in a few projects. Seismic facies cubes are usually available for the reservoir layers. In Fig. 1.9 and 1.10 two example cases from Australia and the North Sea are shown. Seismic facies cubes and maps are used, respectively. Seismic cubes can be given in two–way–time or depth. They require reference horizons to map the corresponding cells from the seismic to the depth model. The resulting facies distribution can be even finer than the major model grid. The invasion percolation method, which is used for modeling of migration, works on a sub–gridding of the cells and takes high resolution features into account (Chap. 6). Capillary entry pressures from the finer scale seismic facies control migration and accumulation.

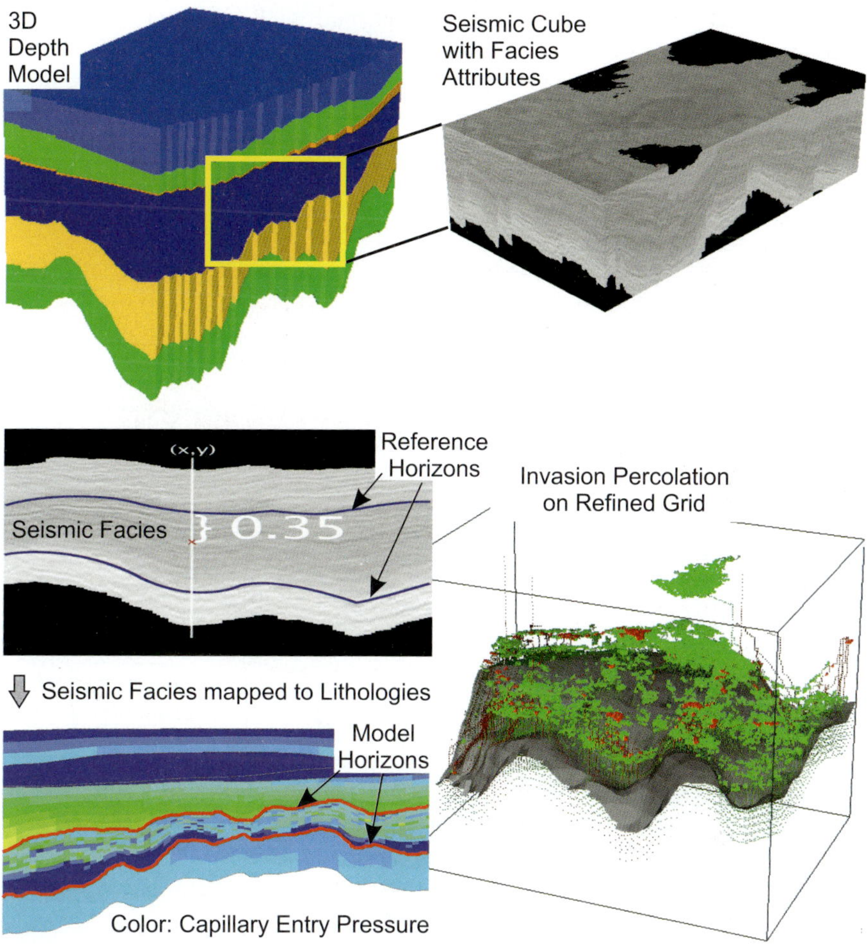

Fig. 1.9. Seismic cube with facies attributes and migration with invasion percolation. The attributes are mapped via reference and model horizons to the 3D model. For example, a point which lies at 35 % vertical distance between two reference horizons is here assumed to lie on the same relative position between the corresponding model horizons

The North Sea petroleum migration example (Fig. 1.10) is mainly restricted to two layers only: the upper Jurassic layer, and the overlaying chalk layer. The Jurassic layer contains high organic content shale and sandstone. It

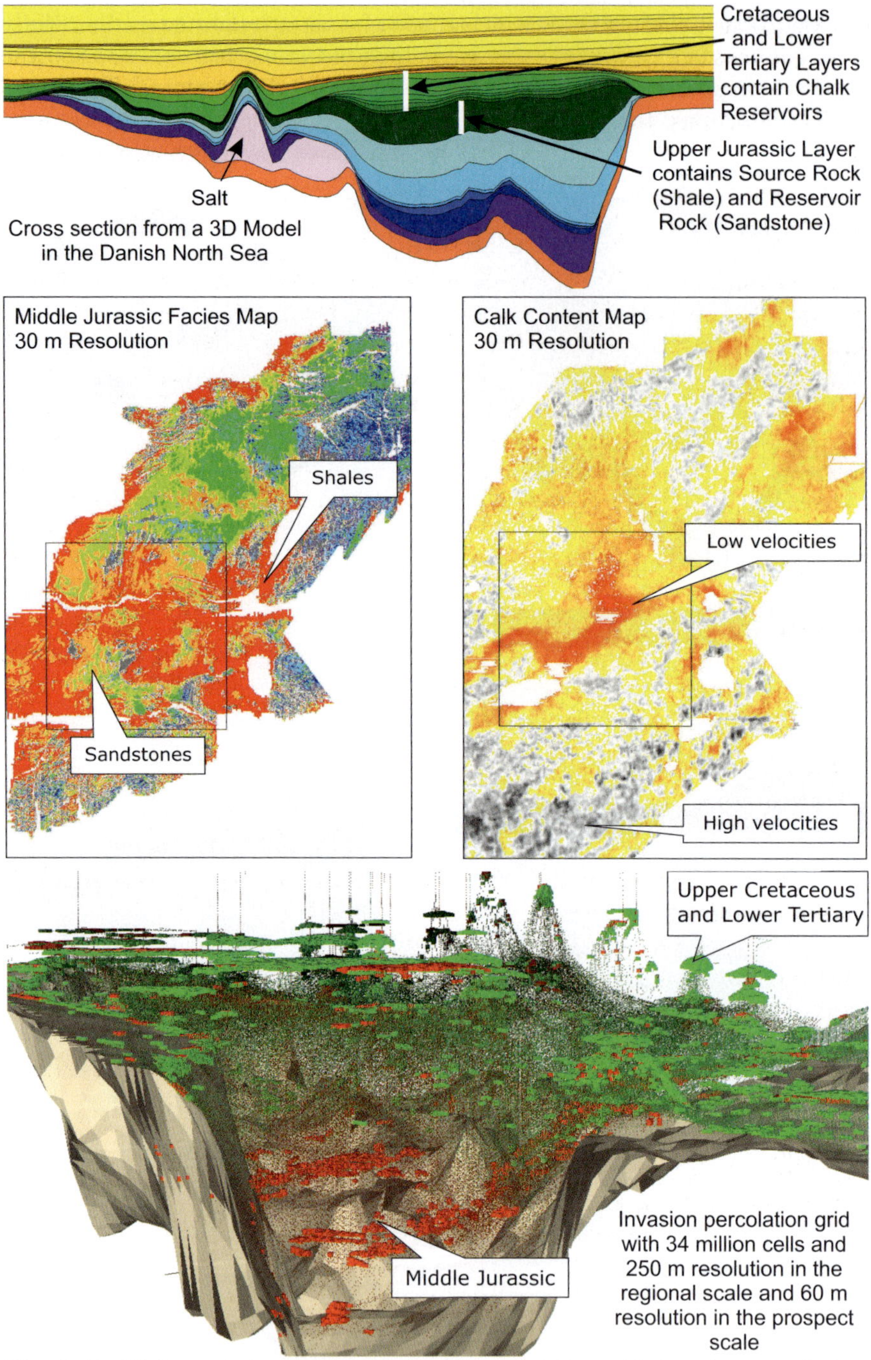

Fig. 1.10. High resolution maps and migration modeling with invasion percolation. The figures are courtesy of MAERSK

is both, a source and a reservoir layer. The chalk layer also contains petroleum accumulations and it is sealed by a dense overlaying shale. The two seismic attribute maps are applied to the layers without any further subdivision in vertical directions. In this case, the invasion percolation method is especially suitable, as high resolution is important and the migration distances are short.

Discretization of a Model

A continuum approach is commonly applied for the general description of heat and fluid flow processes on a macroscopic scale. Practical solutions can, on the other hand, only be obtained for discretized models. A mesher generates grids with the cells as the smallest volumetric units of the geological model. The basin or region of interest is assumed to be covered continuously with cells. Every physical or geological quantity such as temperature, pressure, saturation, concentration, permeability, thermal conductivity, etc. is well defined in the cell as a single, effective or average value. Furthermore, the value can vary continuously from cell to cell at least within parts of the structure. Each cell is used as a finite element or finite volume within the mathematical solvers.

The approach requires that the size of the cell must be small compared to the system being modeled (basin scale) but, at the same time, large compared to the pore scale and grain size. Typical scale sizes are

Molecular Scale:	$10^{-9} \dots 10^{-8}\,\mathrm{m}$
Pore Scale:	$10^{-6} \dots 10^{-3}\,\mathrm{m}$
Bulk Continuum:	$10^{-3} \dots 10^{-2}\,\mathrm{m}$
Cells of the Grid:	$10^{0} \dots 10^{2}\,\mathrm{m}$
Basin Scale:	$10^{3} \dots 10^{5}\,\mathrm{m}$

with cells which are much larger than the pore scale and grain sizes and much smaller than the basin scale.[1]

However, modern simulation programs might contain different grid scales and even different basin scales for the modeling of different geological processes. Such multigrids are typically created with sampled and refined representations of a master grid. Optimal methods can then be applied for each geological process. For example, heat flow is often modeled on the full basin scale with grid cells seldom smaller than 100 m, whereas petroleum systems modeling is sometimes restricted to smaller areas of source rock expulsion and active migration pathways with corresponding grid cells, which can become very small. However, sophisticated up- and downscaling functions (e.g. for fractal saturation patterns) may be required.

Many quantities can be defined as gridded maps at certain events. Alternatively, geological time dependent trend functions are often specified at

[1] In finite element simulators, a continuous crossover within a cell is modeled and the bulk continuum scale, rather than the cell size of the grid, must be compared with the basin scale. Finite elements therefore often show an implicitly higher resolution than other cell types.

individual well locations. Maps are then generated for each event by spatial interpolation over the whole model area. In both cases maps are the central objects for the creation of a basin model.

Size of a Model

A primary target of basin modeling is the assessment of exploration risk by calculation of generated and accumulated petroleum volumes for different geological migration scenarios. Herein, basin to reservoir scale models are used from a total length of hundreds of kilometers down to only a few kilometers (Fig. 1.11). Another study type concerns resource assessments, which cover even more extensive geographical areas such as entire countries (Fig. 1.11). The total amount of oil and gas resources in several layers is estimated. This task often encompasses source rock maturity studies including volumetrics for migration losses with simplified reservoir distributions. Governmental geological surveys and academic institutes often contribute to such studies.

Typical model dimensions and grid data are shown in Fig. 1.11. In practice, there are in general two requirements, a minimum model resolution to approximate the geological structures of interest and a simulation run time of less than 12 hours. This is a "rule of thumb" of the authors: a simulation must to be able to run in one night.

Computer performance has significantly increased since the introduction of parallelized simulations on computer clusters. The average number of cells for a complete simulation is $1-2$ million cells which corresponds to $200-300$ gridpoints in the horizontal directions. Heat, pressure, and Darcy flow computing times depend almost exponentially on the number of cells. Doubling the number of gridpoints in one direction often increases the computing effort by one order of magnitude. That is why big improvements in computer performance and numerical methods often have only a small effect on the grid resolution. However, computing time is very difficult to estimate as some important controlling parameters, such as the number of hydrocarbon containing cells, average and peak fluid flow rates or the number of migration time steps for good convergence, are not known prior to the special conditions of each simulation.

1.4 Petroleum Systems Modeling

A "Petroleum System" is a geologic system that encompasses the hydrocarbon source rocks and all related oil and gas, and which includes all of the geologic elements and processes that are essential if a hydrocarbon accumulation is to exist (Magoon and Dow, 1994).

A petroleum systems model is a digital data model of a petroleum system in which the interrelated processes and their results can be simulated

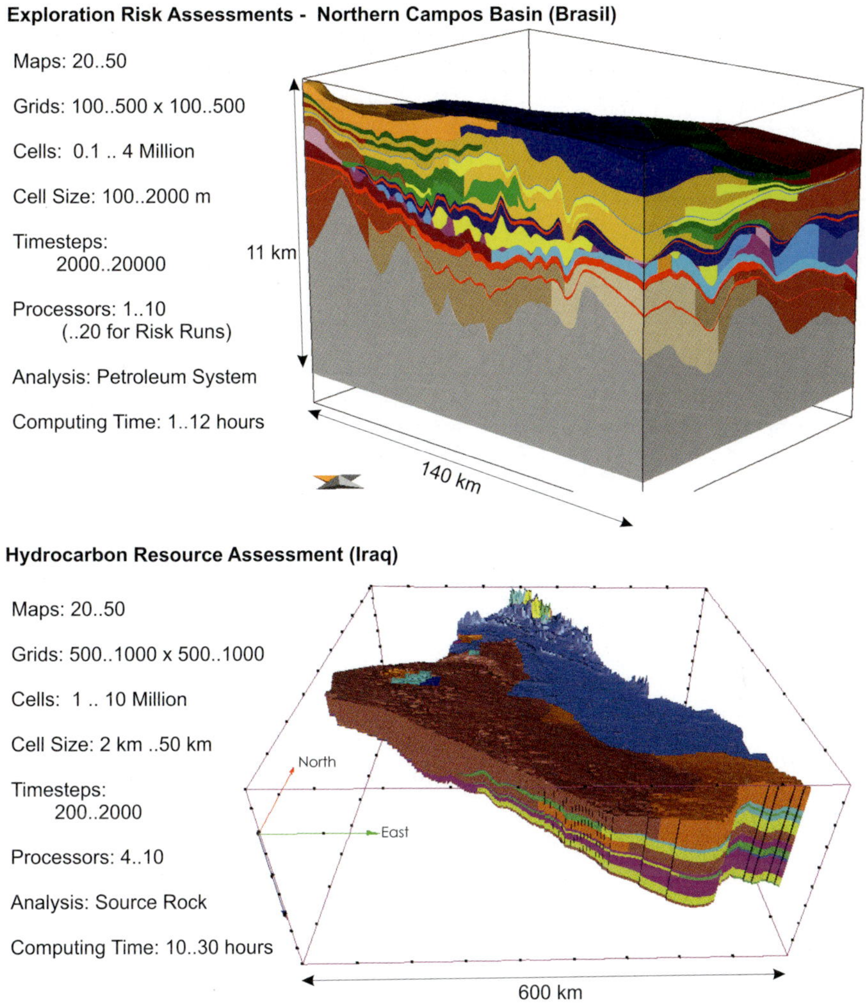

Fig. 1.11. Studies on prospect and regional scales. The figure from Iraq is courtesy of the U.S. Geological Survey and described in Pitman et al. (2003)

in order to understand and predict them. It is a preferably 3D representation of geological data in an area of interest, which can range from a single charge or drainage area to an entire basin. A petroleum systems model is dynamic which means that petroleum systems modeling provides a complete and unique record of the generation, migration, accumulation and loss of oil and gas in a petroleum system through geologic time.

Petroleum systems modeling includes basic assessments such as:

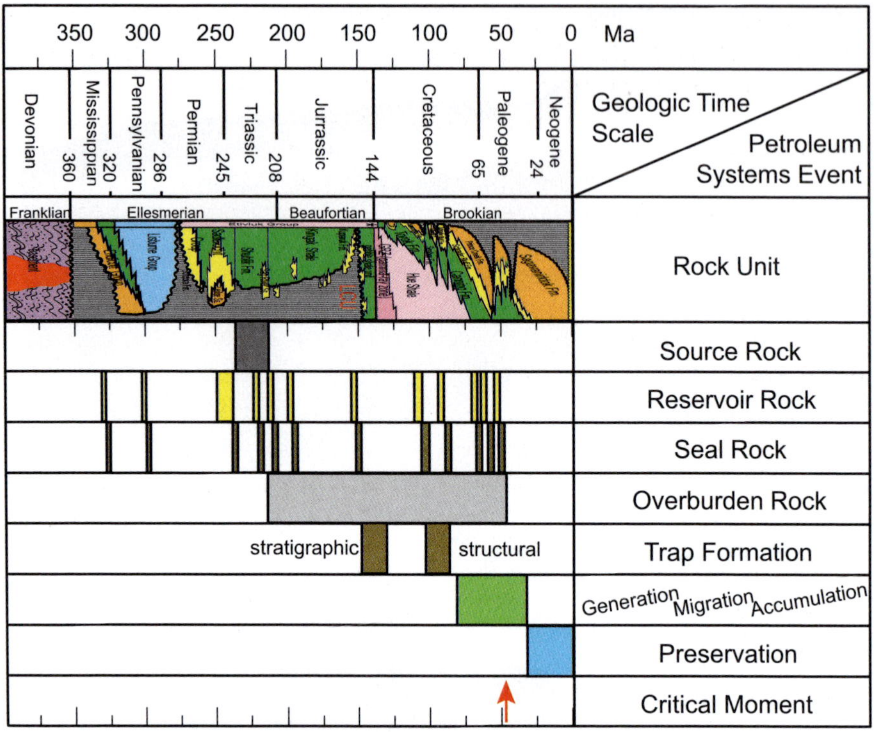

Fig. 1.12. Simplified petroleum system chart of Alaska North Slope after Magoon et al. (2003)

Have hydrocarbons been generated? This includes a full range of services from initial charge risking in frontier areas to regional resource assessments of yet–to–find hydrocarbons.

Where were hydrocarbons generated? If hydrocarbons were generated, their locations can be defined quite accurately so that their possible relationships to prospects can be risked.

When were hydrocarbons generated? There are many clear examples of where basins, plays, and prospects have failed due to timing problems. For example, when oil and gas was generated early and the structures were created much later.

Could hydrocarbons have migrated to my prospect? Modeling of the dynamic process of generation, expulsion, and migration makes it possible to determine if the oil and gas charge could reach the trap.

What are the properties of the hydrocarbons? Modeling of the phase behavior of the hydrocarbons during migration, accumulation and loss makes it possible to determine oil vs. gas probabilities and even predict properties such as API gravities and GORs.

Petroleum systems modeling can be interpreted as a sub-group of basin models, which model the full hydrocarbon lifecycle. It covers the most sophisticated targets of basin modeling.

Each source rock develops its own petroleum system. The petroleum system elements are facies, which contained, transported or sealed the generated petroleum from one source rock. These facies were named according to their function as source rock, carrier rock or seal. All the distributed petroleum of one petroleum system is more or less connected with rest saturation drops, migration stringers and accumulation bodies (Fig. 1.10) and is usually mixed with other petroleum systems from the same basin. The petroleum system chart shows the timing of the petroleum systems elements and allows a first assessment of the process chain (Fig. 1.12).

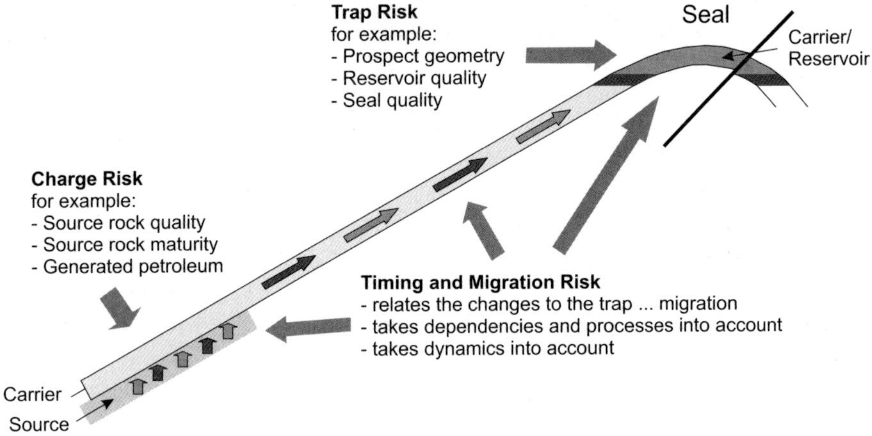

Fig. 1.13. Risk factors of petroleum systems modeling

A primary target of petroleum systems modeling are hydrocarbon exploration risk factors (Figs. 1.13, 1.14). They are the hydrocarbon charge, the reservoir quality, the trap capabilities and the timing relationship between the charge, reservoir, and seal (Fig. 1.13). Exploration risk commissions often evaluate the risk related to charge, reservoir, and seal, separately and subdivided into several factors (Fig. 1.14). Obviously, most of these risk factors can be assessed from a well designed basin model with special emphasis to the charge factors. Probability analysis methods (Chap. 7) allow the total risk to be quantified as a result of special uncertainties of the single risk factors and also take into account the timing relationships. Thus, basin modeling combined with probability analysis can be used as a decision support system for exploration risk assessment.

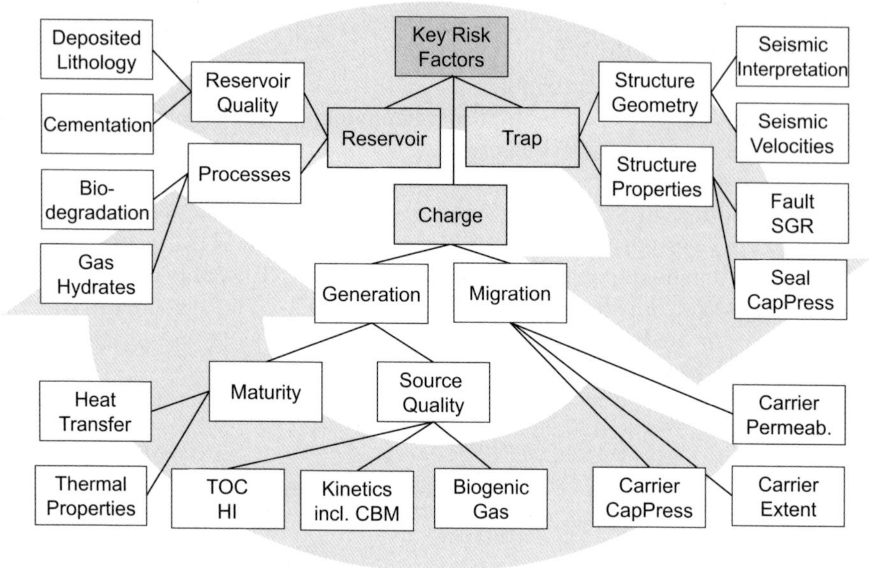

Fig. 1.14. Petroleum Systems Modeling as a Decision Support System

1.5 Modeling Workflows

The employment of some geological processes is optional and sometimes modeling of pressure or migration is not needed. It also makes sense to completely decouple pressure, temperature, and hydrocarbon fluid flow modeling from each other especially for pressure and temperature calibrations or if several migration scenarios should be tested. The following schemes for source rock analysis, reservoir volumetrics, and migration modeling demonstrate some common workflows (Fig. 1.15).

Source Rock Maturation Study

This type of study is performed when knowledge about the basin is sparse or when project deadlines are near. Large uncertainties in the data may not allow a sophisticated modeling. Only basic facts are investigated and emphasis is put on small simulation times.

In the initial step a model is calibrated for pressure and then again for temperature.[2] Both calibrations are performed fully decoupled. Feedback of temperature effects on compaction are not taken into account. This enhances the performance of the procedure drastically. Possible errors are neglected.

After the calibration, generated hydrocarbon masses give a first idea about source rock maturity, peak expulsion times and maximum reservoir fillings.

[2] It is important to perform the pressure calibration before the heat flow analysis since pressure formation influences the paleo–geometry which can have a significant effect on temperature history.

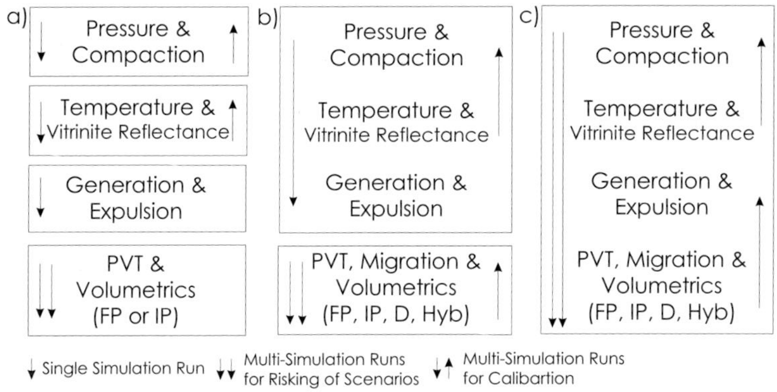

Fig. 1.15. Modeling Workflows for (a) Source Rock Maturation Study. (b) Decoupled Migration Study. (c) Petroleum Migration Study. (FP..Flowpath Modeling, IP..Invasion Percolation, D..Darcy Flow Modeling, Hyb..Hybrid Flow Modeling)

Drainage areas of interest in the reservoir are mapped to the source rocks with some simple procedures and the corresponding expulsion amounts are collected for the volumetrics. Flowpath modeling and invasion percolation techniques can be used additionally in a more advanced manner to consider losses, spill and seal break through amounts.

Multi–simulation runs are often performed for calibration and inversion. Statistical methods can be used to improve the calibration workflow. Histograms with generated amounts are often evaluated as functions of uncertain parameters, such as basal heat flow or SWI temperatures. However, this type of modeling is too crude for risking of individual accumulations.

Decoupled Migration Study

A decoupled migration study is typically performed when multiple migration scenarios are studied. It is often not reasonable to recalculate compaction and temperature for each migration scenario anew because feedback effects between migration and compaction or temperature are usually very small. On the other hand, a lot of simulation time is saved when the pressure and temperature field is not recalculated for each simulation run.

Migration and accumulation are performed on the most sophisticated level. They are considered in more detail than in a source rock maturation study. The selection of the migration model depends on the type of the geological migration process, the model input, the available computer soft- and hardware and the output preferences of the user. Very often different migration methods are tested for their performance in a given basin under certain geological conditions. Darcy flow with time control is often applied, especially as a part of the hybrid migration method. For example, a petroleum system very often consists of several sources. The interaction of different reservoir layers can play an important role. Especially the charging of traps can be studied with the hybrid migration method. A large part of this book deals with explanations and comparisons of the different migration modeling methods.

Multi–simulation runs are performed to explore the range of calculated reservoir fillings dependent on unknown input parameters of the petroleum system. Statistical models are often applied to quantify the risk assessment procedure (Chap. 7).

Coupled Migration Study

The decoupled mode ignores the influences of the petroleum system on temperature and pressure, such as gas generation pressure or oil and gas influences on thermal conductivities. Coupled scenarios ensure modeling of the full interaction. Of course, high resolution 3D models need a lot of computer power for such fully integrated runs, especially when multi–simulation runs are needed for calibration and risk assessment. The calibration of the petroleum system is also part of the procedure when information about known accumulations is available. It cannot be done automatically since there are too many uncertain input parameters which affect the resulting accumulation pattern.

Workflows for modeling geological processes are numerous and most people have their own preferred data and workflows to achieve the desired results. There is no doubt that many of the controlling geological factors involved in these processes are not very well known and difficult to quantify, and that this limits the numerical accuracy of the models. For example, it is still unclear how short–term thermal events ("heat spikes") influence the kinetics of petroleum formation, or how significant errors in the heat flow history that result from insufficient knowledge of the intensity and time of erosional phases can be avoided. Additional restrictions are our limited knowledge of factors affecting carbonate diagenesis (early or late diagenetic cementation?), and subsequent inaccurate estimates of thermal conductivities at the respective diagenetic stages. This list can surely be extended. In many cases, it can be assumed that uncertainties resulting from missing knowledge about uncertain processes are often larger than small errors due to a missing feedback effect. More conceptual models with less coupled processes can be understood, calibrated, and studied more easily. For example, due to higher simulation performance more uncertain parameters can be varied to assess their influence on the modeling results.

1.6 Structural Restoration

Structural restoration deals with the determination of the shape of geological structures at paleo times. Overthrusting and faulting are the main topics. It is often performed with a backstripping approach which is mainly based on the mass and volume balances of rock material.

Structural restoration is tightly linked to basin modeling as the shapes of layers and faults are often used as inputs in basin modeling. Optimization procedures for geometry calculations can then be omitted. However, multiple simulation runs cannot be avoided if porosity is to be calibrated. Fully restored geometries of basins at certain events are needed and extensive restorations have to be performed (Chap. 2).

Structural modeling, geomechanics, and tectonics incorporate the modeling of stresses and strains. They are needed when fault properties, fracturing, and lateral effects on compaction are of interest.

1.7 Comparison with Reservoir Modeling

A role, similar to that of basin modeling in exploration, has traditionally been played in production by reservoir modeling (Aziz and Settari, 1979). There are many fundamental similarities between reservoir modeling and basin modeling, as both technologies are used to model transport processes for hydrocarbon fluid flow in geologic models in order to provide an improved understanding, so that better predictions of possible results can be made.

The scaling of basin and petroleum systems models is however completely different than that of reservoir models, as dynamic geologic processes are considered in basin modeling. Sedimentary basins evolve through geologic time with significant changes in their geometries due to burial subsidence and compaction, uplift, and erosion, and structural complexities. Additionally, the size of sedimentary basins is also orders of magnitude larger than typical field sizes. For example, mega-regional models cover areas the size of the Gulf of Mexico and include the entire sedimentary sequence up to depths of 10 km and more. As a result, pressure and temperature conditions in sedimentary basins vary over a much wider range.

Besides this there are some other fundamental differences which are less important from a technical viewpoint. For example, reservoir modeling deals with forecasts of future production.[3] The Influence of humans on the results, e.g. due to the injection of steam, play a central role. In contrast basin modeling is performed for geological times only. Human influences on the basin are obviously of no interest. Likewise an optimization routine, which is not found in reservoir modeling, is necessary for calibration of the present day geometry. Despite all these differences, basin modeling has benefited greatly from reservoir modeling. For example, fluid analysis was first applied in reservoir modeling and has now evolved to become a sophisticated addition to basin modeling.

[3] History matching is similar to calibration in basin modeling. It is performed to improve the quality of future predictions.

1.8 Outlook

Future trends in basin modeling will involve the refinement of the implementation of all the above listed geological processes. As already mentioned there are, for example, developments for the integration of stresses and strains into simulators. This example is an enhancement of compaction and pore pressure prediction.

Besides this there are other developments which try to incorporate seismic information more directly into basin models. For example, invasion percolation models have a higher resolution than other processes in basin models. The resolution approaches almost the resolution of seismic data. A direct incorporation of seismic data is therefore desired.

Seismic data can also be used in general for facies and lithology assignment. However, appropriate attribute analyses and upscaling laws must be developed.

Summary: Basin modeling is dynamic forward modeling of geological processes in sedimentary basins over geological time spans. It incorporates deposition, pore pressure calculation and compaction, heat flow analysis and temperature determination, the kinetics of calibration parameters such as vitrinite reflectance or biomarkers, modeling of hydrocarbon generation, adsorption and expulsion processes, fluid analysis, and finally migration.

Transport processes for water (pore pressure and compaction), heat (temperature calculation), and petroleum (migration and diffusion) can be formulated in terms of flow equations with appropriate conservation equations for mass or energy which finally yield diffusion type differential equations.

A sedimentary basin is a sequence of geological layers. Each layer was deposited in a given stratigraphical event and is subdivided into regions of similar facies. A facies type specifies the lithological rock type and the organic facies. The lithology includes quantities such as permeability, compaction parameters, heat capacities, thermal conductivities and so on. The organic facies contain the total organic carbon content (TOC), the hydrogen index (HI) and the specification of the kinetic for petroleum generation. Boundary conditions must also be defined. Basal heat flow can be determined from crustal models for basin evolution.

Migration is the most sophisticated process in modeling. Due to its uncertain nature and extensive computing requirements different modeling approaches exist. Hybrid simulators combine the advantages of all approaches.

Additionally, a basin model contains special submodels concerning faults and fault properties, cementation, thermal calibration parameters, salt movement, intrusions, fluid phase properties, secondary cracking, and so on.

Basin models typically cover areas about 10 x 10 km up to 1000 x 1000 km and to a depth of 10 km. They are gridded into volume elements with up to 500 gridpoints in the lateral directions and up to 50 layers. Each volume element contains a constant facies in a bulk continuum approximation. Appropriate upscaling of physical properties from core to grid size might be necessary.

In practice, different workflows for risk evaluation and calibration exist. Dependent on the quality of the data, the geological processes are modeled decoupled, partially coupled or fully coupled. Source rock maturation studies are typically decoupled and petroleum migration studies are fully coupled. In between, decoupled migration studies are performed for risking, when, for example, the migration pathways are not known and different migration scenarios are tested.

Structural restoration yields valuable information about overthrusted layers and faulted geometries. It is an important step for modeling many of the world's basins.

Basin modeling has been performed since about 1980 and became fully three dimensional in respect to all important processes around 1998 when sophisticated 3D-simulators with migration were published.

References

P. A. Allen and J. R. Allen. *Basin Analysis.* Blackwell Publishing, second edition, 2005.

K. Aziz and A. Settari. *Petroleum Reservoir Simulation.* Elsevier, 1979.

H. Bahlburg and C Breitkreuz. *Grundlagen der Geology.* Elsevier GmbH, Muenchen, second edition, 2004.

J. Gluyas and R. Swarbrick. *Petroleum Geoscience.* Blackwell Publishing, 2004.

C. Hermanrud. Basin modeling techniques – an overview. Basin Modelling: Advances and Applications, pages 1–34. Norwegian Petroleum Society (NPF), Special Publication No. 3, Elsevier, 1993.

J. M. Hunt. *Petroleum Geochemistry and Geology.* W. H. Freeman and Company, New York, second edition, 1996.

L. B. Magoon and W. G. Dow. *The petroleum system from source to trap.* AAPG Memoir, 60, 1994.

L. B. Magoon, P. G. Lillis, K. J. Bird, C. Lampe, and K. E. Peters. Alaska North Slope Petroleum Systems: U.S. Geological Survey open file report. Technical report, 2003. URL `geopubs.wr.usgs.gov/open-file/of03-324`.

K. E. Peters, C. C. Walters, and J. M. Moldowan. *The Biomarker Guide*, volume 1 and 2. Cambridge University Press, second edition, 2005.

J. K. Pitman, D. W. Steinshouver, and M. D. Lewan. A 2 1/2 and 3D modeling study of Jurrasic source rock: U.S. Geological Survey open file report. Technical report, 2003. URL `pubs.usgs.gov/of/2003/ofr-03-192`.

B. P. Tissot and D. H. Welte. *Petroleum Formation and Occurrence.* Springer–Verlag, Berlin, second edition, 1984.

P. Ungerer, J. Burrus, B. Doligez, P. Y. Chenet, and F. Bessis. Basin evaluation by integrated two–dimensional modeling of heat transfer, fluid flow, hydrocarbon gerneration and migration. *AAPG Bulletin*, 74:309–335, 1990.

M. A. Yükler, C. Cornford, and D. Welte. Simulation of geologic, hydrodynamic, and thermodynamic development of a sediment basin – a quantitative approach. In U. von Rad, W. B. F. Ryan, and al., editors, *Initial Reports of the Deep Sea Drilling Project*, pages 761–771, 1979.

2

Pore Pressure, Compaction and Tectonics

2.1 Introduction

Most physical transport and related processes depend on both, temperature and pressure. Pressure is one of the fundamental physical values. It is a scalar, which is represented with a single value in each location. The term pressure has only a real meaning for fluids and not solids. In porous media, pressure is often introduced as the pressure within the fluids in the pores, the pore pressure. The equivalent physical entity in solids is the stress tensor, which is a symmetrical 3x3 tensor with six independent values (Sec. 8.2). It can be illustrated with an ellipsoid, whose axes represent the principal stresses in size and direction. Usually, only single components or invariants of the stress tensor are important. Both, rock stress and pore pressure describe the response of the material to an external load. The "average" stress of the porous volume element is called bulk stress. It is therefore a superposition or mixture of pore pressure and rock stress and it has to be in equilibrium with all external loads.

The primary pressure and stress causing process is sedimentation with subsidence, which produces overburden load on the subsurface rocks. Stresses and pore pressures generally increase with depth. Rock stresses and fluid pressures interact with compaction and porosity reduction. The main mechanisms for compaction are rearrangement of the grains to denser packages and cementation, which are called mechanical and chemical compaction, respectively. In summary, three main ingredients needed to formulate a model for the mechanics of the porous sediments, are the concepts of bulk stress, pore pressure and compaction. Additional effects, like mineral transformation, aquathermal pressuring, and kerogen cracking or fracturing, should also taken into account.

2.1.1 Bulk Stresses

A homogeneous body under a constant load from above deforms horizontally and vertically as shown in Fig. 2.1.a. The vertical stress in each location is

T. Hantschel, A.I. Kauerauf, *Fundamentals of Basin and Petroleum Systems Modeling*, DOI 10.1007/978-3-540-72318-9_2,
© Springer-Verlag Berlin Heidelberg 2009

then equal to the top load and the horizontal stress is equal to zero. This stress state is called uniaxial. If the side boundaries of the bodies are fixed (Fig. 2.1.b), the horizontal stress components are compressive as well and equal to a fixed ratio of the top load, namely $\sigma_h/\sigma_v = \nu/(1-\nu)$. The Poisson ratio ν is a material constant and sediments have numbers of $0.1 \ldots 0.4$, which yield stress ratios of $0.11 \ldots 0.67$. Exceptions are salt and unconsolidated sands with Poisson ratios close to 0.5.

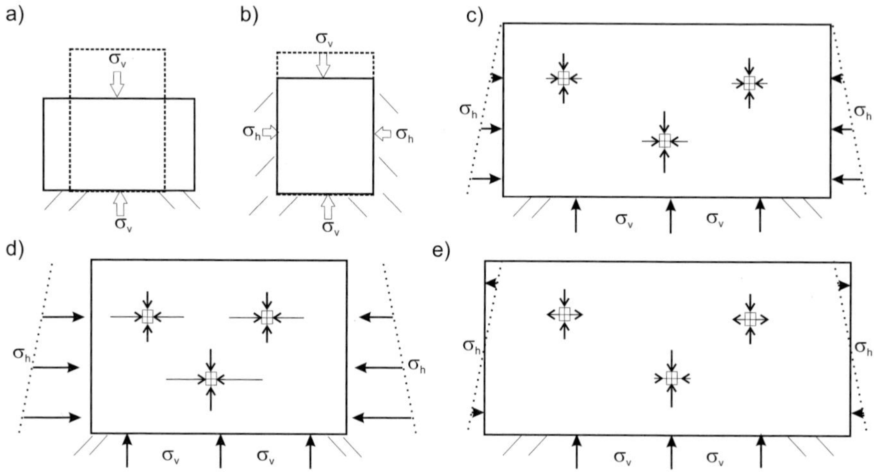

Fig. 2.1. Vertical and horizontal stresses in a homogeneous solid with overburden (a) load on top with free moving sides; (b) load on top with fixed side boundaries (c) gravity loads with fixed side boundaries; (d) together with additional constant compressions on the side boundaries; (e) together with additional constant tensions on the side boundaries

The situation in non–tectonically influenced basins is similar to the fixed solid case (Fig. 2.1.c) with vertical loads increasing approximately linearly with depth. Heterogeneities of geomechanical properties cause different stress ratios and rotation of the main stress axes. Fault planes and salt domes disturb homogeneous trends in stress. Tectonic processes generally add a compressive or tensile stress to the horizontal component (Fig. 2.1.d,e). Extensions of the model to lower horizontal stresses can revert the compressive (positive) stresses into tensile (negative) stresses, while compressive boundaries can increase the horizontal stresses so that they exceed the vertical stresses and become the maximum principal stress.

The stress state in solid grains is mainly controlled by the overburden load of the considered volume element in the case of negligible tectonic forces and homogeneously layered rocks (Fig. 2.1.c). Then, the "lithostatic pressure" approach can be used, which describes the three dimensional stress field by one single value, the lithostatic pressure, assuming the following simplifications:

- The three main stress axes are straight vertically and horizontally directed. This assumption is not valid in heterogeneous layers and salt domes, where the coordinate system of the principal stress components rotates.
- The boundaries of the basins are fixed in terms of displacements, the horizontal stresses are equal in both directions and the stress ratio ($\sigma_\mathrm{h}/\sigma_\mathrm{v}$) is constant. The elastic properties are isotropic and layerwise homogeneous. It also means that the model has no tectonic stresses due to compressional or extensional forces or displacements.
- The vertical stress component is equal to the overburden load. This means that all stresses are conducted straight vertically and will not influence each other.

The vertical component is then equal to the overburden weight, the lithostatic pressure. It represents the 'pressure' state of the porous bulk element.

2.1.2 Pore Pressure Formation and Fluid Flow

The measurable pressure value in the pore fluid is the pore pressure. It is mainly caused by the overburden weight, but fluid flow together with compaction can decrease the overburden induced pressure and the resulting pore pressure is usually smaller than the lithostatic pressure. In a non–compactable porous rock, the lithostatic pressure and the pore pressure are both equal to the overburden load. Fluid outflow allows grain rotation to more compact packages, which decreases pore pressure and porosity. Thus, the difference between lithostatic and pore pressure is a measure of compaction.

Ideal compaction does not reduce pore pressure to zero. Instead a hydrostatic pressure remains, which is equal to the weight of the overlaying water column. Generally, the hydrostatic pressure is defined as the part of the pore pressure which does not contribute to water flow. The hydrostatic zero level can be arbitrarily defined, since only gradients and not absolute values of pressures control pore water flow. The groundwater table is not suitable as a constant reference level, since it varies over basin scale. Instead, the seawater level is used as the hydrostatic zero level. The hydrostatic pressure is then equivalent to the water column weight measured from the seawater level and therefore depends on sea and pore water density. Note that the hydrostatic pressure is not a measurable pressure. It is a theoretical pressure for ideal compactable layers or slow sedimentation.

The difference between the pore pressure and the hydrostatic pressure is the overpressure which directly controls water flow (Fig. 2.2). The pore pressure lies usually between hydrostatic and lithostatic pressure, but there are exceptions. It can be lower than the hydrostatic pressure when high uplift and erosion rates act on deep sand layers which are connected to near surface pressure areas along permeable facies. It can also exceed lithostatic pressures when large overpressures are built up by gas generation or highly permeable facies are connected at large depth levels.

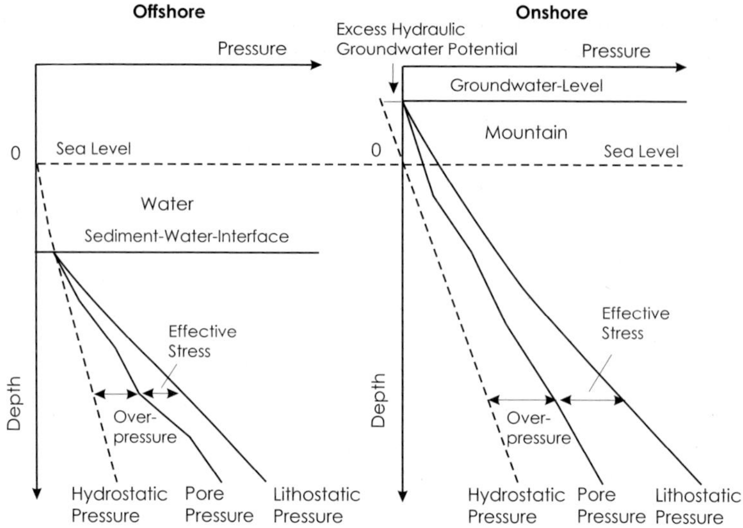

Fig. 2.2. Definitions of pressures and stresses. The groundwater level is often assumed to match the surface in basin modeling. Then, pore and lithostatic pressure have the same zero level, as shown here

One can distinguish between three processes of overpressure build up: overburden load together with mechanical under–compaction, cementation and overpressuring caused by fluid expansion processes (Osborne and Swarbrick, 1997; Swarbrick et al., 2002).

Overburden load induced pore pressure formation due to incomplete sediment compaction, as explained above, is the main process for overpressure formation. Here, compaction is the rearrangement of grains to denser packages with a reduction in pore space related to a decrease in pore throats and connectivity of the pore network. This process of grain rotation, crushing and deformation is called mechanical compaction. Compaction is caused by overburden load. The load acts on the pore fluid and the rock grains according to their compressibilities. Incremental fluid outflow generates a difference between rock stresses and pore pressure, which allows compaction. Compaction in turn changes the ratio between the rock stresses and the fluid pressure, since it decreases the rock and bulk compressibility, enforces further fluid outflow, and decreases the thickness of the solid matrix. The result of this coupled process is always a reduction of overpressure since the outflow from the compacting element is greater then the local increase of overpressure due to the thinning of the solid matrix. This is ensured as the compaction law is formulated in a manner, that relates porosity loss with effective stress increase. Finally, mechanical compaction is considered to be an overpressure reducing process (Fig. 2.3.a). The remaining overpressure could be simply interpreted as a result of incomplete compaction and that is why this process of overpressure formation is called under–compaction.

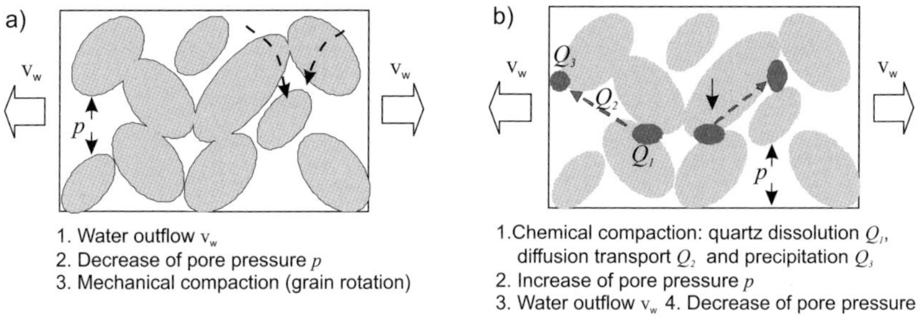

Fig. 2.3. Overpressure and Compaction: (a) mechanical compaction is a result of water outflow, it is always related to decrease in overpressure. (b) Quartz cementation and related compaction transfers lithostatic to pore pressure. It increases pore pressure. Water outflow can partially decrease the overpressure afterwards

Another source for overpressure is chemical compaction due to cementation. Cementation occurs in all sandstones and carbonates. It significantly decreases the porosity, and is mainly responsible for porosity reduction at large depths, where mechanical compaction is almost negligible. Cementation is the result of dissolution of quartz from the horizontal contact areas, diffusive transport within the pore water, and precipitation of a silican cement on free quartz surfaces. Quartz dissolution is mainly stress controlled. Temperature affects the diffusion constant and precipitation rate. Chemical compaction increases overpressure, since rock stress is transfered from the rock matrix to pore pressure. Cementation also drives fluid outflow and compaction with the generated overpressure as the main driving force (Fig. 2.3.b).

The third group of overpressure generating processes encompasses fluid expansion mechanisms: oil and gas generation, oil to gas cracking, aquathermal expansion and mineral changes such as smectite to illite conversion. In all these processes, mass or the density of the fluids changes and yields fluid pressure increase controlled by fluid compressibility. The overpressure increase due to fluid expansion mechanisms is usually small compared to those related to mechanical and chemical compaction.

2.1.3 Compaction and Porosity Reduction

Compaction is the reduction of the sediment bulk volume and is equivalent to volumetric strain $\epsilon_v = V/V_0$, the ratio of a load bearing volume V to the unloaded initial volume V_0. The average of the volumetric charge of a specimen is called the mean stress $\bar{\sigma}$. Stresses and strains are further explained in the Sec. 2.6. A compaction law relates volumetric strain to mean stress changes with an elastic parameter.

	Rock Compaction	Pore Space Compaction
Elastic	Elasticity of the Grains	Elasticity of the Skeleton Elasticity of the Pore Fluid
Plastic	Plasticity of the Grains	Rearrangement of the Grains Pressure Dissolution

Table 2.1. Compaction related Mechanisms

$$C = -\frac{1}{V}\frac{\partial V}{\partial \bar{\sigma}} = \frac{\partial \epsilon_v}{\partial \bar{\sigma}} \,. \tag{2.1}$$

Compaction mainly decreases porosity, but also reduces the grain volume. Generally, the rock and pore volumes are reduced with reversible (elastic) and irreversible (plastic) contributions. Some of the special mechanisms acting on microscopic and mesoscopic scales are listed in Table 2.1 after Schneider et al. (1996).

The compressibility in equation (2.1) is mainly a property of the grain framework and is called bulk compressibility. In the absence of a pore fluid, it relates the bulk volume decrease with the mean total stress. The presence of a pore water retards compaction, which as a first approximation can be described as the introduction of a mean effective stress $\bar{\sigma}' = \bar{\sigma} - p$ instead of the mean total stress in (2.1) with a reduction of the pore pressure p. Terzaghi (1923) confirmed this thesis experimentally, by proving that increasing the mean total stress or decreasing the pore pressure yields the same amount of compaction. Generally, Terzaghi's effective stress can also be introduced as a stress tensor $\boldsymbol{\sigma}'$.

$$\boldsymbol{\sigma}' = \boldsymbol{\sigma} - p\,\boldsymbol{I} \tag{2.2}$$

where $\boldsymbol{I}$ is the unit tensor (Chap. 8).

In practice, compaction laws on the basis of Terzaghi's effective stress definition are written in terms of porosity loss versus the vertical component of the effective stress σ_z'. The usage of porosity change instead of the volumetric bulk strain neglects volume changes of the solid matrix which are small. The restriction to the vertical effective stress means that a fixed ratio between horizontal and vertical stresses is assumed. The corresponding vertical total stress can then be simply approximated by the overburden sediment load pressure p_l.

$$\frac{\partial \phi}{\partial t} = -C_T\,\frac{\partial \sigma_z'}{\partial t} = -C_T\,\frac{\partial (p_l - p)}{\partial t} \,. \tag{2.3}$$

For most rock types, the Terzaghi compressibility C_T decreases rapidly during compaction. This type of compaction law is widely used in basin modeling. However, the formulation with only the vertical components of the stress tensor fails, when active extensional or compressional tectonics occur. Therefore, an extension of the law is proposed in Sec. 2.8.

Biot (1941) worked out a more detailed poro–elastic model for the extension of equation (2.1) for water filled porous rocks, taking into account the effect of the rock compressibility C_r, which yields the following compaction law with the Biot compressibility C_B.

$$\frac{\partial \epsilon_v}{\partial t} = C_B \frac{\partial \bar{\sigma}'}{\partial t} \quad \text{with} \quad \boldsymbol{\sigma}' = \boldsymbol{\sigma} - \alpha p \boldsymbol{I} \quad \text{and} \quad \alpha = 1 - \frac{C_r}{C_B} \,. \tag{2.4}$$

This formulation means, that the retardation of the compaction due to pore pressure drops with lower bulk compressibilities, since the rock compressibilities remain almost constant during compaction. Exceptions are mineral transformations or plastic flow of the grains. In unconfined sediments, $C_r \ll C$ (soil mechanical approach) and $\alpha \approx 1$, while at large depth $C_r \approx C_B \, \phi$ and $\alpha \approx 1 - \phi$ (rock mechanical approach). The case $\alpha = 1$ also means, that the effective stress is equal to the Terzaghi's assumption of negligible rock grain deformations. Note, that these effective stresses are only formal entities and not measurable physical values.

2.2 Terzaghi Type Models

Terzaghi type models are based on the simplifications of the lithostatic stress concept. In these models, overpressure formation related to incomplete mechanical compaction is considered and a fixed relation between porosity reduction and sediment compaction is assumed. The models have been widely used in 1D–Basin modeling programs since the early 90's. The assumptions are as follows:

- The "lithostatic pressure" concept is considered taking into account only the vertical component of the stress tensor as the maximum principal stress. The lithostatic pressure is equal to the overburden weight. The horizontal stresses are fixed ratios of the lithostatic pressure. Additional tectonic stresses, due to compressional or extensional forces, are neglected.
- Pore pressure formation is caused by overburden load. Fluid flow and compaction determine how the pressure is formed and distributed in the basin. Compaction is related to pore fluid outflow and decreases overpressure. One phase fluid flow in a fully saturated rock is considered, which is controlled by permeabilities. Pressure communication within the porous network is assumed.
- Mechanical compaction of the pore space takes into account the rearrangement of the grains to more compact blocks. All compaction is related to porosity reduction caused by pore fluid outflow. This porosity reduction process is controlled by the Terzaghi's effective stress value which is equal to the difference of the lithostatic and the pore pressure: $\sigma' = \sigma_z - p$. A relationship between maximum effective stress and porosity is assumed.
- Water is treated as incompressible.

2.2.1 Basic Formulation

Hydrostatic and Lithostatic Pressure

The hydrostatic pressure p_h at depth h is equal to the weight of a pure water column from sea level with the water density ρ_w.

$$p_h(h) = \int_0^h g\rho_w dz \tag{2.5}$$

with $z = 0$ at sea level. This yields positive values below and negative values above sea level. The negative hydrostatic pressure at groundwater level is the groundwater potential. In basin modeling, the groundwater level is often assumed to be identical to the sediment surface.

The water density varies with changing salinity values, while the dependency on temperature and pressure is relatively small and often neglectable. A further simplification is the assumption of two constant densities for seawater $\rho_{\text{sea}} = 1100\,\text{kg/m}^3$ and pore water $\rho_w = 1040\,\text{kg/m}^3$. This yields piecewise linear curves for hydrostatic pressure versus depth in sediments below sea water (Fig. 2.4).

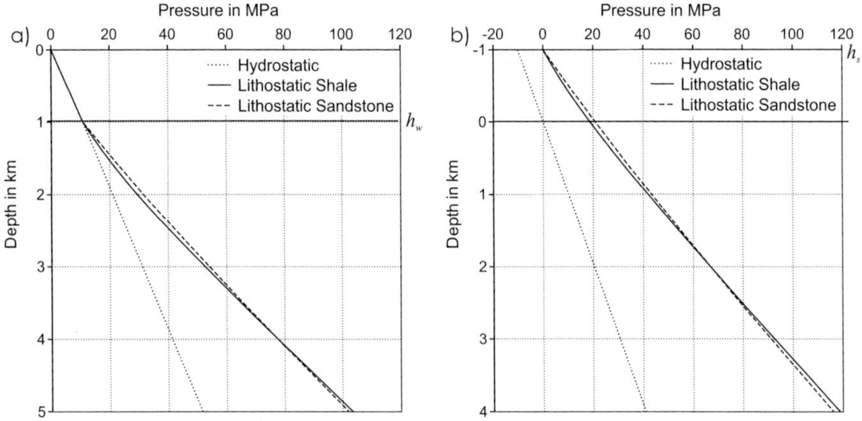

Fig. 2.4. Hydrostatic and lithostatic pressure curves for normal compacted rocks with the following properties: sea water density $\rho_{\text{sea}} = 1100\,\text{kg/m}^3$, pore water density $\rho_w = 1040\,\text{kg/m}^3$, shale density $\rho_s = 2700\,\text{kg/m}^3$, sandstone density $\rho_s = 2720\,\text{kg/m}^3$. (a) Offshore with a water depth $h_w = 1\,\text{km}$. (b) Onshore with a height of $h_s = 1\,\text{km}$. The lithostatic curves cross each other, since shale starts with a higher initial porosity but compacts faster

The lithostatic pressure p_l is equivalent to the total load of the overlaying sediments of bulk density ρ_b and sea water. Lithostatic zero level is the surface onshore and the seawater level offshore.

$$
\begin{aligned}
p_l(h) &= g \int_{h_s}^{h} \rho_b \, dz && \text{onshore} \\
p_l(h) &= g \int_{0}^{h_w} \rho_{\text{sea}} \, dz + g \int_{h_w}^{h} \rho_b \, dz && \text{offshore}
\end{aligned}
\tag{2.6}
$$

where h_s is the sediment surface. The integral over the weight of overburden sediments can be replaced by a sum of the weights of the single layers with thicknesses d_i (i is the layer number), rock densities ρ_{ri}, and and porosities ϕ_i.

$$
p_l(z) = \rho_{\text{sea}} g h_w + g \sum_{i=1}^{n} d_i \left[\rho_w \phi_i + \rho_{ri} \left(1 - \phi_i\right)\right] . \tag{2.7}
$$

For a homogeneous sediment column with a constant rock density ρ_r equation (2.7) can further be simplified as follows:

$$
\begin{aligned}
p_l(h) &= g\rho_{\text{sea}} h_w + g\rho_r (h - h_w) - g(\rho_r - \rho_w) \int_{h_w}^{h} \phi dz \,, && \text{onshore} \\
p_l(h) &= g\rho_r (h - h_s) - g(\rho_r - \rho_w) \int_{h_s}^{h} \phi dz \,, && \text{offshore.}
\end{aligned}
\tag{2.8}
$$

The remaining integral in the above equation is the weight percentage of water in the overlaying sediment column.

At larger depths, the term $(1 - \phi)$ does not significantly change, which means the curve tends toward a straight line for a unique sediment type. Lithostatic pressure curves for shale and sandstone, for hydrostatic compaction with compaction parameters of Fig. 2.8, are shown in Fig. 2.4. The term lithostatic potential u_l is used for the lithostatic pressure minus hydrostatic pressure $u_l = p_l - p_h$.

Pore Pressure Equation

The pore pressure equation is a one phase fluid flow equation based on the mass balance of pore water. A flow equation relates driving forces with flow rates. The driving force for pore water flow is the overpressure gradient. Darcys law establishes a linear relationship between the discharge velocity $\mathbf{v}$ of the pore fluid and the overpressure gradient ∇u assuming relatively slow flow for a Newtonian fluid. The proportionality factor is the mobility $\boldsymbol{\mu} = \boldsymbol{k}/\nu$, which is a function of the rock type dependent permeability $\boldsymbol{k}$ and the fluid dependent viscosity ν.

$$
\mathbf{v} = -\frac{\boldsymbol{k}}{\nu} \nabla u \,. \tag{2.9}
$$

This flow equation is an analogy to Fourier's equation of heat flow, which similarly relates temperature gradient and heat flux with the thermal conductivity tensor. The permeability tensor is often simplified using only two values parallel and perpendicular to the facies layering, named as vertical and horizontal permeabilities.

Mass balance requires, that any fluid discharge from a volume element is compensated by change in the contained fluid mass. The internal fluid mass changes when the fluid density or the fluid volume is modified (App. B).

$$\nabla \cdot \mathbf{v} = -\frac{1}{1-\phi}\frac{\partial \phi}{\partial t} + \frac{1}{\rho}\frac{\partial \rho}{\partial t} . \tag{2.10}$$

Local changes of the fluid densities occur for fluid expansion processes like aquathermal pressuring, mineral transformations or petroleum generation and cracking. Changes in the fluid volume or porosity are related to mechanical and chemical compaction, which are considered as two independent processes.

The porosity reduction due to mechanical compaction is formulated with Terzaghi's compaction law, while chemical compaction induced porosity loss is a temperature and effective stress dependent function $f_c(T, \sigma')$, as specified later in Sec. 2.3.

$$\frac{\partial \phi}{\partial t} = -C\,\frac{\partial \sigma_z'}{\partial t} - f_c(T, \sigma_z') . \tag{2.11}$$

The basic model deals with mechanical compaction only and supposes Terzaghi's effective stress definitions, which yield the following pressure equation.

$$-\nabla \cdot \frac{\boldsymbol{k}}{\nu} \cdot \nabla u = -\frac{1}{1-\phi}\frac{\partial \phi}{\partial t} = \frac{C}{1-\phi}\frac{\partial \sigma_z'}{\partial t} = \frac{C}{1-\phi}\frac{\partial (u_l - u)}{\partial t} . \tag{2.12}$$

Thus,

$$\frac{C}{1-\phi}\frac{\partial u}{\partial t} - \nabla \cdot \frac{\boldsymbol{k}}{\nu} \cdot \nabla u = \frac{C}{1-\phi}\frac{\partial u_l}{\partial t} . \tag{2.13}$$

The equation shows that the overburden load causes overpressure increase and compaction. In the absence of all overpressure generating sources, fluid flow is still admissible, but then the total inflow is equal to the total outflow of each element. Pore water loss is always related to the corresponding overpressure discharge and the grain structure reacts instantaneously with mechanical compaction.

The two lithological parameters, compressibility and permeability control fluid flow and pressure formation. The bulk compressibility describes the ability of the rock framework to compact and it also controls how overburden influences pore pressure. The bulk compressibility in the pressure equation should not be mixed up with pure grain or fluid compressibility, which is orders of magnitude smaller. The higher the compressibility of the element the higher the pore pressure decrease and the smaller the overpressure formation. The permeability controls flow rates, flow paths, and the resulting pore pressure fields. The overpressure in an element cannot decrease if the elements surroundings are impermeable even when the element itself is highly permeable and compressible.

The permeability can vary by several orders of magnitude, ranging from highly permeable facies (sandstone) to low permeability facies (shale) to almost impermeable facies (salt). The two end members of almost impermeable and highly permeable facies are handled with special methods, which will be further discussed later in the 2D– and 3D–pressure examples.

Boundary values of equation (2.13) are overpressures and water flow velocities as illustrated in Fig. 2.5. The upper boundary condition is zero overpressure at the sediment–water–interface offshore and an overpressure equal to the groundwater potential at the sediment surface onshore. The groundwater potential yields topographic driven flow, which is explained in Sec. 2.2.5.

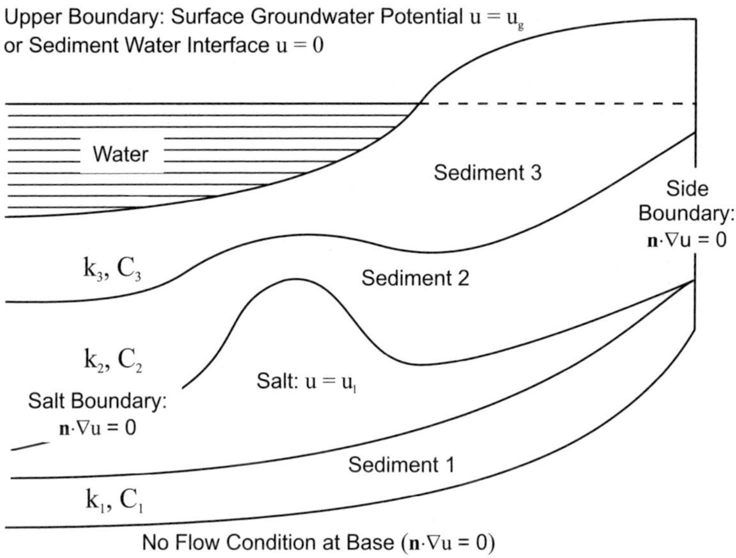

Fig. 2.5. Boundary value problem for overpressure calculation

The lower and side boundaries are no–flow areas, which means the overpressure gradient along the surface normal $\mathbf{n}$ is set to zero $\mathbf{n} \cdot \boldsymbol{\nabla} u = 0$. They are called closed boundaries. In small (prospect) scale models, special overpressures are usually set as side boundary values for some layers. For example, zero overpressure should be set at a permeable layer boundary, if it has a highly permeable connection to a hydrostatic area. To fix an overpressure value as a boundary condition at a certain point, is like injecting or releasing water until the given pressure is achieved. One can also apply a complete pressure array as side boundary values on prospect scale models from precalculated and calibrated basin scale models (Sec. 8.9). Special inner boundary conditions have to be set to impermeable rocks, namely no flow across the boundaries to these areas $\mathbf{n} \cdot \boldsymbol{\nabla} u = 0$ and lithostatic pressure within impermeable regions $u = u_l$.

Compaction and Porosity Reduction

In the basic model, a simple relationship between mechanical compaction and porosity decrease is considered. Hence, the related porosity change is equivalent to the bulk strain and a function of the Terzaghi's effective stress. Several relationships between porosity and effective stress have been developed and they are described in the following section. Although the formulations look different, they are similar to exponential relationships of the following type:

$$\phi \approx k_1 \, e^{-k_2 \, \sigma_z'} \, . \tag{2.14}$$

The compaction in each volume element is usually realized with contraction of its vertical edges when only vertical compaction occurs. The relative decrease in any vertical length is equal to a relative decrease in volume. Then, the actual thickness d is calculated using any previous or initial thickness d_0 from the present and previous porosities ϕ, ϕ_0 as follows:

$$\epsilon_z = \frac{d}{d_0} = \frac{1-\phi_0}{1-\phi} \, . \tag{2.15}$$

2.2.2 Mechanical Compaction

Mechanical compaction is almost irreversible. Hence, porosity is maintained when effective stress is decreased due to uplift, erosion, or an overpressure increase. The general porosity-effective stress relationship (2.14) could then still be used, but with the maximum effective stress value instead of the actual effective stress. This is taken into account when the following compaction laws are formulated in terms of effective stresses. Most mechanical compaction functions are porosity–effective stress relationships with decreasing porosity for increasing effective stress. The lithotype dependent functions can be measured through a triaxial compression test. Soil mechanical models use logarithmic functions between the void ratio $e = \phi/(1-\phi)$ and the effective stress, which yields a similar curve as equation (2.14).

$$e \approx k_1 - k_2 \log(\sigma_z') \, . \tag{2.16}$$

The equivalence of the relationships (2.14) and (2.16) is illustrated in Fig. 2.6. The exponential porosity–effective stress has a wide range of linear porosity versus the logarithm effective stress relationship for most lithologies, and it also behaves almost linearly in the high porosity range when transformed into the corresponding void ratio diagram. Hence, the pure soil mechanical formulation should only be applied to an effective stress of 15 MPa or approximately 1 km.

Compaction curves generally depend on the stress path, but usually only normal compaction curves, with an uniform increase in overburden, are taken into account. Stress release caused by uplift and erosion shows an elastic rebound, usually described with a low incline in the compaction diagram as illustrated in Fig. 2.6.

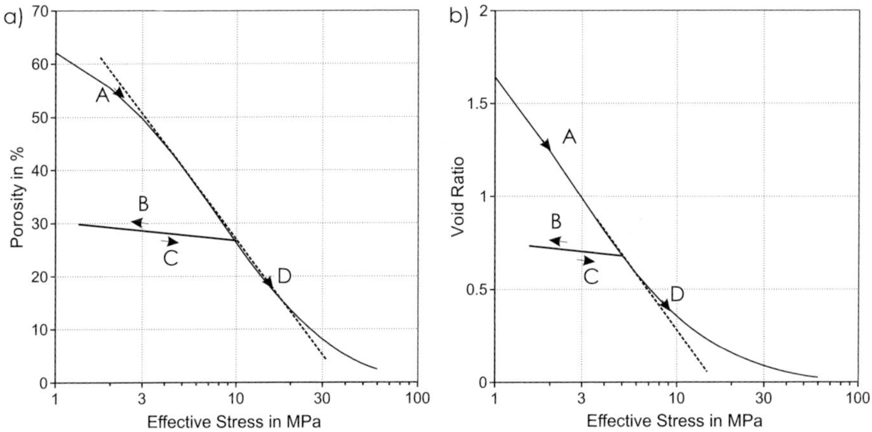

Fig. 2.6. Normal compaction curves (A and D) of a typical shale with an exponential porosity–effective stress relationship. The parameters are given in Fig. 2.8. The paths (B) and (C) represent load removal (erosion and uplift) and reload, respectively. (a) The porosity–stress relationship plotted versus the logarithmic stress axis has a wide range of linear behavior. (b) The void ratio–logarithm effective stress plot also behaves linearly for high porosities. The dashed curves represent linear approximations between porositiy and void ratio and the logarithm of the effective stress

Relationship between Effective Stress, Equivalent Hydrostatic Depth and Compressibility

The following compaction laws relate porosity to either effective stress $\phi(\sigma_z')$, frame compressibility $\phi(C)$, or equivalent hydrostatic depth $\phi(z_e)$, which is the depth of the sample with the same porosity and rock type under hydrostatic pressure conditions. A formulation in terms of one of these independent variables can always be converted analytically or numerically into either of the others. A compaction law has to encompass all three relations during simulation:

- $\phi(\sigma_z')$ determines new porosities after the pore pressure equation yields the effective stresses with the calculated new pore pressures.
- $C(\phi)$ or $C(\sigma_z')$ determines the actual frame compressibilities which are required in the pressure equation. The compressibility is the derivation of the $\phi(\sigma_z')$ function after σ_z'. It defines the slope of the porosity versus the effective stress curve.

- $\phi(z_e)$ is the theoretical porosity versus depth curve assuming hydrostatic pressures and the deposition of the entire column with the same lithotype. Many log and well data are available in terms of porosity versus equivalent hydrostatic depth rather than for porosity versus effective stress data. They are often used to determine the lithotype dependent parameters in the compaction laws.

Effective stress and compressibility based functions can simply be converted to each other by derivation or integration.

$$C(\sigma_z') = -\frac{d\phi(\sigma_z')}{d\sigma_z'}, \qquad \phi(\sigma_z') = -\int_0^{\sigma_z'} C(\sigma)d\sigma \ . \tag{2.17}$$

The hydrostatic porosity–depth function can be derived from the compressibility and porosity–effective stress equations as follows. For hydrostatic conditions, the effective stress change is equal to the change of the lithostatic minus the hydrostatic pressure

$$\frac{d\sigma_z'}{dz_e} = \Delta\rho\, g\,(1-\phi) \tag{2.18}$$

where $\Delta\rho = \rho_r - \rho_w$ is the difference of the rock and water density. Thus,

$$\frac{d\phi}{dz_e} = \frac{d\phi}{d\sigma_z'}\frac{d\sigma_z'}{dz_e} = -\Delta\rho g(1-\phi)C(\phi) \ . \tag{2.19}$$

The porosity–depth function can be analytically expressed if $(1-\phi)C(\phi)$ can be integrated. Analytical porosity–depth functions are very advantageous for the calibration of well data, but some relationships require numerical iteration schemes.

Athy's Law formulated with Effective Stress

Athy (1930) proposed a simple exponential decrease of porosity with depth for a given rock type described only with an initial porosity ϕ_0 and a compaction parameter k. As already explained above, effective stress rather than total depth should be used in the compaction law. A corresponding simple exponential porosity–effective stress function was first proposed by Smith (1971).

$$\phi = \phi_0\, e^{-k\sigma_z'} \ . \tag{2.20}$$

The compressibility function $C(\phi)$ and the hydrostatic porosity–depth function $\phi(z_e)$ are according to (2.17) and (2.19) as follows.

$$C(\phi) = k\phi, \tag{2.21}$$

$$\phi(z_e) = \frac{\phi_0}{\phi_0 + (1-\phi_0)\exp(k\Delta\rho\, g z_e)} \ . \tag{2.22}$$

The exponential function (2.20) is a straight line in the logarithmic porosity versus effective stress diagram with k as the decline angle. Typical compaction curves for clastic rocks are shown in Fig. 2.7. The previous model can be easily extended with consideration of a non–zero minimum porosity ϕ_m.

$$\phi = \phi_m + (\phi_0 - \phi_m)\exp(-k\sigma'_z), \tag{2.23}$$

$$C(\phi) = k(\phi_0 - \phi_m)\exp(-k\sigma'_z) = k(\phi - \phi_m), \tag{2.24}$$

$$\phi(z_e) = \frac{(\phi_0 - \phi_m) + \phi_m(1-\phi_0)\exp(k(1-\phi_m)\Delta\rho\, g z_e)}{(\phi_0 - \phi_m) + (1-\phi_0)\exp(k(1-\phi_m)\Delta\rho\, g z_e)}\ . \tag{2.25}$$

This model is frequently used in basin modeling (Giles et al., 1998), although the use of only one compaction parameter does not give a good match with observed data for many rock types.

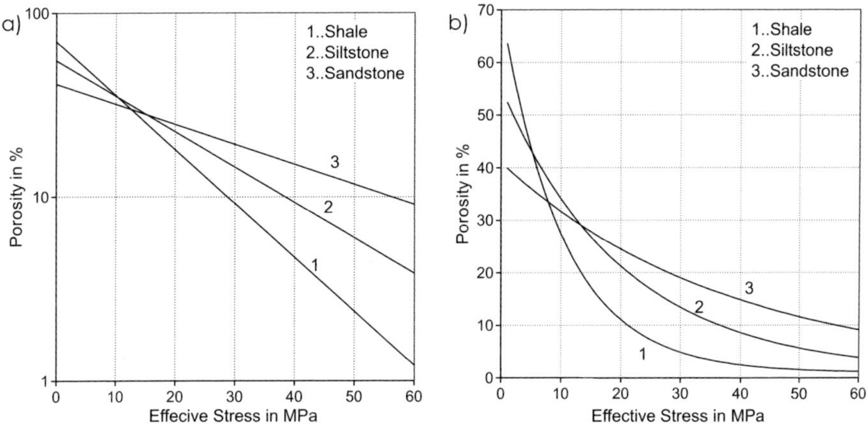

Fig. 2.7. Porosity versus effective stress curves on (a) logarithmic and (b) linear scale for various lithologies using Athy's Effective Stress law with the following parameters: shale $\phi_0 = 0.70; k = 0.096\,\text{MPa}^{-1}$, siltstone $\phi_0 = 0.55; k = 0.049\,\text{MPa}^{-1}$, sandstone $\phi_0 = 0.41; k = 0.0266\,\text{MPa}^{-1}$. The minimum porosity is zero

Athy's Law formulated with Hydrostatic Depth

A depth related porosity law (Athy, 1930) is used with the introduction of an equivalent hydrostatic depth z_e instead of the total depth.

$$\phi = \phi_0 \exp(-kz_e)\ . \tag{2.26}$$

The advantage of this formulation is, that the compaction parameter k can be easily determined when measured porosity versus equivalent depth data is available. The compressibility function $C(\phi)$ and the hydrostatic porosity–depth function $\phi(z_e)$ are according to (2.17) and (2.19) as follows.

$$C = \frac{k}{\Delta\rho\, g} \frac{\phi}{(1-\phi)}, \tag{2.27}$$

$$\sigma_z'(\phi) = \frac{\Delta\rho\, g}{k}(\phi - \phi_0 - \ln\frac{\phi}{\phi_0}) \,. \tag{2.28}$$

The inverse function $\phi(\sigma')$ can be calculated with the Newton iteration method. The resulting porosity–effective stress curves are generally steeper in the high porosity and shallower in low porosity ranges than the Athy versus effective stress functions. Hence, they are more applicable for most rock types even though they are based on only one compaction parameter. The authors prefer this law as a default for most lithologies. Example compaction curves for clastic rocks and carbonates are illustrated in Fig. 2.8.

Schneider Model

An extension of Athy's effective stress law to two exponential terms was proposed by Schneider et al. (1996).

$$\phi = \phi_1 + \phi_a \exp(-k_a \sigma_z') + \phi_b \exp(-k_b \sigma_z') \,. \tag{2.29}$$

Different compaction parameters k_a, k_b for lower and higher porosity ranges are realized with the superposition of two exponential terms. The initial porosity is equal to the sum of the three porosity parameters $\phi_1 + \phi_a + \phi_b$. Both porosities ϕ_a and ϕ_b are usually assumed to be half of the initial porosity value ϕ_0.

The corresponding compressibility function is as follows:

$$C(\sigma_z') = k_a \phi_a \exp(-k_a \sigma_z') + k_b \phi_b \exp(-k_b \sigma_z') \,. \tag{2.30}$$

The hydrostatic porosity versus depth function can be obtained, when equation (2.18) is integrated numerically to get $\sigma_z'(z_e)$, and then $\phi(z_e)$ is calculated with equation (2.29) afterwards. Numerical integration can also be applied to any other model with a given analytical expression for $\phi(\sigma_z')$. The proposed default parameters in App. A yield curves almost identical to those of Athy's depth model (Fig. 2.8).

Compressibility Model

Compressibilities are the derivatives of the porosity versus effective stress and are proportional to the slope of the porosity versus effective stress curves. This model assumes an exponential decrease in the compressibilities from the depositional value C_0 to a value C_m corresponding to the minimum porosity ϕ_m (Fig. 2.9).

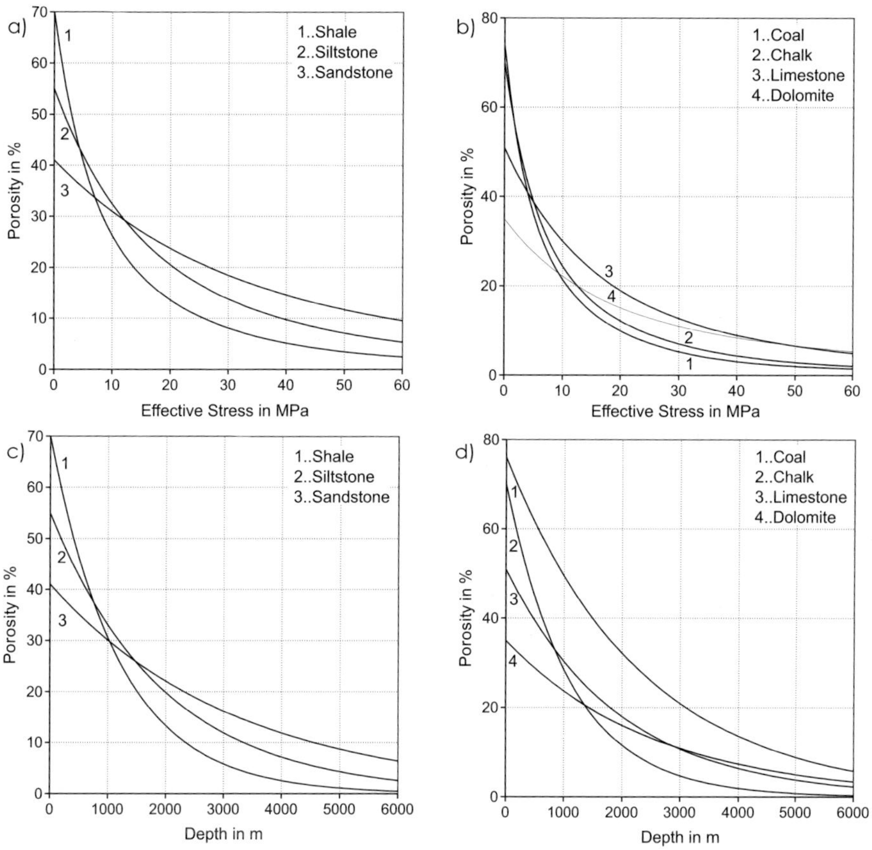

Fig. 2.8. Porosity versus hydrostatic depth and effective stress curves for various lithologies using Athy's depth law with the following parameters: shale $\phi_0 = 0.70, k = 0.83\,\mathrm{km}^{-1}$, siltstone $\phi_0 = 0.55, k = 0.34\,\mathrm{km}^{-1}$, sandstone $\phi_0 = 0.41, k = 0.31\,\mathrm{km}^{-1}$, coal $\phi_0 = 0.76, k = 0.43\,\mathrm{km}^{-1}$, chalk $\phi_0 = 0.70, k = 0.90\,\mathrm{km}^{-1}$, limestone $\phi_0 = 0.51, k = 0.52\,\mathrm{km}^{-1}$, dolomite $\phi_0 = 0.70, k = 0.39\,\mathrm{km}^{-1}$

$$\log C(\phi) = \frac{\phi_0 - \phi}{\phi_0 - \phi_m} \log C_m + \frac{\phi - \phi_m}{\phi_0 - \phi_m} \log C_0 \;. \tag{2.31}$$

This is equivalent to the following expression:

$$\begin{aligned} &C(\phi) = \alpha \exp(\beta\phi) \qquad \text{with} \\ &\ln(\alpha) = \frac{\phi_0 \ln C_m - \phi_m \ln C_0}{\phi_0 - \phi_m}, \quad \beta = \frac{\ln C_0 - \ln C_m}{\phi_0 - \phi_m} \;. \end{aligned} \tag{2.32}$$

Integration of the above exponential function yields the corresponding effective stress correlations.

$$\begin{aligned} \phi(\sigma_z') &= -\frac{1}{\beta}\ln(\alpha\beta\sigma_z' + \exp(-\beta\phi_0)) \\ C(\sigma_z') &= \frac{\alpha}{\alpha\beta\sigma_z' + \exp(-\beta\phi_0)} . \end{aligned} \tag{2.33}$$

Numerical integration of (2.19) can be used to determine the hydrostatic porosity versus depth function $\phi(z_e)$ from $C(\sigma_z')$. Compressibility models generally decrease too fast for low porosities. Default parameters are shown in Appendix A and the curves for clastic rocks are illustrated in Fig. 2.9.

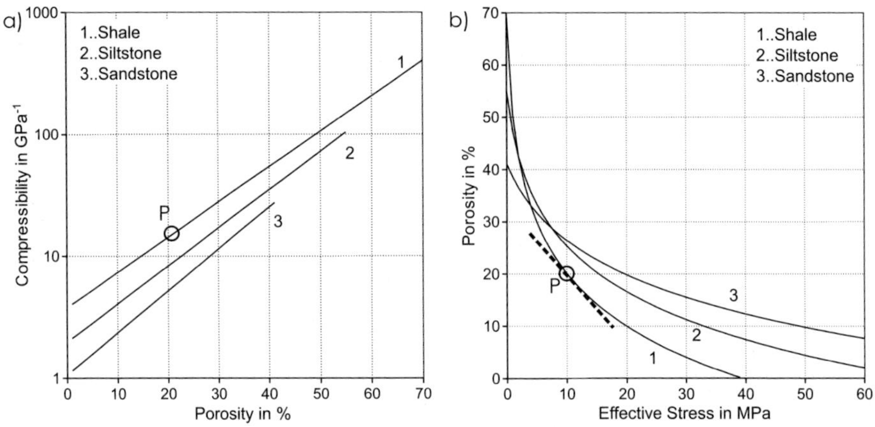

Fig. 2.9. Compaction curves for various lithologies using the compressibility model with the following parameters: Shale $C_0 = 403\,\mathrm{GPa}^{-1}, C_m = 4.03\,\mathrm{GPa}^{-1}$, Siltstone $C_0 = 103\,\mathrm{GPa}^{-1}; C_m = 2.11\,\mathrm{GPa}^{-1}$, Sandstone $C_0 = 27.5\,\mathrm{GPa}^{-1}, C_m = 1.15\,\mathrm{GPa}^{-1}$

Mudstone Model

The following law from soil mechanics is especially applicable for clastic rocks.

$$e = \frac{\phi}{1-\phi} = e_{100} - \beta \log \frac{\sigma_z'}{0.1\,\mathrm{MPa}} . \tag{2.34}$$

The reference void ratio e_{100} at 0.1 MPa can be considered as an initial void ratio $e_{100} = \phi_0/(1-\phi_0)$ although this compaction law yields a singularity $e \to \infty$ for the void ratio at zero stress. The two functions $\phi(\sigma_z')$ and $C(\sigma_z')$ are as follows.

$$\phi(\sigma_z') = \frac{\sigma_z' - \beta \log(\sigma_z'/0.1\,\mathrm{MPa})}{1 + e_{100} - \beta \log(\sigma/0.1\,\mathrm{MPa})}, \tag{2.35}$$

$$C(\sigma_z') = \frac{\beta}{\sigma_z'(1 + e_{100} - \beta \log(\sigma/0.1\,\mathrm{MPa})^2} . \tag{2.36}$$

The above model does not provide an analytical expression for the porosity-depth function. However, with the compressibility model equation (2.19), it can be integrated numerically.

The material dependent constants are the initial void ratio e_{100} and the compressibility β. They can be related to the (volumetric) clay content r for mudstones with the following relationships proposed by Yang and Aplin (2004) Fig. 2.10.

$$\begin{aligned} e_{100}(r) &= 0.3024 + 1.6867\,r + 1.9505\,r^2 \\ \beta(r) &= 0.0937 + 0.5708\,r + 0.8483\,r^2 . \end{aligned} \tag{2.37}$$

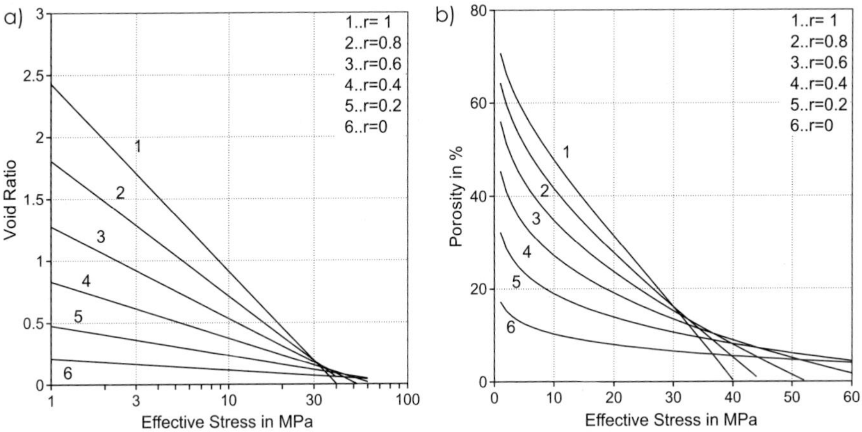

Fig. 2.10. Compaction curves for various lithologies using the mudstone model with the clay dependent functions of Yang and Aplin

Lauvrak (2007) proposed the EasySoil model with the following correlations to sample data for e^*_{100}, β^* and an upscaling to e_{100}, β for natural rocks (Fig. 2.11).

$$\begin{aligned} e^*_{100}(r) &= 0.725 - 0.252\,r + 2.53\,r^2, \\ \beta^*(r) &= 0.218 - 0.119\,r + 1.193\,r^2, \\ e_{100} &= e^*_{100} + 0.76\,\beta^*, \\ \beta &= 1.07\,\beta^* . \end{aligned} \tag{2.38}$$

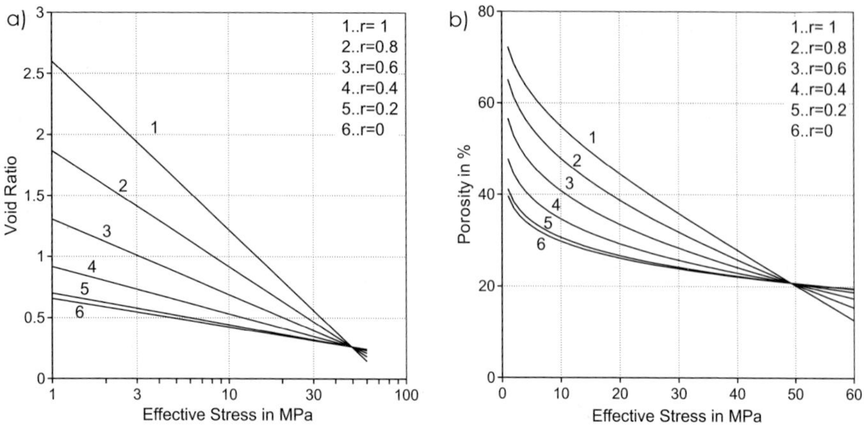

Fig. 2.11. Compaction curves for various lithologies using the mudstone model with the clay dependent functions of Lauvraks EasySoil model

Compressibilities

Bulk compressibilities can be directly derived from the compaction law as previously explained. Example curves for clastic rocks and carbonates are shown in Fig. 2.12 with the parameters of the Athy's hydrostatic depth model. Other compaction models yield similar curves with the proposed default parameters.

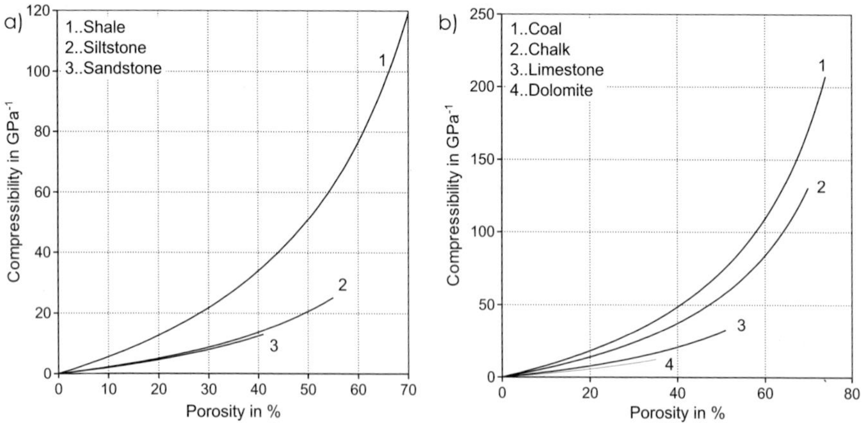

Fig. 2.12. Compressibility curves for various lithologies using Athy's depth law with the parameters of Fig. 2.8

Comparison of Various Lithologies

Although the formulation of the various compaction models look very different from each other, default parameters for most lithologies yield very similar compaction curves. An exception is the mudstone model, which is generally not suitable for approximation of the compaction trend in the lower porosity range. The standard shale curves are shown in Fig. 2.13 for all described models. Compaction parameters are mixed arithmetically. Example curves for shale–sandstone mixtures are shown in Fig. 2.14.

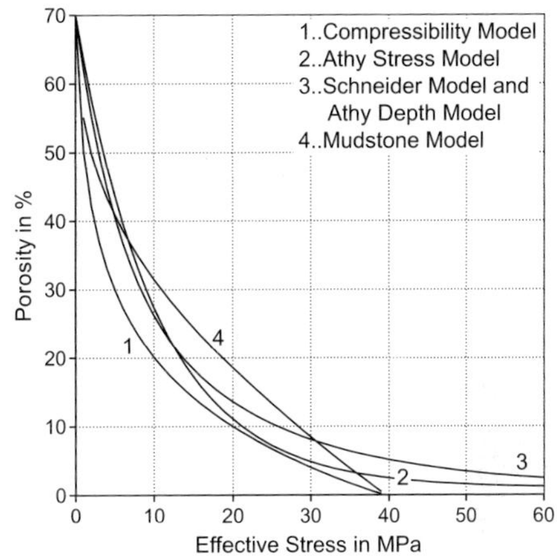

Fig. 2.13. Comparison of different compaction laws for shale: The curves for the Schneider and Athy's depth model are almost identical and plotted in one line. (1) Compressibility model: $C_0 = 403\,\mathrm{GPa}^{-1}$, $C_m = 4.03\,\mathrm{GPa}^{-1}$ (2) Athy's effective stress law: $\phi_0 = 0.70$, $k = 0.096\,\mathrm{MPa}^{-1}$ (3) Schneider model: $\phi_0 = 0.70$, $\phi_a = 0.35$, $k_a = 0.1916\,\mathrm{MPa}^{-1}$, $k_b = 0.0527\,\mathrm{MPa}^{-1}$ and Athy's hydrostatic depth law: $k = 0.83\,\mathrm{km}^{-1}$ (4) Mudstone model: $e_{100} = 1.2889$, $\beta = 0.458$

2.2.3 Permeability and Viscosity

The mobility μ is a measure of the ability of a material to transmit fluids. It includes the rock permeability k and the fluid viscosity ν: $\mu = k/\nu$. In the Darcy law (2.9), the mobility is the proportional factor between pressure gradients and fluid flow velocities. This applies to slow flowing (Newtonian) fluids. The permeability is mainly affected by the pore structure of the rock, and the viscosity describes the internal friction of the moving phase. This indicates that flow velocity rises with rising permeability and reduces with increasing viscosity.

Viscosity

Fluid viscosity is a measure of the resistance of the fluid against flow. It is related to the attractive forces between the molecules. Viscosity generally depends on pressure, temperature, and phase composition. The considered viscosity is the dynamic viscosity ν in contrast to the kinematic viscosity ν/ρ. The unit of the dynamic viscosity is Poise (P, $1\,\mathrm{Pa\,s} = 0.1\,\mathrm{P}$).

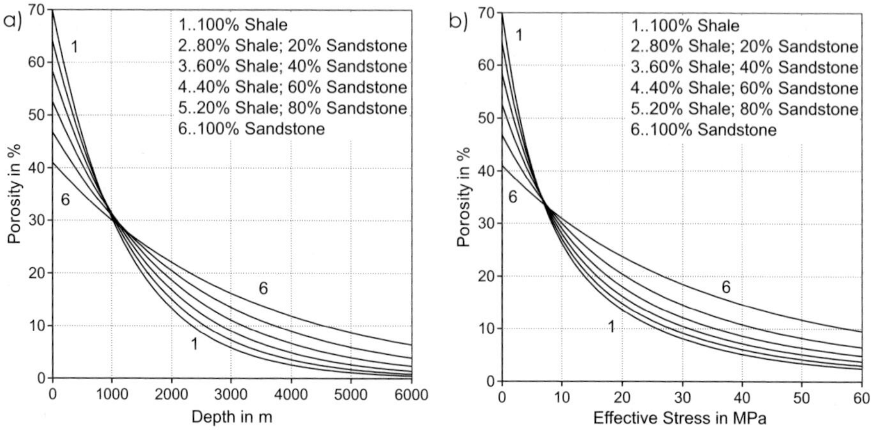

Fig. 2.14. Compaction curves for mixtures of shale and sandstone using arithmetic averages of all compaction parameters. (a) Athy's depth model with shale $\phi_0 = 0.70, k = 0.83\,\mathrm{km}^{-1}$ and sandstone $\phi_0 = 0.41, k = 0.31\,\mathrm{km}^{-1}$. (b) Schneider model with shale $\phi_0 = 0.70, \phi_a = 0.35, k_a = 0.1916\,\mathrm{MPa}^{-1}, k_b = 0.0527\,\mathrm{MPa}^{-1}$ and sandstone $\phi_0 = 0.42, \phi_a = 0.205; k_a = 0.0416\,\mathrm{MPa}^{-1}, k_b = 0.0178\,\mathrm{MPa}^{-1}$

The viscosity of saline water can be estimated from McCain Jr. (1990) in Danesh (1998) as follows:

$$\begin{aligned} \nu &= \nu_T\,(0.9994 + 4.0295 \times 10^{-5}\,P + 3.1062 \times 10^{-9}\,P^2) \\ \nu_T &= T^{-a}\,(109.547 - 8.40564\,s + 0.313314\,s^2 + 8.72213 \times 10^{-3}\,s^3) \\ a &= 1.12166 - 2.63951 \times 10^{-2}\,s + 6.79461 \times 10^{-4}\,s^2 \\ &\quad +5.47119 \times 10^{-5}\,s^3 - 1.55586 \times 10^{-6}\,s^4 \end{aligned} \tag{2.39}$$

with ν in mPa s, T in °F, P in psi, s in mass %, and the validity intervals for ν_T of 38°C $< T <$ 200°C and $s <$ 26%, and for ν of 30°C $< T <$ 75°C, and $P < 100\,\mathrm{MPa}$.

Another formulation was published by Hewlett-Packard (1985) in Mc Dermott et al. (2004).

$$\begin{aligned} \nu &= \nu_0\,[1 - 1.87 \times 10^{-3}\,s^{0.5} + 2.18 \times 10^{-4}\,s^{2.5} \\ &\quad +(T^{0.5} - 0.0135\,T)(2.76 \times 10^{-3}\,s - 3.44 \times 10^{-4}\,s^{1.5})] \\ \nu_0 &= 243.18 \times 10^{-7}\,10^{247.8/(\mathrm{T_K}-140)} \\ &\quad [1 + 1.0467 \times 10^{-6}\,P\,(T_K - 305)] \end{aligned} \tag{2.40}$$

with ν in mPa s, T in °F, T_K in K, P in bar, s in %, and the validity interval 0°C $< T <$ 300°C, $s <$ 25% and $P <$ 430 °C. In the published equation P is the difference between the real and the saturation pressure, but the latter one can be neglected for geological conditions.

Both formulations yield similar results for moderate pressures (Fig. 2.15). The uncertainties of mobilities are controlled more by permeability than viscosity. Hence, the following simplification of equation (2.39) without salinity and pressure dependence is also proposed here (dotted curve inFig. 2.15.a).

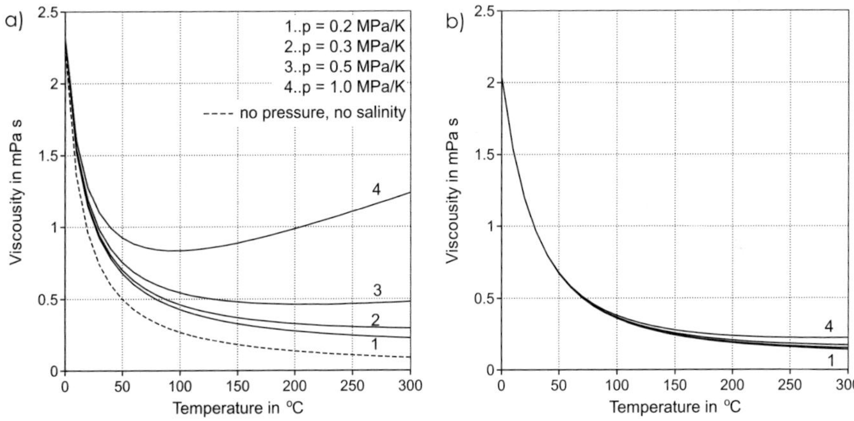

Fig. 2.15. Pressure and temperature dependent water viscosity curves assuming a salinity of 10% after (a) McCain Jr. (1990) in Danesh (1998) and (b) Hewlett-Packard (1985) in Mc Dermott et al. (2004). The pressure and salinity independent curve for the simplified equation (2.41) is the dashed curve in (a)

$$\nu[\text{cp}] = 109.5\, T^{-1.122} \tag{2.41}$$

with temperature T in $°F$. The viscosity of liquid and vapor petroleum is dependent on its composition. Viscosity ranges for standard oils and gases and more sophisticated methods for calculating oil and gas viscosities from compositions are described in (Sec. 5.6.5).

Permeability

Permeability consists of two factors namely rock (intrinsic) permeability and relative permeability, the latter one is further described in Sec. 6.3. The intrinsic permeability k is mainly affected by the pore structure, especially pore throat diameters and pore connectivity. Hence, it is dependent on the compaction state and usually tabulated as a function of porosity (App. A). The unit of permeability is Darcy ($1\,D = 0.98692 \times 10^{-12}\,\text{m}^2$), or millidarcy (mD), but logarithm millidarcy (log mD) is also used, since permeability values often vary over orders of magnitude with decreasing porosity ($1\log\text{mD} = 10\,\text{mD}$, $0\log\text{mD} = 1\,\text{mD}$, $-1\log\text{mD} = 0.1\,\text{mD}$, $-2\log\text{mD} = 0.01\,\text{mD}$, …).

The most commonly used permeability relationship is the Kozeny–Carman relation. A derivation can be drafted from Hagen–Poiseuille's law for fluid flow through a porous structure, which is approximated by a bundle of tubular parallel capillaries.

The flow velocity of a viscous fluid of N parallel tubes of radius r embedded in an rock matrix of bulk area A can be expressed with the Hagen–Poiseuille law as follows.

$$\mathbf{v} = \frac{N}{A} \frac{r^4 \pi}{8\nu} \boldsymbol{\nabla} p \tag{2.42}$$

where $\boldsymbol{\nabla} p$ is the driving pressure gradient along the tubes. The porosity of the considered tubular bundle is $\phi = N\pi r^2/A$, which yields the following fluid velocity.

$$\mathbf{v} = \frac{r^2 \phi}{8\nu} \boldsymbol{\nabla} p \,. \tag{2.43}$$

The comparison with Darcy's Law (2.9) results in a permeability of $k = r^2\phi/8$. The introduction of a tortuosity τ , which is defined as *"the length of the path actually followed between two points divided by the apparent path between these two points"* (Vidal-Beaudet and Charpentier, 2000) or as *"the averaged ratio of path–lengths traveled by a petroleum fluid to the geometrical length of the region of rock considered"* (England et al., 1987) yields

$$k = \frac{r^2 \phi}{8 \tau^2} \,. \tag{2.44}$$

It can be estimated with $\tau = \sqrt{3}$ for many rocks.

This equation can be rewritten to a so called Kozeny–Carman type relationship of the form

$$k = \frac{B \phi^3}{\tau^2 S^2} \tag{2.45}$$

with the specific surface area $S = N 2\pi r/A$ and B a geometrical factor (Mavko et al., 1998).

From consideration of sphere packing it is possible to estimate $S = (3/2)(1-\phi)/d$ with d as grain size. Furthermore, it is common, to replace the porosity ϕ with $(\phi - \phi_c)$ by assuming that the permeability vanishes below a threshold porosity ϕ_c where the pores become unconnected (Mavko et al., 1998).

However, the following revised Kozeny–Carman relationship has been proposed by Ungerer et al. (1990) for practical use in basin modeling.

$$\begin{aligned} k(\phi) &= 2 \times 10^{16}\, \kappa \,\frac{\phi'^5}{S^2 (1-\phi')^2} \quad \text{if} \quad \phi' < 0.1 \\ k(\phi) &= 2 \times 10^{14}\, \kappa \,\frac{\phi'^3}{S^2 (1-\phi')^2} \quad \text{if} \quad \phi' > 0.1 \end{aligned} \tag{2.46}$$

with specific surface area S in m^2/m^3, κ a lithotype dependent scaling factor and ϕ' a corrected porosity $\phi' = \phi - 3.1 \times 10^{-10} S$. Example parameters for clastic rocks are given in Table 2.2.

Lithology	Specific Surface Area in m^2/m^3	Scaling Factor
Shale	10^8	0.01
Siltstone	10^7	0.5
Sandstone	10^6	10.0

Table 2.2. Kozeny–Carman parameters for various lithologies

The Kozeny–Carman type relation (2.46) has two different exponential factors for the high and low porosity range. The permeability decrease for highly porous rocks is mainly caused by the reduction of the pore throat radius, while in the highly compacted rocks, the closure and elimination of pore throats yields a decrease in pore connectivity (coordination number) and a drop in permeability.

However, all Kozeny–Carman type models describe the intrinsic permeability dependent on porosity, pore size, pore throat radii distribution, and coordination number, which is a measure of the pore connectivities. Further considerations based on more complex geometrical models are e.g. given in Vidal-Beaudet and Charpentier (2000), Doyen (1988).

An alternative approach describes the permeability with a piecewise linear function in the log permeability versus porosity diagram. Example curves for many lithologies are tabulated in App. A in terms of three porosity versus log permeability pairs. Some of these curves for clastic rocks and carbonates together with the corresponding Kozeny–Carman curves with the parameters from Table 2.2 are illustrated in Fig. 2.16. Salt, granite and basalt are considered as impermeable.

Permeabilities are mixed geometrically for homogeneous mixtures or lithotypes. In layered mixtures the horizontal values are mixed arithmetically and the vertical values are mixed harmonically.

Generally, permeability $\boldsymbol{k}$ is a symmetrical tensor with six independent components. Similar to the thermal conductivity, it is often approximated with only two independent components: the permeability along the geological layer k_h and permeability across the geological layer k_v with an anisotropy factor $a_k = k_h/k_v$. Typical anisotropy values are $a_k = 2 \ldots 10$ for clastic rocks and $a_k = 1 \ldots 3$ for carbonates, they are tabulated for many rock types in App. A.

The above permeability curves and tabulated permeability values mean vertical permeabilities and equivalent hand-sample values, since most published data and in-house databases in oil companies are derived from hand-sample measurements. Basin scale values for horizontal and vertical perme-

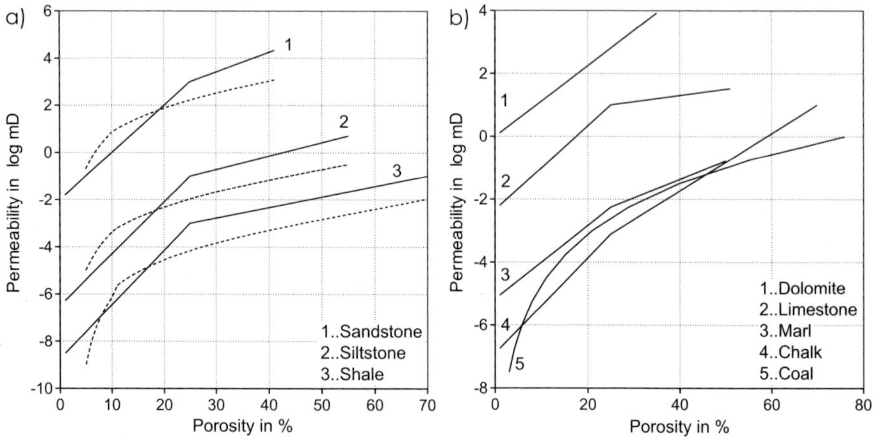

Fig. 2.16. Permeability curves for various lithologies with piecewise linear (solid) and Kozeny–Carman (dashed) relationships. The parameters are from Table 2.2 and from the Appendix A. A special curve is proposed for coal

abilities are calculated from the hand specimen values multiplied with a horizontal and vertical upscaling factor, respectively. The higher values for larger scales are caused by macro-fractures, inhomogeneities and permeable inclusions. Upscaling factors to basin scale elements with lengths greater than 50 m are reported for sandstones: 500 (horizontal) and 10 (vertical) (Schulze-Makuch et al., 1999). Based on the authors' experience, we suggest upscaling factors of 50 (horizontal) and 1 (no upscaling vertical) for all clastic rocks and carbonates, and no upscaling otherwise. Different horizontal to vertical upscaling increases the anisotropy factor in clastic rocks to $a_k = 100 \ldots 500$, respectively.

The permeability of fractured rock is much higher than that of undisturbed samples and is discussed in Sec. 2.6.1.

2.2.4 1D Pressure Solutions

Simplified 1D models can be used to discuss some fundamental processes of overpressure formation and compaction, although 1D solutions are less practical, since most overpressure distributions are strongly influenced by horizontal water flows along highly permeable layers (App. D). In this section, only mechanical compaction is considered in describing the interaction of overburden due to sedimentation, overpressure formation and compaction. The 1D formulation of the general pressure equation (2.13) is as follows.

$$C\frac{\partial u}{\partial t} - \frac{k}{\nu}\frac{\partial u}{\partial z} = C\frac{\partial u_l}{\partial t} \,. \tag{2.47}$$

Pressure curves for a unique rock type deposited with constant sedimentation rates are shown in Fig. 2.17 for shales and siltstones. Shale permeability decreases rapidly during burial, since the log permeability to porosity curve is very steep. Hence, there is only a small sedimentation rate dependent transition zone between the uppermost 1 . . . 3 km and the deeper part, where the pressure gradient is equal to the lithostatic gradient. The transition zone in lower permeable rocks like siltstone occurs over a broader region of sometimes several kilometers. The corresponding porosity curves for homogeneous depositions are shown in Fig. 2.18. The porosity reduction stops in the deep impermeable blocks, when water outflow is near zero.

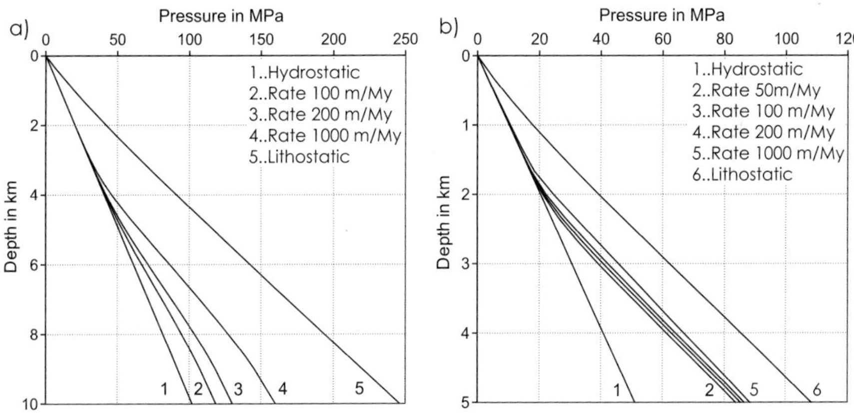

Fig. 2.17. Sedimentation rate dependent overpressure formation of (a) siltstones and (b) shales with piecewise linear permeability curves of Fig. 2.16

The pressure formation in an alternating sandstone–shale sequence is shown in Fig. 2.19. The pressure gradient in sandstone is equal to the hydrostatic gradient, while the pressure in the shale layer returns relatively quickly (after 500 m in the example) to almost the level of the pure shale curve. Hence, the increase of pressure in seals could be much higher than the lithostatic gradients.

The behavior of an impermeable seal is illustrated in Fig. 2.20. All overburden load above the seal is added to the pore pressure of all layers below the seal, since no pore water can cross the impermeable seal. This yields an increase to lithostatic pressure in the seal and a constant offset equal to the overburden load of the overpressure below the seal during the time after the sedimentation of the seal. This displacement with additional pressure is marked with the dotted line in Fig. 2.20.b. The pore pressure (solid line) also includes a small pressure exchange within the block below the seal.

The following calculation shows reservoir pressure decrease by water flow through a permeable seal (Fig. 2.21). The considered reservoir has a thickness

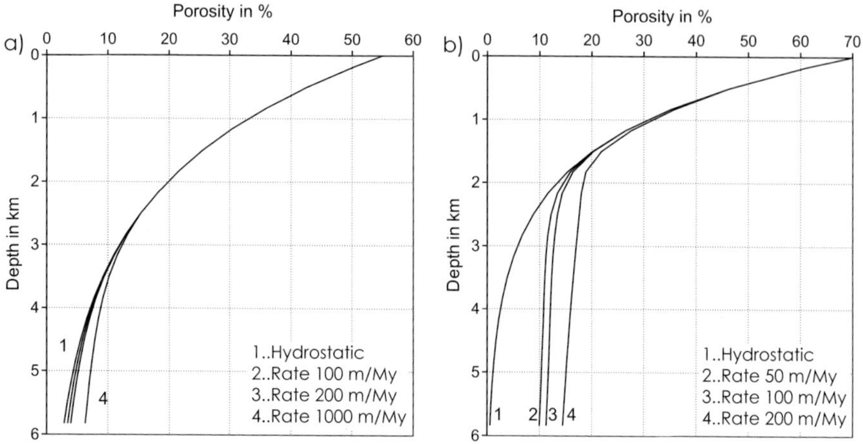

Fig. 2.18. Sedimentation rate dependent compaction of (a) siltstones and (b) shales with compaction curves of Fig. 2.8

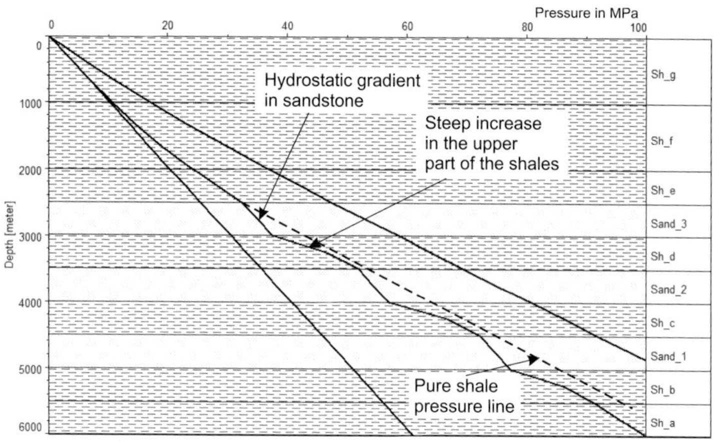

Fig. 2.19. 1D overpressure formation in an alternating sand–shale sequence

h_r, a bulk compressibility C_r and an overpressure u_r. The overpressure in the seal with a permeability k_s and a thickness h_s drops to zero at or near the top of the seal. The flow velocity in the seal is according to the Darcy's law as follows.

$$v = \frac{k_s}{\nu} |\nabla u| \approx \frac{k_s}{\nu} \frac{u_r}{h_s} \; . \tag{2.48}$$

Integration of equation (2.13) over the reservoir area with the assumption of no sedimentation yields the following relationship.

$$\int \mathbf{v} \cdot \mathbf{n} \, dS = \int C \frac{\partial u}{\partial t} dV \; . \tag{2.49}$$

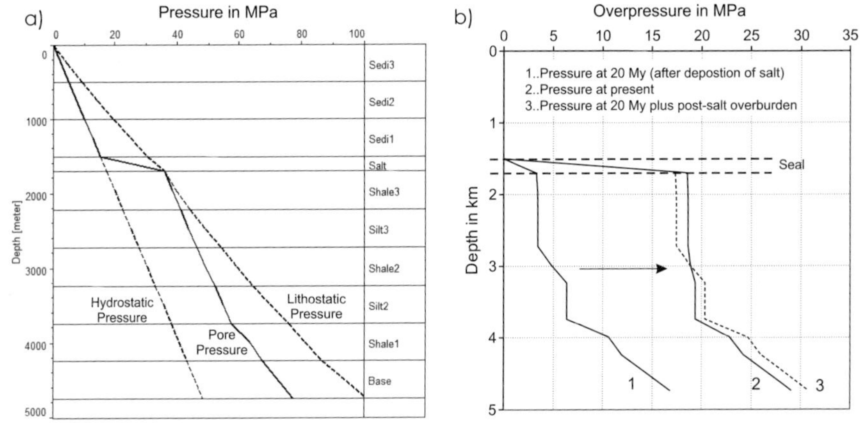

Fig. 2.20. 1D overpressure formation below a perfect seal. (a) Present day pressure versus depth curve. (b) Overpressure versus depth curves for the time of seal deposition and at present day. The pressure curve at seal deposition (20 My) is shifted to the corresponding present day locations to illustrate, that overpressure increase in all layers below the seal from 20 My to present day is a almost the same. The dashed curve is the 20 My curve plus the overburden load after seal deposition. The difference of the dashed curve and the present day overpressure curve is caused by water exchange in the layers below the seal

The outflow of the reservoir is restricted to the reservoir–seal interface with surface A_r. Hence,

$$A_r\, v = -C_r\, V_r\, \frac{\partial u}{\partial t} \tag{2.50}$$

with the reservoir volume $V_r = A_r\, h_r$. Thus,

$$\frac{\partial u_r}{\partial t} = -\frac{k_s}{C_r \nu h_s h_r} u_r \ . \tag{2.51}$$

It yields an exponential decrease in the reservoir pressure as follows.

$$u_r(t) = u_0 \exp(-\frac{t}{\tau}), \quad \tau = \frac{C_r \nu h_s h_r}{k_s} \ . \tag{2.52}$$

with an initial reservoir pressure u_0. The time $t_h = \tau \ln(2)$ when half of the overpressure is dropped is controlled by the permeability of the seal, the bulk compressibility of the reservoir, and the reservoir and seal thicknesses. Typical values for t_h (Fig. 2.22) show, that very low permeabilities are necessary to seal pressure over significant times.

2.2.5 Pressure Solutions in 2D and 3D

Most of the effects discussed in the previous sections are also important in multidimensional pressure calculations: the upper part of the basin is in a

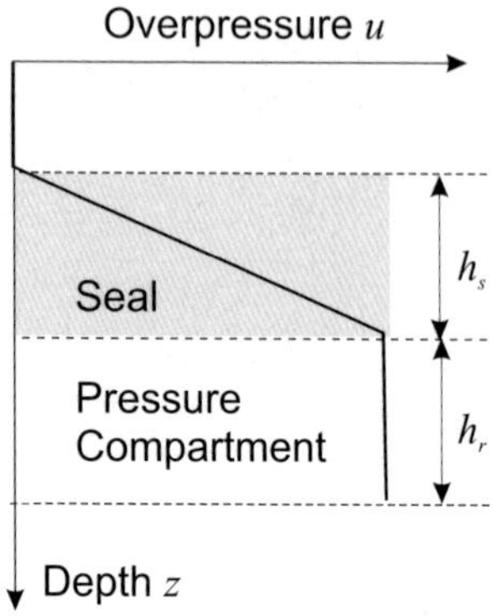

Fig. 2.21. Overpressure formation and decrease below a permeable seal. It is assumed, that the overpressure in a permeable seal drops linearly from reservoir pressure to zero. The overpressure in the compartment is constant. The water flow through the seal and the related decrease in reservoir pressure depend on seal permeability, reservoir compressibility and the thicknesses of the two layers

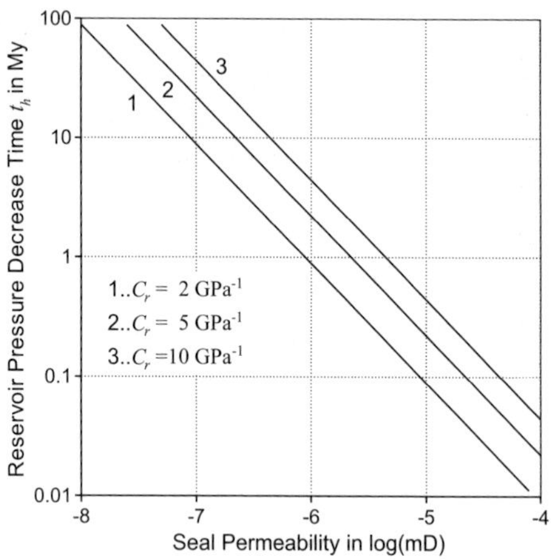

Fig. 2.22. Reservoir pressure decrease times t_h for various seal permeabilities k_s and $h_r = 200\,\mathrm{m}$, $h_s = 200\,\mathrm{m}$, $\nu = 0.5\,\mathrm{mPa\,s}$

hydrostatic state, pressure increases in impermeable layers, and large overpressure areas occur below low permeability seals. Additionally, high permeable layers transmit high water flow rates and yield overpressure equalization in the layer. Even thin high permeable layers affect the multidimensional overpressure field, especially when they are expanded over long distances or large depths. This is illustrated in Fig. 2.23, where high permeability sand layers of very different depths are well connected to each other and yield almost the same overpressure everywhere in the sands. The calculated difference for the connected layers with a permeability of $k = -2.5 \times \log(\mathrm{mD}) = 3.2 \times 10^{-3}\,\mathrm{mD}$ is about $\Delta u = 0.01\,\mathrm{MPa}$.

The pressure difference is higher, when low permeability rocks interrupt the connectivity of the sands. Darcy's law states, that the assumption of the same overall flow rate results in an increase in the pressure difference by one order of magnitude, when the connected permeability decreases by one order of magnitude (10 mD). This example shows how sensitive the multidimensional pressure solution depends on the connectivity of the highly permeable facies.

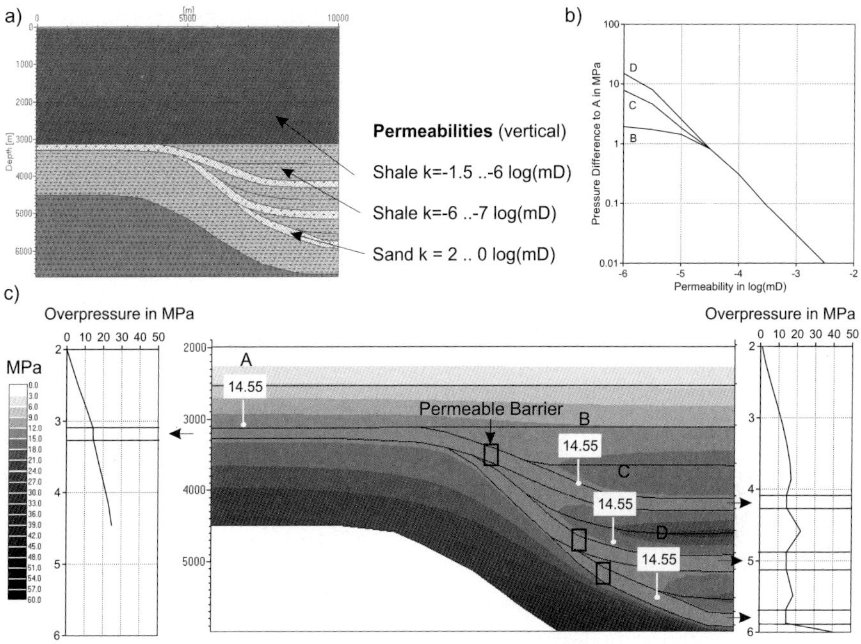

Fig. 2.23. Overpressure equalization along high permeability layers: (a) connected highly permeable sand layers are embedded in a thick shale package and yield almost the same overpressure in the three sandy sub-layers. (b) The pressure difference in the sand layers is almost proportional to the log permeability of the barrier. (c) Pressure solution with highly permeable barriers between the sands

Another similar example is shown in Fig. 2.24. Here a permeable layer connects a highly overpressured area below a thick shale block with a shallow hydrostatic pressure area. This permeable layer is able to discharge the pressure below the shale with resting pressure gradients equivalent to the permeability values of the connecting layer. A 3D example (Fig. 2.25) with a thin permeable layer varying over several kilometers of depth, also shows the pressure equilibration effect along a highly permeable flow avenue. These examples show how the architecture of the sediments in the basin control the pressure distribution.

The overpressure equation (2.13) does not deliver a solution in impermeable facies, such as salt, granite, or basalt, since these permeabilities are equal to zero. The pressure in impermeable structures should be equal to lithostatic pressure, since any fluid inclusion enclosed in an impermeable environment could never drop its pressure due to fluid outflow and must bear the total overburden. Hence, the inner points of impermeable rocks and salt are set as inner boundary conditions with values equal to the lithostatic potential. The overpressure gradient at the top of a salt dome can be a multiple of the lithostatic gradient as the overpressure can increase from a nearly hydrostatic

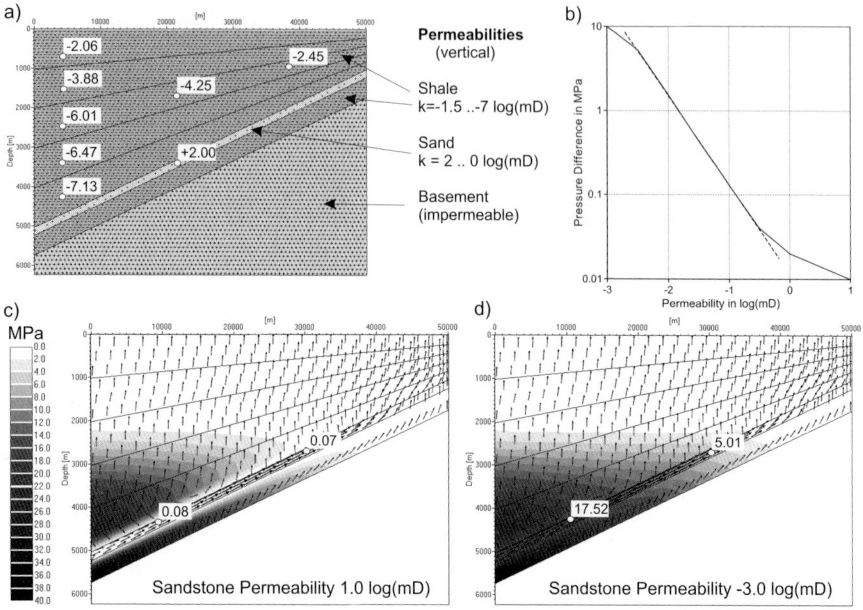

Fig. 2.24. Overpressure discharge in a highly permeable layer: (a) the highly permeable layer connects a high pressure area with a hydrostatic exit. (b) The pressure gradient in the sand layer gradually decreases with increasing permeability. (c), (d) 2D–overpressure fields for two different permeabilities of the permeable layer

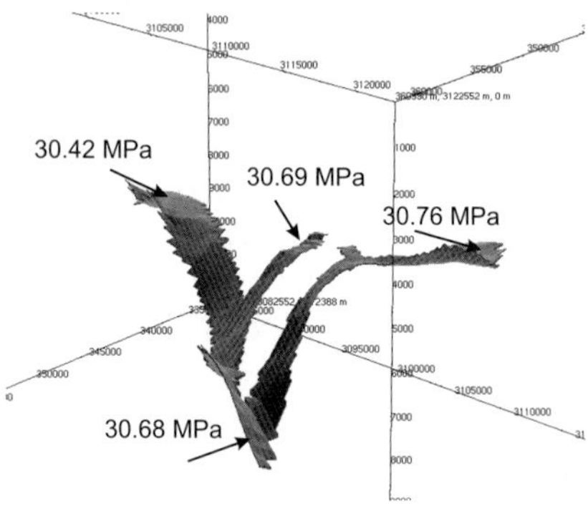

Fig. 2.25. Small overpressure differences along a 100 m thin layer with a permeability of $k = 10^{-2}$ mD

regime in the sediments above to a lithostatic regime in the salt layer over a very short distance.

Uplift, together with erosion, yields overpressure release, since overburden load is decreased, but the porosity is almost maintained and the decompaction path during uplift is different from the normal compaction line (Fig. 2.6). Hence, the compressibility during uplift is much smaller (or close to zero), which results in lower pressure release during erosion when compared with pressure formation during burial.

Some of the multidimensional pressure effects are illustrated in the example calculation of a 2D cross section from the Santos basin offshore Brazil (Fig. 2.26). The pressure is hydrostatic in the shallow area up to the top overpressure surface in 1 to 3 km depth. The pressure is lithostatic in the impermeable salt domes. High overpressure occurs below the thick salt domes, which gradually decrease toward the salt window. The overpressure is much lower below the smaller salt bodies. A thick block of low permeable shale layers also causes overpressure formation, while overpressures are almost equilibrated in the highly permeable facies.

In the above description, the upper pressure boundary condition at the sediment-water-interface is set to zero overpressure. In areas above sea level, the upper boundary is the groundwater surface and the pressure boundary condition is the groundwater potential, which is equal to the weight of the groundwater column above sea level. The pressure variable u in equation (2.13) is named hydraulic potential instead of overpressure in the terminology of groundwater specialists. Both terms are synonymous. The onshore groundwater level far from the coast is usually only some meters beneath the surface. The topographic surface can be taken as the approximate groundwater surface. The groundwater level close to the coast, or in very steep mountains can be significantly decreased, so that the boundary value of the corresponding hydraulic potential must be applied nearer to sea level with a much lower value.

An onshore example with a groundwater potential is shown in Fig. 2.27. The model has an aquifer with a depth of 3 km and a water flow towards the hydrostatic zone. The resulting water flow system and overpressure field of the basin is a superposition of three effects: the topographic driven flow near the surface follows the surface profile, the sedimentation controlled overpressure flow is directed out of the thick sediments, and the aquifer layer transports water toward the hydrostatic area. The water flow is much faster and the overpressure is much smaller, when the mountains are less extended as shown in Fig. 2.27.c.

The formation of mountains is always related with uplift and erosion, which is accompanied by a decrease in the overpressure potential of the uplifted blocks, since overburden is released during erosion. The overpressure release in rapid uplift and erosional periods below the mountains can be so high, that under-pressures arise in the aquifers, and the water flow can redirect toward the mountains as shown in Fig. 2.27.d. An analytical solution of a linearly varying horizontal groundwater potential is described in App. C.

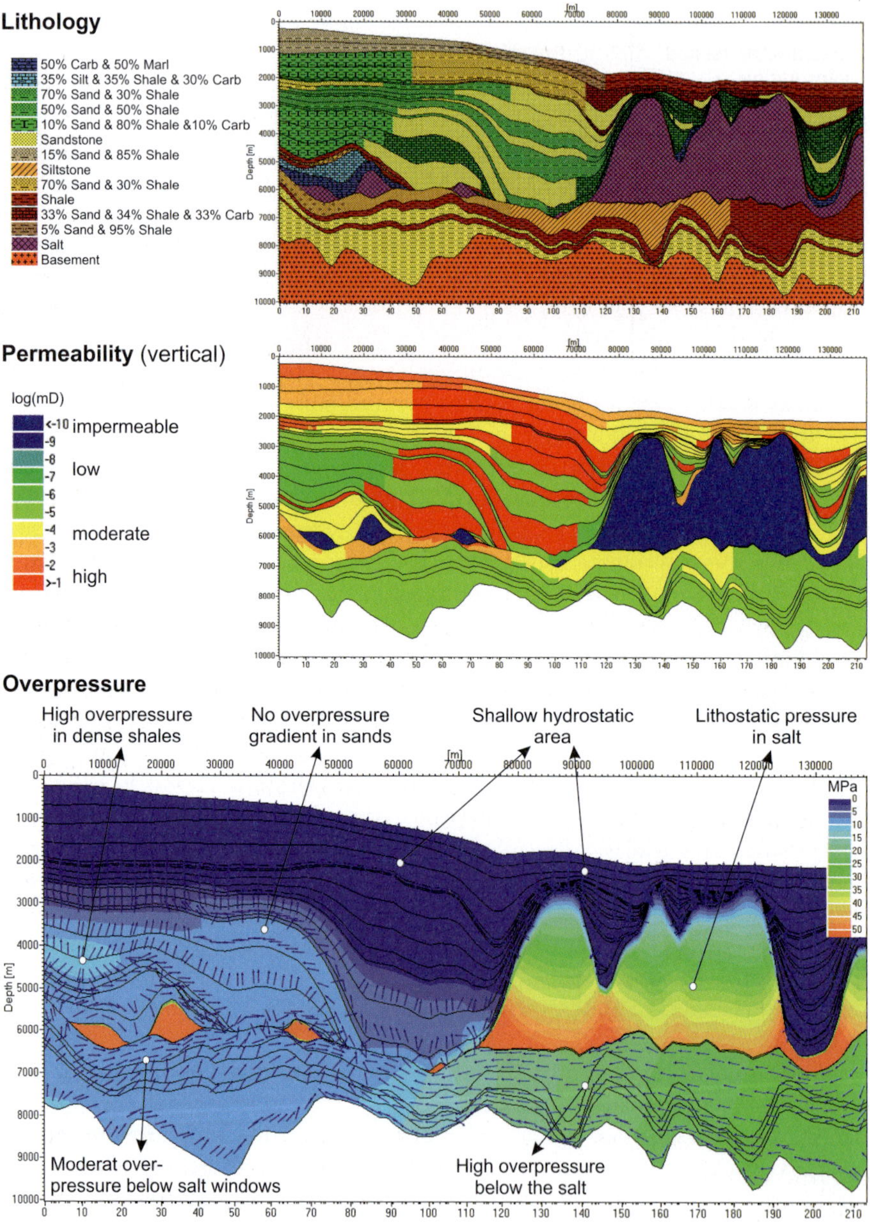

Fig. 2.26. Overpressure formation along a 2D cross section in the Santos basin, Brazil. The blue vectors indicate water flow

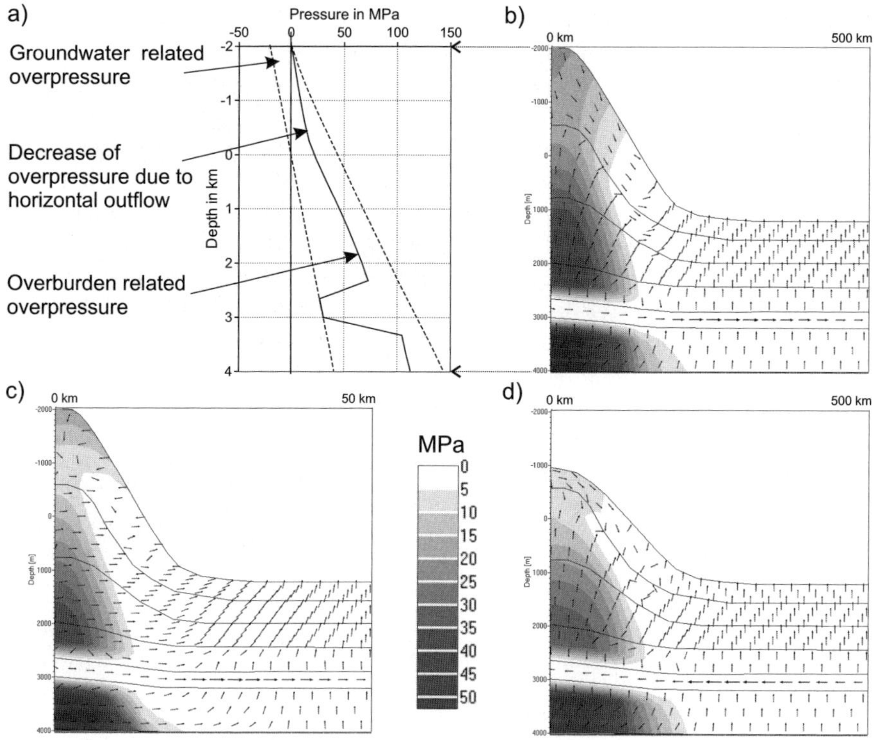

Fig. 2.27. Overpressure formation in an schematic onshore model: (a) 1D–extractions of pressures along a well with an aquifer at 3 km depth. (b) Overpressure formation and water flow vectors at present day. (c) Effect of mountain width, the model length is one tenth compared to model (b). (d) Effect of erosion: uplift is linked with erosion. The water flow vectors in the aquifer change direction from left to right

2.3 Special Processes of Pressure Formation

Special processes of pressure formation are quartz cementation (chemical compaction), aquathermal pressuring, pressure formation due to petroleum generation and cracking, and mineral transformations such as smectite–illite or gypsum–anhydrite.

2.3.1 Chemical Compaction

All sandstones and carbonates are cemented during burial. Quantitative descriptions of cementation processes are proposed by several authors (Walderhaug, 1996, 2000; Walderhaug et al., 2001; Bjørkum, 1996; Bjørkum and Nadenau, 1998; Bjørkum et al., 1998, 2001; Schneider et al., 1996; Schneider and

Hay, 2001; Lander and Walderhaug, 1999). Quartz cementation can be regarded as a three step process: quartz dissolution at grain-grain contacts, transport of the dissolved silica through pore space and precipitation of silica on free quartz grain surfaces (Walderhaug 1996, Figs. 2.28, 2.29). The transport of the solutes is performed via diffusion or pore water flow. All three processes (dissolution, solute transport and precipitation) have different effects on compaction, porosity reduction and pore pressure change. The cementation rate is controlled by the subsurface conditions, the water flow and water chemistry. Subsurface conditions are temperature, total vertical stress, and pore pressure. Water flow rates depend on the permeabilities of adjacent rock, and water chemistry is characterized by the dissolved minerals and the pH–value. It is a common approach to reduce the model to the precipitation process and assume that the other processes always supply enough silica.

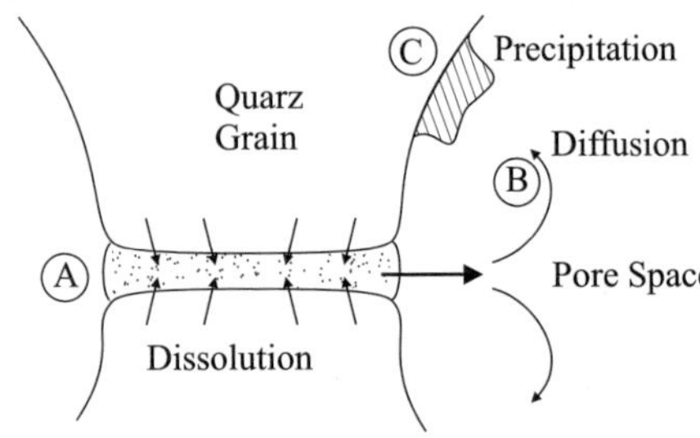

Fig. 2.28. Principal processes of chemical compaction: (A) Pressure dissolution of silica into pore water. (B) Diffusion of dissolved silica within the pore water phase. (C) Precipitation of silica at quartz grains

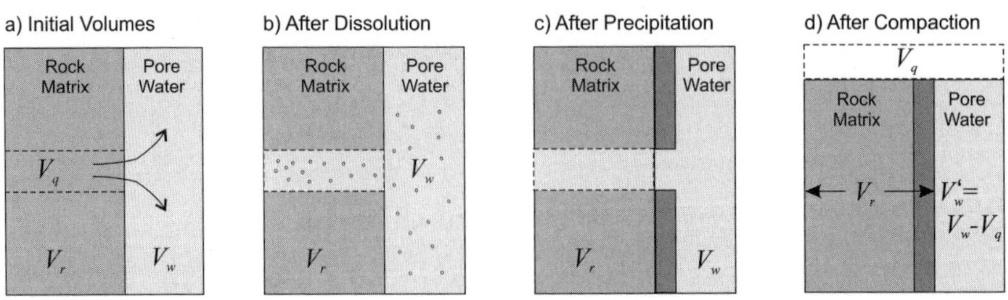

Fig. 2.29. Schematic volume balance for quartz dissolution and precipitation

The volume balance includes changes in the the solid volume (including cement) V_s, the pore fluid volume V_w, the volume of the precipitated cement V_q, and the total volume $V_t = V_s + V_w$ with the porosity $\phi = \mathrm{V_w/V_t}$ and the cementation $\psi = V_q/V_t$.

Dissolution of silica occurs along the grain contacts. The contact zone is a thin film of adsorbed fluids between the rough surface of quartz grains. The dissolution rate C_d is mainly controlled by the effective stress (pressure dissolution) and is dependent on temperature.

$$C_d(\sigma_z', T) = -\frac{1}{V_t}\frac{\partial V_s}{\partial t}\,. \tag{2.53}$$

During dissolution, the solid rock volume is decreased by the amounts of dissolved quartz, while the pore volume is increased by the same amount.

The dissolved silica is transported in water by diffusion and together with water as a separate phase flow. Hence, it depends on the quartz solubility of water, the diffusion rate, the permeabilities and overpressure gradients. The literature distinguishes between an open and closed systems approach, assuming relatively long and short transport paths (Schneider et al., 1996). The closed systems approach is more important, since quartz is usually precipitated near to the location of dissolution. However, the transport of dissolved silica does not influence the porosity, compaction and overpressure. Precipitation of silica as cement occurs on the free grain surfaces with preference to pore throats, which decreases permeability significantly. Precipitation rates C_p are usually temperature dependent (Walderhaug, 1996).

$$C_p(T) = \frac{1}{V_t}\frac{\partial V_s}{\partial t}\,. \tag{2.54}$$

Pure precipitation increases the amount of solid material and reduces the pore space by the same amount. The total balance of quartz dissolution and precipitation is as follows.

$$\frac{\partial V_s}{\partial t} = (C_p - C_d)\,V, \quad \frac{\partial \psi}{\partial t} = C_p\,. \tag{2.55}$$

The total process yields much lower porosities for high effective stresses than pure mechanical compaction would allow. It also increases the pore pressure, since the dissolution of the solid matrix transfers lithostatic pressure to pore pressure.

Closed System Approach

In the closed system approach, short diffusion tracks are assumed with precipitation near to the locations of dissolution. Hence, the precipitation rate is equal to the dissolution rate and the total solid volume remains constant. The ability to drop the porosity by cementation additionally to mechanical compaction requires a change in the compaction law by either increasing the bulk compressibility or adding an additional term f_c as follows.

$$\frac{\partial \phi}{\partial t} = -C\frac{\partial \sigma'}{\partial t} - f_c(T, \sigma') \tag{2.56}$$

with the Terzaghi's compressibility C for mechanical compaction.

The cementation controlled porosity loss is also realized by accompanied water outflow and it is usually almost equal to the relative volume of the precipitated cement ($f_c \approx \partial\psi/\partial t$). The measured relative volumes of silican cement ψ are often used to derive empirical rules for the compaction term f_c.

Empirical laws for the cementation rate are proposed by Walderhaug (2000) and Schneider et al. (1996), named respectively the Walderhaug and Schneider models. The Walderhaug model is a precipitation rate-limited reaction controlled by the temperature and the quartz surface area available for precipitation. Walderhaug argues, that there is usually enough effective stress at large depth to supply enough dissolved quartz and that the effective stress dependency of the chemical compaction can be neglected. He proposed the following relationship with an Arrhenius type temperature dependency:

$$\frac{\partial \phi}{\partial t} = -\frac{M_q}{\rho_q} \frac{6(1-f_q) f_v}{d_q} \frac{\phi}{\phi_0} A \, \mathrm{e}^{-E/RT} \tag{2.57}$$

where R is the gas constant with $R = 8.31447\,\mathrm{Ws/mol/K}$, f_q is the quartz grain coating factor (the fraction of the quartz grain surface that is coated and unsuitable for precipitation), f_v is the quartz grain volume fraction when precipitation starts (the fraction of the detrital grains that are quartz), d_q is the average quartz grain size, and A and E are the frequency factor and activation energy of the quartz precipitation rate. Fixed parameters are $M_q = 0.06009\,\mathrm{kg/mol}$ and $\rho_q = 2650\,\mathrm{kg/m^3}$, the quartz molar mass and density. Default parameters are $f_q = 5$, $f_v = 1$, $d_q = 0.03\,\mathrm{cm}$, $A = 10^{-11}\,\mathrm{mol/cm^2/s}$, and $E = 61\,\mathrm{kJ/mol}$. The activation energy is primarily used for calibration, when sample data are available. The porosity loss described by the Walderhaug model is shown in Fig. 2.30 for various activation energies with smaller cementation rates for higher activation energies.

A viscoplastic type compaction model is proposed by Schneider et al. (1996), who introduced the porosity loss rate proportional to the effective stress σ_z', which represents the quartz supply by pressure induced dissolution. The rate is dropped by a viscosity μ, which decreases with higher temperature according to a Arrhenius type dependency.

$$\frac{\partial \phi}{\partial t} = -(1-\phi)\frac{\sigma_z'}{\mu}, \qquad \mu = \mu_0 \exp\left[\frac{E}{k}\left(\frac{1}{T} - \frac{1}{T_0}\right)\right] \tag{2.58}$$

with the reference temperature $T_0 = 15\,°C$ and viscosity $\mu_0 = 50\,\mathrm{GPa/My}$. The activation energy E can be used for calibration with default values between 16 and 18 kJ/mol. Porosity loss curves for fast and slow sedimentation are shown in Fig. 2.31 for various activation energies. Contrary to the Walderhaug model, porosity rates increase with higher activation energies.

The additional term in the compaction law also appears in the revised pressure equation according to (2.11) as follows:

$$C\frac{\partial u}{\partial t} - \nabla \cdot \frac{\boldsymbol{k}}{\nu} \cdot \nabla u = C\frac{\partial u_l}{\partial t} + f_c(\sigma_z', T) \,. \tag{2.59}$$

The increase in pressure is caused by the transfer of rock stress to pore pressure due to the abbreviation of the vertical stress bearing rock elements, by the

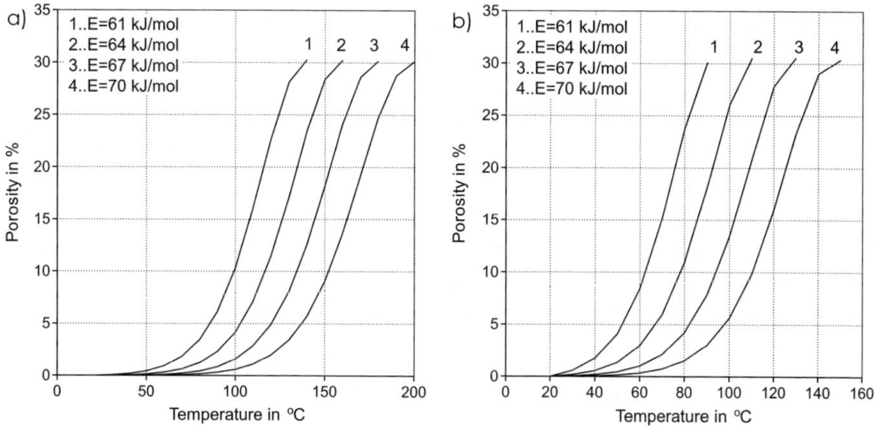

Fig. 2.30. Cemented porosity calculated with the Walderhaug model with coating factor $f_q = 0.5$, quartz grain volume fraction $f = 1$, quartz grain size $d_q = 0.3\,\text{mm}$, initial porosity $\phi_0 = 41\,\%$, frequency factor $A = 10^{-11}\,\text{mol/cm}^2/\text{s}$: (a) with high sedimentation rates $S = 1.0\,\text{km/My}$, (b) with low sedimentation rates $S = 0.1\,\text{km/My}$

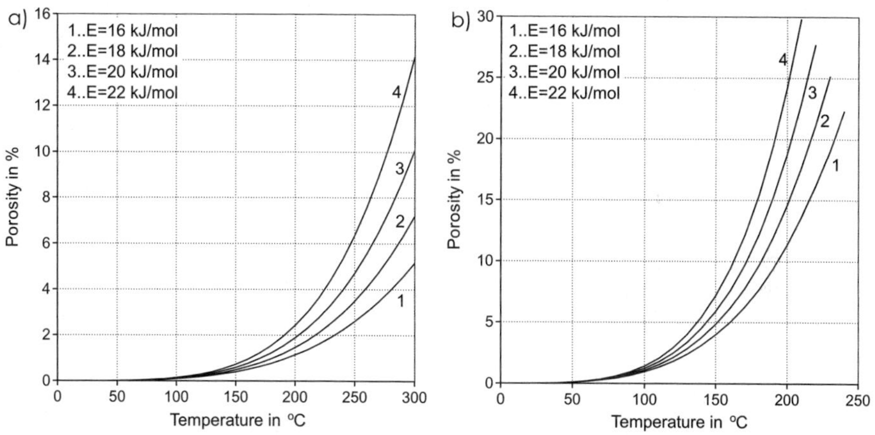

Fig. 2.31. Cemented porosity calculated with the Schneider model with a temperature gradient $dT/dz = 30\,^\circ\text{C/km}$, effective stress gradient $d\sigma'_z/dz = 10\,\text{MPa/km}$: for sedimentation rates (a) $S = 0.1\,\text{km/My}$, (b) $S = 1\,\text{km/My}$

pore space reduction due to precipitated cement, and by the permeability decrease due to thinner pore throats.

An one dimensional example is shown in Fig. 2.32 with alternating shale and sandstone layers. The Walderhaug and Schneider models are compared with respect to porosity loss and additional overpressure generation. The Walderhaug model generally predicts higher cementation rates than the Schneider model. The difference in porosity loss (Fig. 2.32) is very high when the proposed default values are used. The effects are more similar when higher activation energies, rather than the default values, are used in both models. Cementation of the sandy layers yields lower permeability values which significantly influence the dewatering of the shale layers below. Additional overpressures in the sandstones also influence the pressure formation and porosity loss of the overlaying shales.

2.3.2 Fluid Expansion Models

Fluid expansions yield fluid density increases and related overpressure formation, which can be described with an additional source term f_a in the overpressure equation (2.59) as follows:

$$C\frac{\partial u}{\partial t} - \boldsymbol{\nabla} \cdot \frac{\boldsymbol{k}}{\nu} \cdot \boldsymbol{\nabla} u = C\frac{\partial u_l}{\partial t} + f_c(\sigma'_z, T) + f_a(T) \,. \tag{2.60}$$

The source terms in the above equation are understood as the relative pore fluid volume increase over time. The volumetric formulation is obtained, as in the initial mass balance equation (2.10) all the terms were already divided by the pore water density assuming that the water density variations with depth and time are relatively small on the considered scale. The fluid expansion models are here formulated on volume and not on mass balances, although variable fluid densities are considered. They are easier to implement and overview and the differences to more complex formulations are of minor importance (Luo and Vasseur, 1992, 1996). One can also integrate the source terms over the entire burial history for a deep sediment, for example in 5 km depth, to compare the order magnitudes of the different sources for pressure formation with each other. The total source for overburden load F_o is:

$$F_o = \int_t C\,\frac{\partial u_l}{\partial t} dt \approx \bar{C}\,\bar{u}_l \approx 0.75 \tag{2.61}$$

where $\bar{C} \approx 10\,\mathrm{GPa}^{-1}$ is the average bulk compressibility and $\bar{u}_l \approx 75\,\mathrm{MPa}$ is an average total load of a sediment in 5 km depth. Assuming, that in the initial sedimentation phase approximately one third of the overburden was not converted to overpressure, a value of 0.5 is more realistic.

The total load source for chemical compaction F_c is equal to the total porosity reduction $\Delta\phi$ by cementation:

$$F_c = \int_t \frac{\partial \phi}{\partial t} dt \approx \Delta\phi \approx 0.15 \,. \tag{2.62}$$

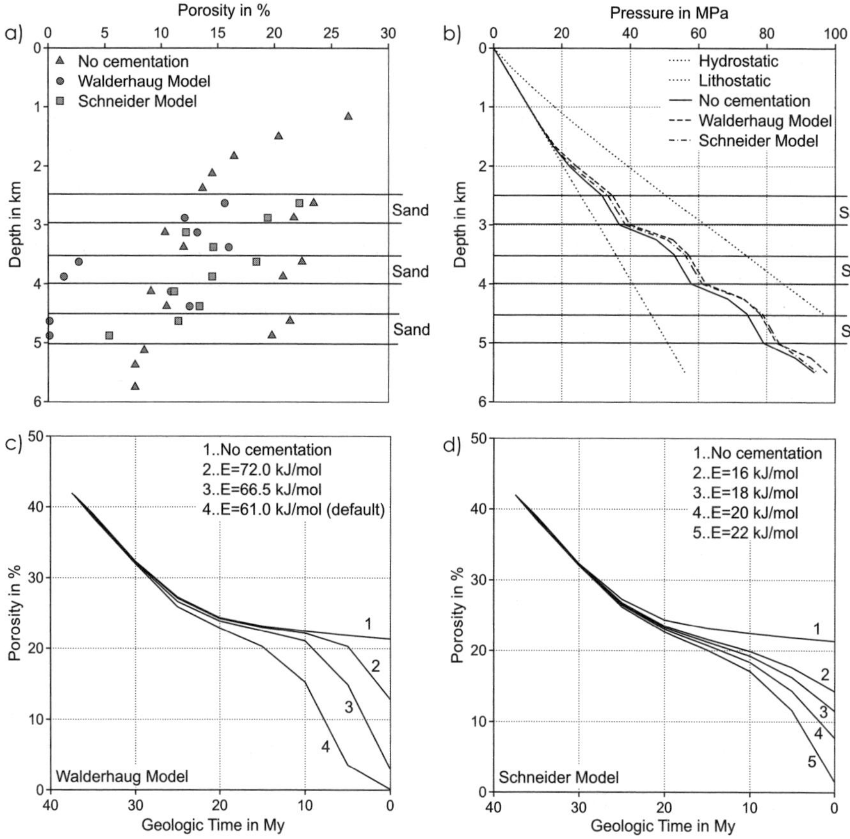

Fig. 2.32. 1D example of three sandstone layers embedded in shale depositions. Porosity loss and overpressure formation is calculated with the Walderhaug and Schneider model for cementation of the sandstone layers and without cementation for comparison. (a) Porosity loss due to mechanical and chemical compaction. The activation energies of the Walderhaug and Schneider model are 61 kJ/mol and 20 kJ/mol. (b) Overpressure formation during cementation. (c), (d) Porosity loss of the lower sandstone layer calculated with the Walderhaug and Schneider model, respectively, for different activation energies

This has a significant effect compared to the overburden load, but is locally restricted to sandstone and carbonate layers only.

Aquathermal Pressuring

Luo and Vasseur (1992) investigated the effect of overpressure formation caused by thermal expansion of the pore water. The additional source term f_a in the overpressure equation depends on the isobaric thermal expansion coefficient $\alpha = \alpha_w - \alpha_r$, which is the difference between the values for water

$\alpha_w = 5 \times 10^{-4}\,\mathrm{K}$ and rock $\alpha_r = 3.3 \times 10^{-5}\,\mathrm{K}$.

$$f_a = \alpha \frac{\partial T}{\partial t} \tag{2.63}$$

which yields the following estimation for the total load source:

$$F_a = \int_t \alpha \frac{\partial T}{\partial t} dt \approx \Delta\phi \approx 0.07 \tag{2.64}$$

with an assumed temperature $T = 150°C$ in a depth of 5 km. Assuming, the overburden is not converted to overpressure for early sedimentation, a value of 0.05 is more realistic. It is one order of magnitude smaller than the effect from the overburden load. The aquathermal pressure formation depends on the heating rate and it becomes higher for fast burial.

Mineral Transformations

Some mineral transformations, such as smectite to illite and anhydrite to gypsum conversion are related to pore fluid volumes changes. The conversion from smectite to illite occurs in all shales and shaly rocks and is described as a complex multi–stage process (Pytte and Reynolds, 1989; Swarbrick et al., 2002). Some of the related processes increase and some decrease the pore water volume with a general release of bound water into pore space and with a total increase of the water relative volume up to 5 %. The volume of the solid matrix is also generally increased, since mainly Na^+ ions are exchanged by K^+ ions with a higher ion radius. This process is controlled by temperature and the availability of K^+ ions in the rock matrix. A widely accepted model was proposed by Pytte and Reynolds (1989) as a fifth–order reaction of the following type.

$$\frac{\partial x}{\partial t} = -x^5\, k_1\, k_2 \tag{2.65}$$

where x is the smectite to illite ratio with an initial value of 0.8, $k_1 = 74.2 \exp(-2490/T[K])$ is the chemical activity of potassium to sodium and $k_2 = 1.64 \times 10^{21}\,\mathrm{My}^{-1} \exp(-16600/T[K])$ is a Arrhenius type temperature dependence. The equation can be written as a usual unimolecular forward reaction type (Chap. 4) as follows:

$$\frac{\partial x}{\partial t} = -x^5\, A\, \mathrm{e}^{-E/RT} \tag{2.66}$$

with the activation energy $E = 37.9\,\mathrm{kcal/mol}$, the frequency factor $A = 1.217 \times 10^{23}\,\mathrm{My}^{-1}$ and the gas constant $R = 8.31447\,\mathrm{Ws/mol/K}$. The conversion depends on the heating rate, which is controlled by sedimentation rates (Fig. 2.33). The simple approach is, that the transformation ratio of the reaction $\mathrm{TR}_s = (x - 0.2)/0.8$ is related with a constant factor $\kappa \approx 0.05$ to the load source f_a as follows:

$$f_a = \kappa \frac{\partial}{\partial t} \mathrm{TR} \,. \tag{2.67}$$

Hence, the overpressure formation caused by smectite to illite conversion also depends on the sedimentation rates with a total source of

$$F_a = \int_t \kappa \frac{\partial}{\partial t} \mathrm{TR} \, dt = \kappa \approx 0.05 \,. \tag{2.68}$$

This effect is one order of magnitude smaller than the overburden load effect. Osborne and Swarbrick (1997) and Swarbrick et al. (2002) proposed the separate consideration of smectite dehydration with the release of interlayer water, but the authors consider this effect to be included in the usual mechanical compaction model for shale.

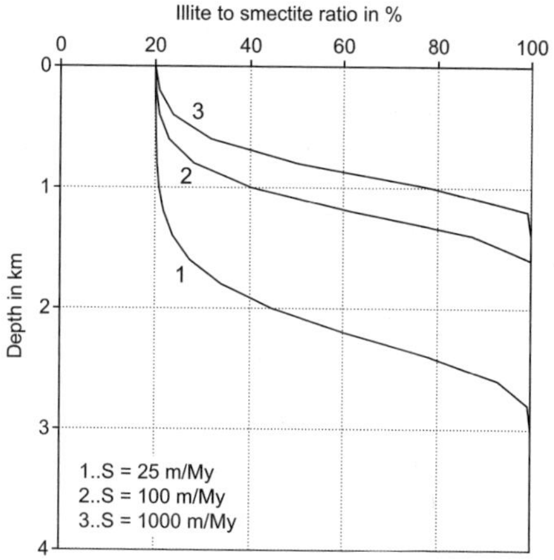

Fig. 2.33. Smectite–Illite after Pytte and Reynolds (1989) calculated for uniform sedimentation with different sedimentation rates

Petroleum Generation Pressure

Catagenetic processes of organic matter change the relative volumes of kerogen, liquid, and vapor petroleum (Fig. 2.34). Here, the primary cracking of kerogen and secondary cracking of the heavier petroleum components are taken into account. The controlling parameters are the changes of the petroleum phase masses and volumes, which result from the generated chemical components and the PVT controlled dissolution into the two petroleum phases (Chaps. 4, 5). The fluid models also provide corresponding modifications of the phase densities, which control the generation driven overpressure formation. The actual densities (mass per volume) of the generated petroleum components, which are dissolved in liquid and vapor phase, are denoted as μ_l and μ_v with the phase densities ρ_l and ρ_v, respectively. Then, the load source for primary generation is as follows:

$$f_a = \left(\frac{1}{\rho_l} - \frac{1}{\rho_k}\right) \frac{\partial \mu_l}{\partial t} + \left(\frac{1}{\rho_v} - \frac{1}{\rho_k}\right) \frac{\partial \mu_v}{\partial t} \, . \tag{2.69}$$

The kerogen density ρ_k has to be taken into account, since the reduction of kerogen volume during cracking is also a significant value. For the estimation of the magnitude of the load source, all generated petroleum with mass density μ_p is assumed to be dissolved in one super–critical petroleum phase with density ρ_p.It follows from (4.8):

$$F_a = \int_t \left(\frac{1}{\rho_p} - \frac{1}{\rho_k}\right) \frac{\partial \mu_p}{\partial t} dt \approx \left(\frac{\rho_r}{\rho_p} - \frac{\rho_r}{\rho_k}\right) \mathrm{TOC}_0 \, \mathrm{HI}_0 \, (1 - \phi) \approx 0.033 \tag{2.70}$$

with an initial total organic carbon content $\mathrm{TOC}_0 = 5\%$, an initial hydrogen index $\mathrm{HI}_0 = 500\,\mathrm{mgHC/gTOC}$, a porosity $\phi = 20\%$, and densities of $\rho_k = 800\,\mathrm{kg/m^3}$, $\rho_p = 500\,\mathrm{kg/m^3}$, $\rho_r = 2200\,\mathrm{kg/m^3}$. This effect is one order of magnitude smaller than the overburden load effect and it is restricted to source rocks only. Exceptions are coals with TOC values higher than 50 % and coal bed methane production, which can form very high overpressures in place. Gas has a much higher compressibility ($100\,\mathrm{GPa^{-1}}$) than the porous framework ($10\,\mathrm{GPa^{-1}}$), which yields a retarded overpressure drop by pore fluid outflow as described in equation (2.52), when high gas saturations occur. This slightly increases the effect of gas generation controlled overpressures.

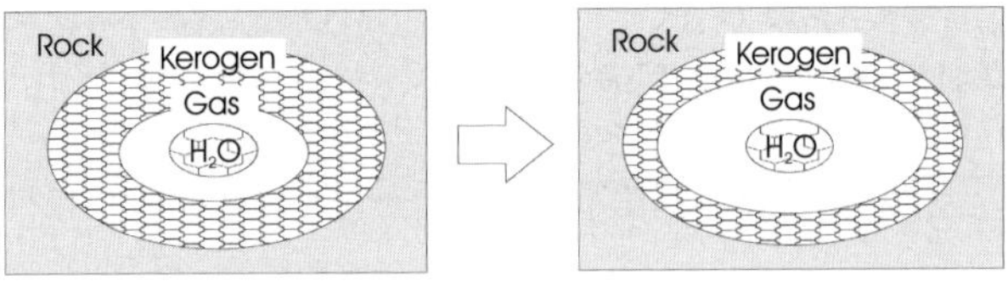

Fig. 2.34. Gas generation from kerogen changes the volumetrics of kerogen and pore fluids

Secondary cracking can also change phase volumes and can be described analogously to primary cracking. The resulting overpressure build up is much smaller, especially because coke with higher density is formed as a by–product.

The generation of petroleum amounts can be more accurately formulated in the multi–phase fluid flow equations (Chap. 2.9) with source terms for the generated masses. Luo and Vasseur (1996) published a detailed analysis for a two-phase system formulation with similar magnitudes for overpressure build-up as described in equation (2.70).

The overpressure equation with all described effects is as follows.

$$C\frac{\partial u}{\partial t} - \nabla \cdot \frac{\boldsymbol{k}}{\nu} \cdot \nabla u = C\frac{\partial u_l}{\partial t} + f_c(\sigma_z', T) + \alpha\frac{\partial T}{\partial t} + \kappa\frac{\partial \mathrm{TR}}{\partial t} + \left(\frac{1}{\rho_l} - \frac{1}{\rho_k}\right)\frac{\partial \mu_l}{\partial t} - \left(\frac{1}{\rho_v} - \frac{1}{\rho_k}\right)\frac{\partial \mu_v}{\partial t}\,. \quad (2.71)$$

2.4 Overpressure Calibration

The overall overpressure is mainly determined by mechanical compaction. Other sources for overpressure, such as chemical compaction or fluid expansion, are often rather localized phenomena and for that reason not included in this section.

Mechanical compaction, as formulated in (2.13), relates pore water flow with porosity reduction and overpressure. An overpressure calibration is therefore a calibration of compressibility and permeability. It can be performed in two major steps. The first step deals with the adjustment of rock compressibilities and the second with permeabilities.

Compressibility is introduced via a relationship of effective stress and porosity in (2.17). Effective stress is defined as the difference $\boldsymbol{\sigma}' = \boldsymbol{\sigma} - p\mathbf{I}$. Relationship (2.17) describes a local property of the rock, and does not contain permeability. If porosity is known, a compaction model such as Athy's law or the Schneider model can be fitted to each lithology in the following way: porosity and pressure value pairs of the same lithology are collected for different depths and locations. It is possible to calculate the corresponding overburden from the basin model for these points. Thus, effective stress can be evaluated and plotted against porosity (Fig. 2.35). Finally, a compaction model with an effective stress versus porosity formulation, is fitted against these data points.

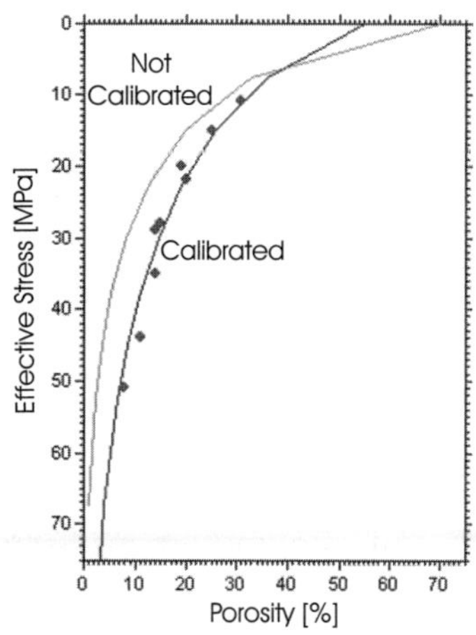

Fig. 2.35. Fit of Athy's law in porosity–effective stress formulation against a few data values

Note that this approach relates effective stress to porosity and not with depth. A porosity depth fit is not achieved until the second major calibration step of permeability adjustment is performed.

The second step consists of a permeability adjustment against overpressure. This step can be sophisticated because overpressure depends in general non–locally and ambiguously on permeability. For example, a pressure drop due to water outflow can often be modeled by different positions and sizes of the "leak". However, some general rules of thumb can be stated. Highly impermeable rocks are not a matter for calibration. If there is no water flow within these rocks, a small change in permeability does not change the overpressure pattern at all. The situation is the other way around for highly permeable sand layers. A change in permeability will not change the overpressure pattern. It remains equilibrated inside these layers. Important for pressure calibration are the layers in which overpressure builds up or is released (Fig. 2.36). Obviously, a permeability variation for these rocks will cause a significant change in the overpressure pattern. Identification of these layers is a key point in overpressure calibration because it drastically reduces the number of calibration parameters.

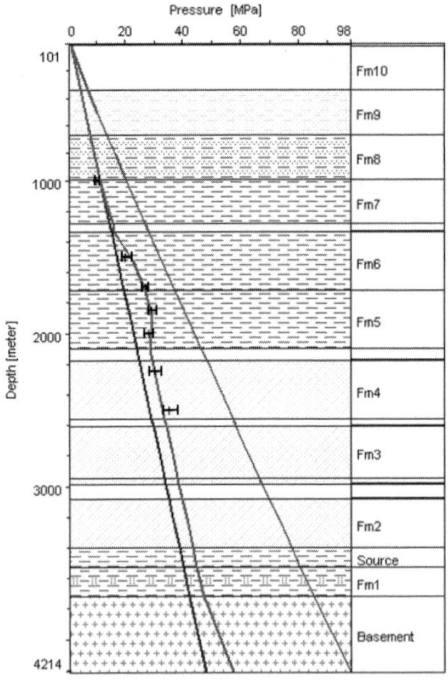

Fig. 2.36. An example of a pressure calibration by adjustment of the permeabilities. Pressure builds up in layer Fm6 and is slightly released in layer Fm5. Hence Fm5 and Fm6 are the key layers in this example. It is possible to calibrate the model by variation of the permeabilities in these layers. Below Fm5 a highly permeable sandstone is located. It transports some water from this region through a slightly higher permeable window area in Fm5/Fm6, far away from this well. Hence it is necessary to incorporate and adjust the permeabilities in this window area for good calibration

Overpressure can, in principal, only be calibrated if the water flow and water balance is adjusted correctly throughout the entire basin. In practice this leads to a situation where the permeabilities of many layers, lithologies, and rock types must be adjusted simultaneously. Due to long range pressure interactions caused by water flow, this is usually very problematic. It is found that it is possible to tackle the problem iteratively. Calibration is first performed on key layers which are connected directly or via other permeable layers to the top of the basin. Water is transported along these pathways out of the basin. Overpressure can be calibrated best if the total amount of water in the basin is adjusted first. According to this picture adjustments of permeabilities in more deeply buried regions will lead to a minor overall adjustment at least on the global water balance.[1] On average, the water flow is upward and therefore calibration should usually be performed from the top down and from more to less permeable lithologies. The procedure can be repeated iteratively until convergence is reached.

This workflow assumes, that recently no erosion with a reduction of overburden appeared. Otherwise, under the assumption of non-decompactable rocks, porosity must be fitted against the maximum effective stress. However, the maximum effective stress might not be calculable as necessary paleo overpressures are possibly unknown. This problem can be overcome with additional overpressure shifts at paleo times, which are also calibrated against the present day overpressure pattern. The whole overpressure calibration procedure including both steps must then iteratively be refined.

2.5 Geomechanical Models

The fundamentals of geomechanics would require a full book in itself. Here only the most important equations, which are needed for basin modeling, are mentioned. Detailed descriptions are given e.g. in Fjaer et al. (1992) and Parry (2004).

Solids conduct forces through the material and react with deformations. The forces and moments acting on each small volume element are described in terms of stress and the deformations are represented in terms of strains. Most materials respond with linear dependent recoverable strains on small stresses, which is called linear elasticity. In practice, stress–strain relations also have terms of non-linearity, irreversibly (plasticity), rate dependency (viscosity or creep) and yield failure, when certain limits of the stress components are exceeded.

The traditional stress–strain concept has been determined for solids. It can be extended to porous media with an introduction of effective stresses, which takes pore pressure into account. The difference between the concepts of rock and soil mechanics is that the first takes into account cement between the grains, while the second refers to unconsolidated rock with loosely connected grain particles.

[1] More deeply buried rocks are usually more compacted and therefore less permeable. This also reduces the capability of water transport and the range of influence on the overpressure pattern.

2.6 Stress and Deformation

The total bulk stress tensor is a superposition of the stress tensors of the grains and the pore pressure of the fluids. Tensors, their principal values and invariants are introduced in Sec. 8.2. The stress tensor σ_{ij} has normal $(i = j)$ and shear $(i \neq j)$ components, which act on surfaces perpendicular to the coordinate axes (Fig. 2.37). Compressional normal components are positive. The principal values are denoted as σ_1, σ_2, and σ_3 with $\sigma_1 \geq \sigma_2 \geq \sigma_3$. The boundary vector $\mathbf{t} = \mathbf{n} \cdot \boldsymbol{\sigma}$ acting on arbitrary area with the normal $\mathbf{n}$ has a normal and a tangential component t_n and t_t.

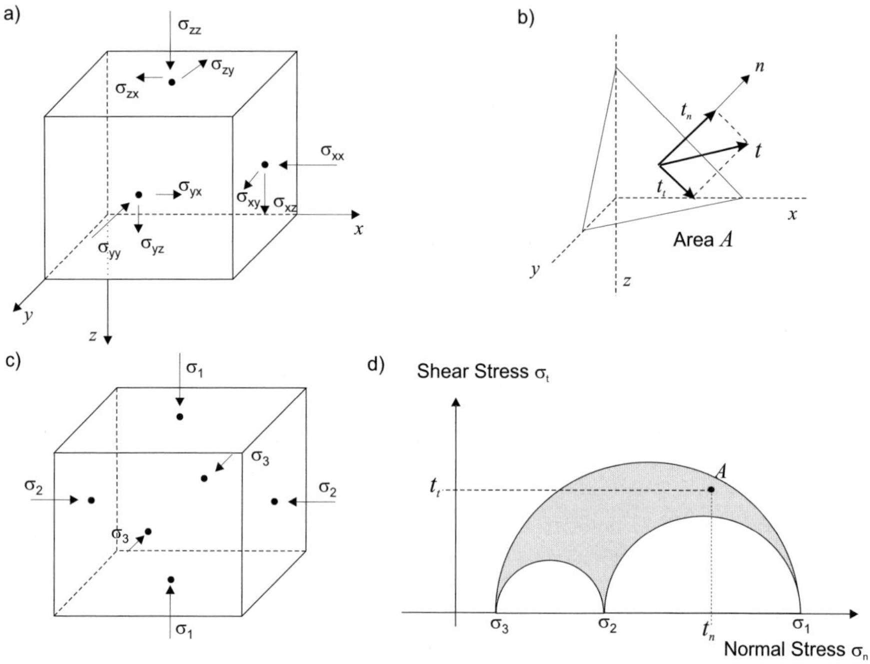

Fig. 2.37. Representations of the stress tensor. (a) Normal and shear components, (b) Boundary stresses at an arbitrary area, (c) Principal Stresses, (d) Mohr circles

There are two important representations of the three dimensional stress state: the Mohr circles used in rock mechanics and the "p-q" plots used in soil mechanics. The Mohr circle construction is based on principal stresses. An arbitrary 3D stress tensor is pictured with three circles (Fig. 2.37) in a normal-shear stress diagram and the area between the circles represents the boundary stress vector acting on any cut-plane of the volume element. The outer circle is important to analyze and illustrates rock failure. Stress in two dimensions is represented with only one circle. Mohr circles for the special cases of biaxial and isotropic stresses, and pure shear are illustrated in Fig. 2.38.

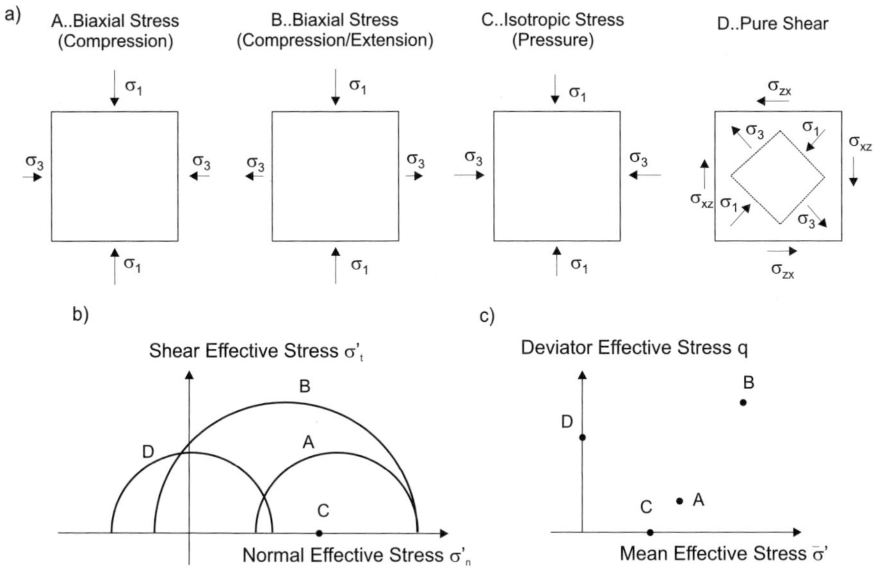

Fig. 2.38. Stress characterization of special load cases (a) Principal stresses for biaxial stresses, (b) Mohr circles, (c) soil mechanical p–q plot

The "p-q" plot uses two characteristic values, the mean stress $\bar{\sigma}$ and the deviatoric stress q.[2] The mean stress is an average volumetric (compressional) stress and the deviatoric stress represents an average shear stress as follows:

$$\begin{aligned} \bar{\sigma} &= \sigma_1 + \sigma_2 + \sigma_3 \\ q &= \tfrac{1}{\sqrt{2}} \left[(\sigma_1 - \sigma_2)^2 + (\sigma_1 - \sigma_3)^2 + (\sigma_2 - \sigma_3)^2 \right]^{1/2} . \end{aligned} \tag{2.72}$$

Any three dimensional stress state is a point in the "p-q" plot as illustrated in Fig. 2.38.

Any movement, rotation and deformation yields a change in the position of the sample particles, which is described with the displacement $\mathbf{u}(\mathbf{r})$. The deformation of a volume element is called strain $\boldsymbol{\epsilon}$ and can be derived from a given displacement vector as follows:

$$\boldsymbol{\epsilon} = \frac{1}{2}(\boldsymbol{\nabla}\mathbf{u} + (\boldsymbol{\nabla}\mathbf{u})^T) . \tag{2.73}$$

This equation is only valid for small deformations. In case of large deformations, additional terms with products of ∇u need to be incorporated in the above equation (Zienkiewicz, 1984). The strain tensor ϵ_{ij} also has normal

[2] The symbol "p" is usually used in soil mechanics terminology instead of $\bar{\sigma}$, but the symbol "p" is already used for pore pressure here.

(or direct) ($i = j$) and shear (or distortion) ($i \neq j$) components. The total volumetric deformation $\epsilon_v = 3\bar{\epsilon}$ is the sum of the principal strains:

$$\epsilon_v = \epsilon_1 + \epsilon_2 + \epsilon_3 \,. \tag{2.74}$$

The two special deformations of pure and simple shear have no volumetric strain $\epsilon_v = 0$ (Fig. 2.39).

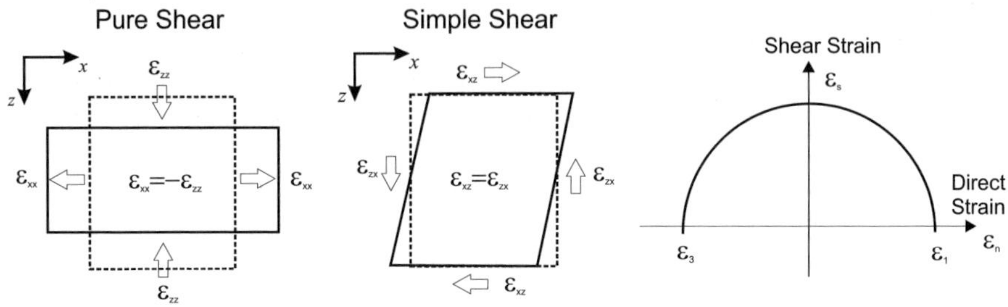

Fig. 2.39. Pure shear and simple shear are special deformations without volumetric strain. They are represented with the same Mohr circle, but they have different orientations of the axes

The elasticity tensor $\boldsymbol{E}$ relates stresses and strains linearly $\boldsymbol{\sigma} = \boldsymbol{E} \cdot \boldsymbol{\epsilon}$ assuming the linear theory of elasticity. It contains only two elastic parameters for isotropic behavior: the shear modulus G and the Poisson's ratio ν.

$$\boldsymbol{\sigma} = 2G\boldsymbol{\epsilon} + \frac{2G\nu}{1-2\nu}\epsilon_v \boldsymbol{I} \tag{2.75}$$

where $\boldsymbol{I}$ is the unit tensor. Alternatively, the Young's modulus E or the bulk modulus K can be used as follows.

$$E = 2G(1-\nu), \quad K = \frac{2G(1+\nu)}{3(1-2\nu)} \,. \tag{2.76}$$

Note, that the inverse of the bulk modulus is the bulk compressibility $C = 1/K$.

The meaning of the elasticity parameters is especially descriptive for uniaxial compression with σ_x and the two resulting strains ϵ_x and ϵ_y, where the elastic properties are $E = \sigma_x/\epsilon_x$, $\nu = -\epsilon_y/\epsilon_x$, and $K = \bar{\sigma}/\epsilon_v$. Anisotropy is described with more then two elastic parameters in the elasticity tensor and non-linear elastic behavior with additional terms of higher order strains.

The principle of force equilibrium states, that any body force $\mathbf{f}$ is compensated by the stress tensor.

$$\boldsymbol{\nabla} \cdot \boldsymbol{\sigma} + \rho \mathbf{f} = 0 \tag{2.77}$$

where ρ is the bulk density. This yields the differential equation based boundary value problem for the model of linear elasticity (2.75) with the gravity (overburden) forces as follows.

$$G\Delta\mathbf{u} + \frac{G}{1-2\nu}\nabla\nabla\cdot\mathbf{u} + \rho g\mathbf{e}_z = 0\,. \tag{2.78}$$

The boundary values are displacements $\mathbf{u}$ and boundary stresses $\mathbf{t} = \mathbf{n}\cdot\boldsymbol{\sigma}$. The differential equation is slightly different when large deformations are taken into account (Zienkiewicz, 1984).

The above theory of stresses and strains has been developed and proofed for pure solids. Extensions for composite media, as needed for pore fluids and rocks, require very complex models for pore pressure and rock stresses. There are simplified models proposed by Terzaghi on an experimental basis and Biot on theoretical derivations, both are based on the idea of introducing an effective stress σ' instead of the total stress σ and using the principal equations of the above concept with some modifications.

$$\boldsymbol{\sigma}' = \boldsymbol{\sigma} - \alpha\, p\, \boldsymbol{I}, \quad \text{with} \quad \alpha = 1 - \frac{K_b}{K_s} \tag{2.79}$$

where K_b, K_s are the bulk moduli for the bulk framework and the solid rock matrix, respectively. Terzaghi's effective stress is defined for $\alpha = 1$, which is a good approximation except at very large depths. The corresponding effective stress based Mohr circles and "p'-q" plots for compacted sediments are illustrated and explained in Fig. 2.40.[3]

2.6.1 Failure Analysis

Elastic material response means, that characteristic stress values like σ_z or $\bar{\sigma}$ increase linearly with the equivalent strain values ϵ_z, ϵ_v. The corresponding stress-strain plots also show non-elastic behavior. The curves are usually obtained by rock mechanical laboratory measurements, such as drained and undrained uni- and triaxial tests. Mean stress and volumetric strain are the important parameters in basin modeling. Typical curves are shown in Fig. 2.41. Usually, elastic and elasto-plastic regions are distinguished, separated by the yield point, the point of maximum stress and the critical state point.

Elastic behavior means no permanent changes. Thus, the rebound curve is identical to the load curve. Beyond the yield point the specimen will not return to the original state, but it still supports increasing loads with yielding. Softening begins at the point of maximum stress. Further deformation yield less ability of the specimen to withstand stress. At the critical state, an instable deformation occurs, like rupture or pore collapse. Failure is usually defined at the point of maximum stress, although it is sometimes used for yielding, since the material structure changes.

[3] Equivalent to the term "p-q" plot, which was introduced as the mean stress versus deviatoric stress diagram, the "p'-q" plot depicts the mean effective versus the deviatoric effective stress diagram. Note that the deviatoric effective stress is equal to the deviatoric total stress for Terzaghi's definition of the effective stress (2.2).

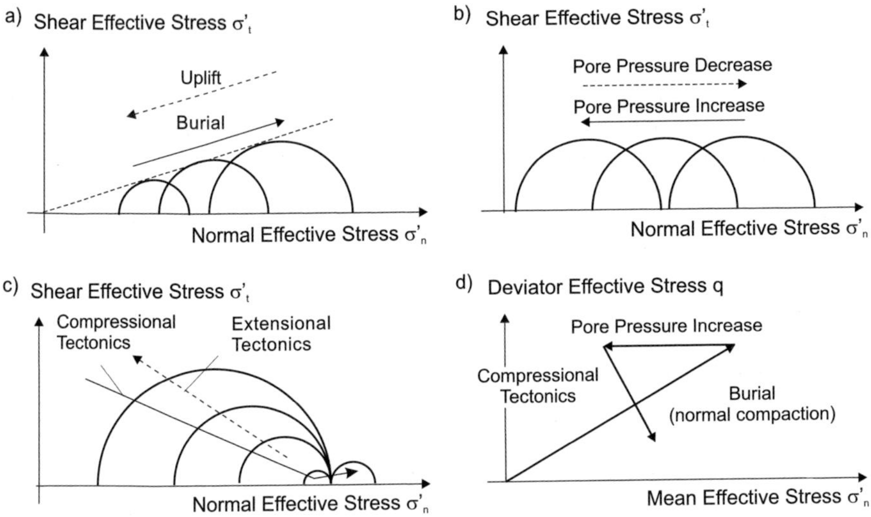

Fig. 2.40. Effective stress based Mohr circles and "p'-q" plots for (a) burial and uplift, (b) pore pressure changes, (c) tectonics. (d) Equivalent "p'-q" plot

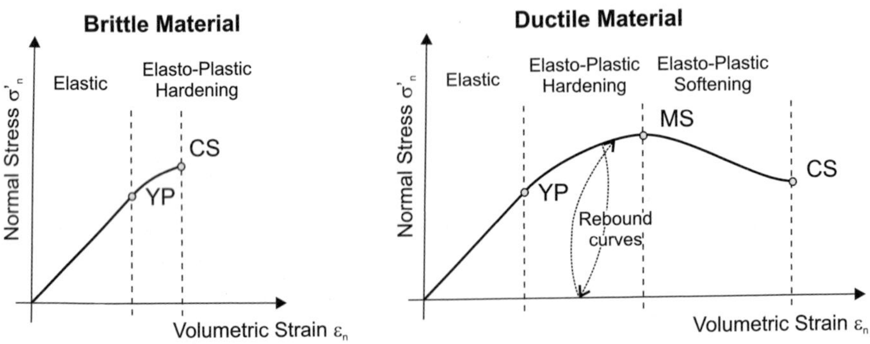

Fig. 2.41. Schematic stress versus strain plot for rocks. Characteristic points are the yield point (YP), the maximum stress point (MS) and the critical state point (CS). Brittle and ductile materials are distinguished by the relative length of the elasto-plastic region. The rebound curve has approximately the same steep angle as the linear elastic curve with a small hysteresis

The behavior of a sample is called ductile or brittle, when the elasto-plastic region is large or small. The curve in the elasto-plastic region of the same sample strongly depends on temperature and the speed of deformation.

Mohr Type Failure

For an arbitrary three dimensional stress state, the failure criterion is a function of all principal stresses, the yield function f:

$$f(\sigma_1, \sigma_2, \sigma_3) = 0\ . \tag{2.80}$$

Mohr proposed a function, which depends only on maximum and minimum principal stresses. This is equivalent to a curve in the Mohr diagram (Fig. 2.42). Failure occurs when the Mohr circle intersects the failure line. The often used Mohr failure curve is a straight line with cohesion C and internal friction μ as offset and steep angle of the line, respectively, which is called a Mohr–Coulomb failure.

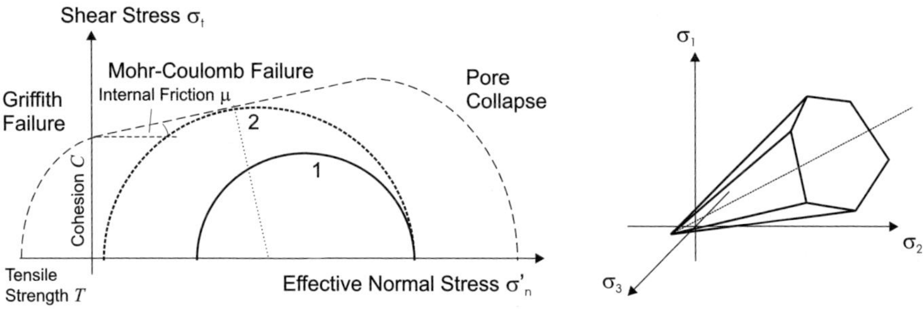

Fig. 2.42. Stresses and pressures in porous rocks

$$\sigma_t = C + \mu\sigma_n\ . \tag{2.81}$$

Mohr–Coulomb failure initiates plastic flow along a failure plane, which is directed at an angle $\beta = (\pi + 2\mu)/4$ between the axes of σ_3 and σ_1. The Mohr–Coulomb failure criterion is also equivalent to the surface of a hexagonal pyramid in the principal stress space (Fig. 2.42). A Mohr-type failure curve can also used to describe pore collapse, but with a negative friction angle, which is equivalent to the "cap" of the Mohr-Coulomb pyramid.

In the extensional region, a Griffith type criterion is usually taken into account. It is derived from a microscopic theory of crack extension in two-dimensional samples. A simple generalization to three dimensional rock samples can be made with a parabolic failure curve in the Mohr diagram (Fig. 2.42), which is equivalent to a parabolic "top" in the principal stress space. Griffith formulated the failure equation for 2D only. The simplest extension to 3D-phenomena is Murell's extension, where the failure surface, in terms of principal stresses, is also a parabolic surface with a simple pyramid on top.

$$\begin{aligned}(\sigma_1 - \sigma_3)^2 + (\sigma_1 - \sigma_2)^2 + (\sigma_2 - \sigma_3)^2 &= 24\,T_0\,(\sigma_1 + \sigma_2 + \sigma_3)\\ \text{or}\quad \sigma_1 = -T_0 \quad \text{or}\quad \sigma_2 = -T_0 \quad &\text{or}\quad \sigma_3 = -T_0\ .\end{aligned} \tag{2.82}$$

Note, that the maximum tensile strength is the only failure parameter for Griffith failure and it is related to cohesion as $C = 12\,T_0$. Griffith failure can also be extended to compressional regions with the same formula. This is usually used for the description of fracturing as explained in Sec. 2.6.1.

Plastic Flow and Critical State

The term critical state indicates damage and the inability of the specimen to support stresses. Usually, the same type of failure condition is used as for yielding, but with different material parameters C_0, T_0 and μ. Plastic flow occurs from the onset of yielding until the critical state is reached. Then, the strain consists of elastic and plastic parts (Fig. 2.43),

$$\boldsymbol{\epsilon} = \boldsymbol{\epsilon}_e + \boldsymbol{\epsilon}_p \; . \tag{2.83}$$

The elastic strain is still related to the effective stress tensor with the elastic modules, and the plastic deformations are directed perpendicular to the failure surface in the principal stress space.

$$d\epsilon_{p,ij} = d\lambda \frac{\partial f}{\partial \sigma'_{ij}} \tag{2.84}$$

where λ is the hardening parameter. In the above equation the plastic flow vector $d\epsilon_{p,ij}$ is directed perpendicular to the yield surface in the σ–space. In porous media, a constant angle between flow vector and yield surface is usually assumed. In general, the function f can be different from the yield function and it is then called plastic potential. It is very important, to note, that the direction of the plastic flow is controlled by the failure parameters C_0, T_0 and μ. Fault planes are directed along fixed angles between the minimum and maximum principal stresses. The formulation of the corresponding elasto–plastic boundary problem takes into account the plastic hardening law (2.84) and the yield condition (2.81). Detailed descriptions are given in Zienkiewicz (1984).

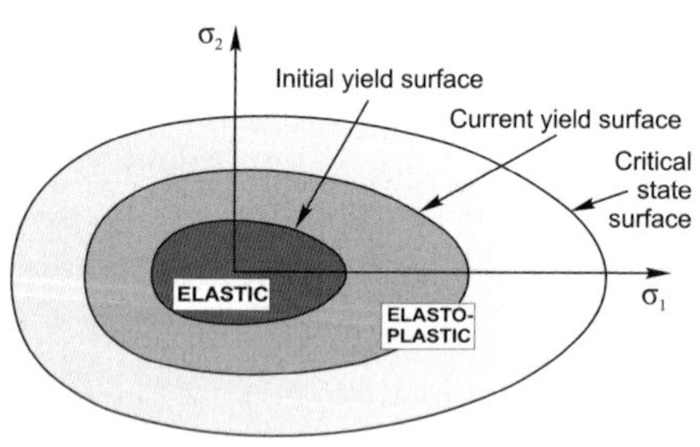

Fig. 2.43. Failure surfaces for yielding and critical state after Fjaer et al. (1992). The process of yielding is equivalent to hardening until the critical state is reached and damage occurs

The soil mechanical equivalent of the yield surface is the the Roscoe and Hvorslev surface in the "p–q–v" diagram (Fig. 2.44), for normally consolidated and overconsolidated rocks, respectively.[4] When the effective stress state in a rock intersects the yield surfaces, further compaction with increasing effective stress occurs along both failure surfaces until the critical state line.

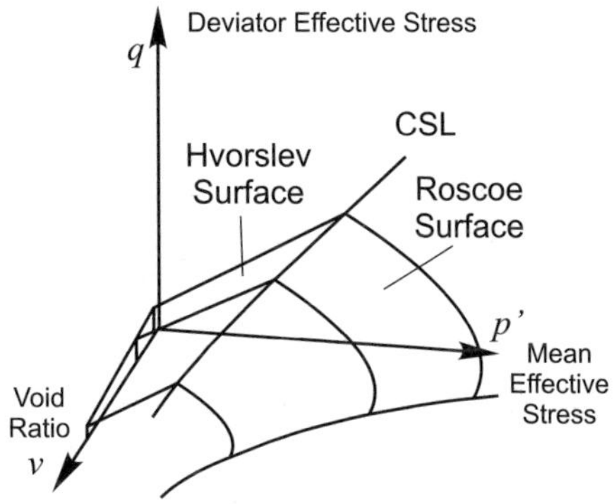

Fig. 2.44. Failure surfaces in a soil mechanics "p'–q–v" diagram. When the compaction controlled effective stress paths hits the failure surface, further compaction follows the failure surfaces until the critical state line (CSL)

Fracturing

Fracturing is another type of failure, which is the formation and growth of microfractures in rocks. Most fracture models are based on the Griffith theory, which defines a brittle failure surface in principal effective stress diagrams.

The most common type of fractures are tensile fractures. They are formed when traction exceeds the tensile strength T_0. Following the usal convention, traction is negative stress. Hence the maximum traction is equal to the minimum principal effective stress σ_3' and the condition for the initiation of fractures is

$$\sigma_3 - p = T_0 \,. \tag{2.85}$$

Obviously, the above condition is valid when the Mohr circle contacts the failure line on the left side (Fig. 2.42). The Mohr circle moves to the left mainly by overpressuring. The minimum overpressure, which is needed to initiate fracturing for a given stress state, is called the fracturing pressure. The fracturing pressure can be drawn in the pressure–depth space to illustrate the threshold pore pressure for fracturing (Fig. 2.42). It is a very common simplification in basin modeling programs to describe the fracturing condition with a fixed fracturing pressure versus depth curve (Fig. 2.45). The tensile strength differs for the different rock types. Hilgers et al. (2006) reported $T_0 = 10\,\mathrm{MPa}$ for sandstone and a significant smaller value for shale. Thus, the fracturing pressure gradient alternates within a shale–sand sequence.

Fracturing increases the rock permeabilities and drops the capillary entry pressures as described with the following relationship.

[4] Normally consolidated rocks are under the maximum effective stress, while overconsolidated rocks have a lower effective stress than at maximum burial

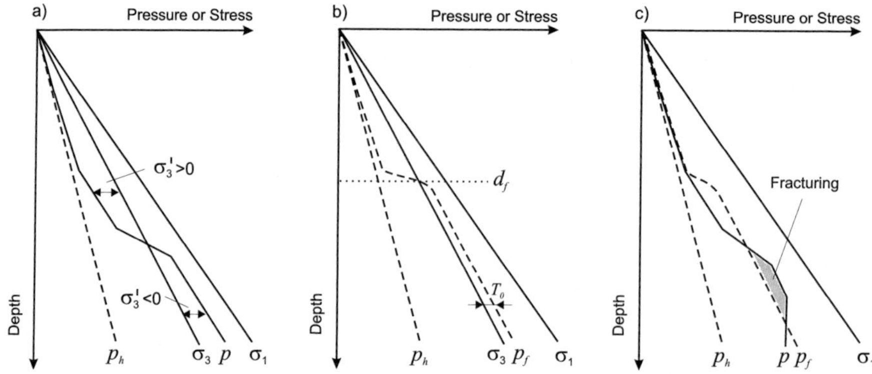

Fig. 2.45. Pressure and stress versus depth diagram for a fracturing model with hydrostatic pressure p_h, pore pressure p, fracturing pressure p_f, principal bulk stresses σ_1, σ_3 and effective stresses σ_1', σ_3'. The maximum stress is almost equal to the lithostatic pressure and the minimum stress is assumed to be a fixed fraction of the lithostatic pressure. (a) The difference between the minimum and maximum effective stress increases with depth. (b) The Griffith model defines the fracturing pressure as the required pore pressure to initiate fracturing. (c) Fracturing occurs when the pore pressure exceeds the fracturing pressure

$$\log k_f = \log k + \lambda_k \frac{p - p_f}{p_f}, \qquad p_c = p_{cf} - \lambda_c \frac{p - p_f}{p_f} \tag{2.86}$$

where k and k_f are the permeability of the unfractured and fractured rock, p_c and p_{cf} are the capillary entry pressures of the unfractured and fractured rock, and λ_k and λ_c are the fracturing parameters. For clastic rocks, the fracturing parameters of λ_k=3 log mD and $\lambda_c = 3$ MPa are frequently used. Fractures can partially anneal when the overpressure decreases below the minimum effective stress, so that the tension turns into compression. It is usually not necessary to exceed tensile strength again when fractures are re-opened, that means that the pore pressure has to be equal to the minimum effective stress. In some models, the simplified fracturing condition, that the fracturing pressure is equal to the minimum principal stress, is used. However, the inclusion of multiple closing and opening behaviors requires hysteresis effects with different fracturing pressures and permeabilities.

2.7 Faults

Faults occur in most basins with large variations in length, thickness, throws, gouge content and related properties. They play an important role in fluid flow and pressure formation. Faults are initiated in consolidated sediments due to extensional and compressional forces mainly caused by plate tectonics. The process of fault formation and growth can be described and modeled with kinematic approaches. This is usually not part of basin modeling, instead the fault geometry and main properties at present and paleo–times are given as a predefined input. Geologists distinguish between normal, reverse, transform and strike–slip faults, although most of the faults are mixed mode faults. The fault type depends on stress conditions in the of formation Fig. 2.46. The fault properties can be predicted by structural or fault seal analysis methods.

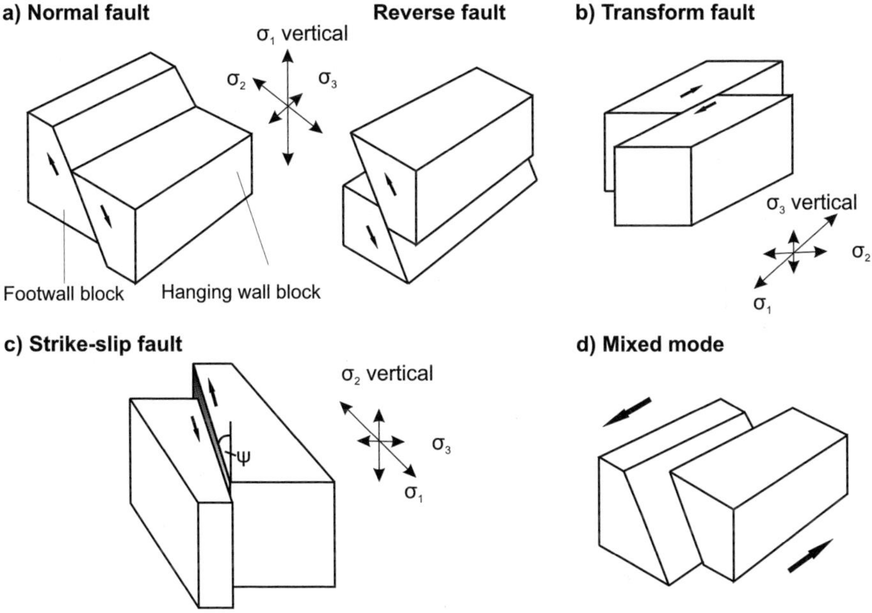

Fig. 2.46. Fault types formed under different stress conditions: (a) maximum principal stress in a vertical direction cause normal and reverse faults. (b) Minimum principal stress in a vertical direction causes transform faults. (c) Medium principal stress in a vertical direction causes strike-slip faults. (d) Most faults in nature are mixed modes. The pictures are from Bahlburg and Breitkreuz (2004)

Fault extensions often exceed several hundreds of meters. Fault zones are often in the range of several meters and much smaller than gridcells of basin scale models. Location and orientation of faults are thus geometrically described with fault planes in 3D–models and lines in 2D–models. Fault lines and planes can be approximated with boundary elements along cell faces and edges in cellular models as illustrated in Fig. 2.47 and Fig. 2.55. The fault planes in 3D–models are constructed from fault lines, which are usually interpreted from seismic at the surface of horizon maps.

Faults can act as preferred migration avenues (in-fault flow) or as hydrocarbon seals which hold column heights of hydrocarbons. The two related flow properties are permeability and capillary entry pressure. Boundary fault

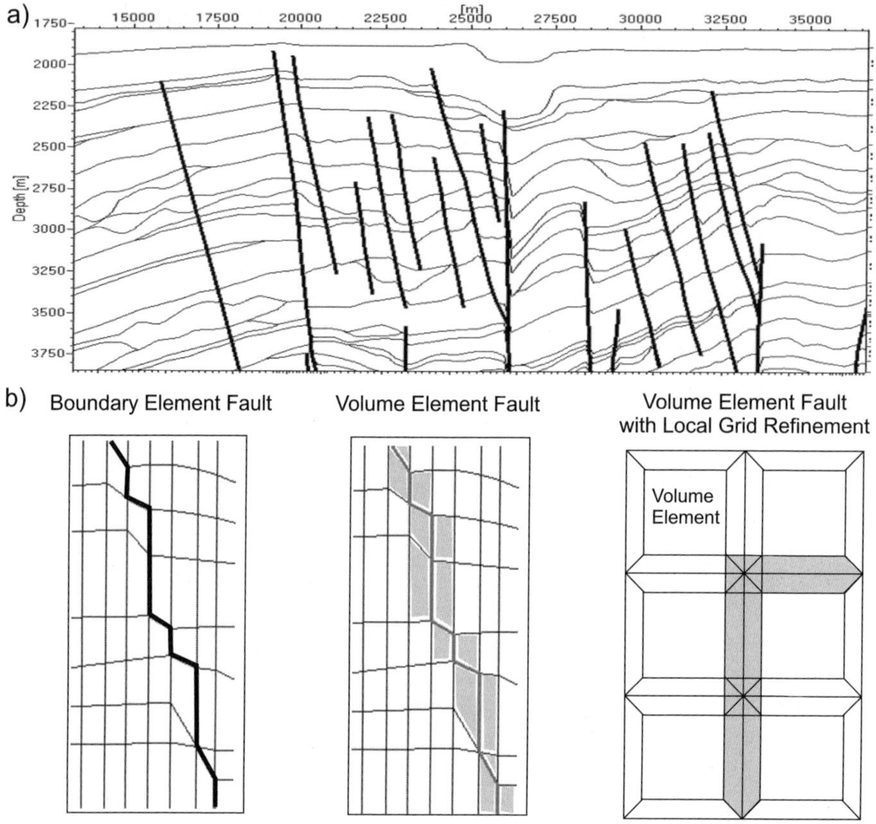

Fig. 2.47. 2D–fault models. (a) Fault lines in a 2D–cross section. (b) Fault line approximation with boundary elements, adjacent volume elements, or with locally refined grid cells

elements can be used for modeling petroleum migration and accumulation. They act between two volumetric elements with a zero volume. Capillary entry pressure can be defined for cells with an infinitesimally small volume, but permeabilities cannot assigned to boundary elements without a volume. This yields instantaneous flow for in–fault flow, which is assumed in some petroleum migration models anyway.

Permeabilities can not be neglected for pore pressure calculations, if fault gouge material with low permeability causes pressure contrasts and compartmentalization. Hence, the inclusion of fault permeabilities for pressure modeling requires the consideration of a fault volume.

The simplest method to work with volumetric fault elements is to define all cells adjacent to the fault plane as fault cells and assign the corresponding fault permeabilities. Obviously, fault zones can be overestimated with this approach, which can yield large errors in the calculated pressures. This is especially problematic for very low permeability faults, which have to be continuously connected to model compartments. Double or triple bands of cells are necessary when topologically regular grids are considered. Irregular gridline spacing with higher gridline density in the vicinity of faults can be used to lower the effect, but this is usually only applied in 2D–models, since the number of gridcells increases significantly.

A good solution to the problem is the introduction of locally refined elements around faults (Figs. 2.47, 8.13), where the real fault width can be taken into account. This method requires significant effort for the development of automatic meshers in 3D, especially for the special cases of layer pinch–outs and dendritic fault segments. An example pressure calculation with locally refined elements around faults, fault widths of 10 m and fault permeabilities of 10^{-6} mD is shown in Fig. 2.56. The pressure is constant within the sandstone compartments and varies in the sandy shales.

The fault permeabilities and capillary pressures are mainly determined by the gouge composition. Very thin faults can be handled as neutral or juxtaposition faults without any property assignment.

The gouge composition is a mixture of the rocks of all layers, which slipped along the location during faulting. An important parameter of the composition is the clay content. Various indicators are proposed (Yielding et al., 1997; Fulljames et al., 1996), such as the shale smear factor or the shale gouge ratio (Fig. 2.48). All of them pay attention to the juxtaposition of sediments between the foot and the hanging wall and depend on fault distance or throw. The shale gouge ratio (SGR) is the volumetric ratio of grains smaller then 100 nm to the larger grains assuming that the value at an actual location is simply the arithmetic average of all the material that slipped since fault movement began. In this approach, it is not considered that different rock types have different supply rates to the gouge. An advantage of the SGR concept is, that the values can be calculated by simple volumetrics of the fault adjacent sediments for each point on the fault surface.

Yielding (2002) proposed simple relations to convert SGR values to capillary entry pressures and permeabilities. Capillary entry pressures control column heights at sealing faults. They are given as mercury-air values and can be converted to the present petroleum-water system via in-situ interfacial tension values of the compositional dependent petroleum phases. The fault capillary pressures (FCP) are therefore capillary entry pressures for the mercury-air displacement. Yielding found linearly increasing FCP values for SGR larger than a threshold SGR, with different ascent angles but unique minimum SGR of 18% in most of the samples. The average value for the parameter k in the following equation is 50 MPa.

$$p_c = k\,(\mathrm{SGR} - 0.18)\,. \tag{2.87}$$

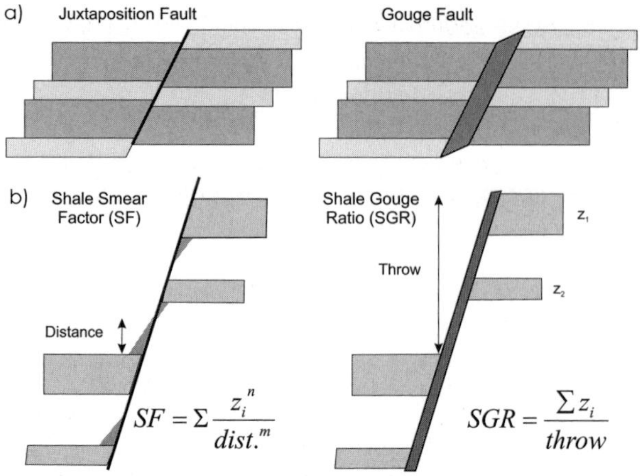

Fig. 2.48. Definition of fault properties after Yielding et al. (1997): (a) juxtaposition and gouge fault, (b) definition of shale smear factor (SF) and shale gouge ratio (SGR)

The permeability value controls in-fault flow, once the accumulation is able to break into the fault, but this seems to be less important as the distances are usually short before an exit to more conductive sediments is found.

It is obvious, that the fault properties (SGR, FCP) experience large variations through geological time. Thus, they have to be specified or precalculated for several time periods. A common simplification is the introduction of special fault properties: ideal open (SGR $<$ 18%, FCP $<$ 0.1 MPa) and ideal closed (SGR $>$ 95%, FCP $>$ 50 MPa) to define faults as completely open or completely closed or via special FCPs or SGRs as in Fig. 2.57.

Diagenetic processes or cataclasis in faults can be described by additional temperature or effective stress dependent corrections of the SGR values.

2.8 Paleo–Models

In a basin with low faulting, throw and tectonics, back-stripping of the present day geometry under consideration of decompaction, erosion and paleo–thickness corrections is a good approximation of the paleo–geometries. Decompaction and erosion are typically vertical phenomena, which do not take into account any horizontal movements and changes in the total length of the layer. Horizontal movements of single layers like salt domes are described with paleo–thickness corrections based on rock volume balances which results in layer squeezing and stretching. Complex tectonic events often yield strongly deformed geometries, which usually overstretch the possibilities of backstripping. Complete paleo–geometries are alternatively used as input for the simulation. They are constructed from structural modeling methods before basin modeling is performed. The simulator then jumps from predefined paleo–geometry to paleo–geometry in the analysis. It has to identify the new location of each single facies and has to take into account facies movements and deformations. In compressional tectonics, overthrusted layers can multiply be defined along a depth line, which is handled in practice with the implementation of a block concept for the thrust belts. Each block represents a compartment treated as a separate unit which can be moved along and against any other block.

2.8.1 Event–Stepping

Backstripping is also called event stepping, since the paleo–geometries are reconstructed from the present day geometry due to given "geological events" with a suitable set of sophisticated rules, which yields topologically similar paleo–models. Decompaction of a layer from present day thickness d_p to depositional thickness d_0 is calculated with the assumption of the conservation of the solid matrix volume according to

$$d_0\,(1-\phi_0) = d_p\,(1-\phi_p) \tag{2.88}$$

with present day and depositional porosities ϕ_p and ϕ_0, respectively. The present day porosity is not known prior to analysis, since it depends on the pore pressure development. Hence, the decompaction in the first simulation run can only be made with an estimation of the present day porosities, used as the steady state values for hydrostatic pressure conditions. The forward simulation then yields calculated present day geometry based on pore pressure controlled compaction, which usually differs from the given present day geometry (Fig. 2.49). This difference is much smaller in the next simulation run, when the calculated present day porosity can be derived for decompaction instead of the estimated steady state values. This optimization procedure can be applied multiple times, but usually two or three loops yield good results.

Modeling of erosion requires the definition of the eroded thicknesses and the erosion ages. Eroded thicknesses can be given with virtual horizons or thicknesses at the time of deposition, at present day or any other geological event (Fig. 2.50). Multiple erosions of one layer and one erosion on multiple layers can also easily be recognized with virtual horizons. The interpretation of eroded thickness is often easier to perform on a backstripped and decompacted paleo–geometry. Herein, the porosity at the erosion age has also to be considered for decompaction of overconsolidated rocks. The eroded thickness and the compaction history of the layer before erosion has to be taken into account in the optimization procedure.

Horizontal movements of layers like salt can be described with additional thickness maps during doming. The changes are realized by layer stretching and thinning. The additional salt thickness layer should be calculated under the assumption of total volume conservation. The simplest model considers a homogeneous depositional layer with the total volume equal to the total

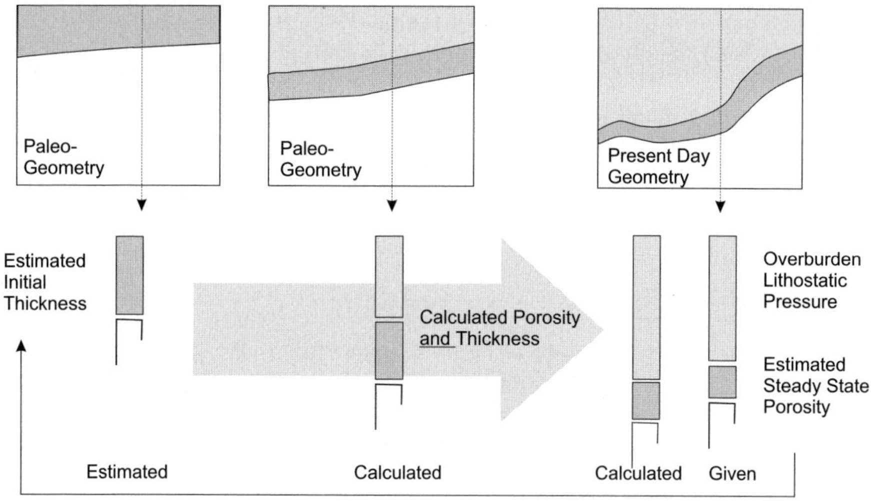

Fig. 2.49. Backstripping with decompaction is based on estimated present day porosities. The calculated porosities of the forward simulation usually improve backstripping in the next run

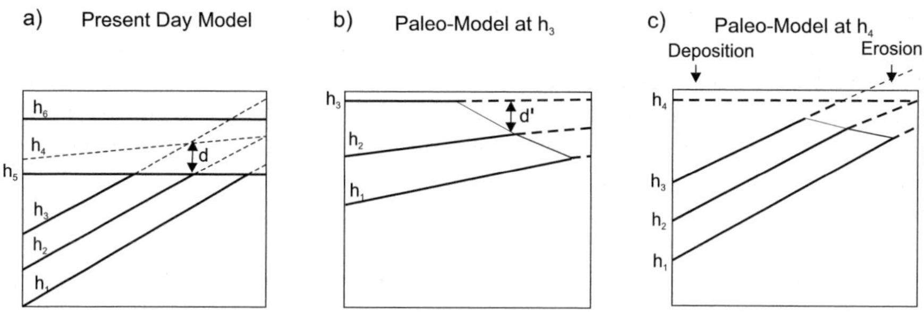

Fig. 2.50. Definition of erosional thicknesses: (a) with virtual horizons at the present day geometry, (b) with additional thicknesses at the time of sedimentation, (c) with virtual horizons at any geological event

volume of the present day salt domes and the definition of the doming ages. A linear interpolation between the initial and the final salt thicknesses can then be realized during doming. The opening of the salt windows should be described with an additional salt map, since the salt windows would otherwise open only during the last time step of the doming (Fig. 2.51).

Structural geologists often provide salt maps for various geological events based on kinematic models, which also can be considered during simulation. Additional thickness maps can be used for the thinning of the salt adjacent layers or for doming of other lithologies, e.g. shale. High overburden can also yield reverse domes and single salt pillows as illustrated in Fig. 2.52. This

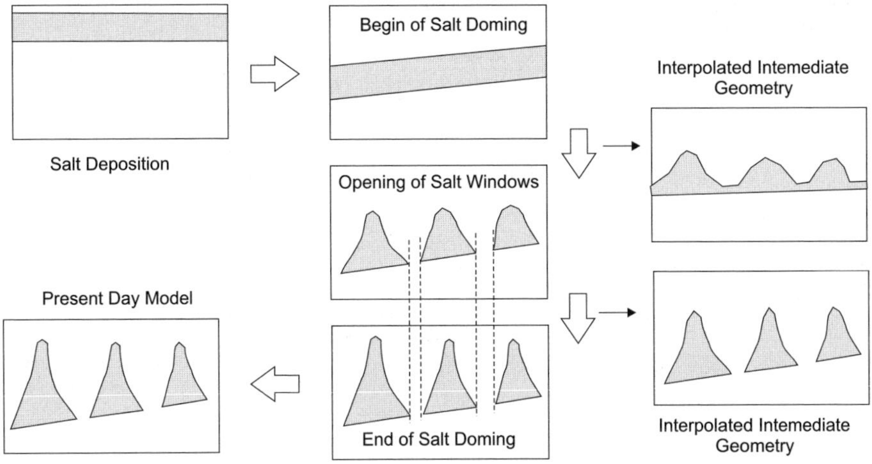

Fig. 2.51. A simple geometrical model with a linear interpolation of the salt thicknesses between the geometries of salt sedimentation, salt window opening and the final doming

requires the introduction of several layers to avoid multiple occurrences of one layer along a depth–line. Another method for handling salt intrusions into the overburden layers is to exchange the lithology of the intruded layers with salt. This is recommended when the intruded layers have big gaps in the present geometry. In very complex basins, pre–computed paleo-geometries might be necessary. This is described in the next subsection.

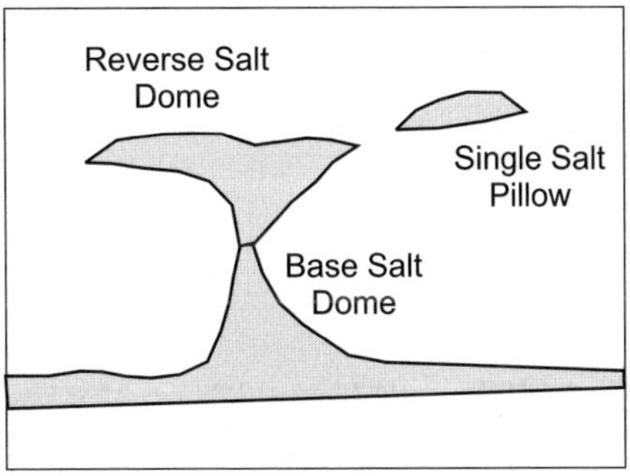

Fig. 2.52. Reverse salt domes and salt pillowing require multiple layer definitions

2.8.2 Paleo–Stepping

The introduction of complete geometrical models for certain paleo–times requires the recognition of facies locations together with the corresponding types of movements and deformations during stepping from one paleo–geometry to the next. A section of a layer can be folded, migrated and/or otherwise stretched so that location and shape in two successive time steps might be very different (Fig. 2.53). A meshing algorithm based on pre-defined gridpoints and sublayers can yield new volumetric cells which are no longer related to the same rock of the previous time step. The consequence is that all bulk rock properties have to be transferred according to the new location.

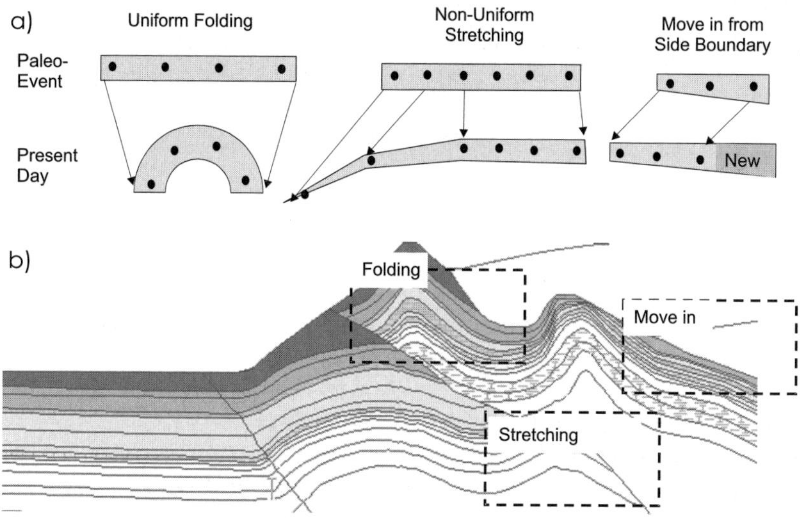

Fig. 2.53. Deformation types of facies during tectonics

In most cases, the deformation is uniform stretching or thinning, which can be achieved with linear mapping operations. Any non-uniform deformations have to be specified manually between paleo–geometries. For moving–in layers, the side boundary values can be taken as the values for the previous time step.

In the following a method is described, how the pressure and compaction problem is solved, when the compaction has already been predefined via paleo–geometries. Both the pressure and the compaction equations can be solved in the usual way. The change in the overburden load of each layer is calculated from one paleo–geometry to the next. The transient equation for overpressure (2.13) can then be solved with the transformed cell values of the previous time step. The results are a change in the overpressure as well as a reduction in the porosity. The only difference from the usual procedure is, that the porosity change is not converted into the new layer thicknesses, since they are already predefined with given paleo–geometries. Hence, porosity reduction and compaction are decoupled processes here and it is accepted that the volume rock matrix is no longer conserved.

Backstripping or event–stepping is applied before the first occurrence of a paleo–geometry with the usual method for optimization (Fig. 2.54).

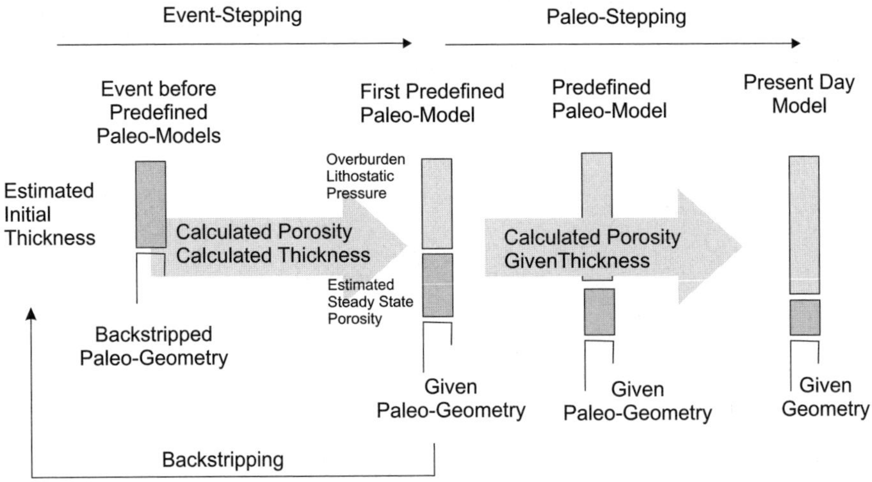

Fig. 2.54. Decoupling of compaction and porosity calculation during paleo–stepping

A difficult problem is the automatic generation of additional paleo–models for time steps between the interpretations. The simplest idea is to use linear interpolations so that the thickness values of each gridpoint are interpolated, but this often yields unsuitable connections in steep faults. Another method is to directly jump to the next paleo–model, and use intermediate time steps for the solution heat and fluid flow equations, but with the same geometry for the whole geological event.

The above procedure clearly separates structural reconstruction from forward basin modeling analysis by work flow and by data. The advantage of this decoupled link is that it is possible to use advanced special tools for both structural and basin modeling and the functionality of both tools are retained. Due to decoupling of processes information is lost. Feedback between modeled processes as well as coupled tools, handling structural and basin modeling together, are principally possible.

2.8.3 Overthrusting

Blocks for overthrust belts are introduced to avoid multiple layer occurrences along one depth–line (Fig. 2.58). Each block is then treated like a "single basin model" with suitable and varying coupling conditions between the block boundaries. The number of blocks can vary during paleosteps, since the breaking of a so called super block into separate pieces leads to the development of complex block substructures from a homogeneous initial model. A hierarchy of block heritages has to be specified as a model input. Splitting of a super block into subblocks also has to be taken into account when considering the shift of the fundamental layer values according to their new locations.

All block boundaries are faults. They can be treated as neutral (juxtaposition), partially or ideally open or closed faults with capillary entry pressures and permeabilities (Sec. 2.7).

Compression or extension yields a total section abbreviation or stretching, which yields an increase or decrease in horizontal stress and causes additional or retarded overpressuring and compaction. Then, compaction should be controlled by mean effective stress instead of vertical effective stress components with the following modified compaction law, which replaces equation (2.3).

$$\frac{\partial \phi}{\partial t} = -C_v \frac{\partial \bar{\sigma}'}{\partial t} = -C_v \frac{\partial(\bar{\sigma} - p)}{\partial t} == -C_v \frac{\partial(\bar{\sigma} - p_h - u)}{\partial t} \tag{2.89}$$

where C_v is the volumetric bulk compressibility, which is related to the Terzaghi compressibility C_T as follows:[5]

$$C_v = C_T \frac{3(1-\nu)}{1+\nu} \tag{2.90}$$

which yields a factor of 1.28 . . . 2.45 for Poisson ratios of ν of 0.1 . . . 0.4. The pressure equation (2.13) is modified as follows.

$$\frac{C_v}{1-\phi} \frac{\partial u}{\partial t} - \nabla \cdot \frac{\boldsymbol{k}}{\nu} \cdot \nabla u = \frac{C_v}{1-\phi} \frac{\partial(\bar{\sigma} - p_h)}{\partial t} \tag{2.91}$$

where p_h is the hydrostatic pressure and $\bar{\sigma}$ is the mean total stress. Assuming, that the total stress differs from Terzaghi's lithostatic pressure assumption only by an additional horizontal stress component the tectonic stress σ_t, the total mean stress is related with the overburden weight pressure p_l as follows:

$$\bar{\sigma} = \sigma_v + 2\,\sigma_h = \frac{1+\nu}{3(1-\nu)} p_l + 2\sigma_t\ . \tag{2.92}$$

The additional assumption of a uniform compression $\sigma_t > 0$ or extension $\sigma_t < 0$ with a constant tectonic stress σ_t yields a simple extension of the pressure equation and compaction law which includes tectonic processes.

[5] In the case without tectonics, it is $\sigma_h/\sigma_v = \nu/(1-\nu)$ and with $\bar{\sigma} = (1/3)(\sigma_v + 2\,\sigma_h)$ it follows that $\sigma_v/\bar{\sigma} = 3(1-\nu)/(1+\nu)$.

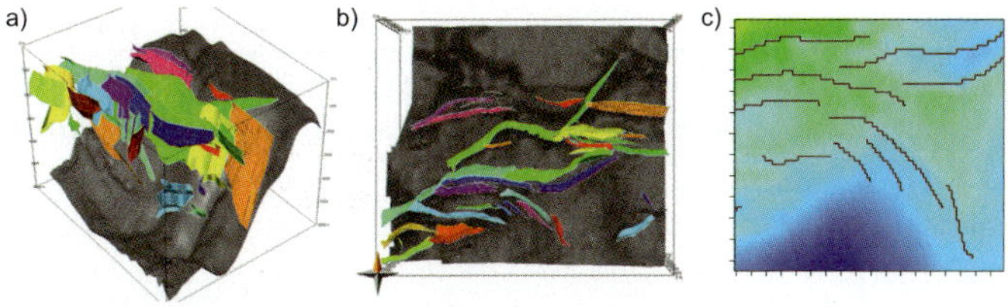

Fig. 2.55. (a) Fault approximation with boundary elements in 3D. (b) Vertical view with horizontal fault elements. (c) Map view of cutout with fault traces

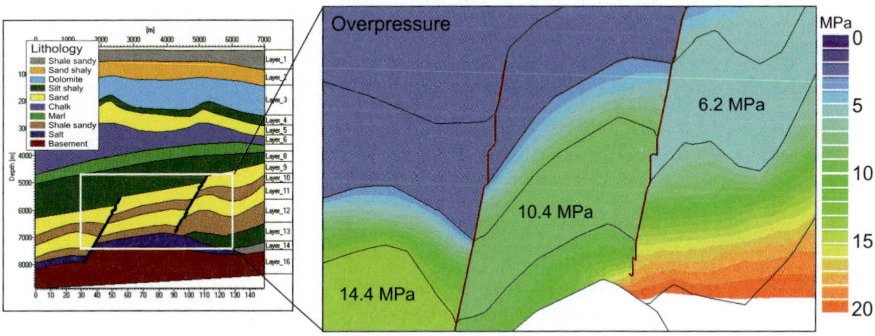

Fig. 2.56. Overpressure example with locally refined elements around faults

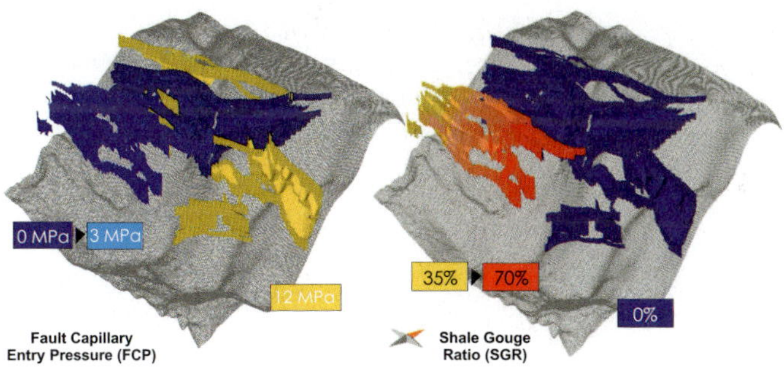

Fig. 2.57. Capillary pressures and SGR values on fault planes in a 3D–model

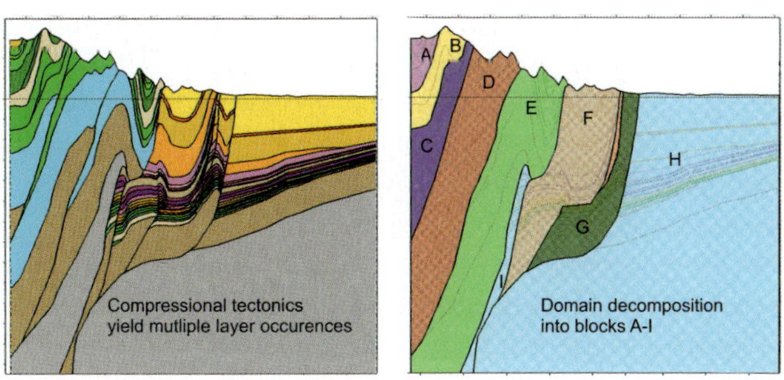

Fig. 2.58. Introduction of blocks for compressional tectonics

Summary: Overburden load and tectonic stresses cause rock stresses, fluid pressure formation, and sediment compaction. An external load on a bulk volume element is balanced partially by the rock skeleton and partially by the pore water. Rock stresses and pore water pressure equalize overburden and external tectonic stresses.

Many geomechanical processes are formulated with overpressure instead of pore pressure and effective stress instead of total rock stress. Overpressure is pore pressure minus hydrostatic pressure, which is the weight of the overlaying pure water column (plus a depth independent shift for zero level adjustment). Effective stress is the total stress minus pore pressure.

Water is mobile. Overpressure gradients causes pore fluid flow, which is mainly controlled by the rock permeabilities. This allows for further rock compaction with reordering of grains. The rock becomes more dense and its internal stress rises as overpressure is usually reduced.

All basic effects of mechanical compaction and overpressure formation can be modeled quite accurately with a Terzaghi–type approach. It is based on the assumptions that rock grains and water are incompressible and that rock compaction is a function of the vertical effective stress only, which is called lithostatic pressure. Water flow is modeled with Darcy's law. Overall mass conservation is taken into account. Appropriate conditions for water in– and outflow at model boundaries must be defined.

Various models for compaction vs. effective stress are proposed. The main characteristic is a logarithmic dependency of effective stress on porosity. The related compaction or bulk compressibilities functions are well known over a wide porosity range for various lithotypes.

Overpressure calibration is a two step process. Firstly, the material parameters of the compaction law must be fitted locally to suitable porosity vs. effective stress relationships. Secondly, permeabilities of relevant layers, which control the overall water flow, must be adjusted. The second step is rather sophisticated and relies on full simulation runs due to possibilities of long range lateral water flows.

Besides pure mechanical compaction, pressure effects due to cementation of pore space, aquathermal expansion, mineral transformations and petroleum generation are found as locally significant.

Alternatively to a calculation of the geometry from compaction laws (event–stepping) the geometry might be imported from purely structural analysis (paleo–stepping). However, overpressures and effective stresses are simulated in any case with similar algorithms.

Faults can be approximated with special volumetric and boundary elements. The main mechanical properties are fault transmissibilities and capillary entry pressures, which can be derived from measured or calculated shale gouge ratios. The inclusion into pressure and fluid flow analysis requires sophisticated numerical models.

References

L. F. Athy. Density, porosity and compaction of sedimentary rocks. *American Association of Petroleum Geophysicists Bulletin*, (14):1–24, 1930.

H. Bahlburg and C Breitkreuz. *Grundlagen der Geology.* Elsevier GmbH, Muenchen, second edition, 2004.

M. A. Biot. General theory of three-dimensional consolidation. *Journal of Applied Physics*, (12):155–164, 1941.

P. A. Bjørkum. How improtant is pressure in causing dissolution of quartz in sandstones. *Journal of Sedimentary Research*, 66(1):147–154, 1996.

P. A. Bjørkum and P. H. Nadenau. Temperature Controlled Porosity/Permeability Reduction, Fluid Migration, and Petroleum Exploration in Sedimentary Basins. *APPEA Journal*, 38(Part 1):452–464, 1998.

P. A. Bjørkum, E. H. Oelkers, P. H. Nadeau, O. Walderhaug, and W. M. Murphy. Porosity Prediction in Quartzose Sandstones as a Function of Time, Temperature, Depth, Stylolite Frequency, and Hydrocarbon Saturation. *AAPG Bulletin*, 82(4):637–648, 1998.

P. A. Bjørkum, O. Walderhaug, and P. H. Nadeau. Thermally driven porosity reduction: impact on basin subsidence. In *The Petroleum Exploration of Ireland's Offshore Basins*, volume 188 of *Special Publication*, pages 385–392. Geological Society of London, 2001.

A. Danesh. *PVT and Phase Behaviour of Petroleum Reservoir Fluids.* Number 47 in Developments in petroleum science. Elsevier, 1998.

P. M. Doyen. Permability, Conductivity, and Pore Geometry of Sandstone. *Journal of Geophysical Research*, 93(B7):7729–7740, 1988.

W. A. England, A. S. MacKenzie, D. M. Mann, and T. M. Quigley. The movement and entrapment of petroleum fluids in the subsurface. *Journal of the Geological Society, London*, 144:327–347, 1987.

E. Fjaer, R. M. Holt, P. Horsrud, A. M. Raan, and Risnes R. *Petroleum related rock mechanics.* Elsevier, 1992.

J. R. Fulljames, L. J. J. Zijerveld, R. C. M. W. Franssen, G. M. Ingram, and P. D. Richard. Fault seal processes. In Norwegian Petroleum Society, editor, *Hydrocarbon Seals - Importance for Exploration and Production*, page 5. Norwegian Petroleum Society, Oslo, 1996.

M. R. Giles, L. Indrelid, and D. M. D. James. Compaction – the great unknown in basin modelling. In S. J. Düppenbecker and J. E. Iliffe, editors, *Basin Modelling: Practice and Progress*, number 141 in Special Publication, pages 15–43. Geological Society of London, 1998.

Hewlett-Packard. Petroleum fluids, manual. Technical Report HP-41C, 1985.

C. Hilgers, S. Nollet, J. Schönherr, and J. L. Urai. Paleo–overpressure formation and dissipation in reservoir rocks. *OIL GAS European Magazine*, (2): 68–73, 2006.

R. H. Lander and O. Walderhaug. Predicting Porosity through Simulating Sandstone Compaction and Quartz Cementation. *AAPG Bulletin*, 83(3): 433–449, 1999.

O. Lauvrak. Personal communication, 2007.

X. Luo and G. Vasseur. Contributions of compaction and aquathermal pressuring to geopressure and the influence of environmental conditions. *AAPG Bulletin*, 76(10):1550–1559, 1992.

X. Luo and G. Vasseur. Geopressuring mechanism of organic matter cracking: Numerical modeling. *AAPG Bulletin*, 80(6):856–874, 1996.

G. Mavko, T. Mukerji, and J. Dvorkin. *The Rock Physics Handbook*. Cambridge University Press, 1998.

C. I. Mc Dermott, A. L. Randriamantjatosoa, and Kolditz O. Pressure dependent hydraulic flow, heat transport and geo-thermo-mechanical deformation in hdr crystalline geothermal systems: Preliminary application to identify energy recovery schemes at urach spa. Technical report, Universitaet Tuebingen, Lehrstuhl fuer Angewandte Geologie, 2004.

W. D. McCain Jr. *The Properties of Petroleum Fluids*. Pennwell Books, second edition, 1990.

M. J. Osborne and R. E. Swarbrick. Mechanisms for generating overpressure in sedimentary basins: A re–evaluation. *AAPG Bulletin*, 81:1023–1041, 1997.

R. H. G. Parry. *Mohr Circles, Stresspaths and Geotechnics*. Spon Press, second edition, 2004.

A. M. Pytte and R. C. Reynolds. The thermal transformation of smectite to illite. In N. D. Naeser and T. H. McCulloh, editors, *Thermal History of Sedimentary Basins: Methods and Case Histories*, pages 133–140. Springer–Verlag, 1989.

F. Schneider and S. Hay. Compaction model for quartzose sandstones application to the Garn Formation, Haltenbanken, Mid–Norwegian Continental Shelf. *Marine and Petroleum Geology*, 18:833–848, 2001.

F. Schneider, J. L. Potdevin, S. Wolf, and I. Faille. Mechanical and chemical compaction model for sedimentary basin simulators. *Tectonophysics*, 263: 307–313, 1996.

D. Schulze-Makuch, D. S. Carlson, D. S. Cherkauer, and Malik P. Scale dependency of hydraulic conductivity in heterogeneous media. *Groundwater*, 37:904–919, 1999.

J. E. Smith. The dynamics of shale compaction and evolution of pore fluid pressure. *Mathematical Geology*, (3):239–263, 1971.

R. E. Swarbrick, M. J. Osborne, and Gareth S. Yardley. Comparison of Overpressure Magnitude Resulting from the Main Generating Mechanisms. In A. R. Huffmann and G. L. Bowers, editors, *Pressure regimes in sedimentary basins and their prediction*, volume 76, pages 1–12. AAPG Memoir, 2002.

K. Terzaghi. Die Berechnung der Duerchlässigkeitsziffer des Tones im Verlauf der hydrodynamischen Spannungserscheinungen. *Szber Akademie Wissenschaft Vienna, Math–naturwissenschaft Klasse IIa*, (132):125–138, 1923.

P. Ungerer, J. Burrus, B. Doligez, P. Y. Chenet, and F. Bessis. Basin evaluation by integrated two–dimensional modeling of heat transfer, fluid flow, hydrocarbon gerneration and migration. *AAPG Bulletin*, 74:309–335, 1990.

L. Vidal-Beaudet and S. Charpentier. Percolation theory and hydrodynamics of soil–peat mixtures. *Soil Sci. Soc. AM. J.*, 64:827–835, 2000.

O. Walderhaug. Modeling quartz cementation and porosity in middle jurassic brent group sandstones of the Kvitenbjoern field, northern North Sea. *AAPG Bulletin*, 84:1325–1339, 2000.

O. Walderhaug. Kinetic modelling of quartz cementation and porosity loss in deeply buried sandstone reservoirs. *AAPG Bulletin*, 5:80, 1996.

O. Walderhaug, P. A. Bjørkum, P. H. Nadeau, and O. Langnes. Quantitative modelling of basin subsidence caused by temperature–driven silicia dissolution and reprecipitation. *Petroleum Geoscience*, 7:107–113, 2001.

A. Y. Yang and A. C. Aplin. Definition and practical application of mudstone porosity-effective stress relationships. *Petroleum Geoscience*, 10:153–162, 2004.

G. Yielding. Shale Gouge Ratio – calibration by geohistory. In A. G. Koestler and R. Hunsdale, editors, *Hydrocarbon Seal Quantification*, number 11 in NPF Special Publication, pages 1–15. Elsevier Science B.V., Amsterdam, 2002.

G. Yielding, B. Freeman, and D. T. Needham. Quantitative Fault Seal Prediction. *AAPG Bulletin*, 81(6):897–917, 1997.

O. C. Zienkiewicz. *Methode der finiten Elemente*. Carl Hanser, second edition, 1984.

3

Heat Flow Analysis

3.1 Introduction

Heat can be transferred by conduction, convection, and radiation in sediments (Beardsmore and Cull, 2001). The sediment–water–interface temperature and the basal heat flow are the main boundary conditions for heat flow analysis in sediments. Magnitude, orientation and distribution of the heat inflow at the base of the sediments are determined by mechanical and thermal processes of the crust and mantle (Allen and Allen, 2005). Two processes result in permanent heat flow from the Earth's interior to its surface: earth cooling and radiogenic heat production with a ratio of 17% to 83% respectively (Turcotte, 1980).

Heat conduction is defined as the transfer of thermal energy by contact according to thermal gradients. It is the primary process in the shallow lithosphere. The controlling lithological parameter is the thermal conductivity. It usually decreases from solids to liquids to gases. Generally, heat conduction is more effective with higher density.

Heat convection is thermal energy transported with the movement of a fluid or solid. In sedimentary basins, it is mainly related to fluid flow of pore water, liquid petroleum and gas. Convection can be more efficient than conduction when flow rates are high, e.g. in permeable layers or in fractures. It is the dominant thermal transport mechanism in the asthenosphere. Fluid movements can either add or remove thermal energy from a sedimentary sequence and can considerably distort conductive heat transfer systems. Solid convection occurs for example during overthrusting and salt doming. It requires very high thrusting rates to be of significance. A special type of combined conduction and convection is advection, which describes, for example, the heating of grains by groundwater flow.

Heat radiation is thermal transport via electromagnetic waves usually with wavelengths of 800 nm to 1 mm. The amount of thermal energy is proportional to the fourth power of temperature. Therefore, only heat transfer from very

T. Hantschel, A.I. Kauerauf, *Fundamentals of Basin and Petroleum Systems Modeling*, DOI 10.1007/978-3-540-72318-9_3,
© Springer-Verlag Berlin Heidelberg 2009

hot areas requires attention. It is negligible in sediments, but should be considered in deep parts of the lithosphere or asthenosphere.

The heat conductivity law states, that a temperature difference between two locations causes a heat flow $\mathbf{q}$. Its magnitude depends on the thermal conductivity of the material and the distance between these locations. In mathematical notation it becomes

$$\mathbf{q} = -\boldsymbol{\lambda} \cdot \boldsymbol{\nabla} T \tag{3.1}$$

with the temperature gradient $\boldsymbol{\nabla} T$ and the thermal conductivity tensor $\boldsymbol{\lambda}$.

The tensor $\boldsymbol{\lambda}$ is often assumed to have only two independent components: the conductivity along a geological layer λ_h and the conductivity across a geological layer λ_v. The heat flow vector at any location is mainly directed along the steepest decrease of temperature from a given location. In the lithosphere, it is mainly caused by the difference between its top and base temperatures: the surface temperature or sediment–water–interface (SWI) temperature at the top and the asthenosphere–lithosphere boundary temperature at its base. Hence, the resulting heat flow is mainly vertically directed when the two boundary surfaces are almost spherical and when the lateral variations of the boundary temperatures are small. The average thermal conductivity and the thickness of the mantle and crustal layer mainly control the heat in–flux into the sediments. This heat flow at the base of the sediments defines the lower boundary condition for the heat flow analysis in the sediments.

In practice, heat flow analysis is commonly subdivided into two problems: the consideration of the crustal model to calculate the heat in–flux into the sediments and the temperature calculation in the sediments afterwards (Fig. 3.1).

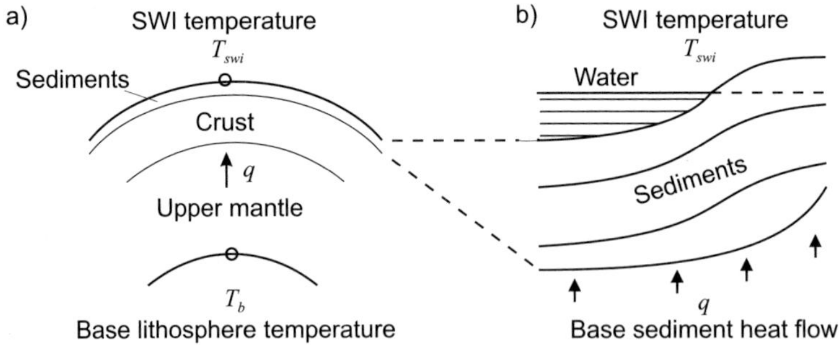

Fig. 3.1. Boundary value problem for a heat flow analysis (a) of the lithosphere and (b) in the sediments

3.2 One Dimensional (1D) Models

In the 1D approach, it is assumed that all heat flow vectors are directed vertically. 1D solutions often provide a good estimate for temperatures since the boundary values define radial core to surface aligned paths. They are especially used for well–based calibrations of basal heat flow trends. Exceptions, which cannot be modeled with 1D approaches are local areas of extraordinarily high thermal conductivities like salt domes, which bundle heat flow vectors from adjacent areas along highly conductive avenues.

3.2.1 Steady State Models

The most simple 1D models are steady state solutions in which all time dependent terms such as transient or convection effects are neglected. In the absence of radioactivity, the heat flow q is constant throughout the sediments and the temperature gradient in a layer is higher the lower the thermal conductivity. Multilayer solutions can then be directly derived from the heat flow equation (3.1) with the assumption that the average bulk thermal conductivity λ_b of a layer sequence is equal to the harmonic average of the corresponding single layer bulk thermal conductivities λ_i. The temperature controlled boundary value problem of the lithosphere yields the following 1D steady state solution with vertical thermal conductivity λ_b and thickness h_l of the lithosphere, and the corresponding properties of the upper mantle λ_m, h_m, the crust λ_c, h_c and sediments λ_s, h_s.

$$q = \lambda_b \frac{T_b - T_{\text{swi}}}{h_l}, \quad \frac{h_l}{\lambda_b} = \frac{h_m}{\lambda_m} + \frac{h_c}{\lambda_c} + \frac{h_s}{\lambda_s} \; . \tag{3.2}$$

The advanced solution of the lithosphere problem, taking into account radioactive heat production, transient effects and convection caused by stretching, is discussed in Sec. 3.8. An equivalent steady state solution for the n–layer model of sediments with a base sediment heat flow q_{bs} can be derived from equation (3.1) for the temperature at base of the sediments T_{bs} and a temperature increase ΔT_i within a layer i as follows.

$$T_{bs} = T_{\text{swi}} + \sum_{i=1}^{n} \Delta T_i, \quad \Delta T_i = q_{bs} \frac{h_i}{\lambda_i} \; . \tag{3.3}$$

The temperatures at all layer boundaries can be calculated from the surface temperature down to the base of section. This algorithm is illustrated in Fig. 3.2 for a simple sediment column consisting of only three lithotypes: shale, sandstone and limestone. The bulk conductivity for each layer λ_i is here approximated by a geometric average of the values for water λ_w and rock λ_r with the porosity ϕ as follows.

$$\lambda_i = \lambda_r^{(1-\phi)} \, \lambda_w^{\phi} \; . \tag{3.4}$$

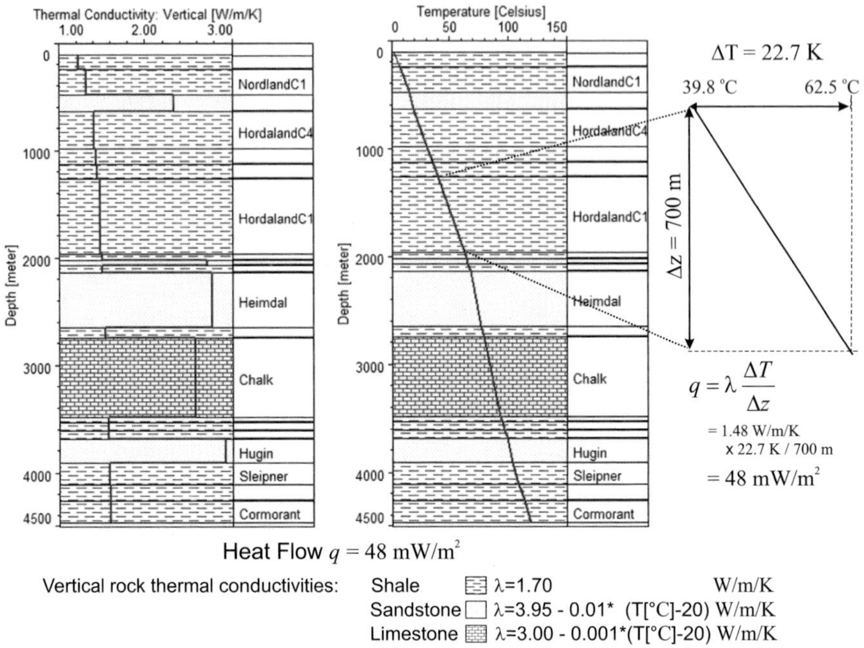

Fig. 3.2. 1D steady state example of a simplified model of a North Sea well. The bulk conductivity of the Hordaland Shale (1.48 W/m/K) is geometrically averaged from rock with $\lambda_r = 1.70\,\mathrm{W/m/K}$ and water with $\lambda_w = 0.7\,\mathrm{W/m/K}$ at a porosity of 15.6 %

The temperature–versus–depth curve clearly shows intervals of steep and low increases due to low and high values of bulk thermal conductivities. This simple steady state argumentation shows that a heat flow analysis which based on a constant thermal gradient is a very rough approximation.

However, the heat flow from the base upward does not remain constant because sediments contain radioactive elements like uranium, thorium and potassium. Radioactivity causes additional heat production, which increases the heat flow through the sediments. Thus, the surface heat flow is higher than the basal value by the amount of generated heat. Each of the radioactive elements generates gamma rays with radiogenic heat production rates Q_r estimated by Rybach (1973) as follows:

$$Q_r = 0.01\,\rho_r\,(9.52\,\mathrm{U} + 2.56\,\mathrm{Th} + 3.48\,\mathrm{K}) \tag{3.5}$$

where ρ_r is the rock density in kg/m^3, U and Th are the concentration of uranium and thorium in ppm, K is the concentration of potassium in % and Q_r is in μW/m^3. Pore fluids do not contribute to radioactive heat production. The resulting increase of vertical heat flow Δq in a layer of thickness h due to a rock heat production rate Q_r is as follows:

$$\Delta q = (1 - \phi)\, h\, Q_r \,. \tag{3.6}$$

The simple layer sequence example of Fig. 3.3 shows an average increase of the heat flow of about 1 mW/m^2 per km sediment, which is a good general estimate.

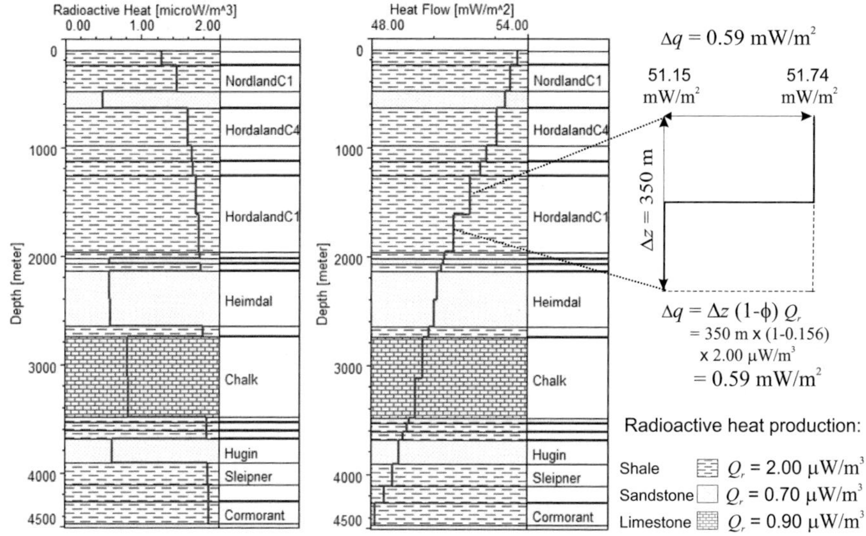

Fig. 3.3. Heat flow increase for the North Sea well with an example calculation in the Hordaland layer. The total increase of 5 mW/m^2 from base to top is about 1 mW/m^2 per km sediment

3.2.2 Transient Effect

A system is in thermal steady state when the heat flow is constant everywhere. Any change of a thermal boundary condition, geometry, properties or temperatures yields non-steady or transient state. The system will gradually return to a new flow equilibrium, when the new conditions do not change any more. The transition time depends generally on the ratio of the transported heat and the inner thermal energy, which is controlled by the size of the system, and ratio of the heat capacity and the thermal conductivity.

The transient 1D temperature distribution along the downward directed z–axis is the solution of the following differential equation:

$$\rho c \frac{\partial T}{\partial t} - \frac{\partial}{\partial z}\left(\lambda \frac{\partial T}{\partial z}\right) = Q_r \tag{3.7}$$

where c and ρ are bulk values, which should be arithmetically mixed from the pore fluid and rock values corresponding to the actual porosity. The transient effect is important during deposition and erosion and when thermal boundary conditions, namely SWI temperatures or basal heat flow, change rapidly.

Deposition results in deeper burial and higher temperatures of underlaying sediments. The actual sediment temperatures are lower than for the steady state solution. Heat is absorbed for heating of the layer and heat flow values decrease towards the surface. In case of constant deposition without compaction the heat flow decreases linearly in vertical direction (Fig. 3.4.a and b). It can indeed be analytically proven, that a constant deposition rate S yields a constant heat flow gradient after a relatively short time of deposition, according to (F.24) with

$$\frac{\partial q}{\partial z} \approx \frac{qS\rho c}{\lambda} \, . \tag{3.8}$$

The example values used in Fig. 3.4 result in a gradient of

$$\frac{\partial q}{\partial z}[\mathrm{mW/m^2/km}] = 3.678\, S[\mathrm{km/My}] \, . \tag{3.9}$$

In the case of erosion the effect is reverse: the sediment temperatures are higher than the steady state solution and the heat flow increases toward the surface. The above rule (3.8) can also be applied to estimate the magnitude of erosion–induced heat change (Fig. 3.4.c). During a hiatus the heat flow gradually returns to the steady state solution, with a constant heat flow value in regions without radioactivity.

The transient effect of an instantaneous heat flow change is shown in Fig. 3.5. Herein, the basal heat flow jumps from $40\,\mathrm{mW/m^2}$ to $60\,\mathrm{mW/m^2}$ while the near surface heat flow change is delayed for more than 5 Ma. Similar examples can be calculated analytically (App. F).

A change in SWI temperature also yields transient heat flow curves (Fig. 3.6). Typical SWI temperature variations yield much lower magnitudes compared to those caused by basal heat flow changes. Here, SWI or surface temperatures are average values over a range of 1000 to 10000 years. Seasonal cycles or short time changes are not taken into account. Sudden surface temperature rises lower the temperature difference between top and base and yield lower surface heat flow values. This effect occurs in glacial intervals of the quaternary period as shown in Fig. 3.21. The presented model uses small time steps of 2000 years for the approximation of the surface temperature variation. The temperature increases are especially steep and cause downward trends in the surface heat flow. Generally, near surface heat flow is correlated with the average surface temperatures of the previous thousands of years.

The heat flow is constant from base to top for 1D solutions without radioactivity and transient effects. Erosion and radiogenic sediments increases the surface heat flow, while deposition lowers surface heat flow. A transient solution of the 1D example model in Fig. 3.2 and Fig. 3.3 is shown in Fig. 3.7.

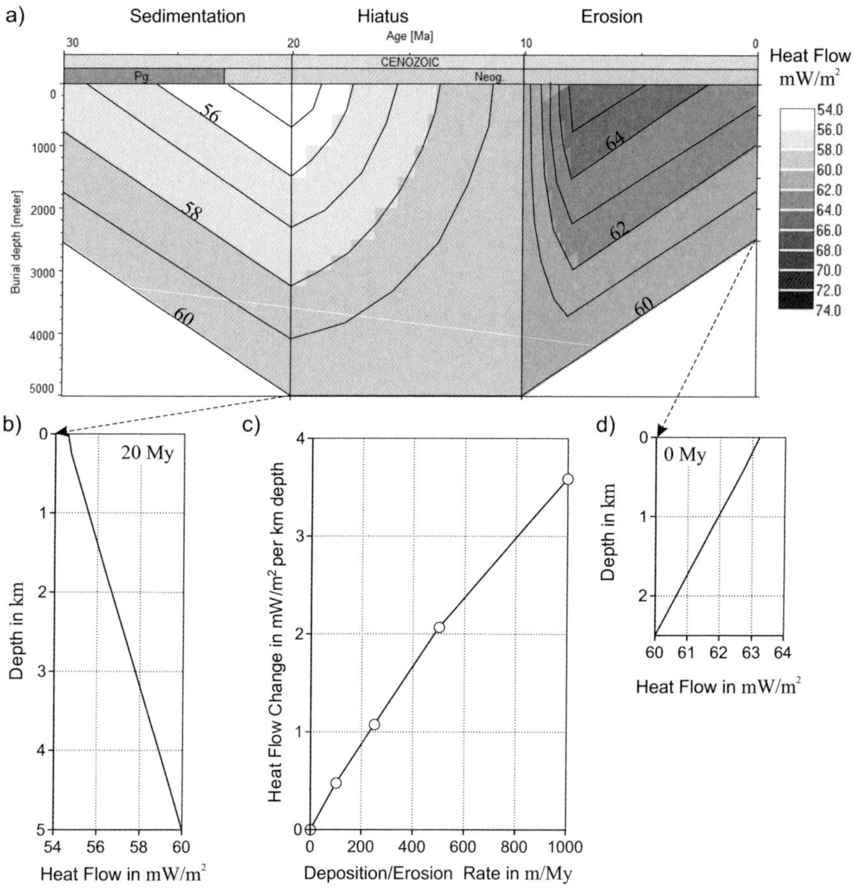

Fig. 3.4. Transient effect of deposition and erosion: (a) burial history with heat flow overlay for periods of uniform deposition, hiatus and uniform erosion. (b) Heat flow vs. depth after deposition. (c) Rate dependent heat flow change per 1000 m sediment. (d) Heat flow vs. depth after erosion. The example is calculated with a constant basal heat flow. Compaction is neglected

Herein, the basal heat flow variation through geological time is the main control on the heat flow in the well. However, the heat flow isolines are not perfectly vertical lines as radioactivity causes a significant slope. Figure 3.7.a shows calculated heat flow values for present day, without radioactivity, to quantify the transient effects. The transient effect of $0.8\,\mathrm{mW/m^2}$ is mainly controlled by deposition and it is much smaller than the increase of $6\,\mathrm{mW/m^2}$ caused by radioactivity.

It is often of interest to evaluate the difference between non–steady and steady–state thermal conditions in a basin. This can be done by solving the heat flow equation twice, once with and once without the transient term. The

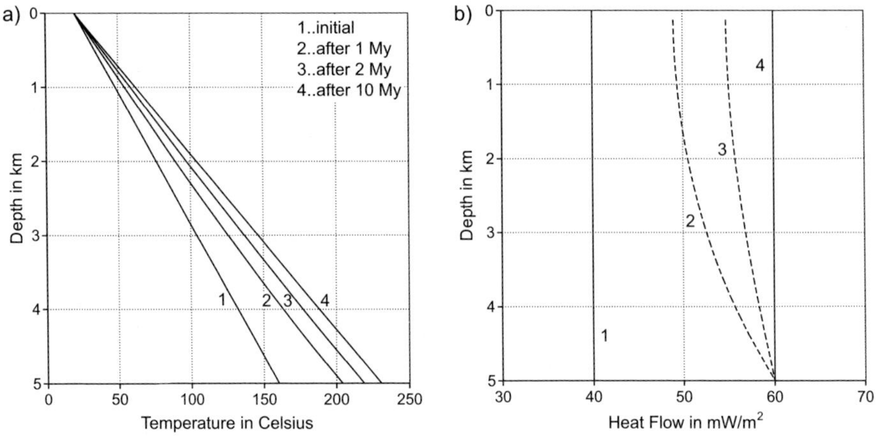

Fig. 3.5. Transient effect due to a basal heat flow switch

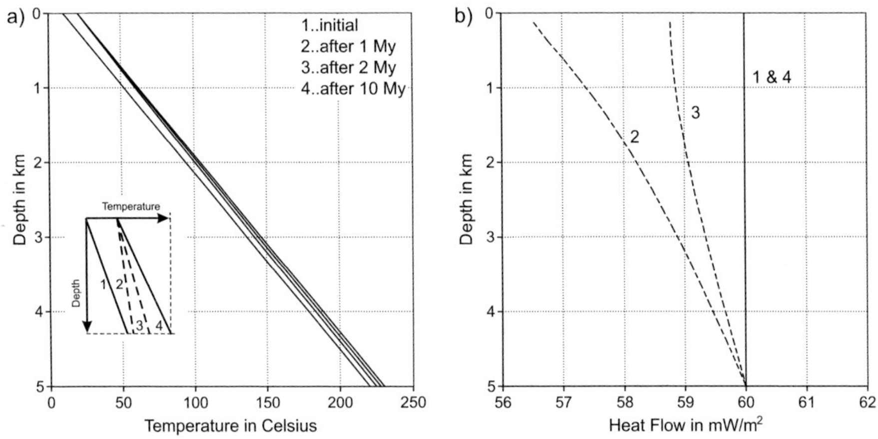

Fig. 3.6. Transient effect of SWI temperature change from 15°C to 25°C

difference between the two types of calculated temperature fields is called the "thermal disequilibrium indicator".

3.3 Thermal Conductivity

Thermal conductivity describes the ability of material to transport thermal energy via conduction. For a given temperature difference a good heat conductor induces a high heat flow, or a given heat flow maintains a small temperature difference. Steep temperature gradients occur in layers with low thermal conductivities. The unit for thermal conductivity is W/m/K.

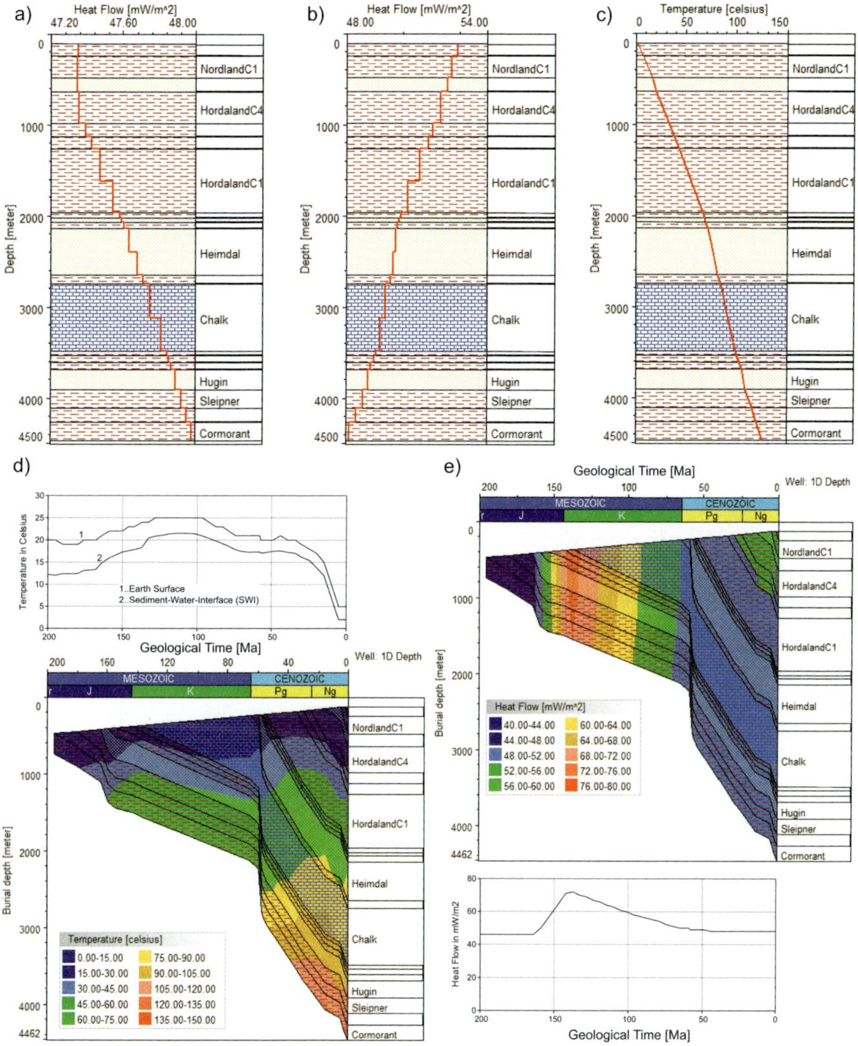

Fig. 3.7. 1D transient solution of a simplified model of a well in the North Sea. (a) Present day heat flow without radioactivity. (b),(c) Present day heat flow and temperature with radioactivity. (d) Temperatures during burial history and SWI temperature trend with radioactivity. (e) Heat flow values during burial history and basal heat flow trend with radioactivity

The bulk thermal conductivity is controlled by conductivity values of rock and fluid components. Mixing rules for rock and fluid components are generally complex and depend on whether the mixture is homogeneous or layered (Sec. 8.3). Sedimentary rocks are anisotropic with higher horizontal than vertical thermal conductivities. Generally, the thermal conductivity $\boldsymbol{\lambda}$ is a symmetrical tensor with six independent components. It is often considered to have only two independent components: the conductivity along the geological layer λ_h and the conductivity across the geological layer λ_v with an anisotropy factor $a_\lambda = \lambda_\mathrm{h}/\lambda_\mathrm{v}$.

3.3.1 Rock and Mineral Functions

Thermal conductivities commonly depend on temperature and vary widely according to the type of rock. Some 20°C values of vertical conductivities λ_v^{20} and anisotropy factors $a_\lambda^{20} = \lambda_\mathrm{h}^{20}/\lambda_\mathrm{v}^{20}$ are given in App. E.

Sekiguchi–Waples Model

The temperature dependence of matrix conductivity of any mineral, lithology, kerogen or coal can be calculated using the following equations adapted from Sekiguchi (1984) and plotted in Fig. 3.8.a.

$$\lambda_i(T) = 358 \times (1.0227\,\lambda_i^{20} - 1.882) \times \left(\frac{1}{T} - 0.00068\right) + 1.84 \qquad (3.10)$$

with $i = v, h, \lambda$ in W/m/K and T in K.

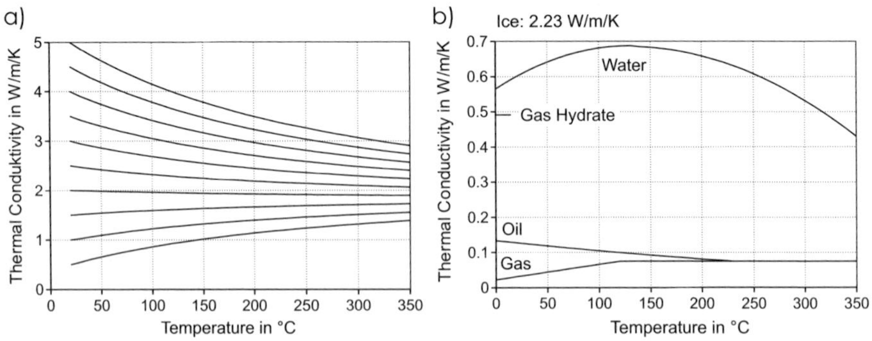

Fig. 3.8. Thermal conductivity functions: (a) rocks, (b) fluids

Most rocks and minerals do not experience significant changes in their anisotropy factors during compaction. Exceptions are claystones with significant dependency on compaction states. The effect can be described with a factor f defining the ratio of the depositional vertical conductivity λ_{vd} to the vertical conductivity of rock with zero porosity λ_{v0}. The latter is an extrapolated value for a fully compacted rock. Horizontal conductivities are calculated using the principle of Waples and Tirsgaard (2002) assuming that the mean thermal conductivity value of the clay minerals $\lambda_m = \lambda_v + 2\,\lambda_h$ remains constant during compaction. It is further assumed that the decrease in vertical conductivity with compaction is exponential as follows:

$$\lambda_v(\phi) = \lambda_{v0} f^{\phi/\phi_0} \tag{3.11}$$

where λ_{v0} is the vertical conductivity of an ideally compacted rock with $\phi = 0$, ϕ_0 is the depositional porosity and f is the grain rotation factor with $f = 1$ for porosity independent anisotropy.[1] The principle of constant mean conductivity yields a porosity dependent horizontal conductivity $\lambda_h(\phi)$ from the horizontal conductivity of an ideally compacted rock λ_{h0} as follows:

$$\lambda_h(\phi) = \lambda_{h0} + \frac{1}{2}\left(\lambda_{v0} - \lambda_v(\phi)\right) . \tag{3.12}$$

Thus, the conductivity values of a rock with any porosity and temperature are calculated in two steps. First, the porosity related corrections are made using (3.12) for claystones only and second, the temperature corrections are made using (3.10) for both (vertical and horizontal) values separately.

In the following example, the matrix thermal conductivity values of a shale with $\lambda_{v0}^{20} = 1.64\,\mathrm{W/m/K}$, $a_\lambda = 1.6$, $\phi_0 = 70\%$ and $f = 1.38$ are calculated at $\phi = 30\%$ and $T = 80°\mathrm{C}$:

$$\begin{aligned}
\lambda_{h0}^{20} &= 1.60 \times 1.64 = 2.624 \\
\lambda_{v\phi}^{20} &= 1.64 \times 1.38^{0.3/0.7} = 1.883 \\
\lambda_{h\phi}^{20} &= (1.64 + 2 \times 2.624 - 1.883)/2 = 2.520 \\
\lambda_v &= 358 \times (1.0227 \times 1.883 - 1.882) \times (0.00283 - 0.00068) + 1.84 = 1.876 \\
\lambda_h &= 358 \times (1.0227 \times 2.520 - 1.882) \times (0.00283 - 0.00068) + 1.84 = 2.402
\end{aligned}$$

The complete thermal conductivity versus porosity functions of the above example shale are shown in Fig. 3.9.

Linear Dependency Model

A simplified alternative model for rock matrix conductivities is based on the assumption of linearly temperature dependent values only.

---

[1] Anisotropy values are often given by the depositional conductivity λ_{vd}, the fully compacted conductivity λ_{v0} at $\phi = 0$ and anisotropy factors $a_{\lambda d} = \lambda_{hd}/\lambda_{vd}$, $a_{\lambda 0} = \lambda_{h0}/\lambda_{v0}$. The constant sum of horizontal and vertical conductivities becomes $\lambda_{vd}(1 + 2a_{\lambda d}) = \lambda_{v0}(1 + 2a_{\lambda 0})$ and therefore $f = (2a_{\lambda 0} + 1)/(2a_{\lambda d} + 1)$.

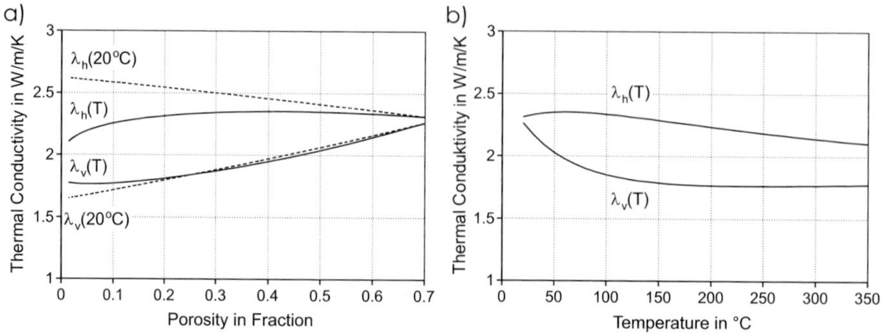

Fig. 3.9. Rock thermal conductivity functions of a typical shale under hydrostatic compaction and a constant temperature gradient of 30°C/1000 m with a temperature $T = 20\,^\circ C$ at deposition. (a) Porosity dependency (b) Temperature dependency

$$\lambda_v(T) = \lambda_v^{20} + \lambda_T\,(T - 20°\mathrm{C}) \tag{3.13}$$

where λ_v^{20} is the thermal conductivity at 20°C and λ_T is the conductivity increase per degree temperature. Default values for some lithologies are listed in App. E. Such a linear temperature dependency is often assumed in an interval from −20°C to 300°C.

Rock Mixtures

Mineral component based matrix conductivity values are calculated with the geometric average of their component minerals when the minerals are homogeneously distributed in the rock (Fig. 8.4). The geometrical average is also used for rocks consisting of homogeneously distributed lithological components. In the case of layered structures the horizontal conductivities are calculated with arithmetic averages and the vertical conductivities with harmonic averages.

3.3.2 Pore Fluid Functions

Thermal conductivities of fluids are isotropic and depend on temperature (Fig. 3.8). Deming and Chapman (1989) published the following formulas for water:

$$\lambda_w = \begin{cases} 0.565 - 1.88 \times 10^{-3}\,T - 7.23 \times 10^{-6}\,T^2 & \text{for} \quad T < 137°\mathrm{C} \\ 0.602 - 1.31 \times 10^{-3}\,T - 5.14 \times 10^{-6}\,T^2 & \text{for} \quad T > 137°\mathrm{C} \end{cases} \tag{3.14}$$

with λ_w in W/m/K and T in Kelvin.

The following equations for conductivities of liquid and vapor petroleum are based on Luo et al. (1994) and a personal communication between Ming Luo and Doug Waples.

$$\begin{aligned} \lambda_o &= 0.2389 - 4.593 \times 10^{-4}\,T + 2.676 \times 10^{-7}\,T^2 && \text{for} \quad T < 240°\text{C} \\ \lambda_g &= -0.0969 + 4.37 \times 10^{-4}\,T && \text{for} \quad T < 120°\text{C} \\ \lambda_o &= \lambda_g = 0.075 && \text{else} \end{aligned} \tag{3.15}$$

with λ_o and λ_g in W/m/K and T in Kelvin. Most hydrocarbon components approach conductivities of about 0.075 W/m/K at high temperatures (Poling et al., 2001). Similarly, methane conductivities also approach values close to 0.075 W/m/K (Lide, 2006). These results seem intuitively reasonable, since under those conditions both phases are supercritical and similarities of the properties dominate.

The thermal conductivity of solid pore substances, such as gas hydrates (clathrates) and ice, are 0.49 W/m/K, and 2.23 W/m/K, respectively (Sloan, 1998). All the above thermal conductivities are shown together in Fig. 3.8.b. Water is a superior thermal conductor, while gas has the lowest conductivity which yields isotherm bending below gas accumulations (Fig. 3.10).

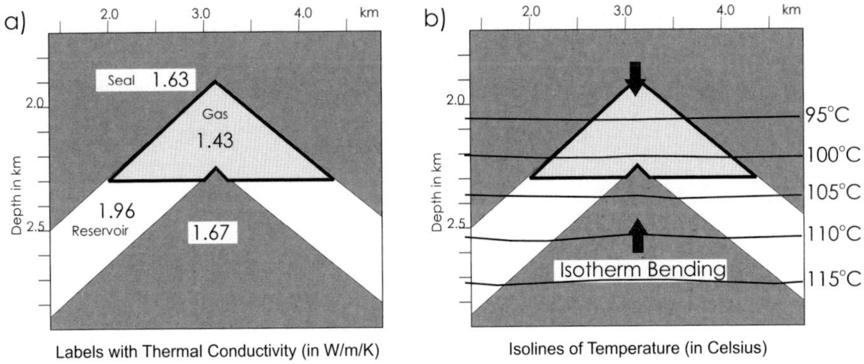

Fig. 3.10. Thermal effects of gas accumulations: (a) vertical bulk thermal conductivities, (b) isotherm bending

If the pore filling is a mixture of several (fluid or solid) phases then the geometrical average of the phase values is used. For water, liquid petroleum and gas it is

$$\lambda_p = (\lambda_w)^{S_w}\,(\lambda_o)^{S_o}\,(\lambda_g)^{S_g} \tag{3.16}$$

where λ_p is the pore fluid thermal conductivity, λ_w, λ_o, λ_g are the thermal conductivities of water, oil, and gas phases and S_w, S_o, S_g are water, oil, and gas saturations. The bulk thermal conductivity is obtained by averaging the rock matrix and pore values with the geometrical average

$$\lambda = \lambda_r^{\,(1-\phi)}\,\lambda_p^{\,\phi}\,. \tag{3.17}$$

A better but more complicated mixing rule is proposed by Buntebarth and Schopper (1998) based on mixing rules for spherical voids in a matrix.

$$\lambda = \lambda_r \frac{1 - E\phi}{1 + \alpha E\phi} \quad \text{with} \quad E = \frac{1 - Z}{1 + \alpha Z} \quad \text{and} \quad Z = \frac{\lambda_p}{\lambda_r} . \tag{3.18}$$

The authors propose a value of $\alpha = 5$ for water. This yields the following simplified law:

$$\lambda = \frac{\lambda_r + 5\lambda_p + \phi(\lambda_p - \lambda_r)}{\lambda_r + 5\lambda_p + 5\phi(\lambda_p - \lambda_r)} . \tag{3.19}$$

The whole procedure of calculating bulk conductivity values is summarized in Fig. 3.11.

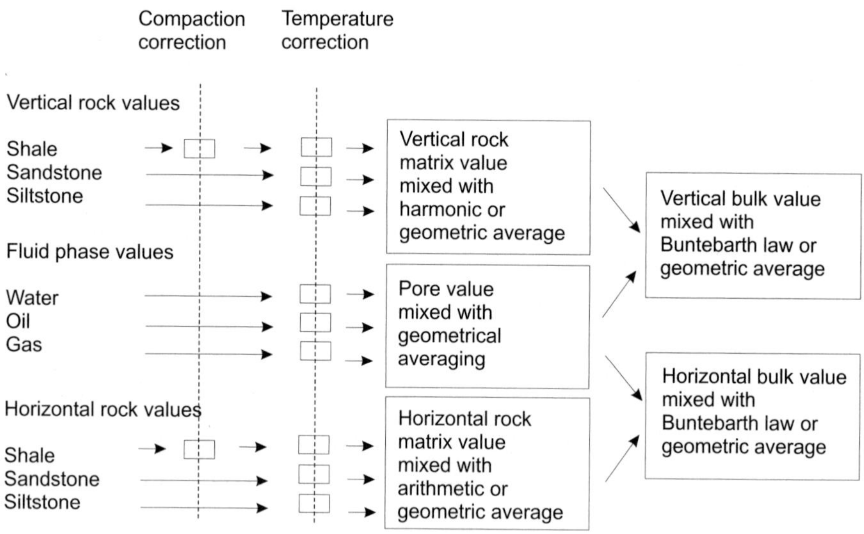

Fig. 3.11. Mixing of bulk thermal conductivities. The compaction and temperature corrections are described, e.g. with equations (3.11), (3.12), and (3.10)

3.4 Specific Heat Capacity

The heat capacity at constant pressure C_p is the ratio of a small (infinitesimal) amount of heat ΔQ absorbed from a body which increases the temperature by ΔT

$$C_p = \left(\frac{\Delta Q}{\Delta T} \right)_p . \tag{3.20}$$

The specific heat capacity or specific heat is defined as heat capacity per mass $c = C_p/m$.

The unit of specific heat is J/kg/K. The specific heat of a rock sample is measured by determining the temperature changes and the corresponding amount of heat entering or leaving the sample. The specific heat capacity is therefore the storage capacity for heat energy per unit mass. The ratio of the heat capacity and the thermal conductivity is a measure of the transient effect.

Heat capacity also controls the magnitude of convection as it determines how much of the stored heat can be moved together with a moving body. The specific heat capacity is a volumetric type property (Sec. 8.3). Rock and fluid component values are mixed arithmetically.

3.4.1 Rock and Mineral Functions

Specific heat capacities depend on temperature. The 20°C values c_{20} for minerals and some standard lithotypes are tabulated in App. E.

Waples Model

The temperature dependency of the heat capacity for any mineral, lithology, or rock value except kerogen and coal can be calculated using the following equation, which has been adopted from Waples and Waples (2004a) and is shown in Fig. 3.12.a.

$$c(T) = c_{20}\,(0.953 + 2.29\times10^{-3}\,T - 2.835\times10^{-6}\,T^2 + 1.191\times10^{-9}\,T^3) \quad (3.21)$$

where c_{20} is the heat capacity at 20°C and the temperature is given in °C. Waples and Waples (2004a) also proposed a special function for heat capacity of kerogen and coal as follows:

$$\begin{aligned} c(T)[\mathrm{J/kg/m}] = {} & 1214.3 + 6.2657\,T - 0.12345\,T^2 + 1.7165\times10^{-3}\,T^3 \\ & - 1.1491\times10^{-5}\,T^4 + 3.5686\times10^{-8}\,T^5 - 4.1208\times10^{-11}\,T^6\,. \end{aligned} \quad (3.22)$$

Linear Dependency Model

The temperature dependency can be approximated with a simple linear function.

$$c(T) = c_{20} + c_T\,(T - 20°\mathrm{C}) \quad (3.23)$$

where c_{20} is the heat capacity at 20°C, c_T is the heat capacity increase per degree temperature. Default lithological values for c_{20} and c_T are tabulated in App. E for a temperature interval from −20°C to 300°C.

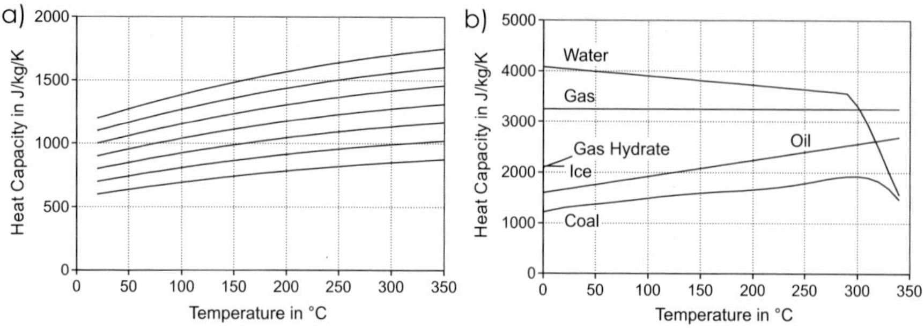

Fig. 3.12. Heat capacity for: (a) rocks, (b) coal and fluids

3.4.2 Pore Fluid Functions

Somerton (1992) developed equations for the specific heat capacity of pure water c_w as a function of temperature. For a relatively constant density the heat capacity decreases linearly to 290°C followed by a strong decrease at higher temperatures according to

$$\rho\, c_w = \begin{cases} 4245 - 1.841\, T & \text{for} \quad T < 290°\text{C} \\ 3703 \exp[-0.00481\,(T-290)+ & \\ \quad 2.34 \times 10^{-4}\,(T-290)^2] & \text{for} \quad T < 290°\text{C} \end{cases} \tag{3.24}$$

with c_w in J/m/K, ρ in kg/m^3 and T in °C. The above described temperature dependent function is shown in Fig. 3.12.b assuming a constant water density of 1040 kg/m^3. According to Kobranova (1989) the specific heat of saline water is slightly lower than that of pure water.

Gambill (1957) published the equation below for the specific heat capacity of oil c_o as a function of temperature and density:

$$c_o = (1684 + 3.389\, T)\sqrt{\rho} \tag{3.25}$$

with c_o in J/m/K, ρ in kg/m^3, and T in °C. It is satisfactory to use a value for the specific heat capacity of natural gas for all compositions at all temperatures and pressures. A reasonable value is 3250 J/kg/K (Waples and Waples, 2004b). Waples also proposed the following equation for the heat capacity of gas hydrates $c_h$in the same paper.

$$c_h = 2097 + 7.235\, T + 0.0199\, T^2 \tag{3.26}$$

with c_h in J/m/K and T in °C. The specific heat capacity of ice (2115 J/kg/K) is about half that of liquid water for temperatures around 0 °C. The pore and bulk values of heat capacity values are mixed arithmetically.

3.5 Radiogenic Heat

Some minerals and rocks contain traces of the radioactive elements uranium (U), thorium (Th) and potassium (K) which are additional heat sources (3.5). Measurements and data catalogs are sparse, but D. Waples developed a data base for most rocks and minerals based on modeling experience and some literature data (App. E). These values should be used carefully especially since the effects on heat flow calculations are tremendous as was discussed in Sec. 3.2.1. Radiogenic heat values for lithologies are given as rock matrix values and are converted to bulk values during simulation by multiplying by $(1 - \phi)$. They can be derived from the following data sources:

- uranium, thorium and potassium concentrations from spectral gamma ray measurements. These values are bulk values with a core sample porosity ϕ_c. One has to use Rybachs law (3.5) and divide by $(1 - \phi_c)$ to get the corresponding matrix heat flow production rate Q_r.
- Gamma Ray API values. Buecker and Rybach (1996) proposed the following law to convert Gamma Ray APIs to matrix heat flow production values:

$$Q_r[\mu\mathrm{W/m^3}] = 0.0158\,(\mathrm{API} - 0.8)\;. \tag{3.27}$$

Concentrations of radioactive elements are present-day values U_0, T_0, K_0. Paleo-values are higher corresponding to their half-lives as follows (with t in My).

$$\begin{aligned} \mathrm{U} &= \mathrm{U}_0\,(1 + 2.77 \times 10^{-4}\,t - 7.82 \times 10^{-8}\,t^2 + 4.53 \times 10^{-12}\,t^3) \\ \mathrm{Th} &= \mathrm{Th}_0\,\exp(0.00005\,t) \\ \mathrm{K} &= \mathrm{K}_0\,\exp(0.000555\,t)\;. \end{aligned} \tag{3.28}$$

Uranium consists of two isotopes having different half–lives. The sum of the two exponential functions is approximated by a third order polynomial. The time correction is small on geological time scales.

3.6 Three Dimensional Heat Flow Equation

In 1D models, heat flows vertically upward from base to top. In multi–dimensional problems, heat flow can also laterally divert to follow layers of high thermal conductivity. The formulation of the multi-dimensional heat flow problem yields a transport–type differential equation with temperature as the field variable and heat flow as the corresponding flow variable. The heat transport equation is based on energy balances, which means that the temperature induced internal energy change in a volume element is equal to the heat conducted into or out of the volume element plus the heat transferred by convection plus radiogenic heat production. It is

$$\rho c \frac{\partial T}{\partial t} - \boldsymbol{\nabla} \cdot \boldsymbol{\lambda} \cdot \boldsymbol{\nabla} T = \rho_p c_p \boldsymbol{\nabla} \cdot (\mathbf{v}_p T) + Q_r \tag{3.29}$$

where $\boldsymbol{\lambda}$, ρ, c are the bulk thermal conductivity tensor, bulk density, and bulk specific heat capacity, v_p, ρ_p, c_p are the pore fluid velocity vectors, density, and specific heat capacity and Q_r is the bulk radioactive heat production.

The full heat flow problem with boundary conditions is shown in Fig. 3.13. The four main terms in the heat flow equation describe the transient effect, heat conduction, heat convection, and the influence of heat sources, respectively. Two material parameters control the magnitude of the effects: thermal conductivity for heat conduction and heat capacity for transient effects and convection. The same temperatures for rock and pore fluid are assumed. For fast moving pore fluids, different temperatures have to be considered instead.

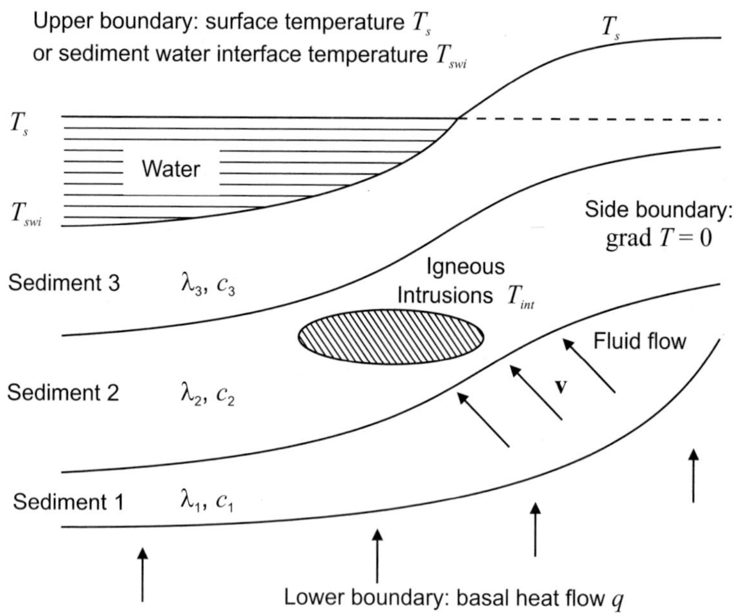

Fig. 3.13. Boundary value problem for the heat flow analysis

Temperatures or heat flow values have to be defined at all model boundaries. Common thermal boundary conditions are surface temperature (onshore) or sediment–water–interface (SWI) temperature (offshore) at the top of the sediments $T = T_{\mathrm{swi}}$, basal heat flow at the sediment base $q = q_b$ and no heat flow $\mathbf{n} \cdot \boldsymbol{\nabla} T = 0$ at the basin sides (Fig. 3.14.a). All boundary conditions have to be defined through geological time as paleo SWI temperatures and paleo heat flow trends, respectively. The lower boundary condition can alternatively be defined partially or completely with fixed base temperatures (Fig. 3.14.b and c). A deep isotherm map for the definition of the lower thermal boundary can also be applied. Then, the model is subdivided into two domains and each of them is separately solved (Fig. 3.14.d).

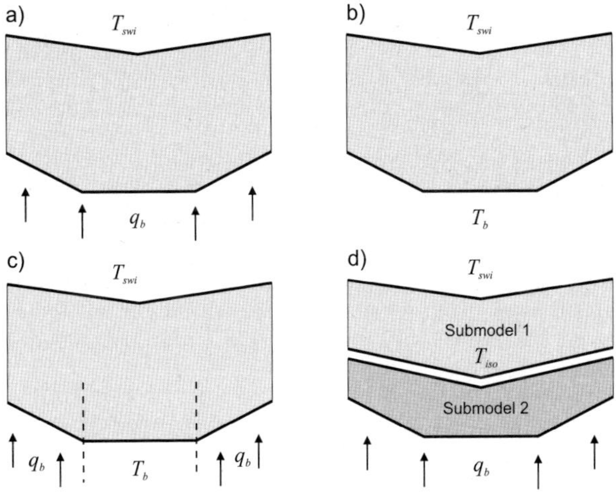

Fig. 3.14. Various types of boundary conditions

Three dimensional effects are important when large variations of the thermal conductivities occur. Salt has much higher conductivities than most other sediments. Thus, salt domes bundle heat flow as shown in Fig. 3.15.a. The preservation of the total energy requires that the heat flow adjacent to the dome has a corresponding lower value. The surface heat flow reflects this effect as well. The corresponding isotherms (Fig. 3.15.b.) bend down at the base of the salt due to higher salt conductivities. They also form a bow upwards at the top of the salt dome due to the higher heat flow values which is a typical 3D-effect and cannot be found in multi–1D models.

Calculated temperature and heat flow distributions of a multi–dimensional model are shown in Figs. 3.16 and 3.17. Thermal conductivities of the clastic rocks increase with depth according to the lower content of pore water, while salt domes are zones of high thermal conductivities. Shales correlate with very high radioactive heat sources. Heat flow from base to top generally increases due to radioactive heat production and it is concentrated along the high conductive salt domes and causes the bending of the isotherms as previously discussed. The higher surface heat flow above the salt domes can be clearly observed in the 3D–model. In such complex situations multi–1D models fail and show errors of more than 50°C (Fig. 3.17).

Some multi–dimensional schematic models can be solved analytically and the results can be used for "benchmarking" of simulation programs. The solutions of the following problems are given and discussed in the Appendix: influence of radiogenic heat production on steady state temperature (App. F.1),

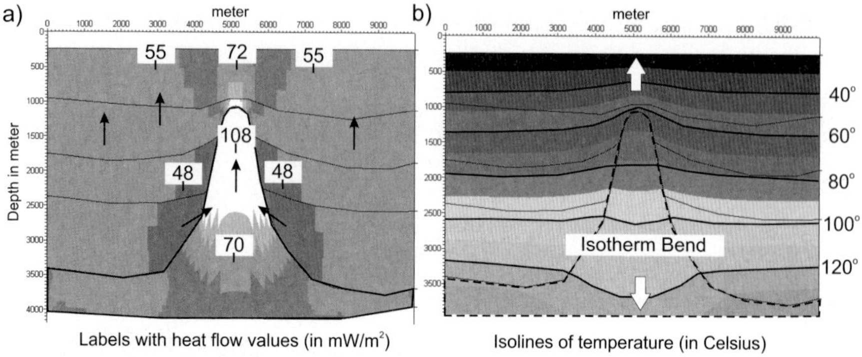

Fig. 3.15. Heat flow through a salt dome: the basal heat flow is $60\,\mathrm{mW/m^2}$ along the entire sediment base. The actual heat flow within the sediments increases to $108\,\mathrm{mW/m^2}$ within the dome. Note that heat flows near but outside the dome are lower than the basal heat flow

influence of lateral basal heat jump on temperature (App. F.2), influence of SWI temperature jump on temperature profiles (App. F.3), the steady state temperature field for a two block model (App. F.4), the transient temperature field of a model with basal heat flow jump (App. F.5), the transient temperature field of a model with SWI temperature jump (App. F.6).

3.6.1 Heat Convection

Heat convection is related to moving masses, solid and liquid. In sediments, heat convection is mainly caused by water flow. Water velocities are calculated when solving the pressure–compaction equations and so the convection term couples heat and fluid flow calculations. The amount of heat ΔQ transfered between two points with a temperature difference ΔT for a moving water mass m_w is

$$\Delta Q = c_w\, m_w\, \Delta T = c_w\, \rho_w\, V_w\, \Delta T \tag{3.30}$$

where c_w, ρ_w and V_w are the specific heat capacity, the density, and transported volume of the water. The corresponding convective heat flow $\mathbf{q}_v$ of water moving trough a cell with a length l, a slice–plane A, a bulk volume $V = A\, l$, and a velocity $\mathbf{v}_w$ is as follows:

$$q_v = \frac{\Delta Q}{A \Delta t} = \frac{c_w\, \rho_w\, \phi\, V\, \Delta T}{A \Delta t} = c_w\, \rho_w v_w\, \Delta T\;. \tag{3.31}$$

The above equation yields a convective heat flow value of $q_v = 0.04\,\mathrm{mW/m^2}$ with a water velocity of $1\,\mathrm{mm/y}$, a temperature difference of $1°\mathrm{C}$ between the flow boundaries, a porosity $\phi = 0.3$, a heat capacity of $c_w = 4186\,\mathrm{J/kg/K}$, and a water density $\rho_w = 1035\,\mathrm{kg/m^3}$. Compaction and overpressure driven water velocities are much smaller. Hence compaction driven convection can

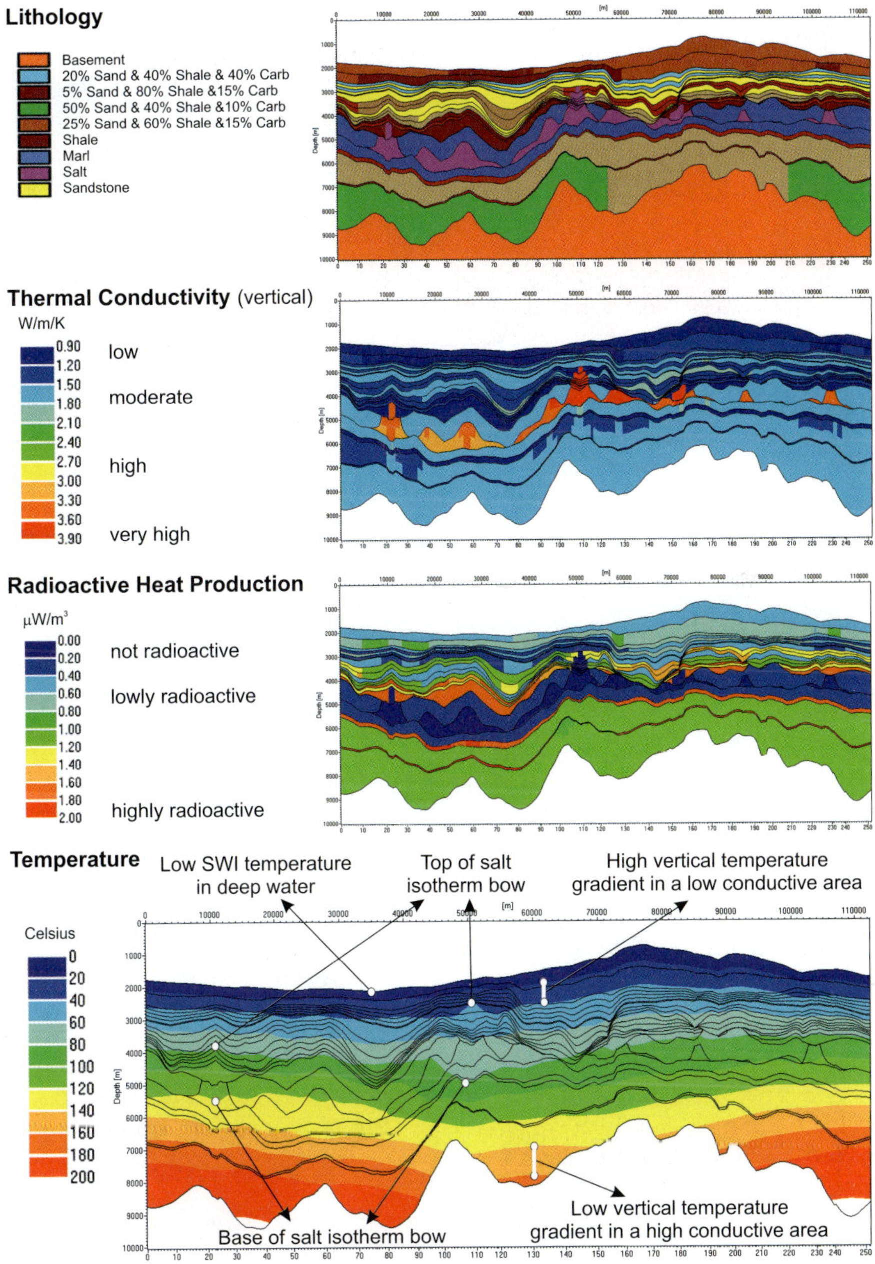

Fig. 3.16. Multi–dimensional heat flow analysis part I, cross-section from Campos Basin, Brazil. The present day temperature distribution shows several multi–dimensional effects like isotherm bending around salt, SWI temperature variation in deep water and temperature gradient variations according to the thermal conductivities

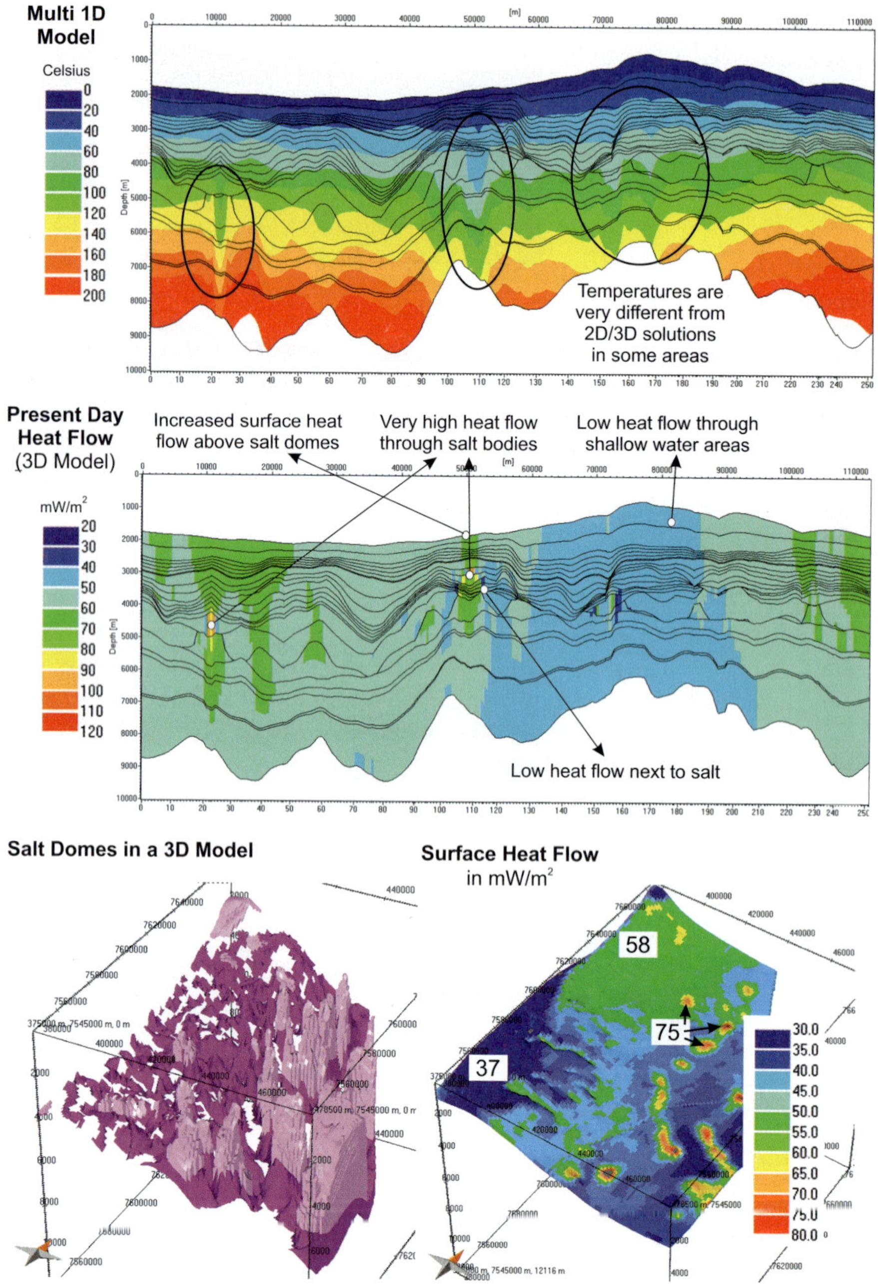

Fig. 3.17. Multi–dimensional heat flow analysis part II, multi–1D temperature model, multi–dimensional heat flow distribution at present day and surface heat flow anomalies above salt domes for a 3D model

be neglected in the thermal budget. Topographically driven aquifer flow and flow of hot water through high permeable fractures and faults can have higher flow velocities and must for that reason be taken into account.

3.6.2 Magmatic Intrusions

Magmatic intrusions can have substantial effects on paleo–temperatures and all thermal calibration parameters. Although the duration of such events is relatively short, extremely high temperatures can trigger rapid chemical reactions in the adjacent environment. Igneous intrusions are modeled with the magmatic temperature as inner boundary condition at the location and time of the intrusion. In subsequent time steps, the temperature decreases in both the intrusion and the surrounding layers. Then, hot liquid magma crystallizes to solid rock. The related crystallization heat is important and has to be taken into account in the heat balance. The principal processes together with some typical values according to Delaney (1988) are shown in Fig. 3.18. Here, it is necessary to switch the lithological properties of the intrusion volume elements twice, at the time of intrusion and again at the time of solidification.

The temperature development during cooling in a simple example is shown in Fig. 3.19, where effects on temperature can still be seen 100,000 years after the time of intrusion. Older intrusions can be recognized in vitrinite reflectance peaks in the vicinity of the intrusions. The use of smaller time steps after the time of intrusion is necessary. Time steps of 500 y, 1 000 y, 2 000 y, 5 000 y, 10 000 y, 20 000 y, 50 000 y, 100 000 y yield suitable results.

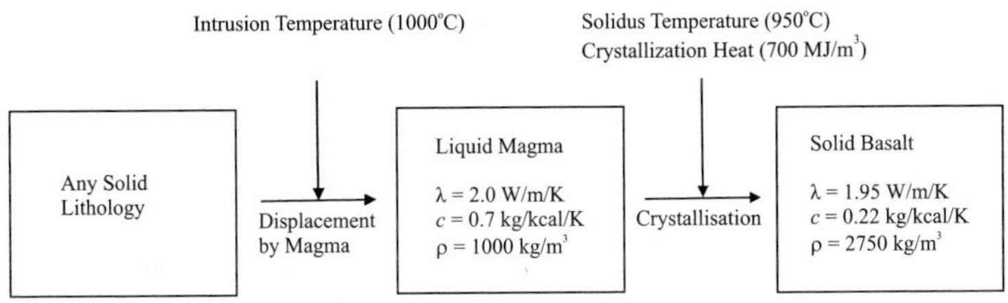

Fig. 3.18. Intrusion model and default values from Delaney (1988)

3.6.3 Permafrost

Modeling permafrost requires the introduction of permafrost lithologies with ice in the pores instead of water. Furthermore, additional heat sources and sinks for ice solidification and melting have to be taken into account. The trigger parameter for converting a lithology into a permafrost lithology is a temperature of 0.7 °C. Hence, temperatures below permafrost are much lower

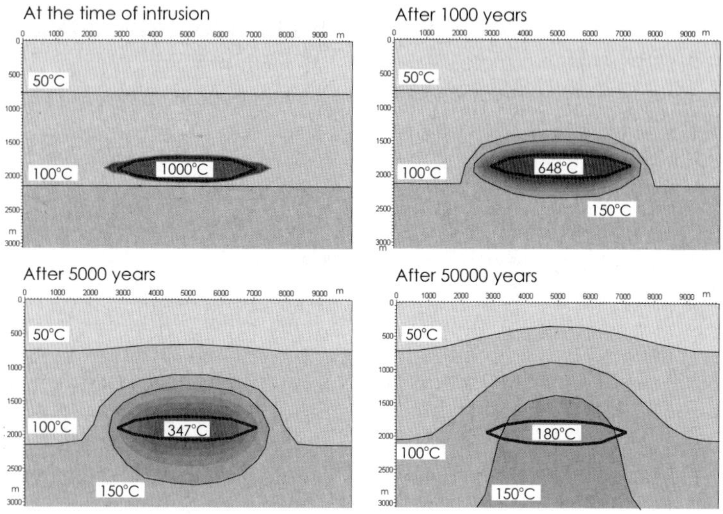

Fig. 3.19. Temperature development around an intrusion of size 300 m × 3000 m

compared to ice–free periods (Fig. 3.20). The high thermal conductivity of ice $\lambda = 2.33\,\mathrm{W/m/K}$ compared to liquid water yields low temperature gradients (Fig. 3.20) and supports the cooling effect. The cooling is further increased by the solidification heat of ice $Q_s = 335\,\mathrm{J/kg}$, which is removed from the permafrost environments. The specific heat capacity of ice (0.502 J/kg/m) is relatively small compared to water.

Modeling the sequence of interglacial periods such as in the Pleistocene, requires the use of very small time steps of about 1000 years to get an appropriate solution for the fluctuations in surface temperature and the corresponding surface heat flow (Fig. 3.21). The surface heat flow peaks generally coincide with the steep changes in surface temperatures, which especially occurred during the change from cold to warm periods.

Ice loading can also be simulated in permafrost periods. Then, an uppermost layer is introduced with the thermal and mechanical properties of pure ice. This yields special characteristics of pore- and lithostatic pressure curves as shown in Fig. 3.21.

3.7 SWI Temperatures

The sediment–water–interface temperature T_{swi} or bottom–water–temperature is the upper boundary for the heat flow problem. It can be determined with estimated paleo mean surface or air temperatures T_s and corrections for water depths. The annual mean ground surface temperature is primarily obtained from mean air temperatures (www.worldclimate.com), which depends

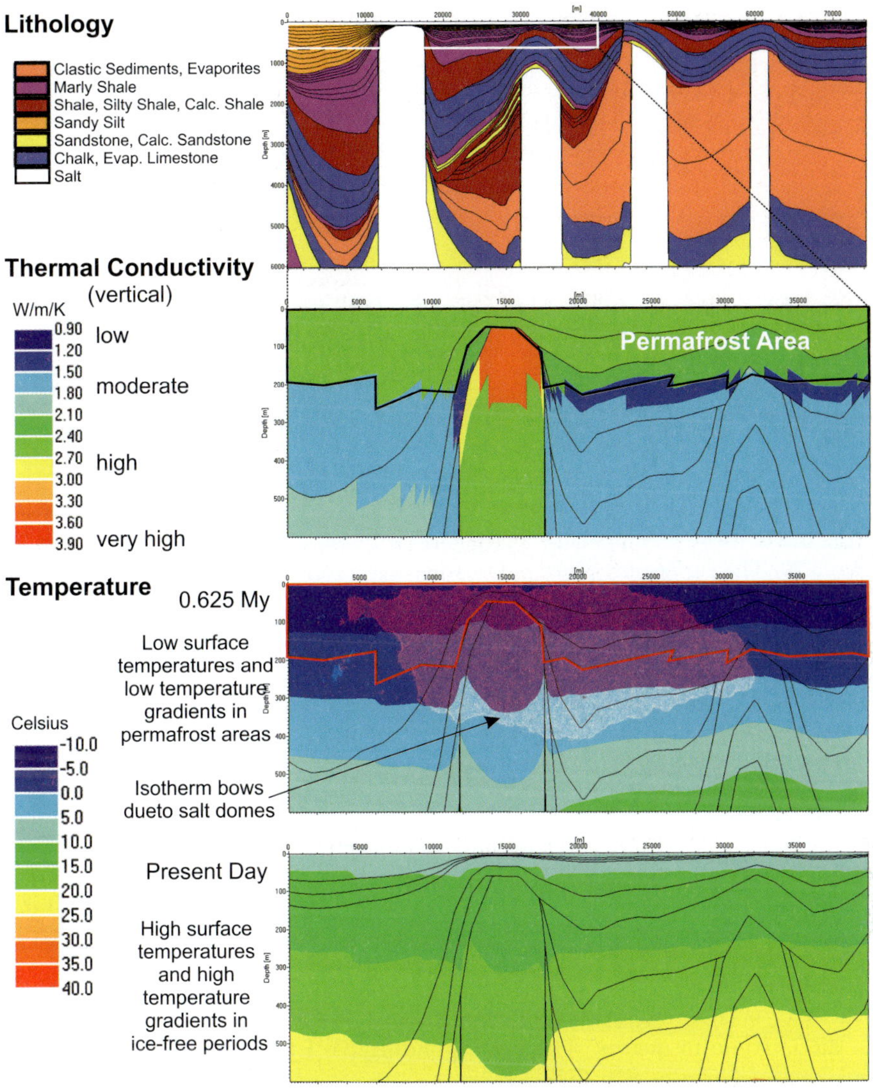

Fig. 3.20. Heat flow analysis in a permafrost area, sample cross–section from the Lower–Saxony Basin, Germany (Grassmann et al., 2005; Delisle et al., 2007). The model has a 200 m thick permafrost layer with very high conductivities at 0.625 My. The resulting temperature gradients differ significantly from the ice free present day temperatures. The thermal conductivities in the salt are generally very high

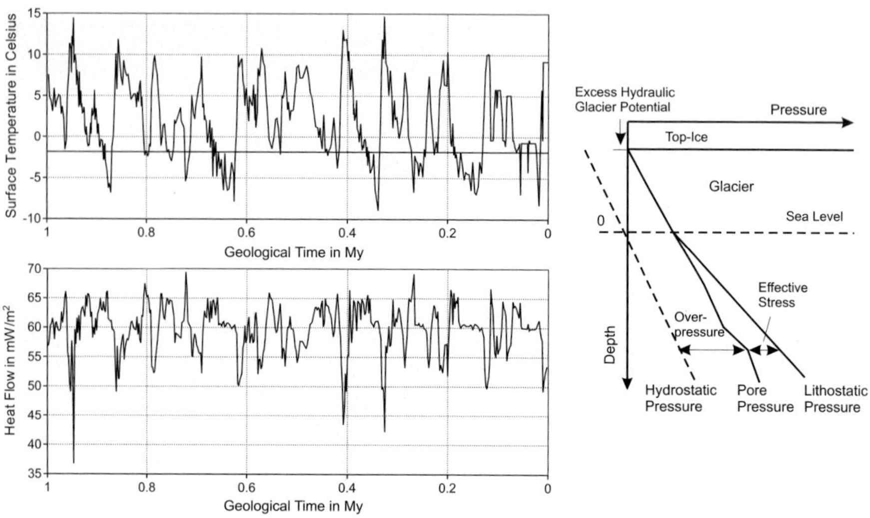

Fig. 3.21. Heat and pressure analysis in a Pleistocene sample cross–section of Fig. 3.20. The surface heat flow is low when the surface temperature increases and visa versa. The glacier causes an excess hydraulic potential on the top glacier surface

on latitude. Beardsmore and Cull (2001) proposed the following latitude and water depth dependent equation for the present day sediment–water–interface temperature with an error bar of $2\,°C$.

$$
\begin{aligned}
\ln(T_{\text{swi}} - T_f) &= a + b \ln z\,, \\
T_f &= -1.90 - 7.64 \times 10^{-4}\, z\,, \\
a &= 4.63 + 8.84 \times 10^{-4}\, L - 7.24 \times 10^{-4}\, L^2\,, \\
b &= -0.32 + 1.04 \times 10^{-4}\, L - 7.08 \times 10^{-5}\, L^2
\end{aligned}
\tag{3.32}
$$

where T_f is the freezing temperature in $°C$, z is the water depth in m, and L is the latitude in degree. The corresponding SWI temperature versus depth curves are shown in Fig. 3.22.

An average air surface temperature history is given in Fig. 3.23 for different latitudes (Wygrala, 1989). Knowledge of the paleo latitude changes through geological time is therefore necessary to be able to derive paleo surface temperatures. This is shown in Fig. 3.24 for several continental areas.

The derivation of the paleo–SWI temperatures from average surface temperature is very difficult to estimate. Wygrala (1989) proposed a decrease of 1.5°C per 100 m in shallow water. The temperature in water deeper than 400 m is primarily controlled by the coldest arctic water temperatures T_n, which are presently affected by polar glaciations. A linear interpolation between the following three fixed points is a common approximation for a water depth based SWI temperature correction.

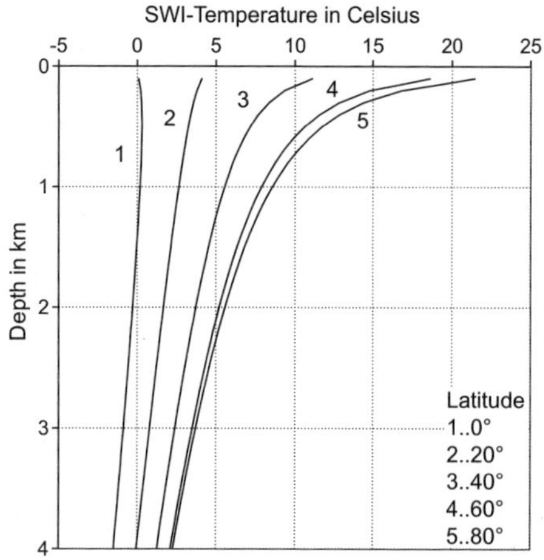

Fig. 3.22. Present day sediment–water–interface curves dependent on latitude and depth according to equation (3.32) after Beardsmore and Cull (2001)

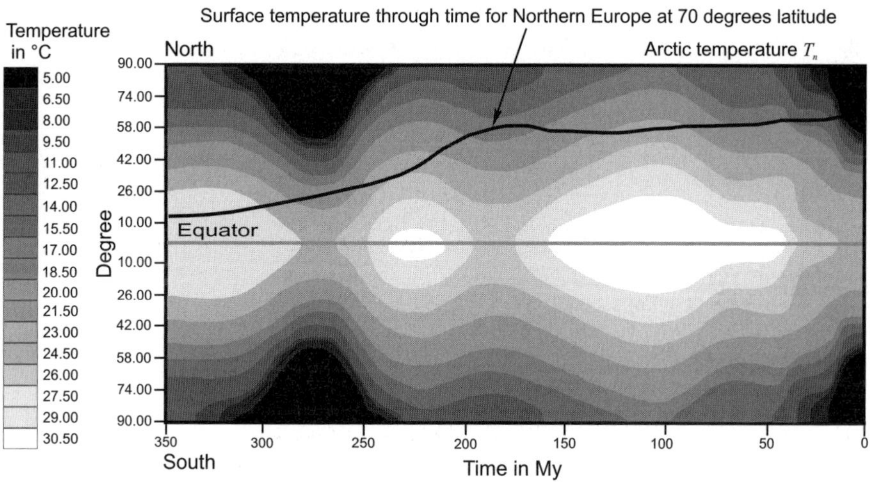

Fig. 3.23. Paleo–surface temperatures

$$T_{\text{swi}}(0\,\text{m}) = T_s, \quad T_{\text{swi}}(200\,\text{m}) = T_s - 3°\text{C}, \quad T_{\text{swi}}(600\,\text{m}) = T_n\,. \tag{3.33}$$

The average arctic temperature is about 4 °C at present, but it was much higher in the past (Fig. 3.23).

3.8 Crustal Models for Basal Heat Flow Prediction

Crustal models describe the mechanical and thermal processes of plate tectonics. In basin modeling, they are used to estimate the basal sediment heat flow as the lower boundary condition in the heat flow analysis (Fig. 3.1). Another

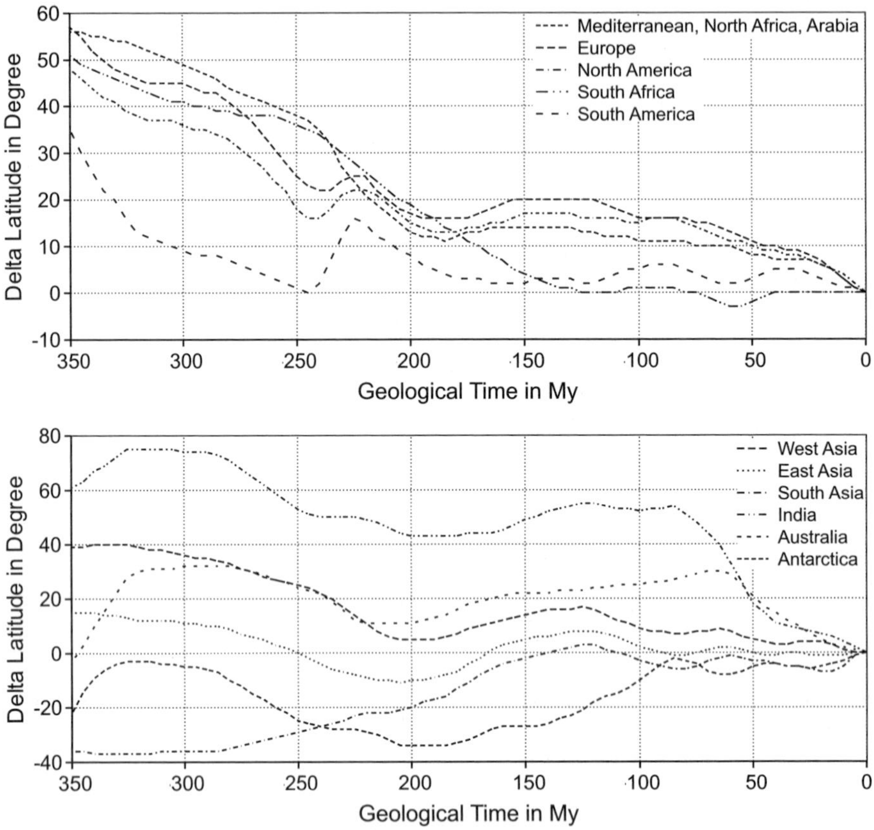

Fig. 3.24. Paleo latitude variations of some continental locations

interesting result of plate tectonic models is subsidence through time, which can be compared with sedimentation rates and paleo–water depths.

The asthenosphere, upper mantle, oceanic crust, lower and upper continental crust, and sediments are usually distinguished based on differences in their chemical compositions and mechanical properties as illustrated in Fig. 3.25. Mechanical behavior is the determining factor to distinguish between the solid lithosphere and the highly viscous (or pseudo–liquid) asthenosphere, which comprises the upper 250 km of the lower mantle. The lithosphere is further divided into the brittle upper crust and the ductile lower crust and upper mantle. Thus, faults are mainly formed in the upper crust during stretching of the lithosphere.

The classification between mantle and crust is based on chemical composition: mantle material mainly consists of mafic silicates, oceanic crust of mafic minerals and feldspar and continental crust of felsic silicates. There is evidence to assume that the entire mantle has a common convection system

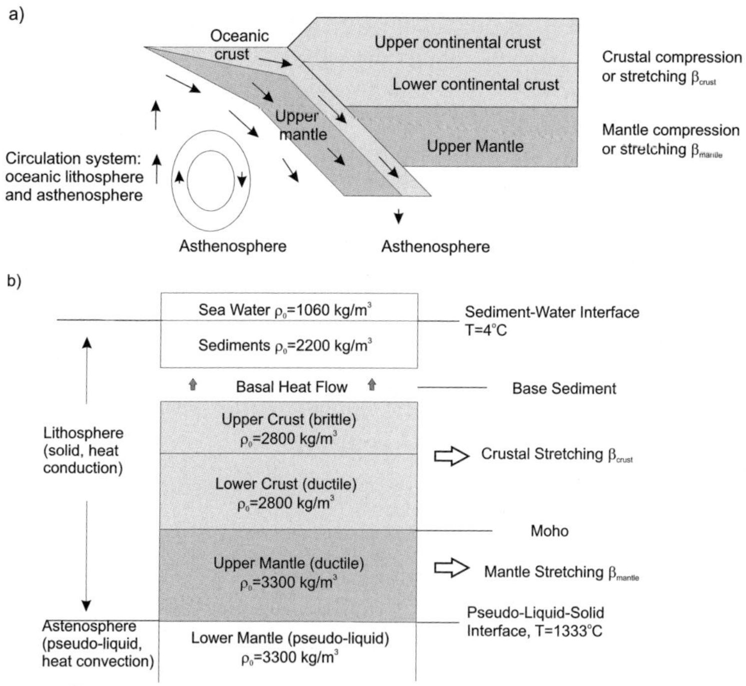

Fig. 3.25. Crust and mantle layer definitions. All rock densities are temperature dependent. The densities values ρ_0 are rock densities at surface conditions

with flow rates of about 10 to 20 cm/y. Oceanic crust, upper mantle and the asthenosphere have similar chemical compositions, since they build a closed circuit with constant formation of oceanic crust material at the mid–ocean ridges and destruction of it in the subduction zones. This circulation system also moves the continental crust pieces causing breakup, stretching, compression and overthrusting. The interface between the upper and lower mantle is the solid to pseudo–liquid boundary with a base lithosphere temperature of $T_a = 1100 - 1350\,^\circ\mathrm{C}$. (Parsons and Sclater, 1977). The value of $T_a = 1333\,^\circ\mathrm{C}$ is used in most publications (McKenzie, 1978), which corresponds to three quarters of the pyrolite melting temperature. The postulate of a fixed and well–known temperature at the base of the lithosphere, is an important assumption in crustal heat flow models.

Sediments are deposited in accommodation spaces as a result of lithospheric stretching and compression with usually different stretching velocities in the lithospheric layers and differences in deformation types. Crustal and mantle layers further differ in densities depending on their compositions and temperatures. Thus, a change in layer thicknesses, affects the weight of the total lithospheric column leading to subsidence with sedimentation on top or uplift with erosion. The depth of the top asthenosphere temperature isosurface and the thermal conductivities of the lithospheric layers primarily control the upward heat flow. In summary, a coupled model of lithospheric stretching, heat flow, and subsidence is necessary to obtain the base sediment heat flow through geologic time.

Plate tectonics yield different stretching, displacement, folding, and subduction processes especially on plate margins, which are related to different phases of basin development. Generalized models have been developed for a stable lithosphere in intra–plate locations, subduction zones at convergent margins and extensional rift–drift phases at divergent margins (Beardsmore and Cull, 2001).

Models of extensional rift basins are established since they can easily be quantified and because they can be applied to many petroleum provinces. The most thoroughly investigated and applied model is the model of uniform stretching, also known as the McKenzie model.

Uniform Stretching Model

This famous model was originally proposed by McKenzie (1978) and it is still frequently used with some minor improvements in basin analysis. It is based on two different periods: an initial stretching phase with constant thinning of the crust and upper mantle and a cooling phase with near or full restoration of the original thickness of the lithosphere. Here uniform stretching is not an uniform decrease of the layer thickness, instead it is a geological time–constant and linearly with depth increasing velocity field. Hence, the base of the lithosphere moves vertically at maximum speed, while the top of the lithosphere is fixed (Fig. 3.26). It also means that the speed of the base lithosphere decreases during uplift, which causes a slowdown in the thinning process.

The total thinning of the layers is described with a stretching factor β, which is the ratio of the initial to the final thickness. For the formulation of the heat transfer problem, it is more important to know the velocity vector of the moving layer element during stretching. The vertical component of the velocity v_z is linearly decreasing from bottom to top. The maximum velocity v_m is the velocity of the rock at the base of the lithosphere at the beginning of the stretching. The relationship between the stretching factor β, the stretching time t_s and the maximum velocity v_m is

$$t_s = \int_{h_0/\beta}^{h_0} \frac{h_0}{z\, v_m} dz = \frac{h_0 \log \beta}{v_m} \tag{3.34}$$

where h_0 is the initial thickness if the lithosphere.

The vertical component of the velocity at any depth $v_z(z)$ of the lithosphere during the entire stretching phase is

$$v_z(z) = \frac{z}{h_0} v_m \ . \tag{3.35}$$

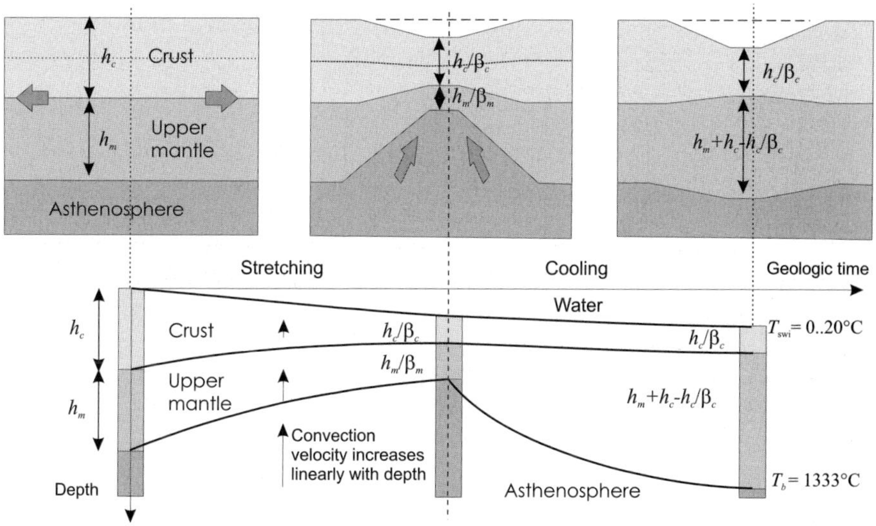

Fig. 3.26. McKenzie model with subsidence due to hydrostatic and isostatic compensation. The stretching velocity is constant in time and increases linearly with depth. This causes thinning of the crust and upper mantle during stretching. The original lithosphere thickness is restored during cooling

In this volume, it is usually dealt with two stretching factors for crust β_c and upper mantle β_m, and multi–1D heat conduction with radioactive heat production in the upper crust is considered.[2]

The model was originally worked out for symmetrical rifting, but it can also be applied to asymmetrical rifting and even compression, when each vertical lithospheric column experiences a phase of uniform thinning, which could be caused by stretching or sliding on detachment faults. Then, the stretching factors vary asymmetrically depending on their location as shown in Fig. 3.27 for the cases of pure shear, simple shear and simple shear–pure shear, respectively.

In the pure shear model, the vertical compressional deformation is equal to the horizontal extensional deformation as explained in Chap. 2. The simple shear model comprises only shear components and in the considered case, one large low angle detachment fault cutting through the entire lithosphere. Instead of extending along this detachment fault, the upper plate slips along the detachment surface. The simple shear–pure shear model considers simple shear in the crust and pure shear in the mantle. The corresponding stretching factor distributions along the section are mixtures of both end member models. The most realistic models are mixtures of the three stretching types, but some

[2] The original McKenzie model deals with one stretching factor only. Multiple stretching factors were introduced by Hellinger and Sclater (1983); Royden and Keen (1980)

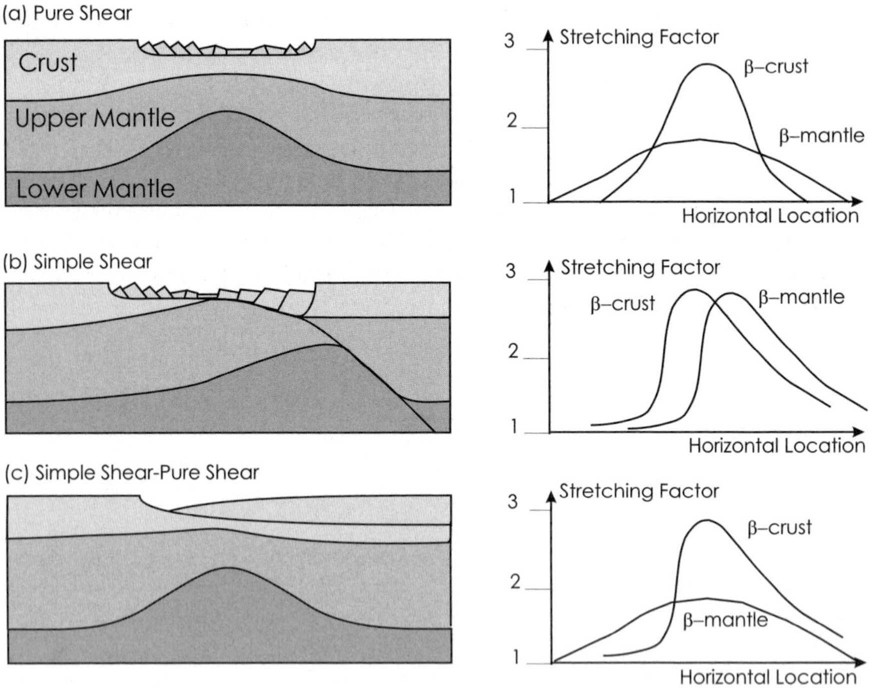

Fig. 3.27. Rift basins, according to Allen and Allen (2005): (a) pure shear, (b) simple shear, (c) simple shear–pure shear

can be approximated by the uniform stretching model when the stretching at each location can be described by only two single stretching parameters.

3.8.1 The Principle of Isostasy

The principle of hydrostatic isostasy states that the weight of all overburden material (lithosphere plus water depth) measured from a reference depth in the asthenosphere is constant. There is gravitational equilibrium between lithosphere and asthenosphere and the elevation depends on the underlying lithosphere column. An increase in lithospheric weight will therefore yield further subsidence so that the additional weight is compensated by the load of lighter water on top and less heavier asthenosphere at the base.

$$\rho_w g h_w + \sum_{i=0}^{n} \rho_{si} g h_{si} + \rho_{cu} g h_{cu} + \rho_{cl} g h_{cl} + \rho_m g h_m + \rho_a g h_a = \text{constant} \quad (3.36)$$

with the subscript indexes w, cu, cl, m, a for water, upper crust, lower crust, upper mantle, and asthenosphere, respectively and si for the ith sediment layer. This equation can be used to calculate water depths or mountain heights from crustal layer thicknesses. The principle is illustrated for two very simple two layer (crust and mantle) models in Fig. 3.28: the Airy and the Platt model. The Airy model supposes a constant density for the entire crust. Thus, the mountain height or water depth is a simple function of the total crust thickness. Hence, the high mountains are above thick crust and large water depths suggest a thin crust below. The Platt model presumes a crust of varying density at the same depth level of the asthenosphere. Thus, the surface elevation is a function of the crustal density only: the higher the mountain the lighter the crust below. These examples illustrate the principle of a hydrostatic lithosphere: each vertical column of lithosphere is able to move independently of the adjacent column to balance itself via its own weight.

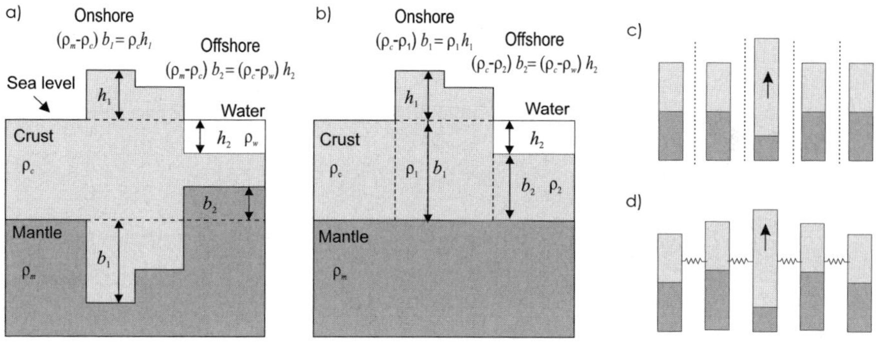

Fig. 3.28. Isostatic compensations: (a) Airy compensation. (b) Platt compensation. (c) Hydrostatic isostasy. (d) Flexural isostasy

In reality, there is an influence from the connected areas, which is described as flexural compensation, but which is not considered here. However, the weakness of the above models is the assumption of constant density with depth, since higher temperatures lower the density by thermal expansion according to

$$\rho(T) = \rho_0 \left[1 - \alpha(T - T_0)\right] \tag{3.37}$$

with the linear expansion factor $\alpha = 3.28 \times 10^{-5}/°\text{C}$ (McKenzie, 1978) and a reference density ρ_0 for surface temperature $T_0 = 20\,°\text{C}$. Another consequence of this equation is that cooling of the lithosphere causes subsidence and warming leads to uplift.

Uniform Stretching Model

The hydrostatic equation of isostasy applied to the uniform stretching model yields the following total mass per unit area for a column of water, crust, upper mantle and asthenosphere above a reference depth in the asthenosphere at any time (Fig. 3.26).

$$
\begin{aligned}
m = \rho_w h_w + \rho_{c0} \int_{h_w}^{d_c} [1 - \alpha_c T(x)]dx + \rho_{m0} \int_{d_c}^{d_m} [1 - \alpha_m T(x)]dx \\
+ \rho_{m0}[1 - \alpha_m (T_a - T_{\mathrm{swi}})]\, h_a
\end{aligned}
\tag{3.38}
$$

where α_c, α_m are the thermal expansions of the crust and mantle, and d_c, d_m are the depths of base crust and mantle, and h_w, h_a are the thicknesses of water and asthenosphere.

The tectonic water depths after instantaneous stretching h_{w_1} and after inifinite cooling h_{w_2} can be analytically calculated with the assumptions of no crustal radioactive heat production, a unique stretching factor β for crust and mantle, equal and constant thermal properties of the crust and mantle. and an linearly increase of the temperature with depth(Jarvis and McKenzie, 1980).

$$
\begin{aligned}
h_{w_1} &= \frac{(h_m + h_c)\left[(\rho_{m0} - \rho_{c0})\frac{h_c}{h_m + h_c}\left(1 - \alpha T_a \frac{h_c}{2h_m + 2h_c}\right) - \frac{\alpha T_a \rho_{m0}}{2}\right]\left(1 - \frac{1}{\beta}\right)}{\rho_{m0}(1 - \alpha T_a) - \rho_w}, \\
h_{w_2} &= \frac{(\rho_{m0} - \rho_{c0}) h_c}{\rho_{m0}(1 - \alpha T_a) - \rho_w}\left[\left(1 - \frac{1}{\beta}\right) - \frac{\alpha T_a h_c}{2h_m + 2h_c}\left(1 - \frac{1}{\beta^2}\right)\right] .
\end{aligned}
\tag{3.39}
$$

The subsidence during stretching is related to an inflow of additional heavy asthenospheric material, while additional subsidence is caused by the cooling of the entire column. Again it should be noted, that the above premises, especially the assumption of instantaneous stretching as seen in equation (3.39) are drastic simplifications. More comprehensive equations should be used instead. Such subsidence curves through geological time are illustrated in Fig. 3.29 for various stretching factors. In Figs. 3.30 and 3.31 radioactive heat production in the crust is taken into account. The subsidence after stretching h_{w_1} can become negative (uplift), when crustal stretching is very small compared to mantle stretching, as shown in the example in Fig. 3.27.a at the margin of the pure shear model and in Fig. 3.32.b.

The subsidence is much larger, when the weight of the sediments is taken into account, in contrast to pure water filled basins. Total subsidence, which is the real subsidence with sediments is different from the tectonic subsidence, which is the theoretical subsidence for water fill only. The relation between tectonic subsidence h_w and total subsidence h_t is as follows:

$$
h_t = \frac{\rho_a - \rho_w}{\rho_a} h_w + \frac{1}{\rho_a} \sum_{i=1}^{n} \rho_{si}\, h_{si} \tag{3.40}
$$

where ρ_{si} and h_{si} are the density and thickness of the i-th sediment layer.

Sometimes, a basement is introduced between the upper crust and the sediments, which is the sediment package before stretching. It has to be determined whether the total and tectonic subsidence is then understood as the top or bottom basement.

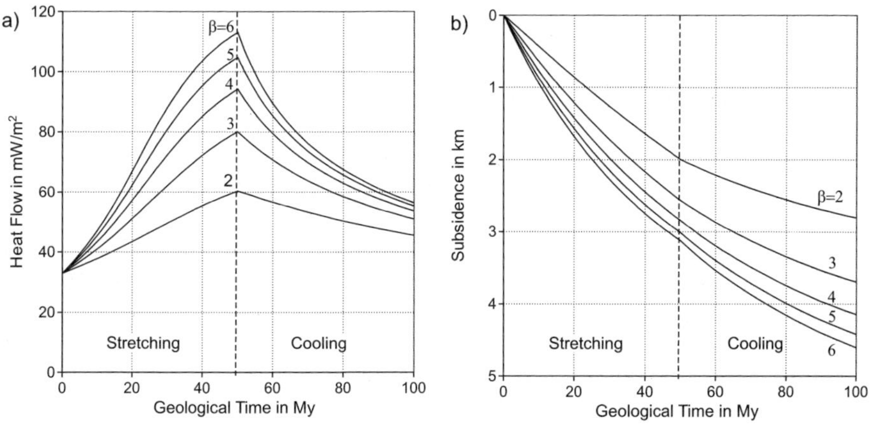

Fig. 3.29. McKenzie model: heat flow and tectonic subsidence for several stretching factors $\beta = \beta_c = \beta_m$ with $\kappa = 0.80410^{-6}\,\mathrm{m}^2/\mathrm{s}^{-1}$, $T_{\mathrm{swi}} = 0°\mathrm{C}$, $T_b = 1333°\mathrm{C}$, $h_c = 30\,\mathrm{km}$, $h_m = 95\,\mathrm{km}$ and $t_s = 50\,\mathrm{My}$

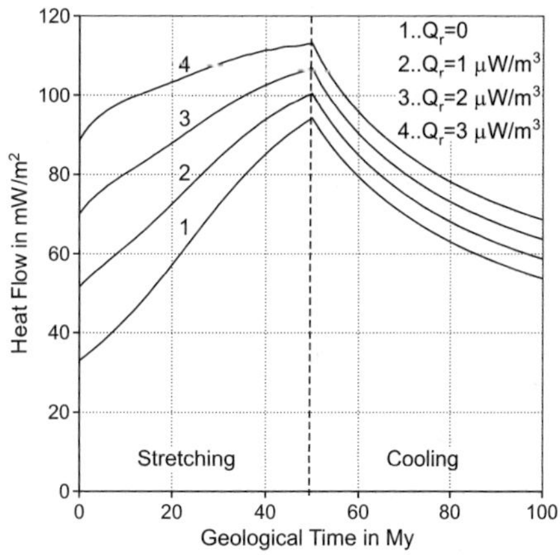

Fig. 3.30. Effect of constant radioactive heat production in the crust. The model parameters are the same as in Fig. 3.29 with the stretching factor $\beta = 4$

The principle of isostasy is universal, it is not restricted to pure shear processes only, e.g. basin subsidence and uplift can also be predicted in cases of underplating or other tectonic processes.

3.8.2 Heat Flow Models

The application of the 3D heat flow equation (3.29) to crustal models requires assumptions for thermal properties, boundary conditions and convection fields. The upper and lower boundary values are usually the sediment water interface and the top of the asthenosphere with the temperatures T_{swi} and T_a. Thermal conductivities and heat capacities of the upper crust depend on temperature as well as on rock composition. Radioactive heat production is

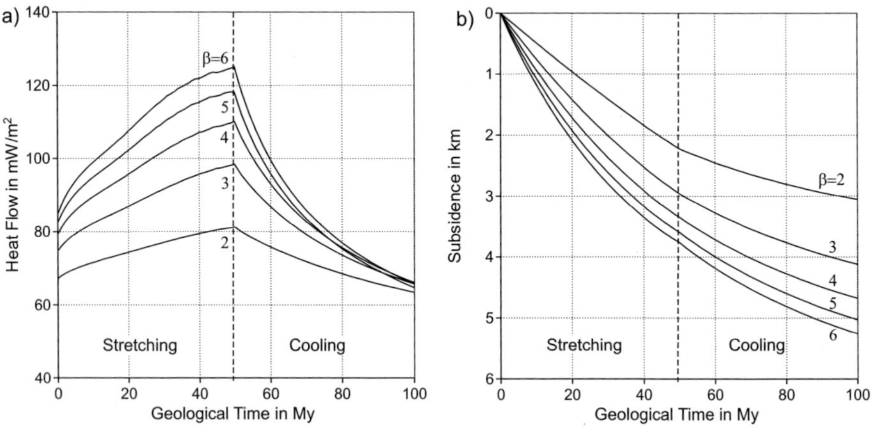

Fig. 3.31. Effect of exponentially decreasing radioactive heat production in the crust. The model parameters are the same as in Fig. 3.29 with $Q_{r0} = 2.5\,\mu\mathrm{W/m}^3$ and $z_h = 7\,\mathrm{km}$

known in the crust as an exponentially decreasing function with depth (Sclater et al., 1980; Allen and Allen, 2005). It can be expressed with the half–value depth z_h which describes the depth at which the concentration of radioactive elements is half the value of the maximum heat production Q_{r0}.

$$Q_r(z) = Q_{r0}\, 2^{-z/z_h} \; . \tag{3.41}$$

The convection term in (3.29) should be used for the moving lithosphere with velocity **v** and can be expressed with stretching factors for the crust and mantle (3.34). The problem is often approximated with a multi–1D solution, since horizontal crustal facies variations, together with extreme thermal conductivities, are assumed to be rather rare.

$$\lambda \frac{\partial^2 T}{\partial z^2} - \rho c \frac{\partial T}{\partial t} + \rho c v_z \frac{\partial T}{\partial z} + Q_r = 0 \; . \tag{3.42}$$

The vertical velocity field $v_z(z)$ can be derived from mechanical models. The stretching velocity increases linearly in the crust and mantle corresponding to the stretching factors, according to (3.34) and (3.35). It is

$$\lambda \frac{\partial^2 T}{\partial z^2} - \rho c \frac{\partial T}{\partial t} + \rho c \frac{\log \beta}{t_s} (h_0 - z) \frac{\partial T}{\partial z} + Q_r = 0 \tag{3.43}$$

where h_0 is the initial lithoshere thickness, and t_s is the stretching time. The thermal diffusivity $\kappa = \lambda/\rho/c$ is often given for the crust and mantle instead of the thermal conductivity λ. Note, that here $Q_r = 0$ in the mantle, $v_z = 0$ during cooling, and β is different for the crust and mantle.

The assumptions of the McKenzie model allow a solution of the heat flow equation with analytical methods. This is not possible when modifications of the model are taken into account such as the introduction of radioactive heat production rates for the upper crust, different and variable thermal properties, and several stretching factors and stretching phases. Then, numerical integration methods such as finite difference methods can be applied.

The upward and downward movement of the highest asthenosphere surface during stretching and cooling yields the typical heat flow peaks for rift basins. Their height and width depends on stretching factors and stretching duration times (Fig. 3.29). Cooling has already an effect during the stretching phase and lowers the heat flow peak in the case of long stretching times. Thus, the maximum peak occurs at instantaneous stretching.

Without radioactivity, the heat flow declines to the initial value after infinite cooling, as the original lithosphere thickness is restored. Radioactive heat production of the crust increases the total heat flow towards the surface, but decreases the relative peak height compared to the initial value, since radioactive heat production decreases with thinning of the crustal layer (Fig. 3.30). This usually yields lower present day heat flow than the original values. Radioactive heat production also results in about 20% higher subsidence curves (Fig. 3.31) compared to models without radioactive heat production (Fig. 3.29). Thus radioactive heat production must not be neglected.

The upper mantle often has a higher stretching rate than the crust because it is more ductile. This also yields lower heat flow peaks and less subsidence compared to higher crustal thinning (Fig. 3.32). Heating the lithosphere with the highest rates at the beginning also has an uplift effect, which is usually balanced by the subsidence caused by crustal thinning. Fast and high mantle stretching and less crustal stretching allows the uplift effect to overcome subsidence, uplifting the basin when stretching starts (Fig. 3.32). Pure mantle stretching always results in uplift during stretching.

The above approach can easily be extended to several phases of uniform stretching with multiple pairs of stretching factors for rifting and cooling periods. Each stretching factor applies the actual thickness of the lithosphere when the new stretching period begins, instead of the initial thickness of the lithosphere (Fig. 3.33).

The linear velocity versus depth curve is independent of the total initial depth, it only depends on the stretching factor and time as expressed in equation (3.34), but the velocity $v_z(d_c)$ of the basal crust layer decreases exponentially in time, since it looses speed during uplift. Stretching factors are inconvenient for the description of any non–uniform stretching behavior. It is possible to work with velocity versus geological time functions instead, and the resulting heat flow equations can still be used in the above manner without any changes.

3.8.3 Workflow Crustal Preprocessing

Crustal models are useful to predict tectonic subsidence and paleo–heat flow when stretching and cooling behavior occurs. Present or paleo–subsidence

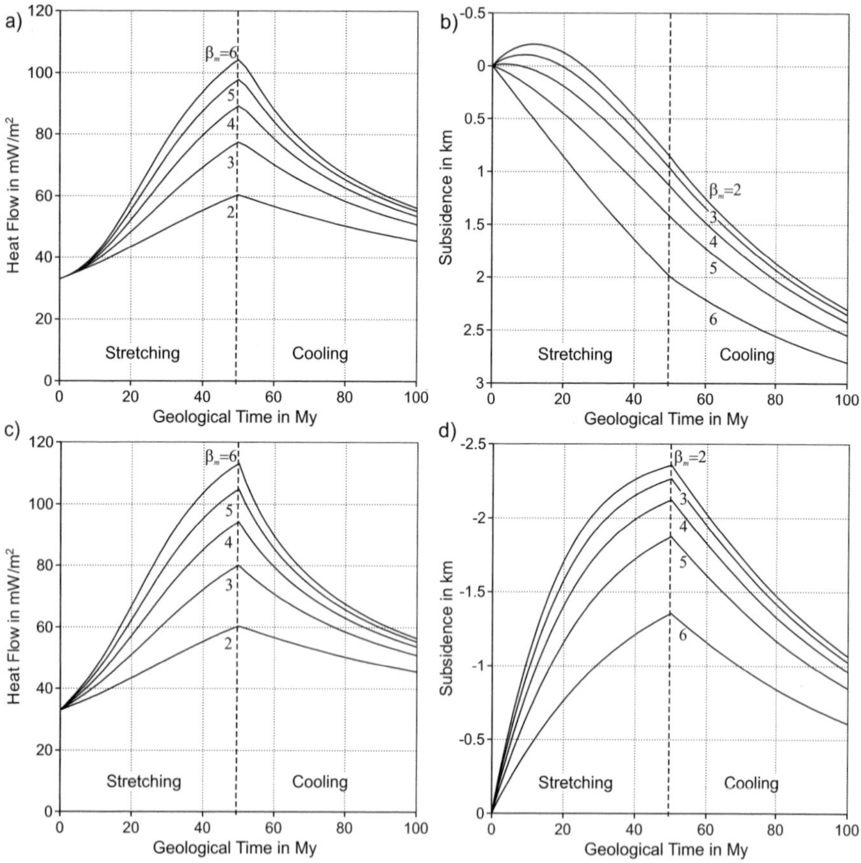

Fig. 3.32. Effect of different stretching factors for crust β_c and mantle β_m: (a),(b) small crustal stretching $\beta_c = 2$, (c),(d) no crustal stretching $\beta_c = 0$

from input geometry and stratigraphy can be used "inversely" to determine the stretching factors. The corresponding paleo–heat flow maps can then be calculated afterwards.

Fig. 3.34 illustrates the workflow for the calculation of paleo–heat flow maps from input geometry with calibrated stretching maps for crust and mantle using the uniform stretching model as described in the previous sections. It is also illustrated in Fig. 3.35.

The workflow starts with the extraction of the total paleo- and present day subsidence maps from the present day input model (Fig. 3.35.c). The corresponding back–stripping routine should also consider estimated paleo–water depth maps, decompaction and salt movement. Then, tectonic subsidence is calculated from total subsidence with the replacement of sediments by water (Eq. 3.40, Fig. 3.35.a). The main computing effort is then needed for inverting

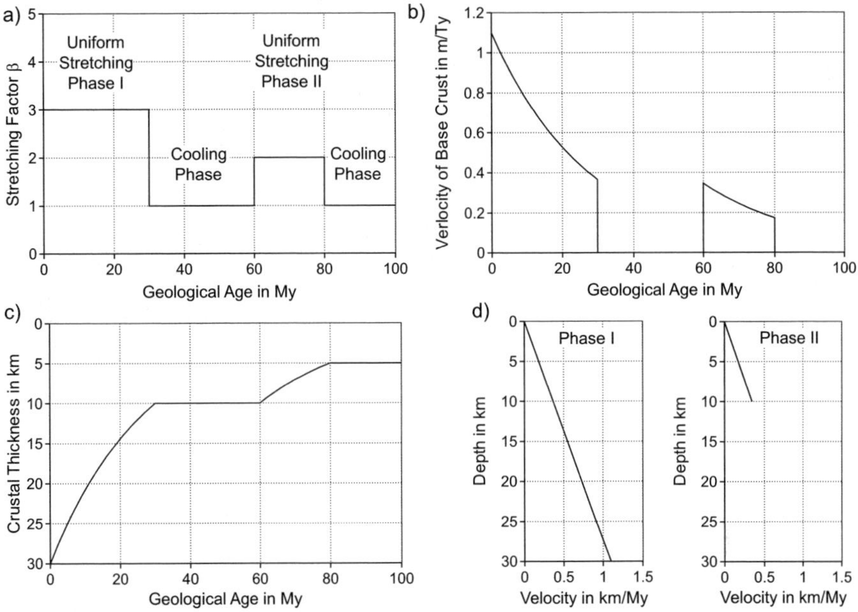

Fig. 3.33. Example with two uniform stretching periods, the initial crust thickness is 30 km: (a) Definition of the stretching and cooling periods. (b) The velocity of the base of the crust decreases exponentially during uplift. (c) The crustal thickness also decreases exponentially. (d) The time–constant velocity versus depth curve in the crust for the two stretching periods

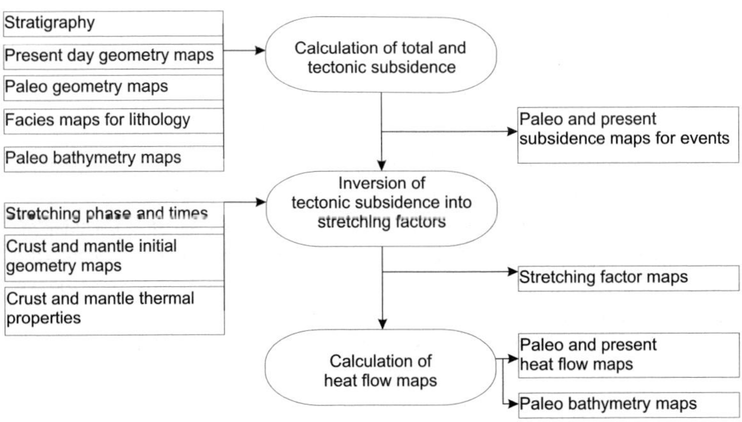

Fig. 3.34. Workflow for (crustal) heat flow preprocessor

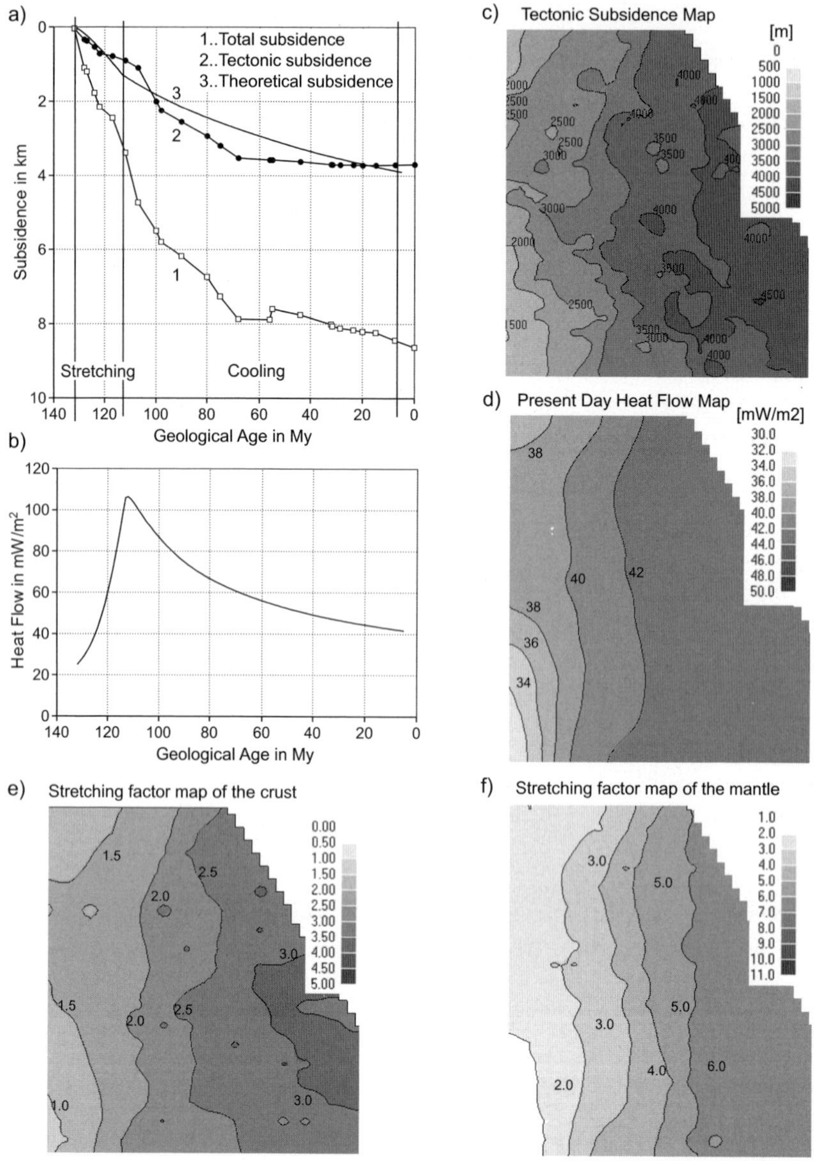

Fig. 3.35. Example from the Northern Campos basin for crustal heat flow analysis with a rift period of $132 - 113$ My, $h_c = 35$ km, $h_m = 95$ km, $Q_{r0} = 2.5\mu\text{W/m}^3$, $z_h = 7$ km: (a) Total and tectonic subsidence from input geometry and calculated theoretical subsidence after calibration at the location of the map midpoint. (b) Calculated heat flow at the basin midpoint with the calibrated stretching factors $\beta_c = 3.0$, $\beta_m = 4.7$. (c) Tectonic subsidence map from input geometry. (d) Calculated present day heat flow map. (e),(f) Maps with calibrated stretching factors for the crust and mantle

the tectonic subsidence maps into stretching factors, since it is an inversion of the heat and mechanical McKenzie type equations 3.43, 3.38. Usually, unique rifting and cooling times and initial crustal and mantle thicknesses are used for the entire map. The only unknowns in the inversion step are two stretching factors, which are calculated for each grid point in a multi–1D approach, so that the main output are two stretching maps for the crust and mantle (Fig. 3.35.e and f). The mantle map should be smoothed afterwards, if there is reason to assume high ductility.

The inversion can be performed for example with the response surface method, which is explained in Chap. 7. For each gridpoint 10 to 100 runs are necessary with the method of nesting intervals, so that there are about one million 1D forward simulation runs. The final stretching maps for crust and mantle can then be used to calculate the paleo and present heat flow maps through time and to recalculate the paleo water depth maps from simulated subsidence values.

This workflow can be extended to more than one rifting event or to define and calibrate other unknowns of the model, such as the initial thickness of the crust. Three stretching maps for the upper and lower crust and the mantle can also be used instead of the presented two layer model, or the two stretching maps can be assigned to upper crust and lower crust/upper mantle, respectively. Gravitational data can also be used for additional calibration parameters, e.g. when the crust geometry directly controls gravity.

Another less accurate and much simpler procedure is use of McKenzie's equilibrium subsidence (3.39) to directly calculate the stretching factor maps from the total present day subsidence map only, and to predict the paleo–heat flow maps and the new water depth maps afterwards.

It is obvious that the McKenzie type models yield only rough estimates of the basal heat flow maps through time and the heat flow maps need to be fine tuned with vitrinite reflectance data and bottom hole temperatures afterwards. A decoupling of the procedures in the two steps of crustal preprocessing and calibration against the thermal markers, allows a better overview and handling of the individual parts and leads to a better understanding of the respective processes.

3.9 Heat Flow Calibration

Heat flow models can be calibrated with measured temperatures from wells and thermal maturity parameters, such as vitrinite reflectance, biomarkers and fission–track annealing data. Thermal maturity parameters are time and temperature dependent. They indicate how long the rock elements remain at certain temperature levels. Thus, single data points of specific thermal markers are often only useful to calibrate a small temperature interval. They cannot be used to specify a total age for the temperature interval. Exceptions are fluid inclusion temperatures, which are often related to paleo–ages.

The most commonly used parameter is vitrinite reflectance, since it is widely available in most sediment types, covers typical oil and gas maturity ranges and is easy and cheap to measure. The importance and main temperature windows of many other thermal markers are compared and explained in more detail in Chap. 4. The most uncertain input parameters are thermal conductivity and paleo- and present basal heat flow values.

There are several workflows and techniques developed to change thermal conductivity and heat flow parameters, when thermal maturity data or measured temperatures, differ from a master run. Recently, more emphasis has been put on the calibration of basal heat flow values as thermal conductivities are much better known. The assumption of a rift–type heat flow peak or any other trend can be obtained as a first estimation from crustal models or other knowledge about geological history. Such trends are typically defined for individual locations and usually calibrated against well data such as bottom hole temperatures or vitrinite reflectance values. The heat flow trends can then be simply shifted entirely or stepwise or other corrections like first order shifting or heat flow peak calibration can be performed until the match with the calibration data is satisfying (Fig. 3.36).

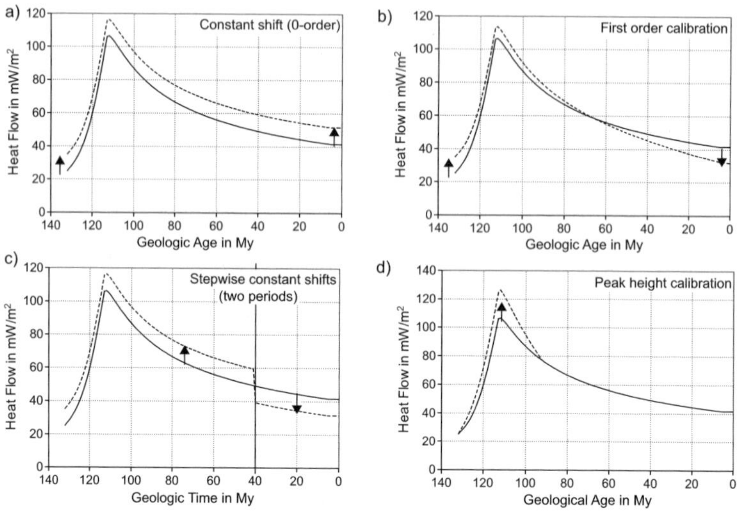

Fig. 3.36. Methods of heat flow trend calibration: (a) Constant shift. (b) First order calibration. (c) Stepwise constant shifts. (d) Special peak height change

Automatic calibration tools can be used, when numerous thermal calibration parameters are available. They allow the definition of time intervals with independent shift corrections, peak corrections or the assumption of additional uncertainties for SWI temperatures or thermal properties. Numerical models used for the automatic calibration or inversion are Monte Carlo simulations, response surface modeling or fast approximated forward simulation techniques, all of which are described in the Chap. 7. One typical workflow of an automatic 3D calibration processor is described below.

The final values of an automatic calibration must make physical and geological sense, rather than simply providing an acceptable mathematical fit between measured and calculated values. This is achieved with fixed parameter ranges or the Bayesian approach (Chap. 7).

3.9.1 Example Workflow for 3D Heat Calibration

The workflow under discussion is illustrated in Fig. 3.37. A fit to calibration data such as temperature and vitrinite reflectance values should obviously be achieved while conserving the shape of the heat flow trend through time. Here, only a constant shift of the heat flow trends is considered. The procedure can be applied to regions of limited size, typically to small areas of interest around a well or a group of wells (Fig. 3.38). The extension of these areas should be large enough to incorporate lateral heat flow effects as they appear e.g. in the vicinity of salt domes. The temperature evolution inside each area can be fitted with its own heat flow shift.

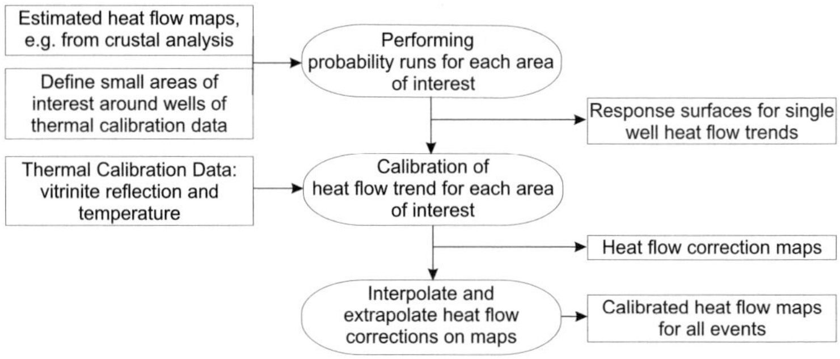

Fig. 3.37. Workflow for heat flow preprocessor

A calibration performed in any area of interest is independent of calibrations in other areas. Therefore, it can, for example, be performed in a separate "mini–model".[3] In practice many wells with calibration data are available and it is advantageous to run all the areas together in one big 3D simulation. The areas around the wells are restricted to a small size. As smaller these sizes as the faster the simulation. This is important in practice, because multiple runs with varying heat flow must be performed.[4] The calibration in each area

[3] If lateral heat flow effects are neglected such a "mini–model" becomes a pure 1D–model.

[4] When sufficient computer resources are available, it is also possible to run the full 3D model without any areal cut–outs. This would reproduce all 3D thermal

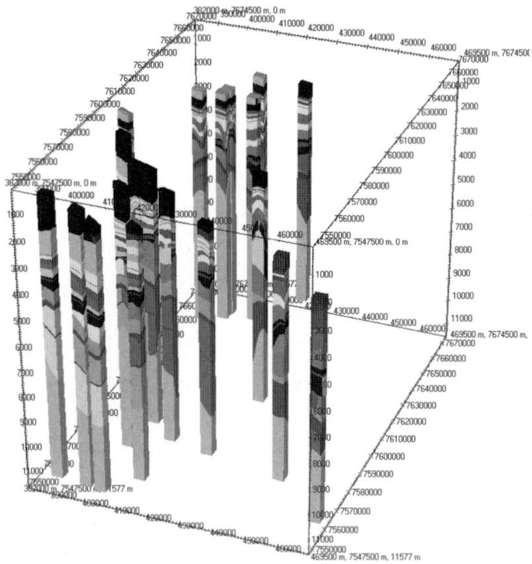

Fig. 3.38. Example of a heat flow calibration in areas around wells with temperature and vitrinite reflectance data. The size of each area is equivalent to the thickness of the corresponding column. These column thicknesses are defined as rather thin here because lateral heat flow effects are small

is performed independently of each other afterwards. Advanced interpolation and extrapolation between different simulation results with, for example, response surfaces as described in section Sec. 7.5.1, yield fast and accurate results (Figs. 3.39, 3.40).

Finally, calibrated heat flow maps can be constructed by spatial interpolation between the areas with calibrated heat flow shifts. If necessary, an additional smoothing of the interpolated heat flow shift can be performed. The results can be tested in a final simulation run.

Crustal heat flow analysis and heat flow calibration can be performed successively (Fig. 3.41). It is thus possible to construct calibrated and geological meaningful heat flow maps from a basin model and additional geological information about stretching phases, crustal structure, bottom hole temperatures and vitrinite reflectance data.

---

effects. However, calibrated heat flow shifts below each calibration well, must be assigned to limited areas for further construction of new structural heat flow maps by interpolation. Additionally, these areas should not be too small. Otherwise in the full 3D model, heat might leave these areas laterally due to lateral temperature gradients. This may yield too low temperatures at the well locations and the calibration might fail.

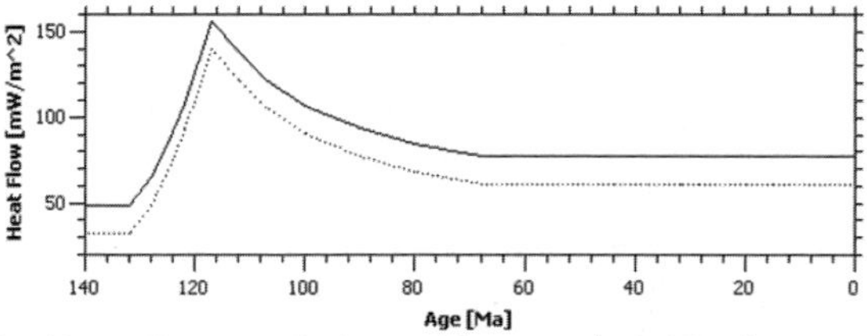

Fig. 3.39. Shifted heat flow trend after calibration (solid line) and before calibration (dotted line)

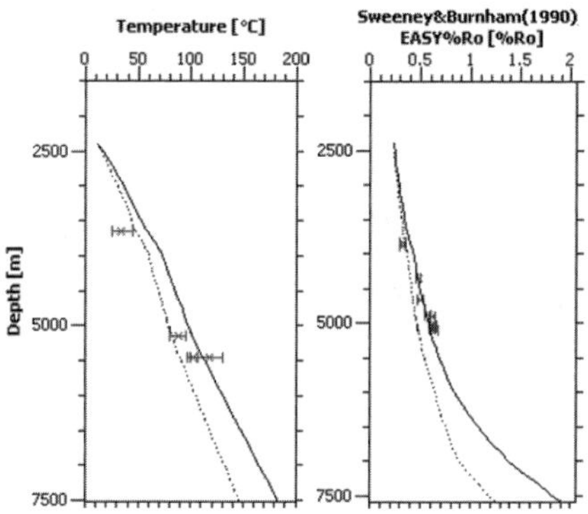

Fig. 3.40. Temperature and vitrinite reflectance in the same well calibrated against data values with error bars. Dotted lines represent results from a crustal model only. The corresponding heat flow shift is shown in Fig. 3.39. Note that a fit against temperature data alone would yield a slightly lower temperature and vitrinite reflectance profile. However, due to small error bars vitrinite reflectance values are treated here with higher importance than temperature values

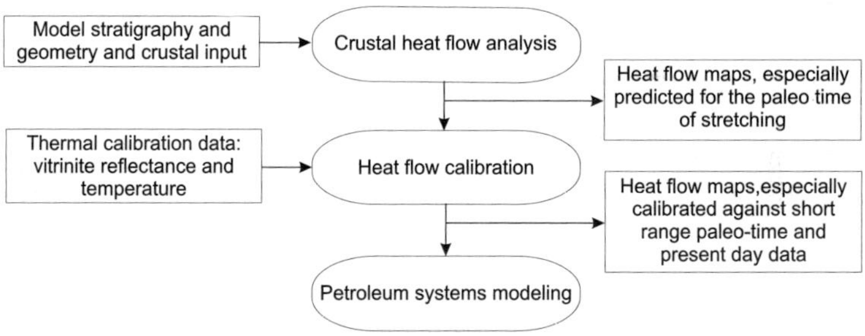

Fig. 3.41. General workflow, which links crustal modeling, heat flow calibration and petroleum systems modeling

Summary: Heat flow analysis is based on a detailed balance of thermal energy that is transported via heat flow through sedimentary basins. Heat flow occurs primarily in form of conduction and convection. The driving forces for conduction are temperature differences. Convection is classified by moving fluid or solid phases that carry their inner thermal energy along. Previously to a detailed energy balance it is necessary to specify heat in– and outflow or alternatively the temperature at the boundary of the sedimentary basin.

The main direction of heat flow in sedimentary basins is vertically upwards. It is thus possible to demonstrate basic effects with crude one dimensional models. Steady state heat flow constitutes the most simple heat flow pattern. Explicit formulas can be calculated. Radioactive heat production can easily be incorporated. The complexity of the system rises with consideration of transient effects which occur during deposition, erosion, and when thermal boundary conditions change. However, some idealized special cases can be solved analytically.

The main thermal properties of the rocks are thermal conductivities, readiogenic heat production, and heat capacities. Detailed specifications of these properties for various lithologies and fluids over wide temperature ranges are well known.

The general formulation of two and three dimensional heat flow problems incorporates heat convection and magmatic intrusions. The quantification of heat flow and temperature boundary conditions is often a major task. SWI temperatures can be derived from paleo climate models. Effects of permafrost require the specification of paleo surface temperatures.

Basal heat flow can be calculated from tectonic stretching and thinning of the crust which causes the evolution of the basin. Models for rift basins are mainly worked out as extensions to the famous McKenzie type crustal models. Basic principles are isostasy and crustal heat flow balance. Finally, the amount of stretching can be calibrated against the known subsidence of the sedimentary package. Comprehensive heat flow trends can be constructed.

These trends can locally be adapted to known temperature histories from well logs and samples, e.g. bottom hole temperatures and vitrinite reflectance measurements. Sophisticated workflows are worked out for fast and efficient calibration procedures.

References

P. A. Allen and J. R. Allen. *Basin Analysis.* Blackwell Publishing, second edition, 2005.

G. R. Beardsmore and J. P. Cull. *Crustal Heat Flow.* Cambridge University Press, 2001.

C. Buecker and L. Rybach. A simple method to determine heat production from gamma logs. *Marine and Petroleum Geology*, (13):373–377, 1996.

G. Buntebarth and J. R. Schopper. Experimental and theoretical investigation on the influence of fluids, solids and interactions between them on thermal properties of porous rocks. *Physics and Chemistry of the Earth*, 23(6): 1141–1146, 1998.

P. T. Delaney. Fortran 77 programs for conductive cooling of dikes with temperature-dependent thermal properties and heat cristallisation. *Computers and Geosciences*, 14:181–212, 1988.

G. Delisle, S. Grassmann, B. Cramer, J. Messner, and J. Winsemann. Estimating episodic permafrost development in Northern Germany during the Pleistocene. *Int. Assoc. Sed. Spec. Publ.*, 39:109–120, 2007.

D. Deming and D. S. Chapman. Thermal histories and hydrocarbon generation: example from Utah–Wyoming trust belt. *AAPG Bulletin*, 73:1455–1471, 1989.

W. R. Gambill. You can predict heat capacities. *Chemical Engineering*, pages 243–248, 1957.

S. Grassmann, B. Cramer, G. Delisle, J. Messner, and J. Winsemann. Geological history and petroleum system of the Mittelplate oil field, Northern Germany. *Int. J. Earth Sci. (Geol. Rundsch.)*, 94:979–989, 2005.

S. J. Hellinger and J. G. Sclater. Some comments on two-layer extensional models for the evolution of sedimetary basins. *Journal of Geophysical Research*, 88:8251–8270, 1983.

G. T. Jarvis and D. P. McKenzie. Sedimentary basin information with finite extension rates. *Earth and Planet. Sci. Lett.*, 48:42–52, 1980.

V. N. Kobranova. *Petrophysics.* Springer–Verlag, 1989.

D. R. Lide. *CRC Handbook of Cemistry and Physics.* 87 edition, 2006.

Ming Luo, J. R. Wood, and L. M. Cathles. Prediction of thermal conductivity in reservoir rocks using fabric theory. *Journal of Applied Geophysics*, 32: 321–334, 1994.

D. McKenzie. Some remarks on the development of sedimentary basins. *Earth and Planet. Sci. Lett.*, 40:25–32, 1978.

B. Parsons and J. G. Sclater. An analysis of the variation of ocean floor bathymetry and heat flow with age. *Journal of Geophysical Research*, 82 (5):803–827, 1977.

B. E. Poling, J. M. Prausnitz, and J. P. O'Connell. *The Properties of Liquids and Gases.* McGraw–Hill, New York, 5th edition, 2001.

L. Royden and C. E. Keen. Rifting processes and thermal evolution of the continental margin of eastern canada determined from subsidence curves. *Earth and Planetary Science Letters*, 51:343–361, 1980.

L. Rybach. Wärmeproduktionsbestimmungen an Gesteinen der Schweizer Alpen (Determinations of heat production in rocks of the Swiss Alps), Beiträge zur Geologie der Schweiz. *Geotechnische Serie*, (51):43, 1973. Kümmerly & Frei.

J. G. Sclater, C. Jaupart, and D. Galson. The heat flow through oceanic and continental crust and the heat loss of the earth. *Journal of Geophysical Research*, 18:269–311, 1980.

K. Sekiguchi. A method for determining terrestrial heat flow in oil basinal areas. In Cerm? V., L. Rybach, and D. S. Chapman, editors, *Terrestrial Heat Flow Studies and the Structure of the Lithosphere*, volume 103 of *Tectonophysics*, pages 67–79. 1984.

E. D. Jr. Sloan. Physical/chemical properties of gas hydrates and application to world margin stability and climate change. In J.-P. Henriet and J. Mienert, editors, *Gas Hydrates: Relevance to World Margin Stability and Climate Change*, volume 137 of *Special Publication*. Geological Society of London, 1998.

W. H. Somerton. *Thermal Properties and Temperature–Related Behavior of Rock/Fluid Systems: Elsevier.* Elsevier, Amsterdam, 1992.

D. L. Turcotte. On the thermal evolution of the earth. *Earth and Planetary Science Letters*, 48:53–58, 1980.

D. W. Waples and H. Tirsgaard. Changes in matrix thermal conductivity of clays and claystones as a function of compaction. *Petroleum Geoscience*, 8: 365–370, 2002.

D. W. Waples and J. S. Waples. A review and evaluation of specific heat capacities of rocks, minerals, and subsurface fluids. Part 1: Minerals and nonporous rocks, natural resources research. 13:97–122, 2004a.

D. W. Waples and J. S. Waples. A review and evaluation of specific heat capacities of rocks, minerals, and subsurface fluids. Part 2: Fluids and porous rocks, natural resources research. 13:123–130, 2004b.

B. P. Wygrala. *Integrated study of an oil field in the southern Po Basin, Northern Italy.* PhD thesis, University of Cologne, Germany, 1989.

4

Petroleum Generation

4.1 Introduction

Modeling of geochemical processes encompasses the generation of petroleum and related maturation parameters, such as vitrinite reflectance, molecular biomarkers, and mineral diagenesis. The transformation and maturation of organic matter can be subdivided into three phases: diagenesis, catagenesis and metagenesis (Tissot and Welte, 1984). The term diagenesis is different from that of the rock types. The formation of petroleum and coal with typical depth and temperature intervals is illustrated in Fig. 4.1.

During diagenesis, most organic particles in the sediments are transformed by microbiological processes into kerogen with a release of volatiles such as CH_4, NH_3 and CO_2. Petroleum is mainly generated during catagenesis, when kerogen is thermally cracked to heavier and lighter hydrocarbons and NSO (nitrogen, sulfur, oxygen) compounds. The transformation rates depend on the organic matter type and the time–temperature history. Heavier petroleum components are generally generated first and they are then cracked into lighter components at higher temperatures, resulting in a so called "oil window" between 1 to 3 km depth. Thermogenic hydrocarbon gas is generated at greater depths.

The maturation of coal also depends on time and temperature. It is mainly described by the change of one of its constituents, the maceral vitrinite. One measure for vitrinite maturation is the intensity of reflected light at a standardized wavelength (546 nm). The occurrence of vitrinite in many sediments enables its usage as a general thermal history marker. Vitrinite reflection can thus be correlated to the maturation of petroleum (Fig. 4.1).

The generation and maturation of HC-components, molecular biomarkers and coal macerals can be quantified by chemical kinetics. Chemical kinetics are formulated using mass balances. It is therefore important to specify and track all chemical reactants of organic matter during the processes of interest. A simple classification of organic matter in sedimentary rocks after Tissot and

T. Hantschel, A.I. Kauerauf, *Fundamentals of Basin and Petroleum Systems Modeling*, DOI 10.1007/978-3-540-72318-9_4,
© Springer-Verlag Berlin Heidelberg 2009

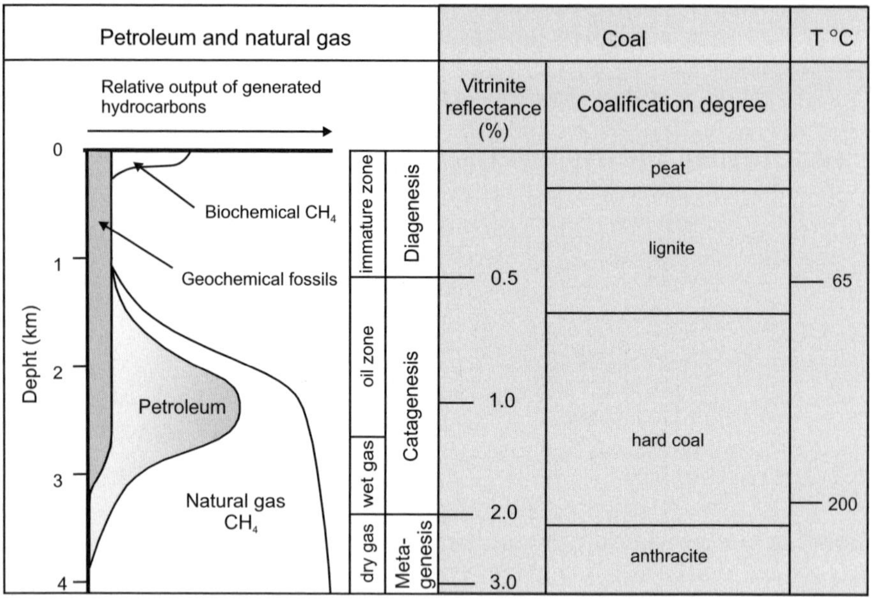

Fig. 4.1. Evolution of organic matter: Diagenetic, catagenetic and metagenetic processes describe the generation of oil and gas and coalification. The picture is from Bahlburg and Breitkreuz (2004). The processes are compared with the relative intensities of light reflected from the coal maceral vitrinite

Welte (1984) is shown in Fig. 4.2. A comprehensive description is given, for example, in Peters et al. (2005).

Diffusion controlled and autocatalytic reactions are not taken into account in this volume, instead decomposition and pseudo–unimolecular kinetics are considered, which can be adequately described by sequential and parallel reactions. This approach is called kinetics of distributed reactivities. Herein, the rate of each reaction is related to an Arrhenius type activation energy and a frequency factor.

4.2 Distributed Reactivity Kinetics

The simplest reaction type, which is used for most sequential and parallel reaction schemes, is the unimolecular forward reaction from an initial reactant X of mass x to the product Y of mass y:

$$\mathrm{X} \xrightarrow{k} \mathrm{Y}, \quad \frac{\partial y}{\partial t} = -\frac{\partial x}{\partial t} = k\, x^{\alpha} \tag{4.1}$$

where α is the reaction order, k is the reaction rate and t is the time. In the following, unit masses are considered with $x_0 = 1$, $x(t \to \infty) = 0$ and $y = 1 - x$.

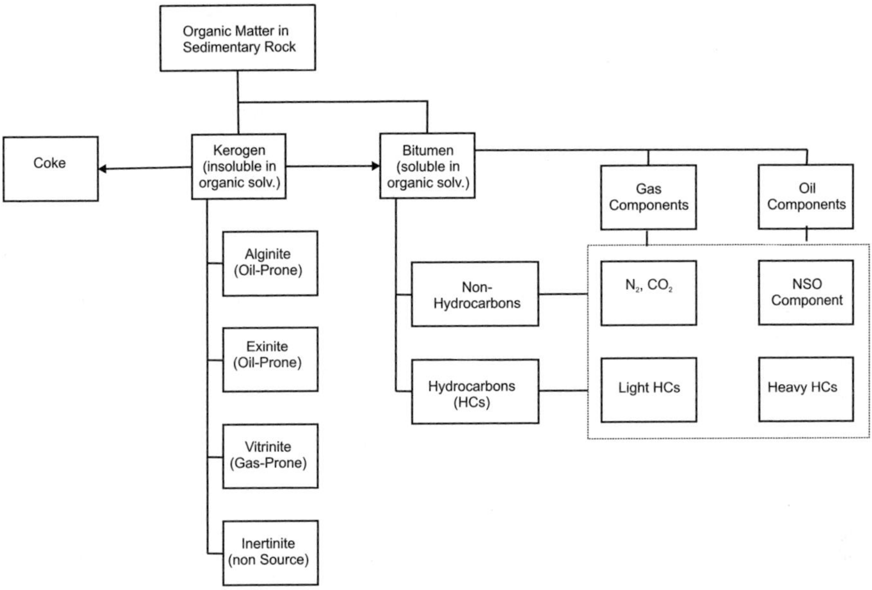

Fig. 4.2. A geochemical fractionation of organic matter

The transformation ratio TR is defined as the converted mass fraction of the initial reactant, which is here TR $= y$. Most geochemical processes are described with first order reactions $\alpha = 1$. Higher or lower reaction orders are used when the transformation rate $d\mathrm{TR}/dt$ has a nonlinear dependency on the reactant's concentration.

The temperature dependency of the reaction rate k is usually described by an Arrhenius law with two parameters, the frequency factor A and the activation energy E:

$$k = A\, e^{-E/RT} \tag{4.2}$$

with the gas constant $R = 8.31447\,\mathrm{Ws/mol/K}$. The frequency (amplitude or pre–exponential) factor represents the frequency at which the molecules will be transformed and the activation energy describes the required threshold energy to initiate the reaction. The Arrhenius law was originally developed as an empirical equation but it is also confirmed from transition theory with a temperature dependent frequency factor (Glasstone et al., 1941; Benson, 1968).

The strong temperature dependence yields significant values for the transformation rate when a threshold temperature is exceeded (Fig. 4.3). The activation energies and frequency factors, which are used in the sample plots, are typical for organic matter decomposition with transformation times of millions of years for heating rates of 10 K/My.

Laboratory measurements are performed with higher heating rates of 1 K/min, which yield peak transformation at much higher temperatures than in nature (Fig. 4.4). Small uncertainties in laboratory based kinetic parameters can result in large effects on predicted transformation rates at geological time scales. Measured transformation ratios versus time can be inverted into (E, A) pairs with a regression line in the "ln k versus $1/T$" Arrhenius plot. Advanced inversion methods are necessary for kinetics with multiple parallel and sequential reactions.

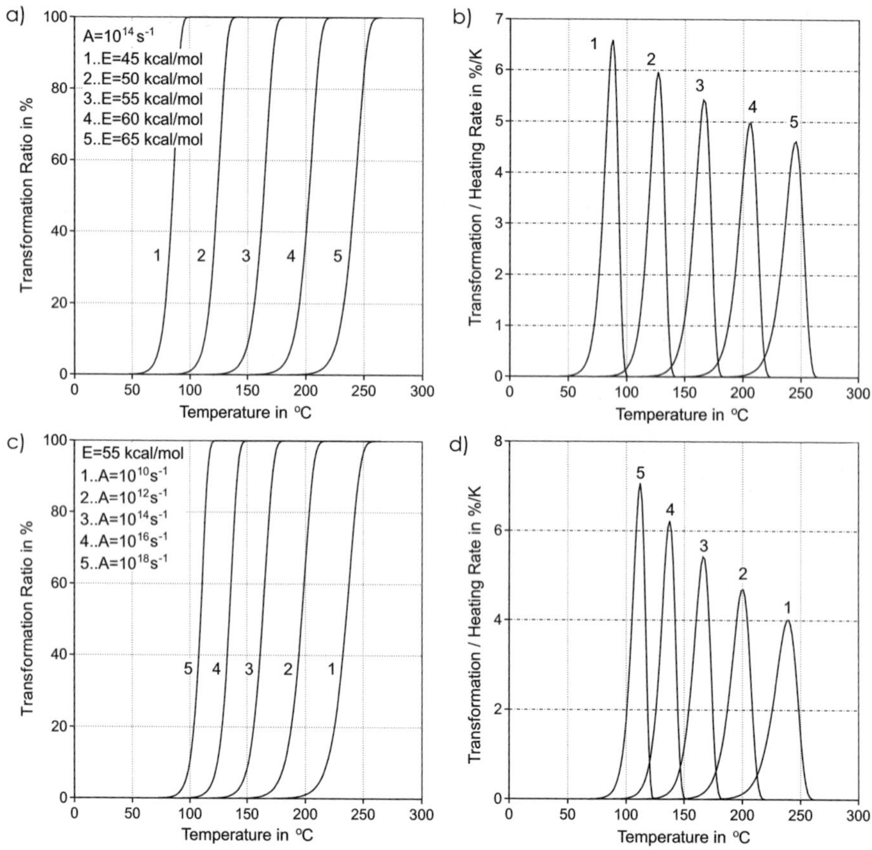

Fig. 4.3. Influence of activation energy and frequency factor on transformation rate (TR) and transformation ratio for first order Arrhenius type reactions with a heating rate of 10 K/My: (a) and (b) fixed frequency ratio and variable activation energy, (c) and (d) fixed activation energy and variable frequency factor

Parallel reactions are used for multiple and not interacting reactants X_i of masses x_i with initial masses x_{0i} and $\sum_i x_{0i} = 1$ which are converted into a product Y as follows.

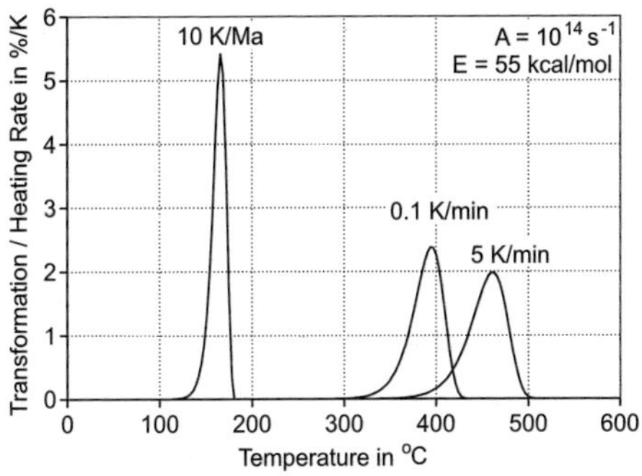

Fig. 4.4. Influence of the heating rate on first order Arrhenius type reactions. The first peak is related to geological heating rates 10 K/My, while the other two peaks are laboratory heating rates with 0.1 K/min and 5 K/min

$$\mathrm{X_i} \xrightarrow{k_i} \mathrm{Y}, \qquad \frac{\partial x_i}{\partial t} = -k_i x^\alpha, \qquad \frac{\partial y}{\partial t} = \sum_{i=1}^{n} k_i x^\alpha . \tag{4.3}$$

Each of the subreactions i can be described with a pair (A_i, E_i), but usually a unique frequency factor is used, which results in a discrete distribution of activation energies $p(E_i)$ equal to the initial masses of the reactants $x_{0i} = p(E_i)$. The transformation ratio is $\mathrm{TR} = 1 - x = y$ with $x = \sum_i x_i$. It also varies between 0 and 1.

Parallel reactions can be used for the decomposition of complex macro–molecules having a wide range of chemical bond strengths to one type of cracking product. Herein, the initial masses x_{0i} correspond to the mass of the reactant related to the cracking of chemical bonds of the corresponding activation energy levels. However, the chemistry of the decomposition of organic matter is generally more complex and the use of parallel reactions is just an empirical formalism for compositional effects of reactants and products.

Arbitrary discrete distributions $p(E_i)$ are mainly used for petroleum and coal formation. Approximations to continuous Gaussian distributions are also popular. Gaussian distributions are described with a mean activation energy μ and a variance σ as follows (Chap. 7):

$$p(E) = \frac{1}{\sigma\sqrt{2\pi}} \exp\left[\frac{-(E-\mu)^2}{2\sigma^2}\right] . \tag{4.4}$$

Burnham and Braun (1999) reported the usage of two other continuous distributions: the Gamma distribution

$$p(E) = \frac{a^\nu \, (E-\gamma)^{\nu-1}}{\Gamma(\nu)} e^{-(E-\gamma)a} \tag{4.5}$$

with parameters ν, a and threshold activation energy γ, and the Weibull distribution

$$p(E) = \frac{\beta}{\eta}\left(\frac{E-\gamma}{\eta}\right)^{\beta-1} e^{-\left[(E-\gamma)/\eta\right]^\beta} \tag{4.6}$$

with the width parameter η, the shape parameter β, and again a threshold activation energy γ. The mean energy of the Weibull distribution is $\bar{E} = \gamma + \eta\Gamma(1/\beta + 1)$, (Fig. 4.5).

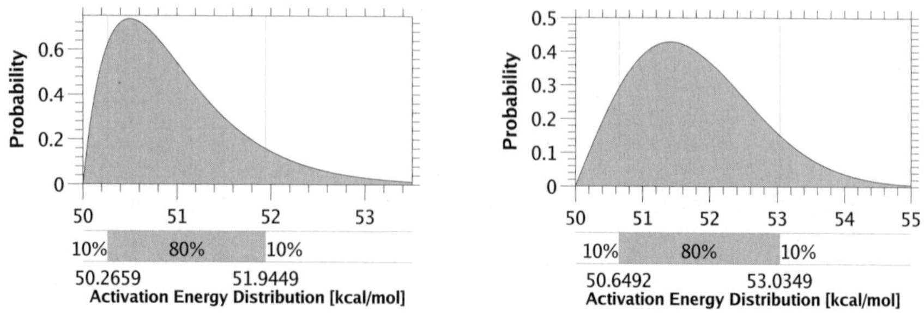

Fig. 4.5. Examples of Gamma (left) and Weibull (right) distributed activation energies. Here it is $\nu = a = \eta = \beta = 2$ and $\gamma = 50\,\text{kcal/mol}$ for both distributions

Equation (4.3) can be numerically integrated in time for stepwise constant heating rates, which yields the following mass reduction from time step l to $l+1$ with duration Δt and temperature $T^{(l)}$:

$$\begin{aligned} x_i^{(l+1)} &= x_i^{(l)} + \frac{\Delta t k_i(T^{(l)})}{\alpha - 1} && \text{for} \quad \alpha \neq 1, \\ x_i^{(l+1)} &= x_i^{(l)} \frac{2 - \Delta t k_i(T^{(l)})}{2 + \Delta t k_i(T^{(l)})} && \text{else}\,. \end{aligned} \tag{4.7}$$

The extension of the above approach to the formation of multiple products is usually performed by superposition of single kinetic reaction schemes, which yields sequential and parallel connected super–schemes. However, the actual processing resembles a bookkeeping of all reactions that are defined in the source rock according to the organic facies. The computing effort is low compared to other processes; e.g. heat and fluid flow calculations. It increases linearly with the number of cells, but the required computer memory can become significant as one value for each of the discrete activation energies of each mass component has to be stored for each cell.

4.3 Petroleum Generation Kinetics

The total content of organic matter (kerogen and bitumen) is usually given in terms of the total organic carbon content (TOC) in mass %. It is the ratio of the mass of all carbon atoms in the organic particles to the total mass of the rock matrix. Hence, it is a concentration value and one needs the total rock mass of the source rock for conversion into generated and expelled petroleum masses.

The generation of petroleum is a decomposition reaction, from heterogeneous mixtures of kerogen macromolecules to lighter petroleum molecules. Petroleum kinetics are distinguished from cracking types (primary or secondary), kerogen types (I — IV) and the number and type of the generated petroleum components (bulk, oil–gas, compositional kinetics).

Kerogen types are chemically classified according to the abundance of the elements carbon (C), hydrogen (H) and oxygen(O). The most common are the H/C and O/C ratios originally used in coal maceral classifications by van Krevelen (1961), which resulted in the definition of three main kerogen types I — III as shown in Fig. 4.6.a after Peters et al. (2005).

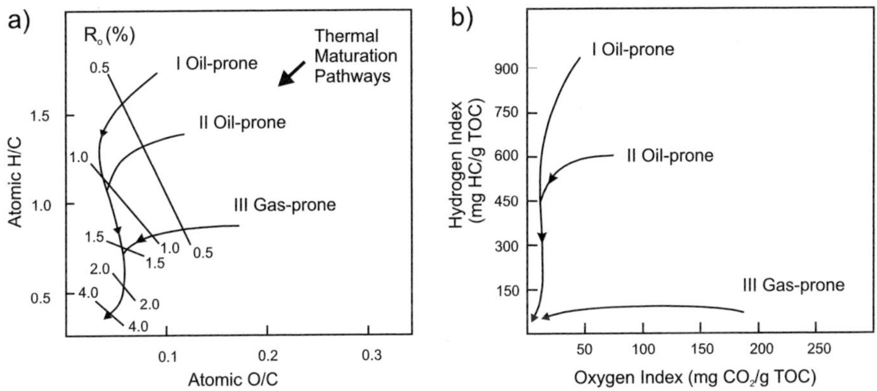

Fig. 4.6. Characterization of kerogen by van–Krevelen diagrams after Peters et al. (2005): (a) according to the abundance of the elements in kerogen in ratios of H/C and O/C, (b) according to the generative amounts of HC and CO_2 in Rock–Eval parameters HI (hydrogen index) and OI (oxygen index) from Fig. 4.7, and vitrinite reflectance R_0 The paths are idealized. Real samples can be different

The kerogen types are weakly linked to depositional environments. Type I is mostly derived from lacustrine algal matter, although some petroleum source rocks deposited in marine settings are also dominated by type I kerogen. Type II is the most widespread type. It is common in marine sediments, indicating deposition of autochtonous organic material in a reducing environment. Type III with the highest relative oxygen content of the main kerogen types, indicates an origin in terrigenous environments, i.e. from plant organic matter. Type IV kerogen has very low HI values. The corresponding maturity path is close to the x-axis in the van–Krevelen diagram. Most low maturity coals contain type III kerogen, although coals dominated by type I, II or IV kerogen also occur. Detailed descriptions of the occurrence and the genesis of the different kerogen types are given in Tissot and Welte (1984) and Peters et al. (2005).

The maturation paths of decreasing H/C and O/C ratios can be related by simple mass balance calculations to the corresponding decrease in the generative masses of HCs and CO_2. The generative HC and CO_2 masses of a kerogen sample are measured by Rock–Eval pyrolysis in terms of the HI and OI potential, which yields the "HI–OI" van-Krevelen diagram as shown in Fig. 4.6.b.

The Rock–Eval method is an open system pyrolysis. The rock sample containing organic matter is heated at approximately 50 K/min and the released masses of HCs and CO_2 are measured (Fig. 4.7). The first HC peak of the thermally distilled HCs (S1) corresponds to the residual bitumen. It is the already generated and not yet expelled mass of HCs in the rock sample. The second peak of pyrolytic generated HC amounts (S2) is the total generative mass of HCs, which is related to the hydrogen index (HI), given in mg/gTOC. Hence, the HI multiplied with the TOC and the rock mass is equal to the total generative mass of HCs in the rock. The peak of the pyrolytic generated CO_2 is analogically related to the oxygen index (OI) measured in mg/gTOC.

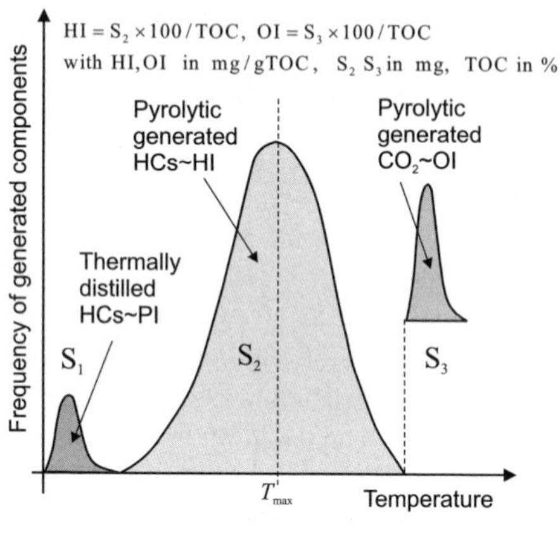

Fig. 4.7. Schematic pyrogram from the Rock–Eval pyrolysis: Hydrogen Index HI, Oxygen Index OI, Production Index PI. The S1 and S2 peaks mainly contain hydrocarbons. They are measured with a flame ionization detector. The S3 peak contains the generated CO_2. It is measured with a thermal conductivity detector after the heating is finished at room temperature

The production index PI=S1/(S1+S2) is a measure of cracked kerogen. It varies between 0 and 1. The PI is equal to the transformation ratio, if no petroleum has been expelled out of the sample.

Another characteristic value of the Rock–Eval method is the oven temperature T_{max} at the maximum HC generation rate for S2 (Fig. 4.7). This temperature can be used as a maturity parameter for the kerogen sample. It can be calculated in source rocks and used as a thermal calibration parameter.

The classification of kerogen into the three van Krevelen types is not sufficient to predict the generated petroleum composition. The type of the generated petroleum generally depends on various factors, such as the biological input, the oxic or anoxic environment, and marine or deltaic facies. Jones

(1987) introduced the term organic facies for an improved classification of the kerogen types according to the generated type of petroleum. Di Primio and Horsfield (2006) proposed a characterization of organic facies with the generated portions of the HC classes ($C_1 - C_5$, $C_6 - C_{14}$ and C_{15+}) and the sulfur content of the source rock (Fig. 4.8). They also proposed a quantitative description of generation kinetics with 14 hydrocarbon components, which allows an improved prediction of HC phases with equation–of–state type flash calculations (Chap. 5). The prediction of phase properties, such as densities, viscosities or phase compositions is more reliable with these organic facies based kinetics than with generalized van–Krevelen type kinetics.

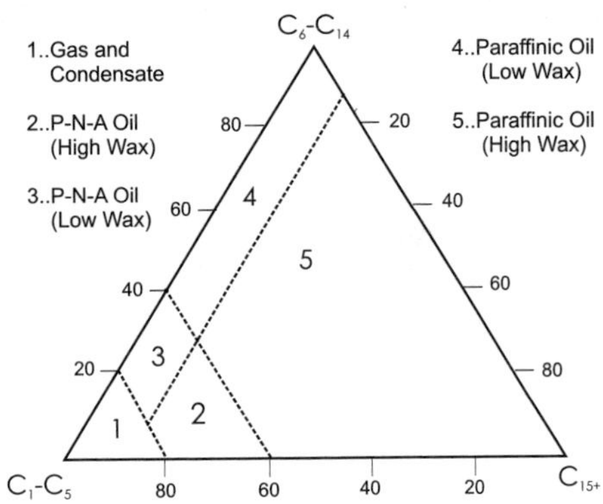

Fig. 4.8. Characterization of kerogen by the generated petroleum type: Five organic facies are defined according to the generative potential for three HC classes ($C_1 - C_5$, $C_6 - C_{14}$, C_{15+}). Facies 3 and 5 are further distinguished into a low and a high sulfur type . P–N–A means paraffinic–naphthenic–aromatic type

The number of considered petroleum components depends on the available sample data, the type of pyrolysis system, and whether phase properties such as APIs and GORs should be calculated.

Many publications describe primary generation kinetics, while secondary cracking from higher into lower molecular weight HCs with coke as a by–product are rather seldom. Many secondary cracking models are only based on methane and coke as products. Some are described as chain reactions each of them with the next lower molecular weight component and coke as products. In the case of two component oil–gas systems, secondary cracking is defined as an oil to gas reaction.

4.3.1 Bulk Kinetics

Bulk kinetics focus on kerogen cracking and do not distinguish between several petroleum components. They are described with n parallel reactions (4.3) from kerogen $X_i, i = 1, \ldots, n$ to petroleum Y. Herein, the "i–th" parallel reaction corresponds to the chemical bonds, which have to be cracked with the activation energy E_i in the kerogen molecules. The x_{0i} and x_i are the initial and actual relative masses of the generative petroleum according to the activation energy E_i with $\sum_{i=1}^{n} x_{i0} = 1$ and $\sum_{i=1}^{n} x_i = x$, and y as the generated relative petroleum mass. The energy distribution $p(E_i)$ is equal to the initial relative mass distribution of the generative petroleum x_{0i}.

Usually, the petroleum potential $y_p = y\,\mathrm{HI}_0$ in gHC/kgTOC is used rather than the relative masses of petroleum y to describe the generated petroleum amounts. Some typical bulk kinetics are shown in Fig. 4.9 for type II and III kerogen after Tegelaar and Noble (1994). The petroleum potential is usually calculated in all layers and not only in the source rock to illustrate the dependency and sensitivity of the source rock generation by depth. Herein, the same HI_0 is assumed for all layers.

The transformation ratio TR = 0.5 defines the critical point of generation. Some example curves for type II kinetics at three different sedimentation rates show the dependency of the generated petroleum on sedimentation or heating rates (Fig. 4.10). The depth of the critical point ranges from 4.5 to 7 km.

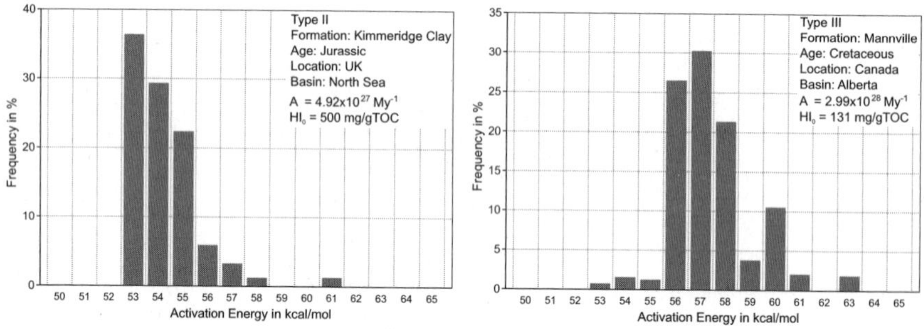

Fig. 4.9. Example of bulk kinetics from Tegelaar and Noble (1994) for a type II and a type III kinetic model. A unique frequency factor is assumed

The relative petroleum masses y can be converted into real masses as follows.

$$m_p = \mathrm{TOC}_0\,\mathrm{HI}_0\,V\,(1-\phi)\,\rho_r\,y \tag{4.8}$$

where TOC_0, HI_0 are the initial TOC and HI values V is the considered source rock volume, ϕ is the porosity, and ρ_r is the rock density.

The initial HI value is usually estimated from the kerogen type, while the initial TOC value is often difficult to determine. When measured HI and TOC values from mature source rock samples are available, the initial TOC value can be reconstructed with the following equation (Peters et al., 2005):

$$\mathrm{TOC}_0 = \frac{p\,\mathrm{HI}\,\mathrm{TOC}}{\mathrm{HI}_0(1-\mathrm{TR})(p-\mathrm{TOC})+\mathrm{HI}\,\mathrm{TOC}} \tag{4.9}$$

where $p = 83\,\%$ is the percentage of carbon in generated petroleum, TOC is given in % and HI in mgHC/gTOC. Peters et al. (2005) also give an equation to calculate TR from measured Rock–Eval data and estimated initial values: HI, HI_0, PI and $\mathrm{PI}_0 \approx 0.02$. A better alternative is to take the TR values from

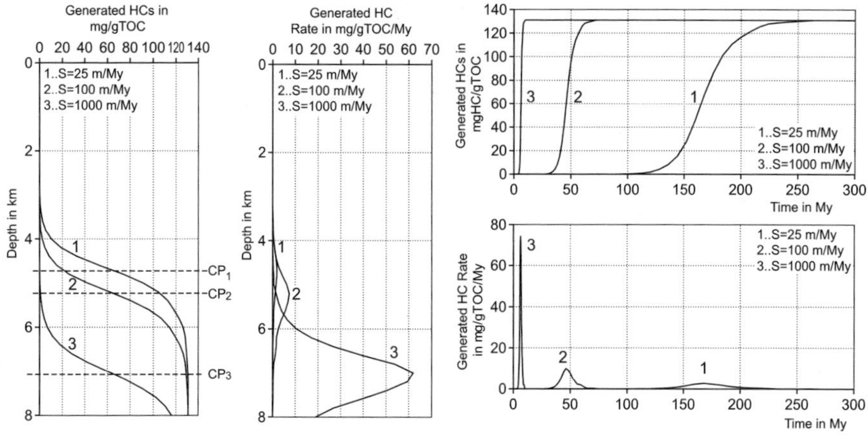

Fig. 4.10. Petroleum generation potential and corresponding rates of the type II bulk kinetics of Fig. 4.9. In each diagram three different sedimentation rates are modeled. The critical points (CP) of the curves correspond to TR=0.5

a simulation with arbitrary HI_0 and TOC_0 as the TR is not dependent on the initial HC mass.

4.3.2 Oil–Gas Kinetics

Oil–gas kinetics are two component models. The gas component lumps together the lighter HCs ($C_1 - C_5$) and the oil components comprehend all heavier HCs (C_{6+}). Oil and gas are almost equal to the liquid and vapor phases at surface conditions. They are very different from the liquid and vapor phases at in–situ conditions, i.e. during petroleum migration and in reservoirs. A two component — two phase model is well established in Darcy type fluid flow models by the name "Black Oil model" (Secs. 5.3, 6.3.5). The reaction scheme describes the two primary cracking reactions from kerogen X into oil Y and gas Z and secondary cracking from oil to gas, each described by a set of first order parallel reactions:

$$X_{1i} \xrightarrow{k_{1i}} Y, \quad Y_i \xrightarrow{k_i'} Z, \quad X_{2i} \xrightarrow{k_{2i}} Z\,. \tag{4.10}$$

The relative masses of generative oil and gas for primary cracking are x_{1i} with $i = 1, \ldots, n_1$, and x_{2i} with $i = 1, \ldots, n_2$, respectively, the initial values x_{01i}, x_{02i} and the number of parallel reactions n_1 and n_2. The total actual and initial generative masses are

$$x = \sum_{i=1}^{n_1} x_{1i} + \sum_{i=2}^{n_2} x_{2i}, \qquad x_0 = x_{01} + x_{02} = \sum_{i=1}^{n_1} x_{01i} + \sum_{i=2}^{n_2} x_{02i} = 1\,. \tag{4.11}$$

The two frequency distributions of the activation energies $p_1(E_i)$, $p_2(E_i)$ for primary cracking of oil and gas are normalized to the initial relative masses x_{01i}, x_{02i}. The total relative amount of generated masses of oil and gas are y and z, respectively, and the frequency distribution over the activation energies for secondary cracking is $p'(E_i)$ with $i = 1, \ldots, n'$. Then, the mass balance of the above kinetics is as follows.

$$\begin{aligned} \frac{\partial x_{1i}}{\partial t} &= -k_{1i}\, x_{1i}, \quad \frac{\partial x_{2i}}{\partial t} = -k_{2i}\, x_{2i}, \\ \frac{\partial y_i}{\partial t} &= p'(E_i) \sum_{j=1}^{n_1} k_{1j}\, x_{1j} - k_i'\, y_i, \\ \frac{\partial z}{\partial t} &= \sum_{i=1}^{n_2} k_{2i}\, x_{2i} + R \sum_{i=1}^{n'} k_i'\, y_i \end{aligned} \tag{4.12}$$

where k_{1i}, k_{2i}, k_i' are the reaction rates for primary kerogen to oil and gas cracking, and secondary oil to gas cracking, respectively. Each generated oil is distributed to portions y_i according to the frequency distribution of the activation energy for secondary cracking.

Secondary cracking of oil to gas yields a reduction of organic mass with coke as a by–product, as described by the reduction factor, where $R = 1$ means no coke and $R = 0$ all coke. R usually ranges between 0.4 and 0.7. The following principal equation results in a reduction factor of $R = 16/28 = 0.57$.

$$\begin{array}{ccccccc} 2\mathrm{CH}_2 & \longrightarrow & 1\mathrm{C} & + & 1\mathrm{CH}_4 \\ 2 \times 14 & \longrightarrow & 12 & + & 16 \end{array} . \tag{4.13}$$

In analogy to bulk kinetics, oil and gas potentials y_p, z_p are defined as $y_p = y\,\mathrm{HI}_0$ and $z_p = z\,\mathrm{HI}_0$ with the unit mgHC/gTOC. It is also necessary to integrate ratio (or percentage) values to describe how the total HI value is subdivided into the HI for kerogen to oil HI_o and the HI for kerogen to gas HI_g reactions.

Relative masses for oil and gas are also used to define principle zones of active oil and gas generation. Herein, the generated masses of oil and gas are compared to the maximum generative masses for oil y_t and gas z_t:

$$y_t = \sum_{i=1}^{n_1} x_{01i}, \quad z_t = \sum_{i=1}^{n_2} x_{02i} + R \sum_{i=1}^{n_1} x_{01i} . \tag{4.14}$$

with

$$\begin{array}{llll} \text{Immature:} & y/y_t < 0.1 & \text{and} & z/z_t < 0.1 \\ \text{Oil Generation:} & y/y_t \geq 0.1 & \text{and} & z/z_t < 0.1 \\ \text{Gas Generation:} & z/z_t \geq 0.1 & \text{and} & z/z_t < 0.9 \\ \text{Overmature:} & z/z_t \geq 0.9 & & \end{array} \tag{4.15}$$

Oil–gas kinetics are mainly used in source rock maturity studies, when petroleum phase properties are of minor importance. Some typical oil–gas kinetics for type I, II and III kerogen after Pepper and Corvi (1995a); Pepper and Dodd (1995) are shown in Fig. 4.11. The calculated oil and gas generation potential versus depth and versus geologic time of the three kerogen types are shown in Fig. 4.12. The effect of the sedimentation rate on the generation potential is illustrated in Fig. 4.13.

4.3.3 Compositional Kinetics

In compositional kinetics, more than two HC–components are considered. Each of the components can be generated by primary cracking from kerogen X, which is described by parallel decomposition reactions. The kinetics for secondary cracking is a triangular scheme where each component can be cracked into all lighter ones. In the formulation of the following kinetic schemes and mass balances, chemical components and distributed activation energies are numbered serially with the indices $i, j, l = 1, \ldots, N$ and $r, s = 1, \ldots, n_i'$, respectively. The petroleum components $\mathrm{Y_i}$ are ordered according to their molar masses, starting with the lowest. The complete reactions scheme encompasses the following N primary and $N(N-1)/2$ secondary reactions:

$$\begin{aligned} \mathrm{X_{ir}} &\xrightarrow{k_{ir}} \mathrm{Y_i} \quad \text{for} \quad i = 1, \ldots, N, \\ \mathrm{Y_i} &\xrightarrow{k'_{ijr}} \mathrm{Y_j} \quad \text{for} \quad i, j = 1, \ldots, N,\; i > j\,. \end{aligned} \tag{4.16}$$

The relative generative component masses of primary cracking are denoted as x_{ir} in analogy to the oil–gas kinetics and the initial values are denoted as x_{0ir}, which are equal to the frequency of the activation energies for primary cracking $p_i(E_r) = x_{0ir}$. This is an array with the dimension of the number of components and discrete activation energies and with

$$\sum_{i=1}^{N} \sum_{r=1}^{n_i} p_i(E_r) = 1 \tag{4.17}$$

(Fig. 4.14). In the most general case of secondary cracking, each petroleum component i can be cracked into any other lighter component j with, $i > j$, which can also be described by a frequency array with dimension of the number of lighter components and activation energies $p'_{ij}(E_r)$ with

$$\sum_{j=1}^{N} \sum_{r=1}^{n_j} p'_{ij}(E_r) = 1 \tag{4.18}$$

for each component i. The mass balance of the coupled reaction is for all $i = 1, \ldots, N$ and $r = 1, \ldots, n_i$

$$\frac{\partial x_{ir}}{\partial t} = -k_{ir}\, x_{ir} \tag{4.19}$$

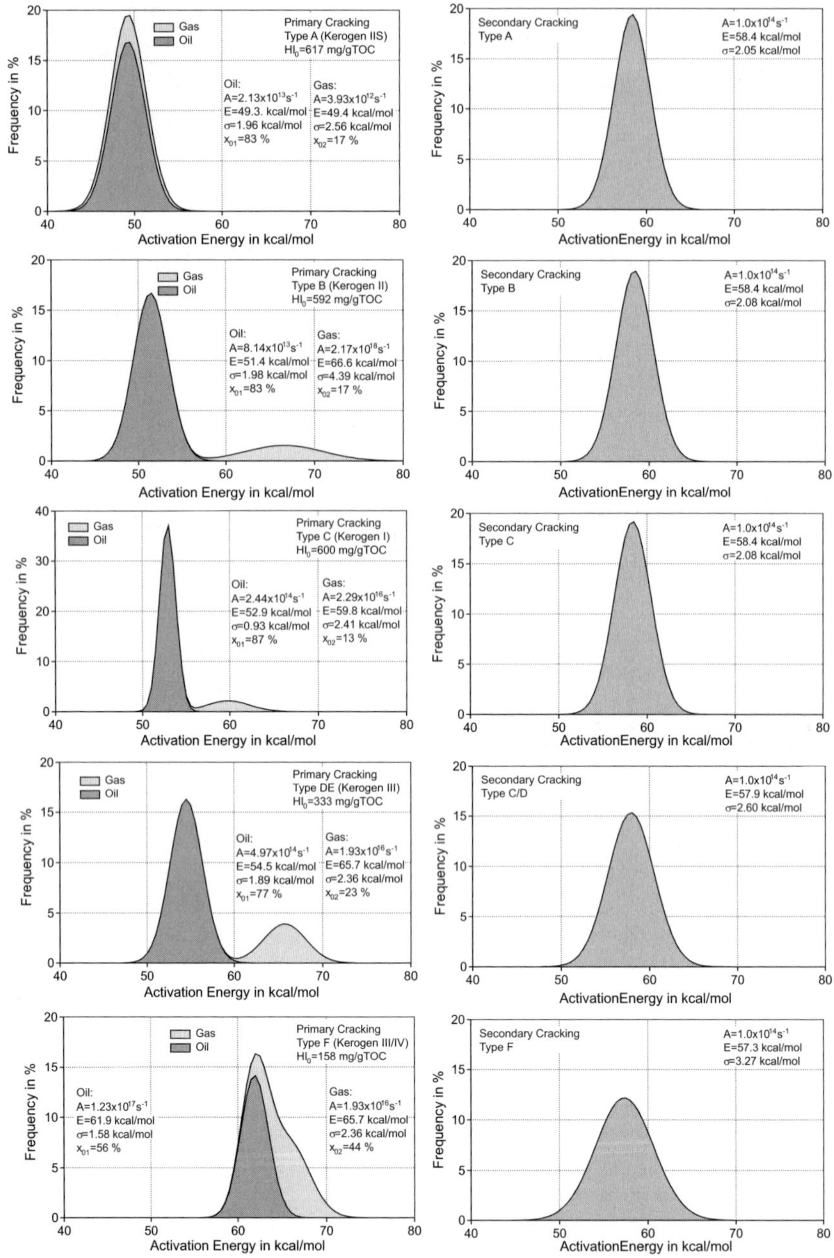

Fig. 4.11. Oil–gas kinetics for typical kerogen types I — III after Pepper and Corvi (1995a); Pepper and Dodd (1995). The kinetics are approximated in this volume by discrete distributions. Note, that the frequency curves for gas are added to the oil curves and that oil and gas are separately normalized to their initial kerogen fraction x_{0i}

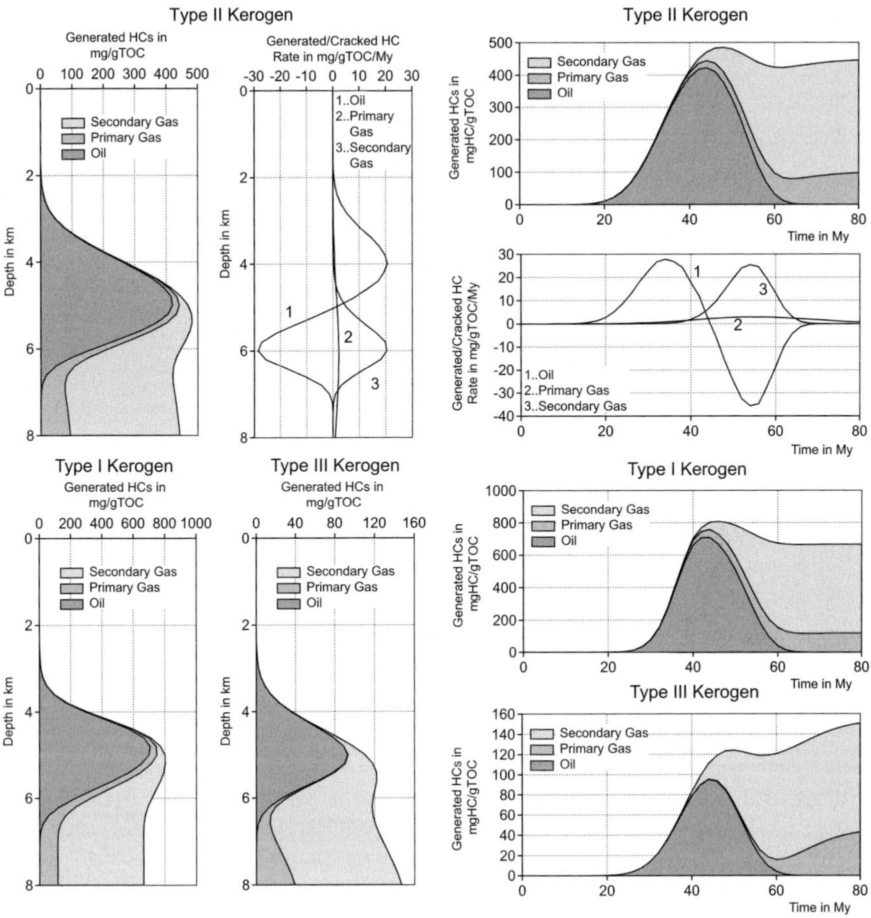

Fig. 4.12. Comparison of kerogen type I, II and III kinetics from Fig. 4.11 with a sedimentation rate of 100 m/My. The gas is further subdivided into primary and secondary cracked gas. The generation potential is calculated for all three kinetic models. Generation rates are also shown for type IIB kerogen

and for all $i = 1, \ldots, N$ and $j = 1, \ldots, (i-1)$ and $r = 1, \ldots, n'_{ij}$

$$\frac{\partial y_{ijr}}{\partial t} = p'_{ij}(E_r) \sum_{s=1}^{n_i} k_{is}\, x_{is} + p'_{ij}(E_r) \sum_{l=i+1}^{N} R_{il} \sum_{s=1}^{n'_{il}} k'_{ils}\, y_{ils} - k'_{ijr}\, y_{ijr} \quad (4.20)$$

where R_{ij} is the reduction factor for the secondary cracking reaction of component i to j. The total masses of the generative kerogen x and the petroleum components y_i are

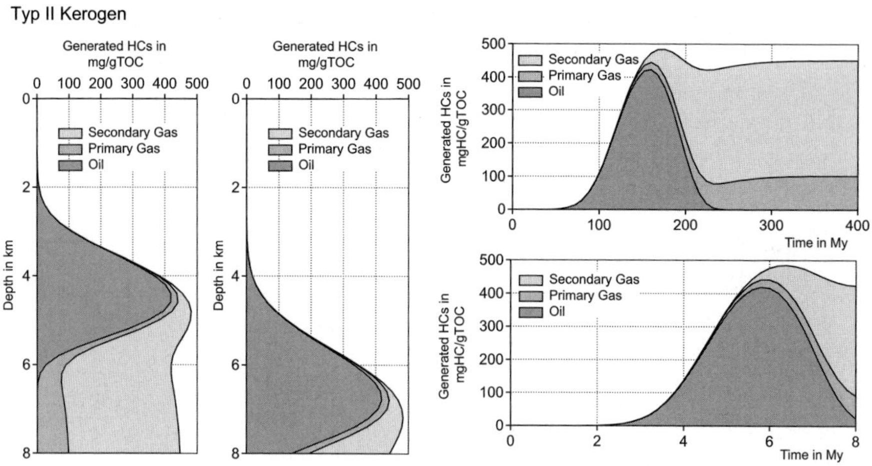

Fig. 4.13. Influence of different sedimentation rates (25 m/My and 1000 m/My) on oil and gas potentials of the type IIB kinetics from Fig. 4.11

$$x = \sum_{i=1}^{N} \sum_{r=1}^{n_i} x_{ir}, \quad y_i = \sum_{j=1}^{i-1} \sum_{r=1}^{n'_{ij}} y_{ijr} \,. \tag{4.21}$$

The three terms of equation (4.20) correspond to generated masses from kerogen, generated masses from the heavier petroleum components and cracked masses into lighter petroleum components. The transformation ratio TR $= 1 - x$ cannot be directly derived from the actual masses y_i, since the mass loss of coke depends on the ratio of primary and secondary cracked components.

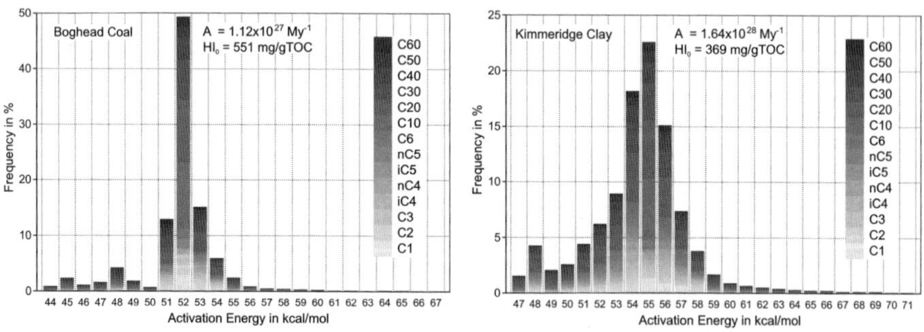

Fig. 4.14. 14 component kinetics for Kimmeridge Clay after di Primio and Horsfield (2006)

A typical simplification of the secondary cracking scheme is that the heavier components are only cracked to methane ($i = 1$) and only with one activation energy. Then all $n'_{ij} = 0$ for $j \neq 1$ and small j's and $n'_{ij} = 1$ for great j's and (4.20) can be simplified as follows.

For methane is

$$\frac{\partial y_1}{\partial t} = \sum_{r=1}^{n_1} k_{1r} \, x_{1r} + \sum_{i=2}^{N} R_i \, k'_i \, y_i, \tag{4.22}$$

and for all other components $i > 1$ is

$$\frac{\partial y_i}{\partial t} = \sum_{r=1}^{n_i} k_{ir} \, x_{ir} - k'_i \, y_i \tag{4.23}$$

where R_i is the reduction factor for secondary cracking of a component i to methane and $k'_i(A_i, E_i)$ is the reaction rate for secondary cracking of the component i based on the pair of frequency factor and activation energy.

Two kinetic schemes are often used in practice, the 4-component approach using boiling point classes and the 14-component scheme which is especially suitable for predicting phase properties such as GOR, API and saturation pressure.

Boiling Point Classes

Espitalie et al. (1988) introduced four HC classes C_1, $C_2 - C_5$, $C_6 - C_{15}$ and C_{15+} for multi–component kinetics, based on pyrolysis data of the components with similar boiling point properties. The classes have been used with small modifications in many publications and databases to describe primary cracking such as Ungerer et al. (1990), Behar et al. (1997) and Vandenbroucke et al. (1999). In all those papers first order parallel reactions are proposed with a unique frequency factor for each reaction. The models for secondary cracking are restricted to methane generation only as described with equation (4.22).

In Fig. 4.15, the kinetics from Behar et al. (1997) for typical type II and type III kerogen are shown, based on anhydrous open and closed pyrolysis. Other kinetics are illustrated in App. G. Note, that the lumped C_{15+} class was split into the three classes, C_{15+} saturates, C_{15+} aromatics and NSOs in the original paper. A separation into higher molecular components is shown in Fig. 4.16 after Abu-Ali et al. (1999). In most boiling point kinetics, the activation energy distributions of the components are approximately normal distributed with increasing deviation and a higher amount of gas generation from type I to type III kerogen, similar to the oil–gas kinetics.

Compositional Phase Kinetics

Di Primio and Horsfield (2006) proposed a 14 component scheme and published data for eight sample kinetic models described with a classification of organic facies as shown in Fig. 4.8. These samples were measured with

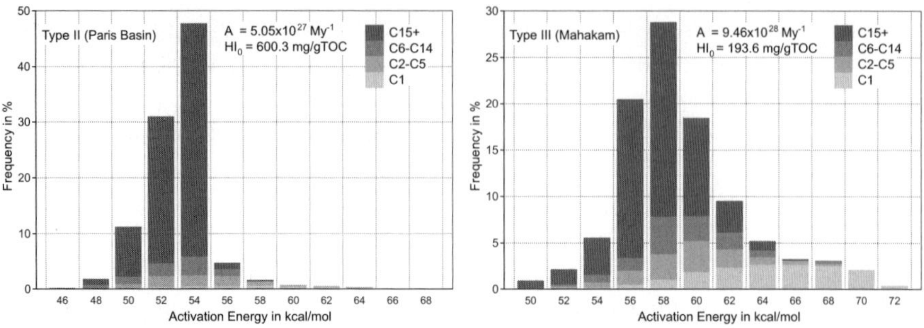

Fig. 4.15. Boiling point class kinetics according to Behar et al. (1997) for type II and type III samples from the Paris basin and the Mahakam delta

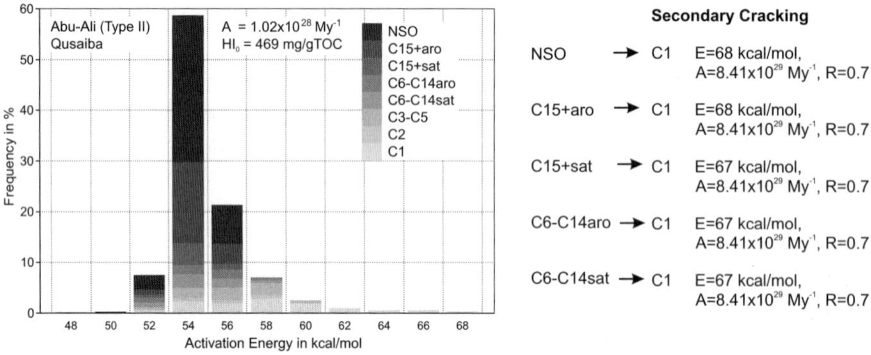

Fig. 4.16. Boiling point kinetics after Abu-Ali et al. (1999) with an additional split of the heavier petroleum components. Simple secondary cracking reactions were added by the authors based on calibration data from projects in the Gulf of Arabia

combined open– and closed–system pyrolysis. A method is also proposed to determine the gas composition from characteristic petroleum properties GOR (gas oil ratio) and saturation pressures.[1] The 14 component scheme takes the following classes into account: C_1, C_2, C_3, iC_4, nC_4, iC_5, nC_5, nC_6, $C_7 - C_{15}$, $C_{16} - C_{25}$, $C_{26} - C_{35}$, $C_{36} - C_{45}$, $C_{46} - C_{55}$ and C_{55+}.[2] The eight example kinetic models for primary cracking are shown in Fig. 4.14 for Kimmeridge Clay and in App. G for Boghead Coal, Tasmanite Shale, Woodford Shale, Tertiary Coal, Teruel Oil Shale, Toarcian Shale and Brown Limestone.

The PVT behavior of a multi-component kinetic is illustrated in Fig. 4.17 and Fig. 4.18 with the Kimmeridge Clay kinetics. In the open source rock approach no secondary cracking is considered. The generated petroleum masses

[1] Gas composition influences phase behavior and pyrolysis alone cannot reproduce gas composition in natural rocks.

[2] In the following description, the symbols C_{10}, C_{20}, C_{30}, ... are used for the higher molecular weight components $C_7 - C_{15}$, $C_{16} - C_{25}$, $C_{26} - C_{35}$, ..., respectively.

and related maturity increase with depth. The differences in the PVT diagrams and petroleum compositions with increasing maturity and GORs and API densities (Sec. 5) are relatively small, e.g. the GOR changes from 85 to 112 kg/kg and densities from 29.4°API to 29.6°API. The phase diagrams are typical for light oil.

The effect of secondary cracking is considered in the closed source rock system approach. In the example kinetics the components C_{10} to C_{60+} are cracked to methane with a normal activation energy distribution. The generated petroleum type changes from a gas condensate for early generated petroleum to dry gas for late generation with much higher GORs and APIs.

4.4 Thermal Calibration Parameters

4.4.1 Vitrinite Reflectance

The most widely used thermal maturation indicator is the reflectance of the vitrinite maceral in coal, coaly particles, or dispersed organic matter. It increases as a function of temperature and time from approximately $R_o = 0.25\%$ at the peat stage to more than $R_o = 4\%$ at the meta-anthracite stage. Vitrinite is a very complex substance and undergoes a complicated series of changes during pyrolysis. The general reaction leads to the assumption that vitrinite is transformed to residual (modified or mature) vitrinite and some condensate.

$$\text{Vitrinite} \xrightarrow{k_i} \text{Residual Vitrinite} + \text{Volatiles} \tag{4.24}$$

Three models were proposed in the 1980s by Waples (1980); Larter (1988); Sweeney and Burnham (1990) and they are still very popular.

Burnham & Sweeney Model

The model from Burnham and Sweeney (1989); Sweeney and Burnham (1990) uses distributions of activation energies for each of the four reactions: the elimination of water, carbon dioxide, methane and higher hydrocarbons. Each of the reactions is described as a parallel decomposition reaction with a set of discrete distributed activation energies.

$$\begin{aligned} \text{vitrinite} &\xrightarrow{k_{1i}} \text{residual vitrinite} + \mathrm{H_2O}\,, \\ \text{vitrinite} &\xrightarrow{k_{2i}} \text{residual vitrinite} + \mathrm{CO_2}\,, \\ \text{vitrinite} &\xrightarrow{k_{3i}} \text{residual vitrinite} + \mathrm{CH_n}\,, \\ \text{vitrinite} &\xrightarrow{k_{4i}} \text{residual vitrinite} + \mathrm{CH_4}\,. \end{aligned} \tag{4.25}$$

The vitrinite reflectance value is then calculated with the four transformation ratios of the reactions. In a simplified version of the model (Easy–Ro model),

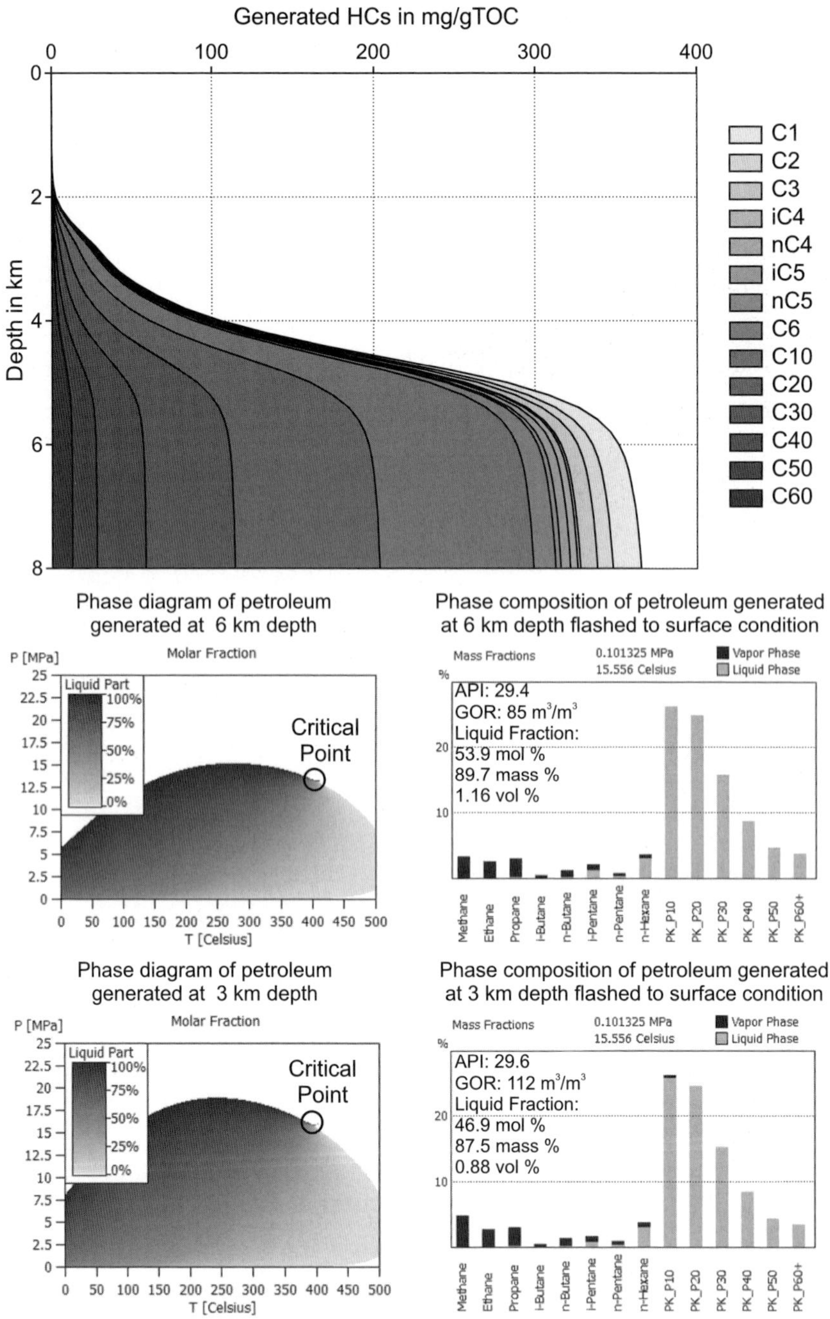

Fig. 4.17. Generated petroleum components and related phase diagrams for Kimmeridge Clay kinetics without cracking (open source rock system)

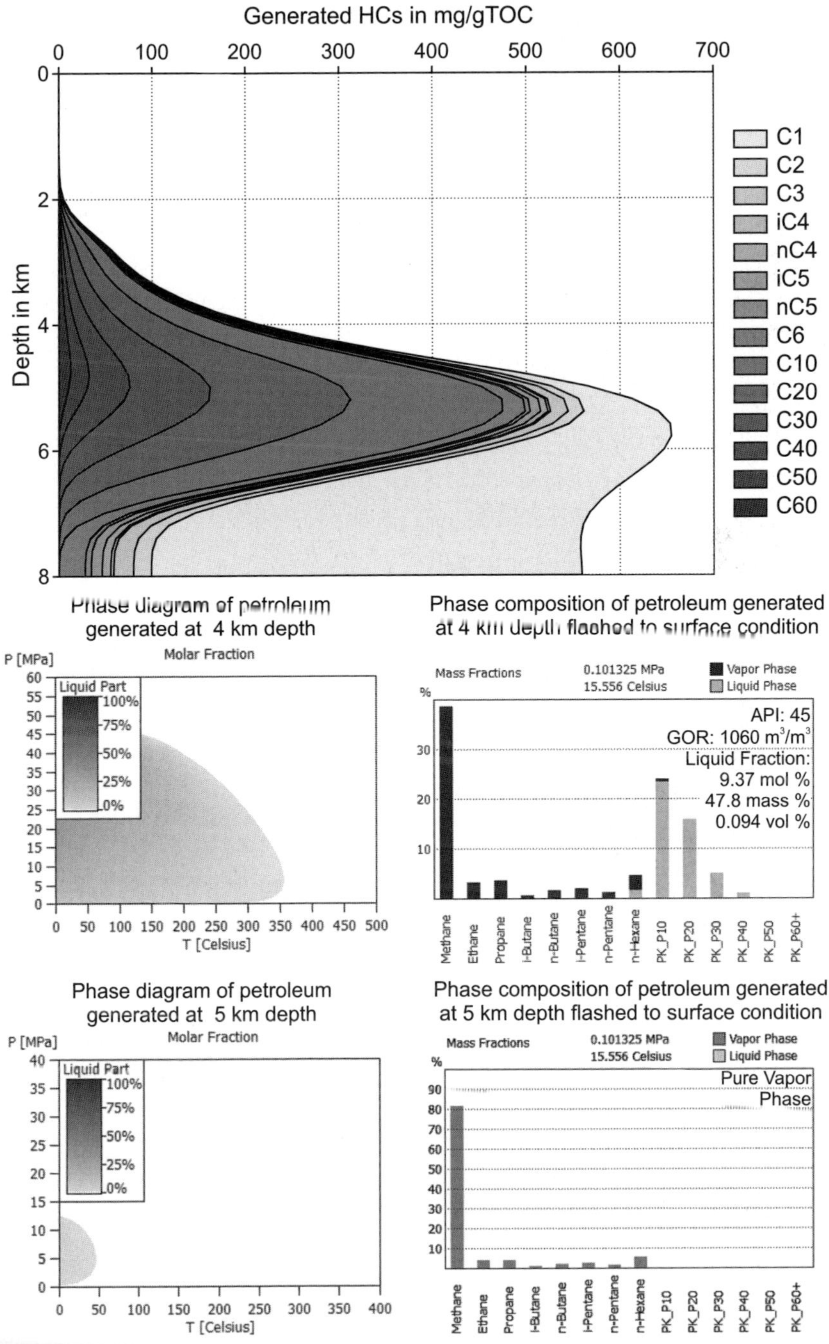

Fig. 4.18. Generated petroleum components and related phase diagrams for Kimmeridge Clay kinetics with cracking (closed source rock system)

the kinetics is approximated by superposition of the four reactions with only one frequency factor. The final distribution of the activation energies is shown in Fig. 4.19. The reflection value of the Easy–Ro model is then exponentially correlated with the TR to the interval [0.20%, 4.66%], as follows:

$$R_o[\%] = 0.20 \left(\frac{4.66}{0.20}\right)^{\mathrm{TR}} . \tag{4.26}$$

Carr (1999) extended the above model to incorporate overpressure u retardation, as there is an expansion of the products due to the generation of volatiles. The pressure dependency is proposed to be included via a modified frequency factor $A(u)$.

$$A(u) = A_h \, e^{-u/c} \tag{4.27}$$

where $A_h = 3.17 \times 10^{26}\,\mathrm{My}^{-1}$ is the original frequency factor from the Easy–Ro model, defined here as the hydrostatic frequency factor, u is the overpressure, and $c = 590\,\mathrm{psi}$ is a scaling factor.

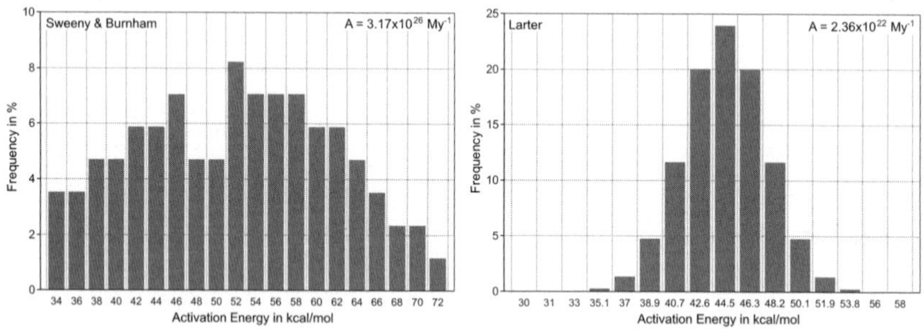

Fig. 4.19. Distributions of the activation energies used in the vitrinite models of Sweeney and Burnham (1990) and Larter (1988)

Larter Model

Larter (1988) utilized quantitative pyrolysis gas chromatographic data of isolated vitrinite kerogen to achieve the concentrations of structurally specific moieties (alkylphenol precursor) in vitrinite as a function of rank. The kinetics describe the phenol precursor loss from vitrinite.

$$\text{precursor} \xrightarrow{k_i} \text{phenol} . \tag{4.28}$$

The pyrolysis data are inverted into a parallel decomposition reaction with normal Gaussian distributed activation energies of mean energy $\bar{E} = 186\,\mathrm{kJ}$, standard deviation $\sigma = 13.0\,\mathrm{kJ}$, and frequency factor $A = 2.36 \times 10^{22}\mathrm{My}^{-1}$. An approximated discrete distribution is shown in Fig. 4.19. The phenol yield is then linearly correlated to the interval [0.45%, 1.60%] with vitrinite reflectance.

$$R_o[\%] = 0.45 + 1.15\,\mathrm{TR} \tag{4.29}$$

where TR is the transformation rate of the parallel reaction. The model limits the reflection value to 1.6% since the study is based on chemical reactions primarily in oil generating zones.

TTI Model

This model was proposed by Waples (1980). It calculates TTI (Time–Temperature–Index) maturity, which is converted to vitrinite reflectance. The definition of the TTI is based on the assumption that the maturity rate of vitrinite almost doubles every 10°C.

$$\mathrm{TTI} = \sum_n (\Delta t_n)\, 2^n \tag{4.30}$$

where the integer n represents a temperature interval with the following scheme:

$$\begin{aligned} n &= -1 \quad \text{for} \quad T = [90°C, 100°C], \\ n &= 0 \quad \text{for} \quad T = [100°C, 110°C], \\ n &= 1 \quad \text{for} \quad T = [110°C, 120°C], \\ n &= 2 \quad \text{for} \quad T = [120°C, 130°C] \ldots . \end{aligned} \tag{4.31}$$

he value Δt_n is the time in My spent by the vitrinite in the corresponding temperature interval. The TTI value is converted to vitrinite reflectance with empirical functions, e.g. with the following rule by Goff (1983).

$$R_o[\%] = 0.06359\,(1444\,\mathrm{TTI})^{0.2012} \,. \tag{4.32}$$

Although this model is only rule based, it is still used.

Vitrinite reflectance is the most common thermal maturity parameter and is generally used for calibration of heat flow history. Herein, paleo-heat flow values as the most uncertain input parameters are changed according to differences between calculated and measured vitrinite reflectance parameters. This requires reliable vitrinite reflectance models and measurements.

Calculated vitrinite reflectance models are shown in Fig. 4.20 for different sedimentation curves. They differ in the initial (immature) value on the surface and further at greater depth. The most commonly used model is the Burnham & Sweeney model.

Vitrinite reflectance curves commonly have offsets at erosional discordances (Fig. 4.21) as the uplifted layers have already experienced a heating period with a corresponding increase in maturity. The offset is typically higher for larger erosional thickness. Thus, jumps in the vitrinite data at a certain depth can be used to estimate the erosion thickness. The maturation process

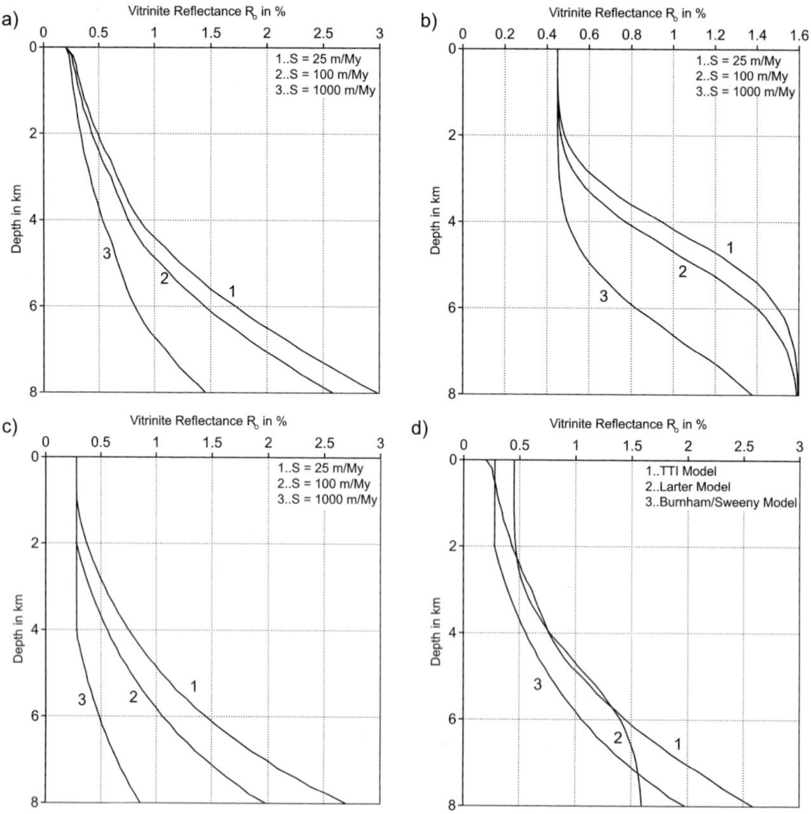

Fig. 4.20. Comparison of calculated vitrine reflectance curves after Burnham & Sweeney (a), Larter (b), and Waples (c) for different sedimentation curves. Figure (d) shows models together with a sedimentation rate of 100 m/My

significantly slows during a hiatus and uplift as there is no further temperature increase, but it does not completely stop.

Vitrinite reflectance can be generally correlated with source rock maturation and the stage of oil and gas generation. The following intervals define empirical relationships for primary oil and gas generation from kerogen of type II and type III: immature ($R_o \leq 0.55\,\%$), early oil($R_o \leq 0.70\,\%$), main oil ($R_o \leq 1.00\,\%$), late oil ($R_o \leq 1.30\,\%$), wet gas ($R_o \leq 2.00\,\%$), dry gas ($R_o \leq 4.00\,\%$), overmature ($R_o \geq 4.00\,\%$). This correlation is not always reliable as the generation histories of different petroleum kinetics can significantly differ from each other. In Fig. 4.22 the correlation of the transformation ratio with the vitrinite reflectance value is shown for the eight different multi–component data sets of Fig. 4.14 and App. G from di Primio and Horsfield (2006) assuming an average sedimentation rate of 100 m/My. The differences

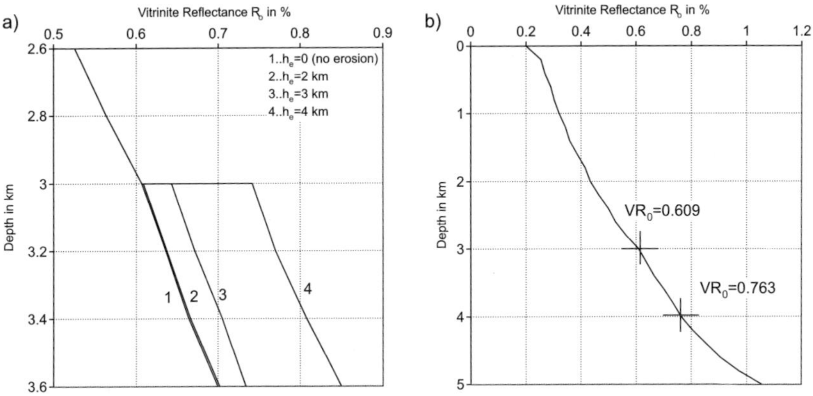

Fig. 4.21. Calculated vitrinite reflectance curves can have significant offsets at erosion horizons (a). A discordance occurs at 3 km depth with eroded thicknesses of 2, 3, and 4 km, which corresponds to a maximum burial depth of the eroded horizon of 2, 3 and 4 km. In case of 4 km erosion, the offset is approximately the difference of the vitrinite reflectance values between 3 and 4 km of uniformly deposited layers (b). The Burnham & Sweeney model is used here and the sedimentation rate is 100 m/My

are obvious, e.g. the R_o interval for main oil corresponds to the TR intervals of (70%, 97%) for Brown Limestone and (5%, 72%) for Alaskan Tasmanite.

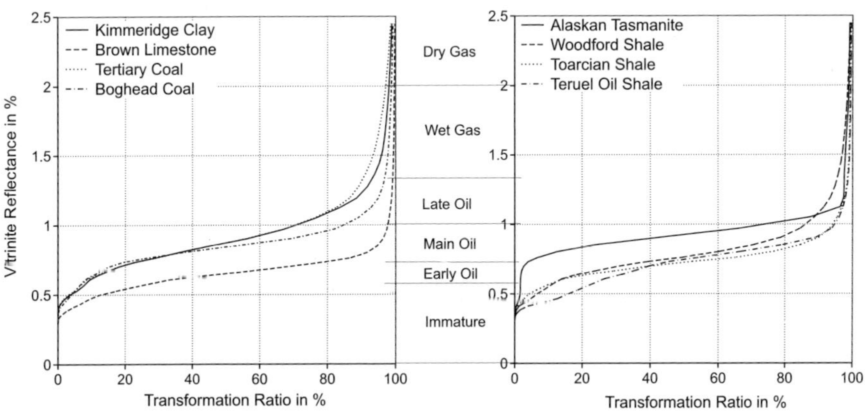

Fig. 4.22. Correlation of transformation ratio and vitrinite reflectance

There are pitfalls for the interpretation of vitrinite reflectance, such as contamination and sampling problems, misidentification of vitrinite, and poor sample preparation, but the main problem is the dependency on kerogen types. CSIRO Australia proposed a more comprehensive method to take into account the fluorescence measurements of multiple macerals (Wilkins et al., 1998). The method is called FAMM (Fluorescence Alteration of Multiple Macerals). The fluorescence intensities of vitrinite, liptinite and inertinite are measured for different maceral fragments. They are plotted on a fluorescence alteration diagram, which can be used to derive an equivalent vitrinite reflectance value.

4.4.2 Molecular Biomarkers

The molecular characteristics of kerogen and petroleum during catagenesis are controlled by the types of deposited organisms, the environmental and preservation conditions and the thermal maturity. Gas chromatography and gas chromatography–mass spectrometry show the relative abundance of individual biomarkers during kerogen pyrolysis (Fig. 4.23). Chromatographic fingerprints and correlations can therefore be used to determine the organic facies types, the thermal maturation state and the degree of biodegradation. Peters et al. (2005) worked out comprehensive lists, which link high concentrations of biomarkers and relative ratios of special biomarkers to the biological origins and environments (App. H). For example, biomarkers in crude oil samples can be used to distinguished an origin from shale or carbonate source rocks. The frequency of some biomarkers can also be used to determine the deposition age and the thermal maturity (App. H). Examples are the isomerization of the C_{20} chiral carbon in steranes and of C_{22} chiral carbon in hopanes and the aromatization of steroid hydrocarbons. These reactions occur before and during the early stages of oil formation. They are commonly described as unimolecular equilibrium reactions with forward and backward reaction rates.

$$\text{Reactant} \overset{k_f;k_b}{\longleftrightarrow} \text{Product}, \quad \frac{\partial x}{\partial t} = -\frac{\partial y}{\partial t} = -k_f\, x + k_b\, y\,. \tag{4.33}$$

The reaction rate of the forward reaction is calculated from a single component Arrhenius law with the activation energy E and the frequency factor A. The ratio between forward and backward reactions $c = k_b/k_f$ is usually independent of temperature and is usually given as an input parameter instead of specifying the backward reaction rate directly. Sample values for A, E and c are experimentally determined and are described by several authors (Sajgo and Lefler, 1986; Mackenzie and McKenzie, 1983) as shown in Table 4.4.2.

Fig. 4.24 shows calculated transformation ratios of the isomerization and aromatization reaction. Similar to all Arrhenius type reactions, the transformation ratio is controlled by the temperature history and therefore by the sedimentation rate. Finally, every equilibrium reaction approaches a transformation ratio equal to $c/(c+1)$. Other molecular biomarker indices are the methylphenanthrene index (MPI) from Radke and Welte (1983), the methyladamantane index (MAI), methyldiamantane index (MDI) from Chen et al. (1996), and the trisnorhopane ratio from Peters et al. (2005).

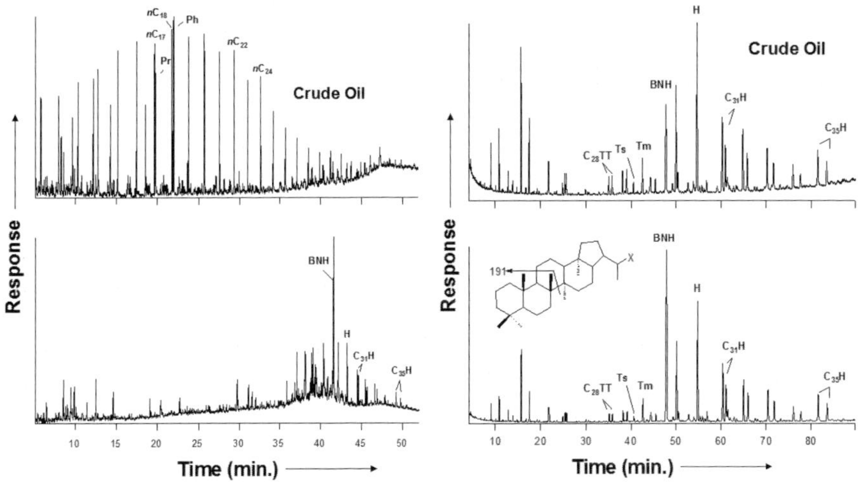

Fig. 4.23. Gas chromatograms

	E in kcal	A in My^{-1}	c
Sterane Aromatization			
Mackenzie and McKenzie, 1983	47.769	5.680×10^{27}	0
Rullkoetter and Marzi, 1988	43.324	1.529×10^{24}	0
Sajgo and Lefler, 1986	29.044	3.345×10^{16}	0
Sterane Isomerization at C_{20}			
Mackenzie and McKenzie, 1982	21.496	1.893×10^{11}	1.174
Rullkoetter and Marzi, 1988	40.363	1.533×10^{22}	1.174
Sajgo and Lefler, 1986	21.878	7.574×10^{10}	1.380
Hopane Isomerization at C_{22}			
Mackenzie and McKenzie, 1982	21.496	5.050×10^{11}	1.564
Rullkoetter and Marzi, 1988	40.124	2.554×10^{22}	1.564
Sajgo and Lefler, 1986	20.971	1.104×10^{12}	1.326

Table 4.1. Kinetic parameters for biomarker models

4.4.3 T_{max} Values

The T_{max} value is an output of the Rock–Eval pyrogram: the oven temperature at the peak S2 (Fig. 4.7). In the Rock–Eval pyrolysis, the contained petroleum (S1–curve) and the remaining kerogen potential (S2–curve) are measured by heating chips of the rock sample until all remaining kerogen is cracked to hydrocarbons.

The remaining kerogen potential x_{ij} can be taken from a simulation. The S2–curve of the Rock–Eval plot can then calculated for each matrix x_{ij} of the remaining kerogen potential and the calculated T_{max} value can be compared to measured T_{max} values from Rock–Eval pyrolysis. The equations (4.3) for bulk kinetics, (4.12) for oil–gas kinetics, (4.19), and for multi–component kinetics

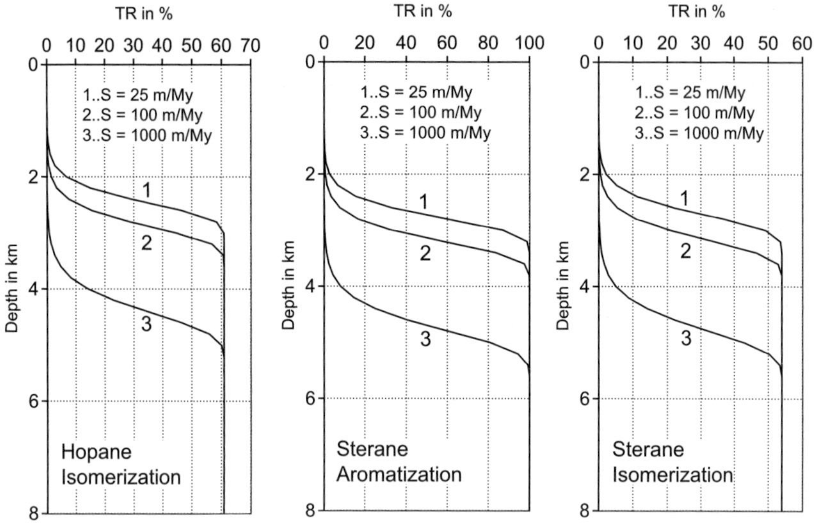

Fig. 4.24. Calculated molecular biomarker ratios with kinetic parameters of Mackenzie and McKenzie (1983) for the isomerization of hopanes and steranes. The conversion curves are shown for three different sedimentation rates S

are used taking a higher heating rate into account. Sweeney and Burnham (1990) proposed to start with 150 °C in time steps of 0.5 °C until the first decrease of the rate occurs. The procedure is more reliable, when the entire S2–curve is calculated, since more complex multi–component kinetics often yield more than one peak. The Rock–Eval curve and T_{max} depend on the heating rate, which is here assumed to be 25 °C/min. Example calculations for the Kimmeridge Clay kinetics (Fig. 4.14) are shown in Fig. 4.25 for different maturation levels.

Measured T_{max} values always increase with higher thermal maturity of the rock sample, since the remaining activation energies are the highest available and higher temperatures are needed to crack the high activation energy bonds. Calculated T_{max} values can drop with increased maturity when different frequency factors are used for the components such as for the oil–gas kinetics of Pepper and Corvi (1995a); Pepper and Dodd (1995) of Fig. 4.11 (kerogen type IIB, Fig. 4.26). Higher frequency factors yield smaller shifts when upscaled from geological to laboratory heating rates (Fig. 4.4). For example, both pairs of activation energies and frequency factors: ($E = 55\,\mathrm{kcal/mol}$, $A = 10^{14}\,\mathrm{s}^{-1}$), ($E = 65\,\mathrm{kcal/mol}$, $A = 2 \times 10^{18}\,\mathrm{s}^{-1}$), yield peak temperatures of $170\,°C$ at a geological heating rate of 10 K/My, but the peak temperatures for laboratory heating rate of 5 K/min are $460\,°C$ and $420\,°C$, respectively. This means, that for the considered Pepper kinetics, the oil peak moves to higher temperatures than the gas peak when transferring to the laboratory scale. Thus, kinetics with different frequency factors for oil and gas (in the very popular

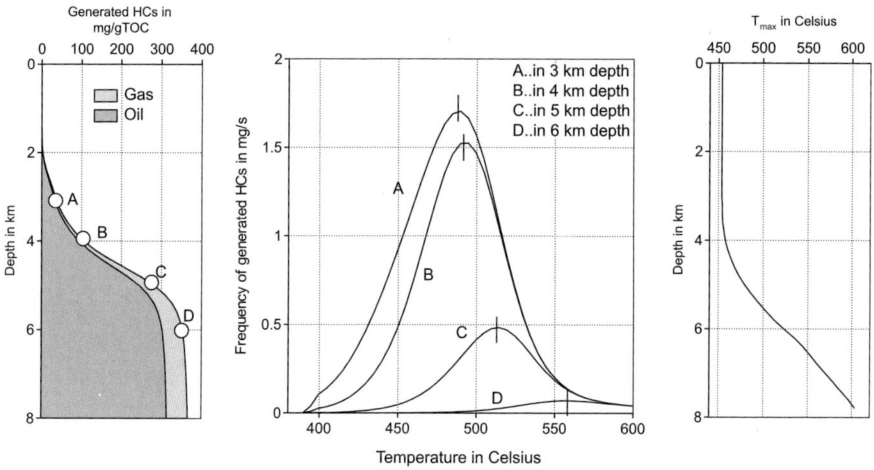

Fig. 4.25. Calculated T_{max} values for the Kimmeridge Clay kinetics of (Fig. 4.14). The left diagram shows the generated masses of gas (C_1 … C_5) and oil (C_6 … C_{60+}). The middle diagram shows the calculated Rock–Eval S2–curves for samples from four different depths with the corresponding T_{max} values. The entire T_{max} versus depth curve is shown in the right diagram

Pepper kinetics the frequency factors differ by four orders of magnitude) are not suitable for the use of calculated Rock–Eval plots.

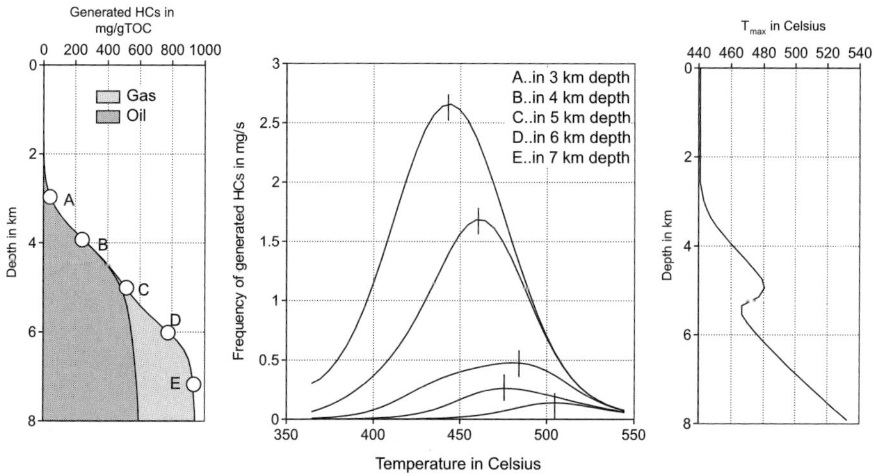

Fig. 4.26. Calculated T_{max} values for the oil–gas kerogen type IIB Pepper and Corvi (1995a); Pepper and Dodd (1995) kinetics of Fig. 4.11

Calculated T_{max} values are very well correlated to the transformation ratios (TR) from the kinetics. In all the multicomponent phase kinetics shown in Fig. 4.14 and App. G, the TR versus T_{max} curves are almost independent of the thermal history. The TR versus T_{max} curves differ by less than 0.1% for sedimentation rates between 20 and 1000 m/My for one organic facies, but the curves vary for different organic facies (Fig. 4.27). Hence, there is no general correlation between vitrinite reflectance and T_{max}.

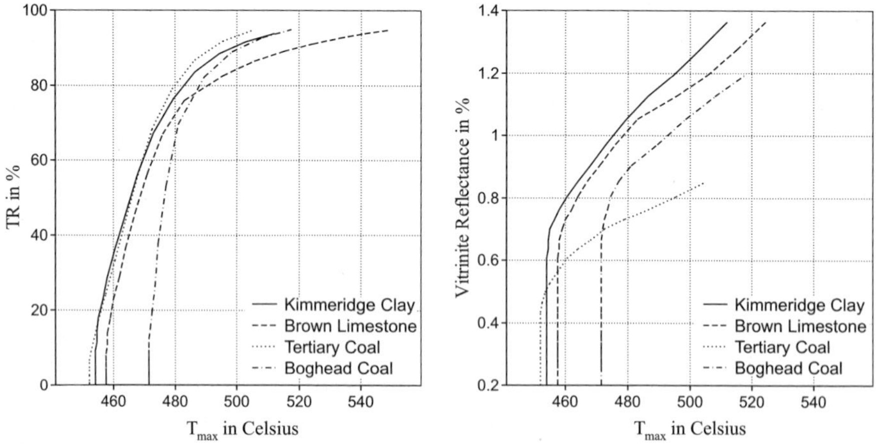

Fig. 4.27. Correlation of vitrinite reflectance and T_{max} for various multi–component kinetic models. The kinetic data for the samples are from Fig. 4.14 and App. G

The production index PI and the hydrogen index HI can also be derived from the calculated Rock–Eval curve and used for calibration of kinetic parameters.

4.4.4 Isotopic Fractionation

Thermally generated gas components become isotopically heavier with increasing maturity, as C^{12}–C^{12} bonds are easier to break than C^{13}–C^{12} bonds. Cramer et al. (2001) analyzed light gases generated from coals by open system pyrolysis, gas chromatography and mass spectrography to measure δ^{13} values of methane, ethane and propane with increasing maturity. The δ^{13} value is defined as follows:

$$\delta^{13} = 1000 \left(\frac{R}{R_0} - 1 \right) = 88990 \, \frac{m^{13}}{m^{12}} - 1000 \tag{4.34}$$

where R is the isotopic ratio, $R_0 = 0.01124$ the standard isotope ratio, and m^{12}, m^{13} are the masses of the C^{12} and C^{13} molecules, respectively. Cramer et al. (2001) proposed to distinguished between C^{12} and C^{13} methane each generated with a slightly different rate described by a constant shift in the activation energy distribution ΔE of 25 cal/mol. The isotopic ratio R of methane can then be tracked in petroleum migration and accumulation analysis. The isotopic fractionation can be used to distinguish between biogenic and thermogenic gases.

4.4.5 Fission–Track Analysis

Fission–tracks are trails of disrupted atoms in the crystal structures of various minerals caused by radioactivity. The trails can be etched using acid so that they become visible under a microscope. For example they are found in apatites and are almost exclusively generated by the spontaneous fission of the uranium isotope ^{238}U. It has a decay constant of $4.92 \times 10^{-18} s^{-1} = 1.55 \times 10^{-4} \mathrm{Ma}^{-1}$, which corresponds to a half life of 4471 Ma. Fission–tracks show a temperature dependent annealing and are thus used to constrain paleo–temperature histories. Introductions to fission–track analysis are given in Gallagher et al. (1998), and Beardsmore and Cull (2001).

Track length distributions can be predicted in a forward modeling approach from temperature histories which can easily be extracted from basin models. These distributions can be compared with sample data for calibration purposes. In the case of good data quality, the temperature history can even be evaluated based on an inversion of forward modeling (Ketcham et al., 2000). This section deals with a short summary of forward modeling of the track length distributions only. More details can be found in Green et al. (1986); Laslett et al. (1987); Duddy et al. (1988); Green et al. (1989a,b); Carlson et al. (1999), and Donelick et al. (1999).

Track length distributions are usually formulated with the reduced track length $r = l/l_0$ with l the confined track length during annealing and l_0 the initial track length. Forward modeling of track length reduction through time is described in Laslett et al. (1987). According to their results it is possible to fit the relationship between time t in seconds and annealing described by r at constant temperature T in Kelvin by a "fanning" fission–track Arrhenius fit of the form

$$\ln t = A(r) + B(r)/T \tag{4.35}$$

with e.g. $A = -28.12$,

$$B(r) = [g(r) + 4.87]/0.000168 \tag{4.36}$$

and

$$g(r) = ([(1 - r^{2.7})/2.7]^{0.35} - 1)/0.35 \,. \tag{4.37}$$

In the case of time dependent temperature $T(t)$ (4.35) can be differentiated and becomes

$$\frac{dr}{dt} = \frac{1}{A'(r) + B'(r)/T(t)} \left(e^{-A(r)-B(r)/T(t)} + \frac{B(r)}{T^2} \frac{dT}{dt} \right) \tag{4.38}$$

with initial condition $r(0) = 1$. The rate equation (4.38) can be solved more easily than the implicit formulation (4.35). However, it must be discretized with caution for a numerical solution.

The uranium isotope ^{238}U is constantly producing new fission tracks with an initial track length distribution of Gaussian form and standard deviation $\sigma = 0.0704\,\mu m$ (Ketcham, 2003). The resulting present day track length distribution is thus a sum of all track length distributions that were generated continuously over time and annealed according to their generation time.

It must be taken into account that the transformation from reduced track length distribution r to observable track length density ρ is biased according to Fig. 4.28. Length distributions must be weighted according to this bias before summing them up.

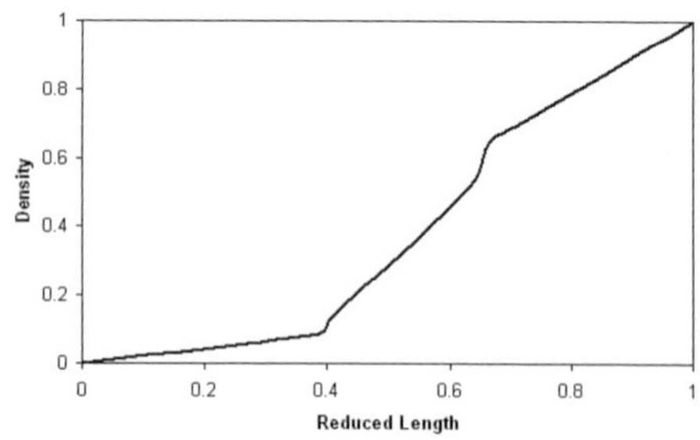

Fig. 4.28. Bias between length r and observable density ρ according to Green (1988); Ketcham (2003)

This bias must also be considered if pooled fission track ages t_a are calculated from reduced track length distributions. According to Ketcham et al. (2000) it is

$$t_a = \frac{1}{\rho_{\mathrm{st}}} \int \rho(t)\, dt \tag{4.39}$$

with the integral limits ranging from sedimentation time to present day. The density ρ_{st} is given by the ratio of present–day spontaneous mean track length to mean induced track length with a typical value of $\rho_{\mathrm{st}} = 0.893$.

Examples of forward modeled track length distributions are shown in Fig. 4.29. Temperature histories can now be calibrated against track length distributions found in real samples. Technically, this can be done with a Kolgomorov–Smirnov test which is designed for comparison of two distributions (Press et al., 2002).

Finally, with the outlined procedure only basic approaches of fission track modeling are covered. Topics, such as the integration of track projections to the crystallographic c–axis into the forward modeling procedure or more advanced models incorporating important factors such as the Cl–content, are not included here. These topics would exceed the capacity of this volume.

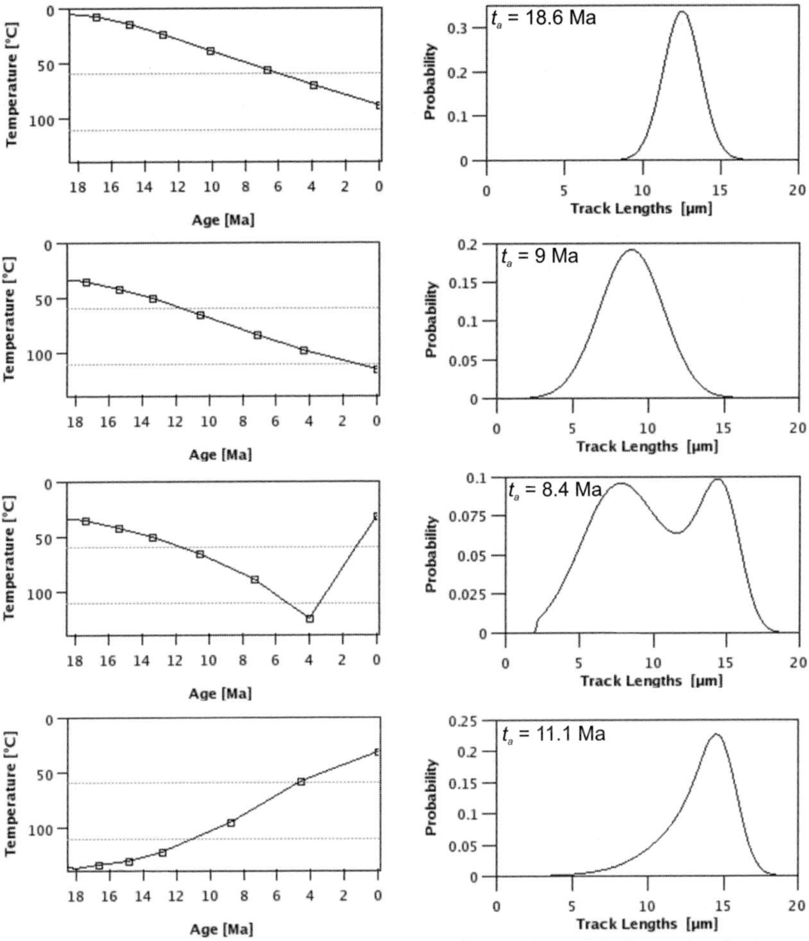

Fig. 4.29. Examples of temperature histories, forward modeled apatite fission track length distributions and pooled ages t_a. The marked temperature interval between 60 °C and 110 °C is called "Partial Annealing Zone"

4.5 Adsorption

Unreacted and inert kerogen can bind the generated petroleum components by adsorption before releasing them into the open pore space of the source rock. The term expulsion is used to specify the amount of petroleum (phases) passing from the source rock to the carrier interface: that means it encompasses all the processes the petroleum molecule is undergoing within the source.[3] The main processes and nomenclature for petroleum migration from source

[3] Generally, the term expulsion is used differently in the literature, for desorption and/or primary migration.

to reservoir are shown in Fig. 4.30 and are considered as a chain of discrete steps.

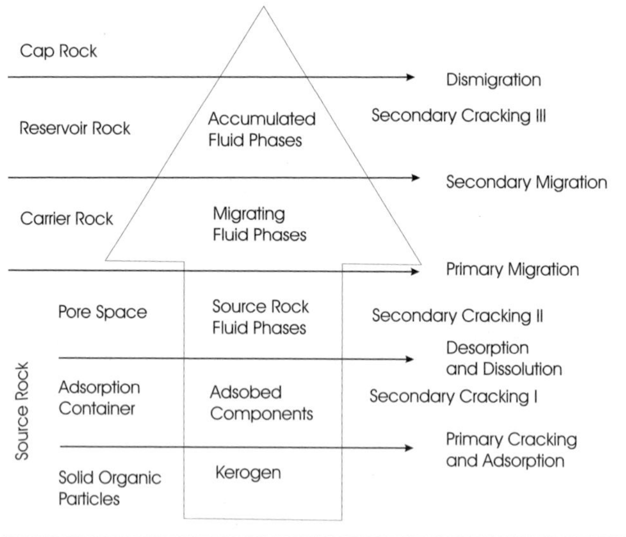

Fig. 4.30. Adsorption and migration processes

- Primary Cracking: kerogen is cracked to petroleum components via first order parallel reactions.
- Adsorption: the primary generated components are adsorbed within the unreacted kerogen and coke in the source rocks. The adsorbed amounts are not included in the petroleum in the pore space. The amount of solid organic particles and coke controls the maximum adsorption amounts. Adsorbed amounts can be further (secondary) cracked following a secondary cracking scheme. The generated coke as a by-product in secondary cracking increases the maximum adsorption amounts.
- Desorption: adsorption amounts can be released into the pore space when the generation amounts exceed the maximum adsorption amounts. The maximum adsorption amounts are therefore modeled as a container that the petroleum must fill completely before entering the pore space.
- Dissolution: the desorbed components are dissolved in the fluid phases. Further transport in the pore space is now controlled by the phase (instead of the component) properties and handled with models for fluid flow.
- Primary Migration: the fluids have to pass through the source rock pore system until their expulsion into the carrier. A critical saturation value (initial oil saturation or endpoints in the capillary entry and relative permeability curves) is used in Darcy type models to quantify the amount

of initial and residual saturation during expulsion (Sec. 6.3). This value therefore acts as a second container for the expulsion of petroleum. In Darcy flow models, the time for the movement through the source network is controlled by permeabilities. Secondary cracking is also modeled for petroleum in the free pore space of source rocks.

- Secondary Migration: all further petroleum transport in carrier rocks, reservoirs and through seals is modeled by Darcy, flowpath and/or invasion percolation models (Chap. 6). There is controversy in the literature about whether the secondary cracking scheme in reservoirs is different from or similar to that in source rocks. Most basin modeling programs allow the use of three different (adsorption, source and reservoir) secondary cracking schemes.

Quantification of the adsorbed masses is the target of different adsorption models. Herein it is assumed that the adsorption capacity is proportional to the amount of available kerogen, which is given by a crackable or reactive and an inert fraction of kerogen

$$m_{\text{ker}} = m_{\text{ker,reac}} + m_{\text{ker,inert}} \, . \tag{4.40}$$

Initially, before cracking started, a fraction r of $m_{\text{ker},0}$ is assumed to be inert and a fraction $1 - r$ is reactive. The inert fraction can directly be evaluated to

$$m_{\text{ker,inert},0} = \frac{r}{1-r} m_{\text{ker,reac},0} \, . \tag{4.41}$$

Substitution of this result and of $m_{\text{ker,reac}} = m_{\text{ker,reac},0}(1 - \text{TR})$ into (4.40) yields

$$m_{\text{ker}} = m_{\text{ker,reac},0} \left(\frac{1}{1-r} - \text{TR} \right) \, . \tag{4.42}$$

with $m_{\text{ker,reac},0} = \text{TOC}_0 \, V(1-\phi) \, \rho_r \text{HI}_0$ where V is the bulk volume, ϕ the porosity and ρ_r the rock density.

Hydrocarbons can adsorb on the surface of kerogen. The hydrocarbons are called adsorbate and the kerogen adsorbent. This process is typically described by three different approaches (en.wikipedia.org/wiki/Freundlich_equation, en.wikipedia.org/wiki/Adsorption):

The Freundlich equation is an empirical equation of the form

$$m_{\text{ads}} = m_{\text{ker}} \, K_F \, c^{1/n} \tag{4.43}$$

where c describes the molar concentration of the adsorbate in the solution and with constants K_F and n for a given combination of adsorbent and adsorbate at a particular temperature.

The Langmuir equation is based on a kinetic approach of adsorbate molecules which adsorb on the surface of the adsorbent. It is further assumed that the adsorbent is uniform, that the adsorbate molecules do not interact and that only one monolayer of adsorbate is formed. This finally yields

$$m_{\text{ads}} = \frac{\text{m}_{\text{ads,max}} K_L c}{1 + K_L c} \tag{4.44}$$

with $\text{m}_{\text{ads,max}}$ as the maximum mass which can be adsorbed in one monolayer on the kerogen surface and $K_L = k_{\text{ads}}/k_{\text{des}}$ as the ratio of reaction rate constants for adsorption and desorption. Obviously, K_L is strongly dependent on the temperature. It may be possible to model k_{ads} and k_{des} with an Arrhenius law.

The BET theory is an enhancement of (4.44) for the adsorption of multiple layers of adsorbates. It becomes

$$m_{\text{ads}} = \frac{\text{m}_{\text{ads,max}} K_B c}{(1-c)\left[1 + (K_B - 1)c\right]} \tag{4.45}$$

with K_B similar to K_L and $m_{ads,max}$ again as the mass of one monolayer.

The Langmuir equation can easily be extended for multiple types of molecules which might be adsorbed with the assumption, that each molecule of each type occupies the same space on the adsorbent. The adsorbed amount of component i becomes

$$m_{\text{ads},i} = \frac{\text{m}_{\text{ads,max,i}} K_{L,i} c_i}{1 + \sum_k K_{L,k} c_k} \tag{4.46}$$

with $K_{L,i}$ the Langmuir factor, $\text{m}_{\text{ads,max,i}}$ the mass of one monolayer, and c_i the concentration of species i.

The mass of one monolayer can be estimated to be

$$\text{m}_{\text{ads,max}} = g \, m_{\text{ker}}^{2/3} \tag{4.47}$$

with a geometrical factor g. It can be assumed that the kerogen is distributed randomly in the sediments so that the exponent 2/3 should be replaced by an exponent from a fractal theory. A value about 1 can be interpreted as an adsorption potential proportional to the number of kerogen atoms, which is consistent with the Freundlich equation.

In practice, temperature and adsorption kinetics are often ignored. Adsorbed masses are equal to maximum adsorption amounts if enough adsorbates are available. The maximum adsorbed mass is assumed to be proportional to the mass of kerogen, which encompasses the actual generative and the inert kerogen. Two different basic approaches are commonly in use.

Pepper and Corvi (1995b) proposed adsorption of each species independent from other species (Fig. 4.31). The following formulation is often used:

$$m_{\text{ads,max},i} = a_i \, m_{\text{ker}} \tag{4.48}$$

where a_i is the adsorption coefficient and $m_{\text{ads,max},i}$ the (maximum) adsorption amount of component i.

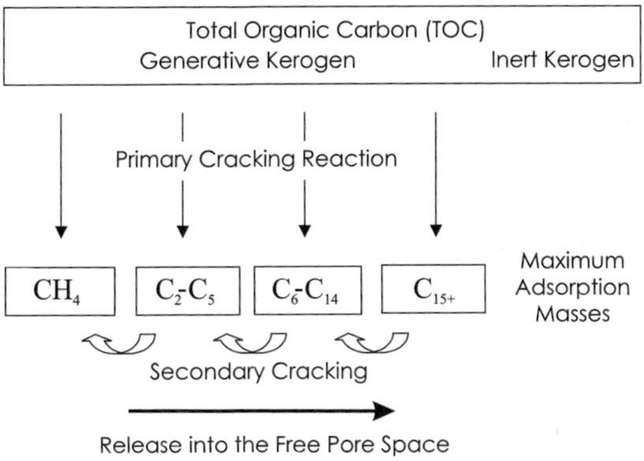

Fig. 4.31. Adsorption model for independent species

Alternative adsorption models exist. Again the total adsorption mass is proportional to the available kerogen mass. The calculation is similar to that of the adsorption model for independent species with only one adsorption coefficient a as

$$m_{\text{ads,max}} = a\, m_{\text{ker}} \; . \tag{4.49}$$

In a common formulation with interacting component adsorption, the adsorption of each component is assumed to be proportional to the relative mass concentration of the adsorbate in the solution. This yields for $m_{s,i}$ as the mass of component i in solution and $\Delta m_{s,i}$ as a small amount of component i which is currently adsorbed

$$\frac{\Delta m_{s,i}}{w_i m_{s,i}} = \frac{\Delta m_{s,k}}{w_k m_{s,k}} \; . \tag{4.50}$$

It is $m_{s,i} = m_{0,i} - m_{a,i}$ with $m_{0,i}$ the total initial petroleum adsorbate mass and $m_{a,i}$ the already adsorbed mass of component i. The factor w_i is an extra weight factor which is introduced for different adsorption amounts of different components. Assuming that there are three components available with weights 1:2:3, then the components are adsorbed in the ratio of 1:2:3 of the unadsorbed relative masses, as long as there are enough generated masses available or until the surface is completely covered (Fig. 4.32).

Equation 4.50 can be rewritten in differential form and integrated for the total adsorption mass of each component. It becomes

$$\left(\frac{m_{s,i}}{m_{0,i}}\right)^{1/w_i} = \left(\frac{m_{s,k}}{m_{0,k}}\right)^{1/w_k} \; . \tag{4.51}$$

If $m_{s,i}$ is known, $m_{s,k}$ can be calculated. Therefore only the dissolved mass $m_{s,i}$ of one component has to be known. Only two cases have to be considered for the full solution. First, $\sum_i m_{0,i} \leq m_{\text{ads,max}}$ which is trivial, because

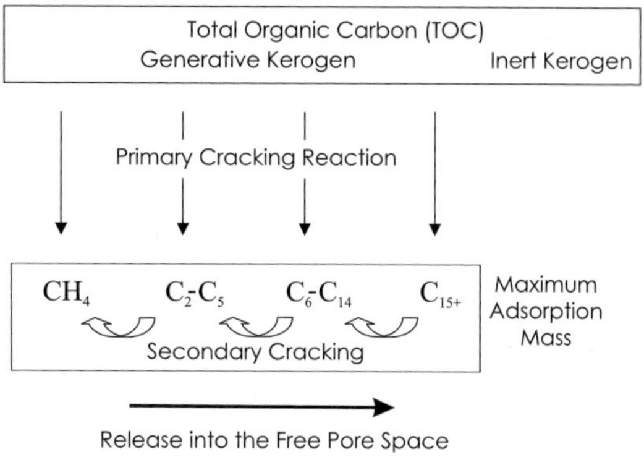

Fig. 4.32. Adsorption model for interacting species

everything is adsorbed. Second, $\sum_i m_{0,i} > m_{\text{ads,max}}$. For $m_0 = \sum_i m_{0,i}$ this yields the condition

$$m_0 - m_{\text{ads,max}} = \sum_i m_{s,i} = \sum_i m_{0,i} \left(\frac{m_{s,k}}{m_{0,k}} \right)^{w_i/w_k} \tag{4.52}$$

for one fixed reference component k. With $m_s = \sum_i m_{s,i} = m_0 - m_{\text{ads,max}}$ one may alternatively write

$$m_s = \sum_i m_{0,i} x_k^{w_i/w_k} \tag{4.53}$$

with $x_k = m_{s,k}/m_{0,k}$. The root x_k of this equation can easily be found, e.g. with the Newton–Raphson method.

Note that the values of the weights are only important in the ratios relative to each other. For comparison with the adsorption model for independent species they can be scaled according to $aw_i = a_i$ so that the total adsorption becomes similar (Fig. 4.33).

Very often there is no data available for a differentiation between generated and expelled compositions. The generation kinetic is thus sometimes calibrated against expulsion. For prevention of a compositional error adsorption must in such a case be identical to the composition of the already generated petroleum. However, wrong results may occur when adsorbed amounts are released later.

4.6 Biodegradation

In shallow reservoirs microbes can transform or biodegrade crude oil. The resulting oil is biodegraded. Its density and viscosity is increased and its composition has changed significantly. The process is assumed to occur mainly at

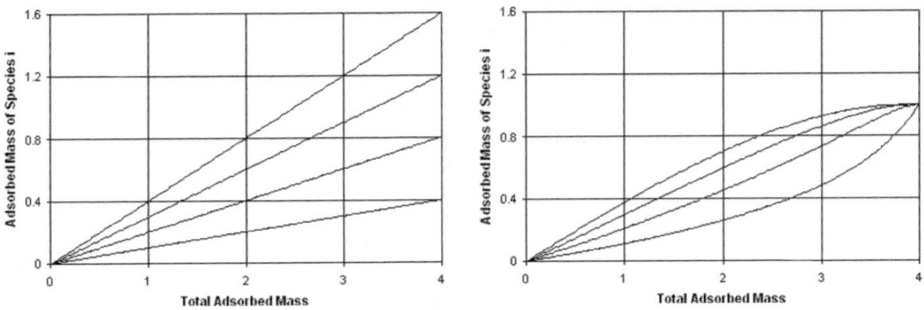

Fig. 4.33. Adsorption model for four independent species on the left and with interaction on the right. The species are adsorbed with a relative weight of 4 : 3 : 2 : 1. The initial masses $m_{0,i}$ are distributed according to these relative weights on the left and according to equal values of one on the right. In a different case with equal initial mass amounts on the left, the adsorption is additionally limited by the available amount of each species. With increasing total adsorbed mass, a bigger part of the remaining species must be adsorbed. However, all cases are becoming the same for small adsorption masses

the free oil water contact (OWC). It is further assumed to be mainly dependent on the size of this area, the temperature, the composition of the oil, the filling history and the supply of additional nutrients which are also needed by the microbes (Fig. 4.34). When the temperature rises above a threshold of about 80°C almost all biodegradation stops. This effect is called paleopasteurization.

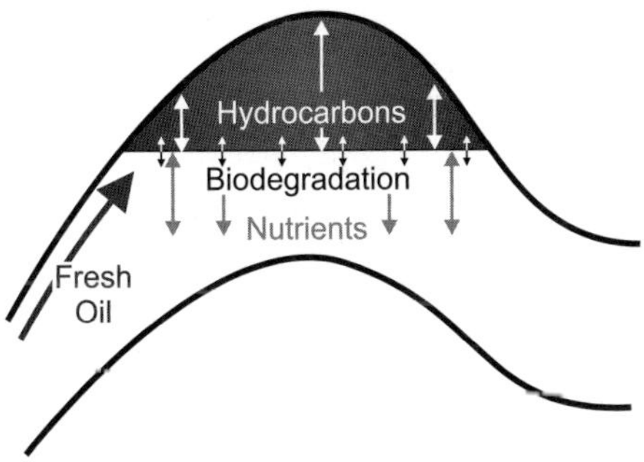

Fig. 4.34. Scheme for biodegradation at oil water contact

The biodegradation process is not fully understood and quantitative approaches are at best approximations. For example, the biodegradation rate of a lumped compound class is often assumed to follow a simple decay law of the form

$$\frac{dm_i}{dt} = -r_i r A \tag{4.54}$$

with m_i as the mass of components in compound class i, r_i the relative degradation rate of class i, r the total degradation rate and A the size of the free OWC area (Fig. 4.35). The total degradation rate r can be approximated by

$$r = \begin{cases} r_{\max}, & T \leq T_1 \\ r_{\max} \exp\left(-\frac{(T-T_1)^2}{2\sigma^2}\right), & T > T_1 \end{cases} \tag{4.55}$$

(Blumenstein et al., 2006; Krooss and di Primio, 2007). The parameters T_1 and σ describe the form of the temperature dependency of the model. Temperature T_1 defines the limit above and σ the temperature range over which biodegradation decreases. Biodegradation is assumed to have a maximum rate $r_{\max}$ below T_1, which is dependent on the composition of the oil and the individual degradation environment particularly the supply of nutrients. Typical values are $T_1 = 50°C$, $\sigma = 10°C$ and $r_{\max} = 100\,\mathrm{kg/m^2/My}$ (Fig. 4.36) and relative degradation rates r_i around 0.8 for light components up to C_7 and down to 0.1 for e.g. C_{56+} (Table 4.2).

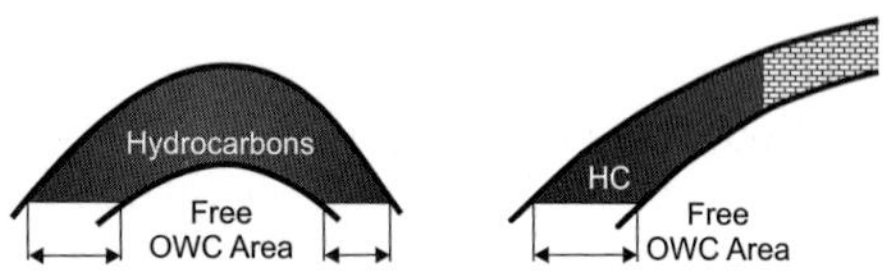

Fig. 4.35. Structural (left) and a stratigraphic (right) traps with their free oil water contact (OWC) area. Free OWC areas can easily be calculated to a high degree of accuracy in reservoir analysis based on flowpath models (Chap. 6.5)

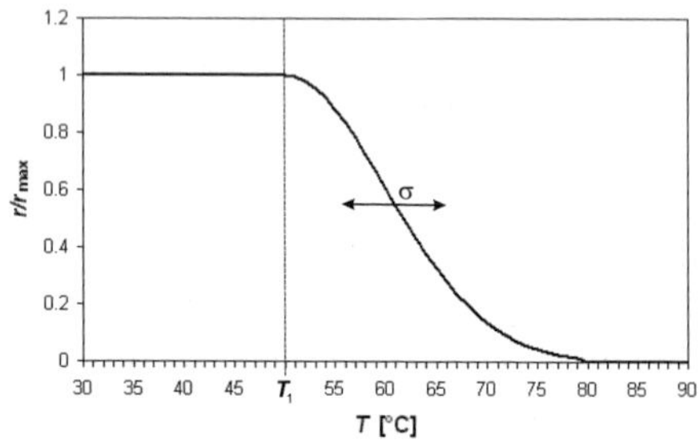

Fig. 4.36. Biodegradation rate according to equation (4.55)

Not every chemical component of oil is degradable. Hence each compound class representing a group of components, is often separated into degradable and non degradable fractions. This fraction of biodegradability is usually assumed to be about 1.0 for all light components and decreasing for heavier compound classes starting at C_6 down to almost 0 for C_{56+} (Table 4.2). As a product of biodegradation, methane is obviously an exception with degradability 0. The concept of degradability fractions limits the degradation within each compound class. This ensures that the mass decrease according to equation (4.54) stops at reasonable values.

Compound Class	Degradation Rate r_i	Degradability
Methane	0.00	0.00
Ethane	0.40	1.00
Propane	1.00	1.00
i–Butane	0.80	1.00
n–Butane	1.00	1.00
i–Pentane	0.70	1.00
n–Pentane	0.80	1.00
n–Hexane	0.80	1.00
C_{7-15}	1.00	0.80
C_{16-25}	1.00	0.60
C_{26-35}	0.80	0.40
C_{36-45}	0.30	0.20
C_{46-55}	0.20	0.10
C_{56+}	0.10	0.02

Table 4.2. Example of degradation rates and degradability fractions according to Blumenstein et al. (2006); Krooss and di Primio (2007)

Biodegradation is assumed to follow an overall reaction scheme such as

$$4\,C_{16}H_{34} + 30\,H_2O \longrightarrow 49\,CH_4 + 15\,CO_2 \tag{4.56}$$

(Zengler et al., 1999; Larter, 2007). The products of biodegradation are primarily methane and carbon dioxide. Other products are generated in rather small amounts and are therefore neglected in quantitative modeling although degraded oils with residual nitrogen or sulfur containing products are found. The generated amount of methane and carbon dioxide is estimated by stoichiometry and a balance of carbon atoms.

An example of biodegradation is shown in Fig. 4.37 and Fig. 4.38. The API decreases by 9°API. All components from ethane up to n–hexane are fully degraded and additional methane is produced. The enhancement of heavy components is a relative effect. Biodegradation generally reduces the amount of oil. Lighter components degrade faster and hence the relative amount of heavier components increases. Besides this effect, outgassing is also taken into account. The class C_{7-15} is exceptional. It is not fully degraded but only slightly reduced in quality.

4.7 Source Rock Analysis

The source rock analysis is an important aspect of petroleum system analysis. The information about generated and expelled petroleum masses, peak gen-

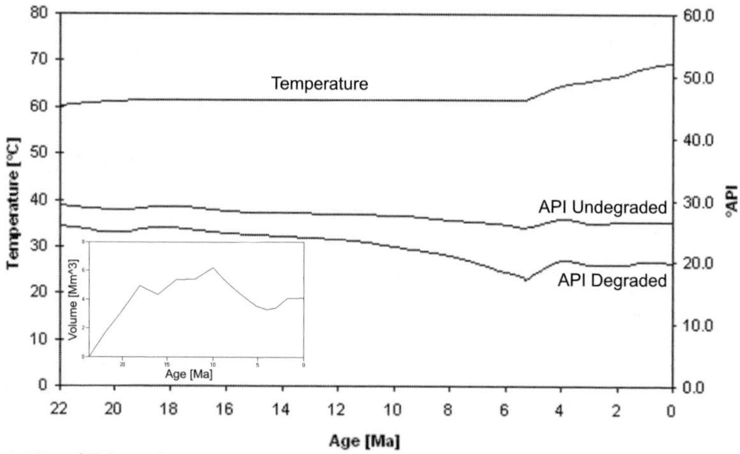

Fig. 4.37. API and temperature of accumulated petroleum with and without biodegradation according to (4.54). The filling history with spilling and partial refilling is shown at the bottom right. It started at 25 Ma. The first API values are available for the following time step at 22 Ma

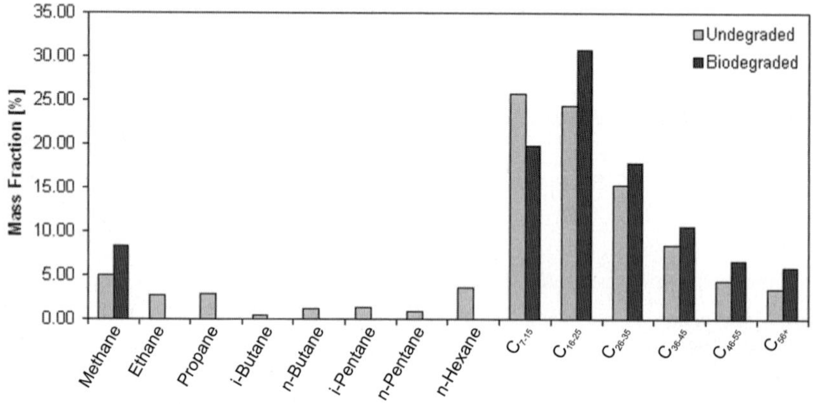

Fig. 4.38. Composition of undegraded and degraded oil at present day. Note that the molar fraction of methane changed from 34% to 61%

eration, and expulsion times already gives a first idea about possible reservoir charge even if migration is not yet considered in detail. The most important input data and results for a source rock analysis are illustrated in Fig. 4.39 and Fig. 4.40 for a model of the Northern Campos basin. Herein, three source rocks are defined in the Bota, Coquina and Lagoe Faia formations, each with primary and secondary kinetics. The source rock kinetics can be based on measured sample data or equivalent default kinetics from a database. Kinetics related adsorption factors should also be defined for each HC component. The kerogen content and quality are defined with TOC and HI maps, which are specified for all three source rocks. Usually, TOC and HI values from well data are used to construct interpolation maps.

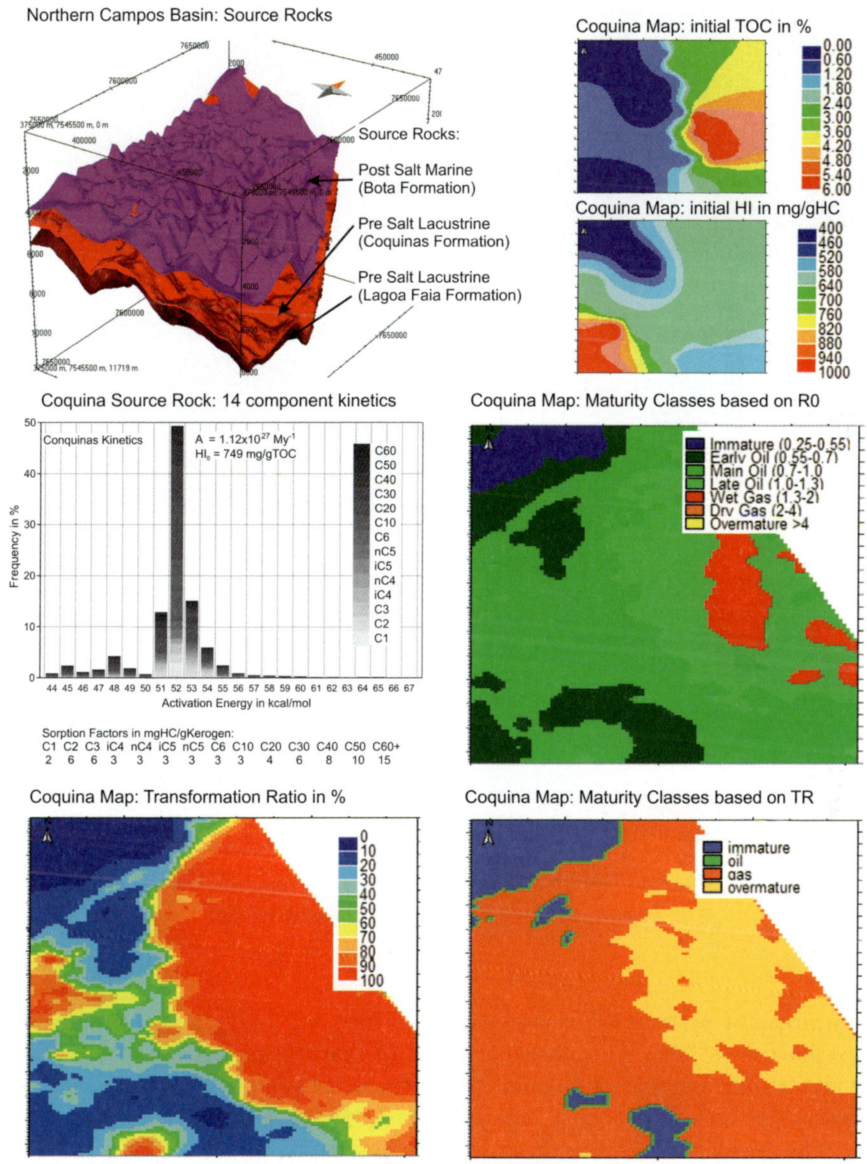

Fig. 4.39. Example: Campos Basin, Brazil

The calculated transformation ratios at present day and paleo-times show how and where most of the petroleum is generated. These can be converted

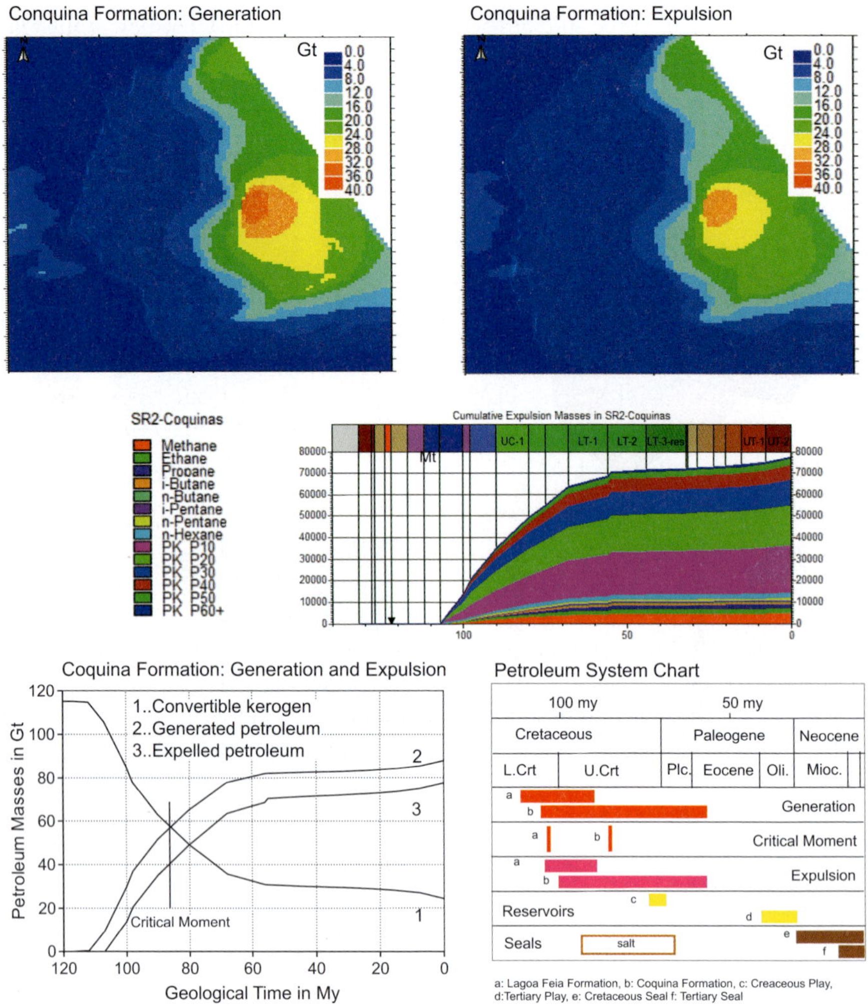

Fig. 4.40. Example: Campos Basin, Brazil

into areas of maturation classes for immature, oil and gas generating, and overmature regions. The principal areas of maturation can also be derived from vitrinite reflectance calculations without consideration of the generated masses, the source rock thickness and the TOC and HI values.

The petroleum systems chart summarizes the main periods of generation and expulsion for each source rock for comparison with the periods of reservoir rock deposition and the occurrence of effective seals. These periods are derived from the generation curves through geologic time and specification of the critical moment of the petroleum system. The petroleum system chart shows whether the petroleum system elements occur in the correct geological sequence and allow accumulation and preservation of petroleum.

Summary: Petroleum generation and coal formation mainly proceed during catagenesis. The maturation processes are time and temperature controlled. Petroleum generation models are generally used for source rock analysis. The amounts and the timing of generation and expulsion of the petroleum system are studied.

Petroleum is generated from kerogen, which can be subdivided according to the van Krevelen diagram into type I — IV corresponding to its hydrogen and oxygen content. Oil generation is often modeled with two–component kinetics for oil and gas. The kinetic data are derived from Rock–Eval or similar open system pyrolysis methods. The quantity and quality of the organic matter which is available for petroleum generation is described with the total organic content (TOC) and the hydrogen index (HI), respectively.

A different and more detailed characterization of the organic matter is based on the generated petroleum product and from five organic facies types. The corresponding data are derived from combined open and closed system pyrolysis. The petroleum composition is usually described using 14-component kinetics, which have to be combined with fluid models to obtain petroleum properties such as density and viscosity.

Quantitative analysis of petroleum generation is based on chemical kinetics for the primary cracking of kerogen and the secondary cracking of petroleum. Chemical kinetics are formulated with mass balances and distributed reactivity kinetics. These kinetic schemes encompass sequential and parallel reactions that are approximated as first order Arrhenius type reactions. Herein, the reaction rate depends exponentially on temperature, described with the activation energy and the frequency factor. The primary cracking of each petroleum component is modeled with an activation energy distribution. Secondary cracking reactions are usually simplified with the cracking of the heavier components into methane and coke, also formulated with an activation energy distribution. The main results of the analysis are the generation potentials and the generated mass of each petroleum component through geologic time.

The most widely used thermal maturation indicator is the reflectance of vitrinite. Other parameters include molecular biomarkers, T_{max} values from Rock–Eval pyrolysis and the annealing of fission tracks. Most of the corresponding kinetic models are also based on distributed kinetic reactions. Thermal maturity parameters can be used to calibrate heat flow histories. They also can be correlated to the maturity of petroleum.

Special kinetic models are used for the adsorption of petroleum in the source rock and the biodegradation of petroleum in the reservoir.

References

M. A. Abu-Ali, J. G. Rudkiewicz, J. G. McGillivray, and F. Behar. Paleozoic petroleum system of central Saudi Arabia. *GeoArabia*, (4):321–336, 1999.

H. Bahlburg and C Breitkreuz. *Grundlagen der Geology.* Elsevier GmbH, Muenchen, second edition, 2004.

G. R. Beardsmore and J. P. Cull. *Crustal Heat Flow.* Cambridge University Press, 2001.

F. Behar, M. Vandenbroucke, Y. Tang, and J. Espitalie. Thermal cracking of kerogen in open and closed systems: determination of kinetic parameters and stoichiometric coefficients for oil and gas generation. *Organic Geochemistry*, 26:321–339, 1997.

S. W. Benson. *Thermodynamical Kinetics.* Wiley, 1968.

I. O. Blumenstein, R. di Primio, W. Rottke, B. M. Krooss, and R. Littke. Application of biodegradation modeling to a 3d–study in N. Germany, 2006.

A. K. Burnham and R. L. Braun. Global kinetic analysis of complex materials. *Energy and Fuels*, 13:1–22, 1999.

A. K. Burnham and J. J. Sweeney. A chemical kinetic model of vitrinite maturation and reflectance. *Geochim. Cosmochim. Acta*, 53:2649–2657, 1989.

W. D. Carlson, R. A. Donelick, and R. A. Ketcham. Variablility of apatite fission–track annealing kinetics: I. Experimental results. *American Mineralogist*, 84:1213–1223, 1999.

A. D. Carr. A vitrinite kinetic model incorporating overpressure retardation. *Marine and Petroleum Geology*, 16:355–377, 1999.

J. Chen, J. Fu, G. Sheng, D. Liu, and J. Zhang. Diamondoid hydrocarbon ratios: novel maturity indices for highly mature crude oils. *Organic Geochemistry*, 25:179–190, 1996.

B. Cramer, E. Faber, P. Gerling, and B. M. Krooss. Reaction kinetics of stable carbon isotopes in natural gas – insights from dry, open system pyrolysis experiments. *Energy and Fuels*, 15(15):517–532, 2001.

R. di Primio and B. Horsfield. From petroleum–type organofacies to hydrocarbon phase prediction. *AAPG Bulletin*, 90:1031–1058, 2006.

R. A. Donelick, R. A. Ketcham, and W. D. Carlson. Variablility of apatite fission–track annealing kinetics: II. Crystallographic orientation effects. *American Mineralogist*, 84:1224–1234, 1999.

I. R. Duddy, P. F. Green, and G. M. Laslett. Thermal annealing of fission tracks in apatite 3. Variable temperature behaviour. *Chemical Geology (Isotope Geoscience Section)*, 73:25–38, 1988.

J. Espitalie, P. Ungerer, I. Irwin, and F. Marquis. Primary cracking of kerogens. experimenting and modelling C1, C2–C5, C6–C15 and C15+. *Organic Geochemistry*, 13:893–899, 1988.

K. Gallagher, R. Brown, and C. Johnson. Fission track analysis and its applications to geological problems. *Annu. Rev. Earth Planet Sci.*, 26:519–572, 1998.

S. Glasstone, K.J. Laidler, and H. Eyring. *The theory of rate processes.* McGraw-Hill, 1941.

J. C. Goff. Hydrocarbon generation and migration from jurassic source rocks in East Shetland Basin and Viking graben of the northern North Sea. *J. Geol. Soc. Lond.*, 140:445–474, 1983.

P. F. Green. The relationship between track shortening and fission track age reduction in apatite: Combined influences of inherent instability, annealing anisotropy, length bias and system calibration. *Earth and Planetary Science Letters*, 89:335–352, 1988.

P. F. Green, I. R. Duddy, A. J. W. Gleadow, P.R. Tingate, and G. M. Laslett. Thermal annealing of fission tracks in apatite 1. A qualitative description. *Chemical Geology (Isotope Geoscience Section)*, 59:237–253, 1986.

P. F. Green, I. R. Duddy, A. J. W. Gleadow, and J. F. Lovering. Apatite fission–track analysis as a paleotemperature indicator for hydrocarbon exploration. In N. D. Naeser and T. H. McCulloh, editors, *Thermal History of Sedimentary Basins, Methods and Case Histories*, pages 181–195. Springer–Verlag, 1989a.

P. F. Green, I. R. Duddy, G. M. Laslett, A. J. W. Gleadow, and J. F. Lovering. Thermal annealing of fission tracks in apatite 1. Quantitative modelling techniques and extension to geological timescales. *Chemical Geology (Isotope Geoscience Section)*, 79:155–182, 1989b.

R. W. Jones. Organic facies. In Academic Press, editor, *Advances in Petroleum Geochemistry*, pages 1–90. 1987.

R. A. Ketcham. Personal communication, 2003.

R. A. Ketcham, R. A. Donelick, and M. B. Donelick. Aftsolve: A program for multi–kinetic modeling of apatite fission–track data. *Geological Materials Research*, 2, No. 1 (electronic), 2000.

B. M. Krooss and R. di Primio. Personal communication, 2007.

S. Larter. Bugs, biodegradation and biochemistry of heavy oil. The 23rd International Meeting on Organic Geochemistry, Torquay, England, 2007.

S. R. Larter. Some pragmatic perspectives in source rock geochemistry. *Marine and Petroleum Geology*, 5:194–204, 1988.

G. M. Laslett, P. F. Green, I. R. Duddy, and A. J. W. Gleadow. Thermal annealing of fission tracks in apatite 2. A quantitative analysis. *Chemical Geology (Isotope Geoscience Section)*, 65:1–13, 1987.

A. S. Mackenzie and D. McKenzie. Isomerization and aromatization of hydrocarbons in sedimentary basins. *Geological Magazine*, 120:417–470, 1983.

A. S. Pepper and P. J. Corvi. Simple kinetic models of petroleum formation. Part I: oil and gas generation from kerogen. *Marine and Petroleum Geology*, 12(3):291–319, 1995a.

A. S. Pepper and P. J. Corvi. Simple kinetic models of petroleum formation. Part III: Modelling an open system. *Marine and Petroleum Geology*, 12(4): 417–452, 1995b.

A. S. Pepper and T. A. Dodd. Simple kinetic models of petroleum formation. Part II: oil – gas cracking. *Marine and Petroleum Geology*, 12(3):321–340, 1995.

K. E. Peters, C. C. Walters, and J. M. Moldowan. *The Biomarker Guide*, volume 1 and 2. Cambridge University Press, second edition, 2005.

W. H. Press, S. A. Teukolsky, W. T. Vetterling, and B. P. Flannery. *Numerical Recipes in C++*. Cambridge University Press, second edition, 2002.

M. Radke and D. H. Welte. The methylphenanthrene index (MPI): A maturity parameter based on aromatic hydrocarbons. In M. Bjoroy et al., editor, *Advances in Organic Geochemistry.* Proceedings of the 10th International Meeting on Organic Geochemistry, University of Bergen, Norway, 14–18 September 1981, Wiley and Sons, 1983.

C. S. Sajgo and J. Lefler. A reaction kinetic approach to the temperature–time history of sedimentary basins. *Lecture Notes in Earth Sciences*, 5:123–151, 1986.

J. J. Sweeney and A. K. Burnham. Evaluation of a simple model of vitrinite reflectance based on chemical kinetics. *AAPG Bulletin*, 74:1559–1570, 1990.

E. W. Tegelaar and R. A. Noble. Kinetics of hydrocarbon generation as a function of the molecular structure of kerogen as revealed by pyrolysis–gas chromatography. *Advances in Organic Geochemistry*, 22(3–5):543–574, 1994.

B. P. Tissot and D. H. Welte. *Petroleum Formation and Occurrence.* Springer–Verlag, Berlin, second edition, 1984.

P. Ungerer, J. Burrus, B. Doligez, P. Y. Chenet, and F. Bessis. Basin evaluation by integrated two–dimensional modeling of heat transfer, fluid flow, hydrocarbon gerneration and migration. *AAPG Bulletin*, 74:309–335, 1990.

D. W. van Krevelen. *Coal. Typology–Chemistry–Physics–Constitution.* Elsevier, 1961.

M. Vandenbroucke, F. Behar, and L. J. Rudkiewicz. Kinetic modelling of petroleum formation and cracking: implications from high pressure, high temperature Elgin Field (UK, North Sea). *Organic Geochemistry*, 30:1105–1125, 1999.

D. W. Waples. Time and temperature in petroleum formation: application of Lopatin's method to petroleum exploration. *AAPG Bulletin*, 64:916–926, 1980.

R. W. T. Wilkins, C. P. Buckingham, N. Sherwood, Russel N. J., M. Faiz, and Kurusingal. The current status of famm thermal maturaty technique for petroleum exploration in australia. *Australian Petroleum Prduction and Exploration Asociation Journal*, 38:421–437, 1998.

K. Zengler, H. H. Richnow, R. Rosselló-Mora, W. Michaelis, and F. Widdel. Methane formation from long–chain alkanes by anaerobic microorganisms. *Nature*, 401:266–269, 1999.

5

Fluid Analysis

5.1 Introduction

On a macroscopic level fluids consist of physically distinct phases which fill empty space in regions with defined boundaries. The amount, composition, and properties of the phases vary with the overall composition of the fluid and external parameters such as pressure, volume, and temperature (PVT). Migration and other sophisticated aspects of basin modeling can only be simulated if the phases and their properties are known. Hence, the subject of this chapter is modeling of phase compositions and properties.

On a molecular level fluids can be grouped into compounds with (approximately) the same physical properties. These compounds are called components and are typically "pure" chemical species such as methane and water or "lumped" species like alkanes with a defined number of carbon atoms e.g. C_nH_{2n+2} (Danesh, 1998). The subject of fluid analysis can be subdivided into three parts, the determination of the coexisting phases, their compositions and their properties.

In basin modeling it is usually assumed that a water phase is present. The polar structure of water causes a separation from the non–polar hydrocarbons, and hence, as a result of this, there are at least two distinct phases. An exception are gas hydrates which occur only under special conditions (Sec. 5.8). It is further assumed that other highly polar components, including salts, dissolve almost completely in the water phase. Small amounts of light HCs can also be dissolved in water. However, usually HCs form their own phases. Furthermore it is commonly assumed that the HC phases exist as liquid and vapor or of only one single phase, a supercritical or undersaturated phase. So, in practice, fluid analysis is reduced to HC phase analysis with dissolution of components in one or the other of two possible phases. From Gibbs' phase rule it can be deduced that the number of phases κ can become $\kappa = N + 2$ where N is the (sometimes large) number of components. It should be kept in mind that more than two HC phases may exist (Sengers and Levelt, 2002; Danesh, 1998; Pedersen et al., 1989; Pedersen and Christensen, 2007).

T. Hantschel, A.I. Kauerauf, *Fundamentals of Basin and Petroleum Systems Modeling*, DOI 10.1007/978-3-540-72318-9_5,
© Springer-Verlag Berlin Heidelberg 2009

From experience it is known that components with higher molecular weights are enriched in the denser liquid phase, whereas lighter components favor the vapor phase. At ambient surface conditions (standard conditions) of 60 °F = 15.556 °C and 1 atm, a rule of thumb can be applied, according to which components heavier than pentane are found almost entirely in the liquid and the remaining lighter components almost entirely in the vapor phase.[1] In petroleum system studies, components found in the liquid phase are referred to as oil components and components found in the vapor as gas components. At surface conditions the compositional analysis is reduced to a distribution of gas components in the vapor phase and oil components in the liquid phase. Properties of the phases can be estimated using empirical methods such as the API method for density predictions (Sec. 5.4). Retaining these phase compositions for arbitrary pressure and temperature conditions, is called the "Fixed Phase Model" (Fig. 5.1). It is obviously not suitable for conditions which differ a lot from surface conditions. Nevertheless it contains concepts that are useful to the following discussion.

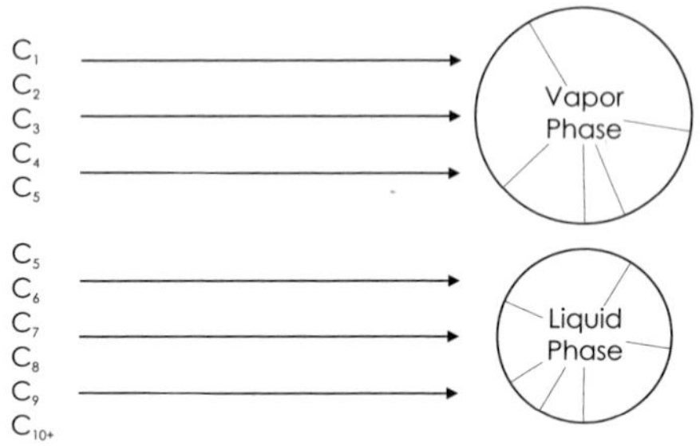

Fig. 5.1. The fixed phase model attributes light components to the vapor and heavy components to the liquid phase. C_5 is an intermediate and can often not be classified unambiguously as a light or a heavy component

The water phase is commonly well separated from the hydrocarbon phases. Methane is an exceptional component which might dissolve in non–negligible parts in the water phase. This must be taken into account prior to further compositional analysis and thus the next section deals with the influence of the water phase. A simple fluid model for limited pressure and temperature ranges is discussed afterwards in Sec. 5.3. Section 5.4 deals more with the theoretical background and advanced equations of state (EOS). This knowledge is the basis for an algorithm for the prediction of phase compositions in Sec. 5.5. The prediction of properties such as density, viscosity or the gas oil ratio (GOR) of fluids and their phases, is the topic of Sec. 5.6. The chapter continues with a discussion about how fluid models are usually integrated, tuned, and calibrated in basin modeling (Sec. 5.7). Finally, the behavior of gas hydrates is briefly introduced in Sec. 5.8.

[1] Pure pentane is a liquid at room temperature with a boiling point of 36°C. In multicomponent systems it is usually enriched in the liquid but also the vapor phase contains often non–negligible amounts.

Especially in this chapter, many examples are plotted for typical pressure temperature paths (PT–paths) in sedimentary basins, which range from cool and overpressured to hot and hydrostatic (Fig. 5.2). The PT–path in a real basin will often not follow a straight line. For example, overpressured regions often start below 3 km depth. Hence steep curves with values above 0.5 MPa/K are often found only above 100°C. However, the range between path 1 and 5 in Fig. 5.2 defines an area of possible pressure temperature points in arbitrary sedimentary basins.

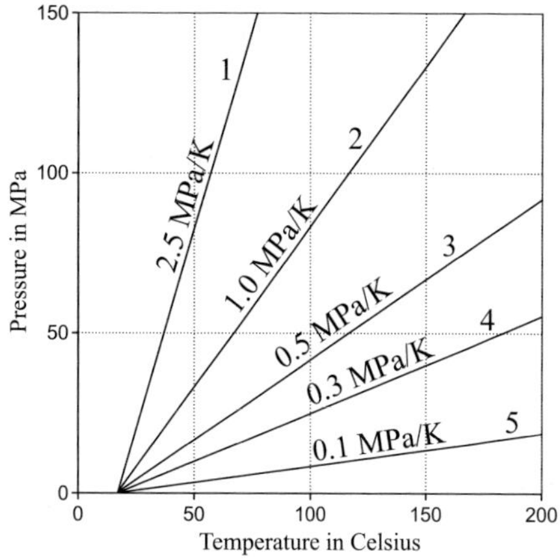

Fig. 5.2. Typical pressure temperature paths (PT–paths) in geological basins. Path 1 is cool and overpressured, path 2 is overpressured, path 3 is a typical path, path 4 is hydrostatic and path 5 is hot and hydrostatic

5.2 Water Phase

In a first rough approximation H_2O is found almost completely in the water phase only. More complex approaches such as models including H_2O as a component partially dissolved in petroleum (Pedersen et al., 1989), are usually not taken into account. The dissolved amount of water is so small that it can be neglected in basin models.

It is also commonly assumed that all HCs, except methane, do not dissolve in water. The dissolved amount of methane in water depends approximately on pressure, temperature, and the overall amount of methane, but not on the properties of the HC phases. Therefore, methane dissolution can be treated independently before HC phase analysis.

Methane solubility has been studied, among others, by Haas (1978) and Battino (1987). The IUPAC (International Union of Pure and Applied Chemistry) recommended the smoothed equation

$$\ln x = -55.8111 + \frac{7478.84}{T} + 20.6794 \ln \frac{T}{100} + 0.753158 \ln p \qquad (5.1)$$

for the molar fraction x of methane in water, where the unit for temperature T is Kelvin and for pressure p is MPa (Nelson and Simmons, 1995). It is drafted in Fig. 5.3.

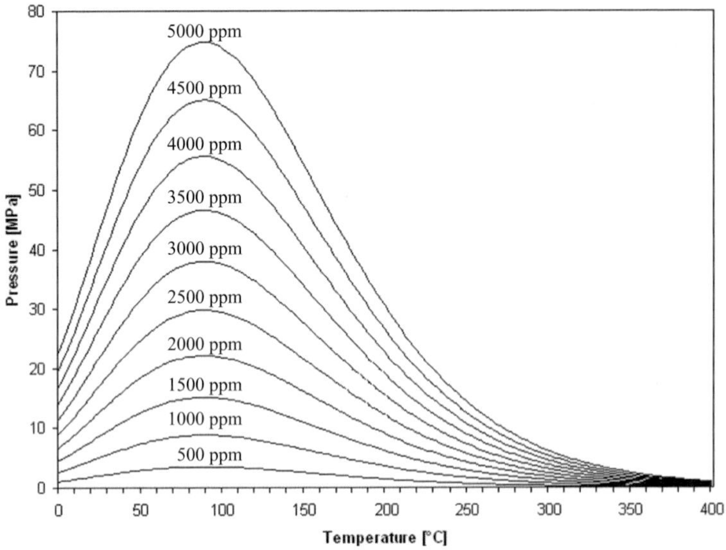

Fig. 5.3. Isolines of mol fraction solubility x for the dissolution of methane in water according to (5.1)

In the subsurface methane dissolves in water only in the immediate neighborhood of hydrocarbon pathways. Due to the size and resolution of a basin model it is often not possible to resolve pathways accurately. Additionally, methane solubility in seawater—equivalent pore water is reduced by approximately 14% (Nelson and Simmons, 1995). Therefore, the average maximum solubility must be lower than given in (5.1).

Another component of special interest concerning petroleum systems modeling is carbon dioxide. It is often found in concentrations up to 2% (molar) in petroleum (di Primio, 2008). The physical properties of carbon dioxide differ form typical hydrocarbons. It is slightly polar and dissolves well in water. In basin modeling it is for that reason often assumed that it is completely dissolved in the water phase. An empirically determined fraction of carbon dioxide is sometimes kept in the petroleum phases. This approach is rather crude, especially at locations of high petroleum saturation e.g. in accumulations where the water content is relatively small and water is not capable of absorbing all the carbon dioxide.

Due to a lack of spatial resolution, basin models are generally not feasible for sophisticated modeling of small concentrations of impurities, which might dissolve in the water or in a hydrocarbon phase. Alternatively, empirically determined concentrations are commonly assumed for each phase.

5.3 Binary Mixtures and Black Oil Models

Binary mixtures are of great interest for modeling since they represent simple systems which show qualitatively the same phase behavior as arbitrary petroleum fluids. Black oil models are binary mixtures which have been adjusted to quantitatively match the phase behavior of petroleum, at least in limited pressure temperature ranges.[2]

A starting point for the discussion of binary mixtures is the phase diagram of a one–component system (Fig. 5.4). The vapor–pressure line defines the conditions at which liquid and vapor coexist. Directly above this line pure liquid and below it pure vapor is found. The line ends at the critical point. Property differences such as a density contrast between the liquid and vapor phase, decrease continuously and finally vanish when approaching the critical point.

Gibbs' phase rule states that the degrees of freedom F equal $F = N - \kappa + 2$. The number of components is given by N and κ is the number of phases. Here it is $N = 1$ and therefore $\kappa = 3 - F$. An area in the pressure–temperature diagram has two degrees of freedom and thus $\kappa = 1$ which means that this area can only be represented by one phase. A line has one degree of freedom so on a line two phases can coexist and on a single point, namely the triple point, even three phases may coexist side by side. The whole pressure–temperature plane is subdivided into areas with one phase, lines with two coexisting phases and one triple point. The solid state behind the melting–point line is with the exceptions of permafrost and gas hydrates of no interest in basin modeling.

Quotation marks in Fig. 5.4 are used for liquid and vapor, since the definition of liquid and vapor becomes ambiguous in a one phase region. A simultaneous temperature and pressure change on a path from a point in the "liquid" region to the "vapor" region, which goes around the critical point through the "supercritical" region and without crossing the vapor–pressure line, causes a continuous change from "liquid" to "vapor". In other words, there is no point with a physical criterion for the distinction between "liquid" and "vapor" and thus the word supercritical is a better match for the description of this one phase area.

An extension to two components causes an "opening" of the vapor–pressure line to an area of coexisting phases (Fig. 5.4). With $N = 2$ the phase rule yields $\kappa = 4 - F$ which allows for two phases with two degrees of freedom. All isolines of constant composition end in the critical point. The cricondentherm is defined as the pressure temperature point with highest possible temperature for two coexisting phases.

---

[2] "Black Oil Model" should not be mistaken for "Black Oil", which describes a class of oils with special properties (Sec. 5.5.1).

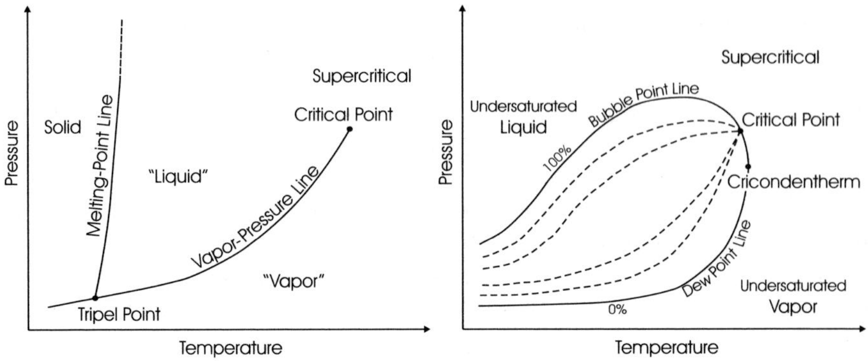

Fig. 5.4. Schematic pressure–temperature diagram of a one–component system on the left and of a two–component system on the right

The one–phase region is often called "undersaturated liquid" or "undersaturated vapor". This naming convention can easily be understood with the help of Fig. 5.5 and the example of a one–phase system with 15% methane at point P. This system would be "liquid–like" since it is above curve 8, which indicates the outline of the 15%–methane two–phase region. An increase of the methane–content to 30% moves the limiting slope to curve 7, which touches point P. The liquid is capable of doubling its methane content until it is finally saturated and cannot absorb anymore methane. The remaining methane will then form a separate vapor phase.

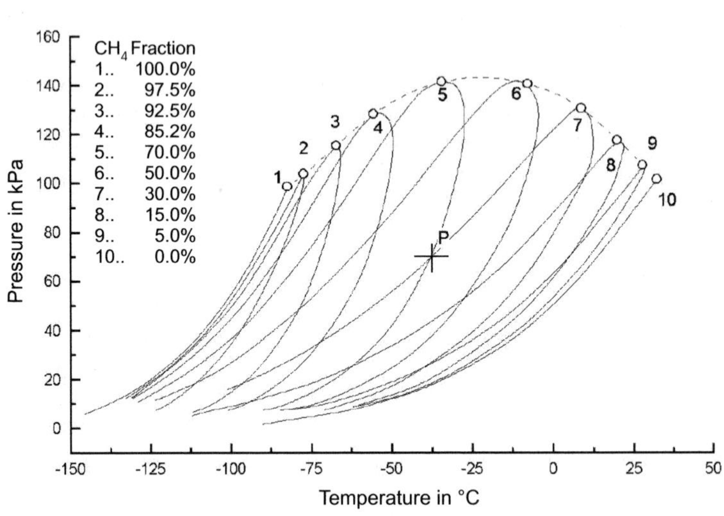

Fig. 5.5. Two phase areas of binary methane–ethane mixtures (McCain Jr., 1990)

In the two–phase region each phase will be saturated. Hence, with a further increase of the methane content a methane–saturated vapor phase appears. The dew point line of this touches point P (Fig. 5.5). This phase consists of 70% methane. At point P the dew point line of mixture 5 (70% methane and 30% ethane) intersects with the bubble point line of mixture 7 (30% methane and 70% ethane), indicating that the co-existing vapor and liquid phases contain 70% and 30% methane, respectively. The total amount of both phases can be calculated with simple material balances:

Let $x_{1,2}$ be the molar ethane and methane fractions in the liquid, $y_{1,2}$ the molar ethane and methane fractions in the vapor and $z_{1,2}$ the total molar ethane and methane fractions.[3] Then

$$x_1 + x_2 = 1, \qquad y_1 + y_2 = 1 \qquad \text{and} \qquad z_1 + z_2 = 1\,. \tag{5.2}$$

With the total fractions, n_l of liquid and n_v of vapor, the balance equations can be formulated as

$$n_l + n_v = 1 \qquad \text{and} \qquad z_1 = x_1 n_l + y_1 n_v, \quad z_2 = x_2 n_l + y_2 n_v. \tag{5.3}$$

It is easy to calculate n_l and n_v with the knowledge of $x_{1,2}$, $y_{1,2}$ and $z_{1,2}$ to

$$n_l = \frac{z_1 - y_1}{x_1 - y_1} \qquad \text{and} \qquad n_v = \frac{z_2 - x_2}{y_2 - x_2}\,. \tag{5.4}$$

Component one is the heavier ethane and it prefers the liquid phase, so $x_1 > y_1$. Methane favors the vapor phase, so $y_2 > x_2$. Thus undersaturated vapor exists if $z_1 < y_1$ because n_l cannot be negative and undersaturated liquid exists if $z_2 < x_2$. This means that undersaturated liquid exists if the total amount of methane is less than the capability of a saturated liquid to dissolve methane. In the other cases a two–phase system emerges.

The result (5.4) can be interpreted graphically (Fig. 5.6) with so called tie lines (McCain Jr., 1990).

Qualitatively, binary mixtures show similar properties to multi–component petroleum. A quantitatively better approximation can now be constructed starting with the idea of gas and oil components from the fixed phase model described in the introduction: All gas components are lumped together into one artificial pseudo gas component and all the oil components into one artificial pseudo oil component. This artificial two component system can then be treated as a binary mixture. The two pseudo components are distributed in both phases. All the data needed for this procedure are the bubble and dew point curves for arbitrary compositions.[4] A "Black Oil Model" is a binary

[3] In this chapter x is used for liquid, y for vapor and z for the overall composition or if none of both is specified. This is not consistent with Chap. 4 where x denotes kerogen, y oil/liquid and z gas/vapor. However, the notation is commonly used in the literature (Danesh, 1998).

[4] It is even more efficient to use two lookup tables for x_1 and y_2.

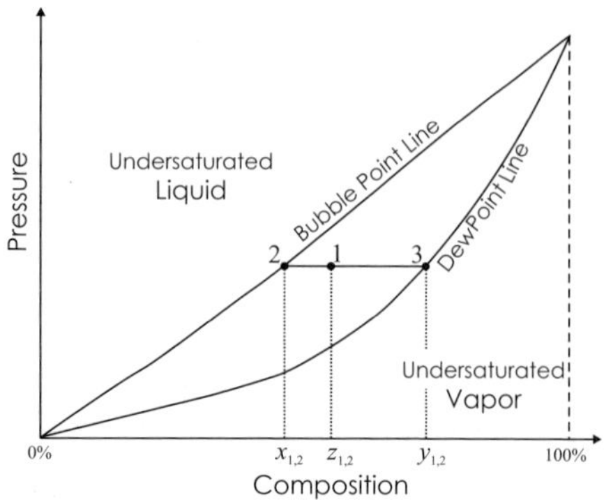

Fig. 5.6. Tie line construction for a binary mixture with overall composition $z_{1,2}$, saturated liquid with composition x_1 and saturated vapor with composition y_2: The vapor and liquid phase amounts $n_{v,l}$ can be calculated with the length of the distances $\overline{12}$, $\overline{13}$ and $\overline{23}$ by $n_v = \overline{12}/\overline{23}$ and $n_l = \overline{13}/\overline{23}$, which is the same as (5.4)

mixture with no solubility of the heavier component in the vapor or $y_1 = 0$ (Peaceman, 1977). Binary mixtures without this restriction are often referred to as "Symmetrical Black Oil Models" (SBO) (Fig. 5.7).

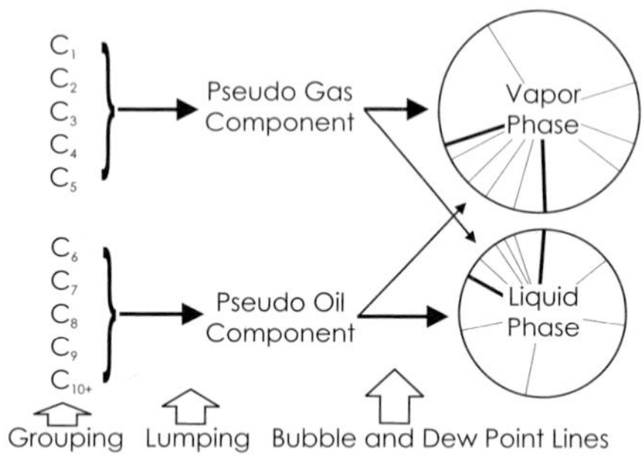

Fig. 5.7. Symmetrical black oil model: Pure components are grouped into light or heavy components and then lumped into artificial gas and oil pseudo components. A binary mixture is used for the distribution of the pseudo components into the liquid and vapor phases. "Un-lumping" of the pseudo components yields the amount of each pure component in any phase

The procedure of grouping gas and oil components, the lumping into pseudo components and the determination of a consistent set of dew and bubble point curves, are dependent on the overall composition of the petroleum under consideration. Usually, components are grouped following the fixed phase model. All the light components are collected in the gas component and the heavy components in the oil component. The dew and bubble point curve data are often calculated using thermodynamic methods, if they are not known from oil sample analysis. Therefore, properties such as the critical temperature or pressure of pseudo components must be known. Various approaches such as simple molar averaging according to

$$T_c = \sum_i z_i T_{ci} \tag{5.5}$$

or special formulas such as Lee–Kesler–Averaging (Danesh, 1998) according to

$$v_c = \frac{1}{8}\sum_{ij} z_i z_j \left(v_{ci}^{1/3} + v_{cj}^{1/3}\right)^3 ,$$

$$T_c = \frac{1}{8v_c}\sum_{ij} z_i z_j \left(T_{ci} T_{cj}\right)^{1/2} \left(v_{ci}^{1/3} + v_{cj}^{1/3}\right)^3 , \tag{5.6}$$

$$Z_c = 0.2905 - 0.085 \sum_i z_i \omega_i$$

are commonly in use. Here T_c is the critical temperature of the lumped component, T_{ci} the critical temperatures of the pure components, z_i the molar fraction of each component, v_c the critical molar volume, ω the acentric factor and Z_c the critical compressibility. The quantities indexed with i are the values of the pure components.

The biggest advantage of black oil models is their high performance during simulation. Grouping and lumping are performed independently before solving the differential equations of basin modeling. Phase changes during migration are calculated just by searching dew and bubble point data in lookup tables and the solution of the simple balance Equations (5.2) – (5.4). The disadvantages however are limited pressure and temperature ranges and difficulties to determine the dew and bubble point curves of these artificially constructed binary mixtures.

5.4 Equations of State (EOS)

Further thermodynamic aspects, which go beyond the scope of a symmetrical black oil model, can only be considered, if more about the properties of the phases are known. The most important property is the density or its inverse, the molar volume. Relationships of pressure and temperature with volume are called "Equations of State" (EOS).

The origin of most EOS is the ideal gas equation

$$pv = RT \tag{5.7}$$

with p as the pressure, v the molar volume, R the universal gas constant and T the temperature. The ideal gas equation is known to be a good approximation to the behavior of dilute gases at low pressures (Fig. 5.9). It can be derived from classical statistical mechanics with assumptions of point–like molecules, which move without interaction (Huang, 1987). The ideal gas equation cannot be used for the description of phase transitions.

For further considerations it is convenient to introduce the compressibility factor Z which is defined as

$$Z = \frac{pv}{RT} . \tag{5.8}$$

The ideal gas equation then becomes[5] $Z = 1$.

[5] Note that Z is not directly related to compressibilities such as introduced in (2.1). For the ideal gas equation it is $C = -(1/V)\,\partial V/\partial p = 1/p$.

A virial expansion can be interpreted as a systematic starting point to derive more realistic EOS. It has the form

$$Z = 1 + \frac{A_1}{v} + \frac{A_2}{v^2} + \frac{A_3}{v^3} + \dots \tag{5.9}$$

with the virial coefficients A_1, A_2, $A_3, \dots$, which are temperature dependent. The expansion can also be motivated by arguments from statistical mechanics (Huang, 1987). An important example is the Benedict–Webb–Rubin EOS (Danesh 1998, App. J).

Many other approaches are used for derivation of EOS. Four families are commonly distinguished (Pedersen et al., 1989):

- van der Waals (vdW)
- Benedict–Webb–Rubin
- Reference–fluid equations
- Augmented-rigid–body equations

The vdW family is well known for its accuracy and simplicity which makes it the method of choice in basic modeling. Although other families are often very precise, they are less general and thus impractical to use.

Van der Waals directly improved the ideal gas equation. He took attractive forces between the molecules into account which are present due to induced dipole–dipole interactions and repulsive forces which originate from the finite volume of the molecules (Becker, 1985; Huang, 1987). The vdW equation has the form

$$p = \frac{RT}{v-b} - \frac{a}{v^2} \, . \tag{5.10}$$

The parameter b is a measure of volume of molecules and called "co-volume" whereas a describes the intermolecular attraction.

With the notation of compressibility (5.10) becomes

$$Z^3 - (1+B)Z^2 + AZ - AB = 0 \tag{5.11}$$

with the dimensionless parameter

$$A = \frac{ap}{(RT)^2} \qquad \text{and} \qquad B = \frac{bp}{RT} \, . \tag{5.12}$$

Therefore, vdW equations are referred to as cubic EOS. The form of their isotherms is illustrated in Fig. 5.8. The isotherms have up to three volume–roots for given pressure and temperature. For that reason vdW equations have the ability to model phase transitions. The smallest root describes the liquid–like and the largest the vapor–like state. It can be shown that the intermediate root describes a thermodynamically instable state. Two phases coexist only if the pressure has special values, which can be found graphically according to the so called Maxwell equal area rule (Fig. 5.8). The restriction to special values of pressure for two phase states corresponds to Gibbs' phase rule.

Fig. 5.8. Pressure–volume diagram with isotherms according to the vdW equation. Below the critical temperature T_c two phases may exist simultaneously. It can be proven thermodynamically that this is only possible for pressures p_b, which cut the temperature isolines in a way that the vertical and horizontal hatched areas are equal in size (Becker, 1985). The region of two phases is situated below the dashed line

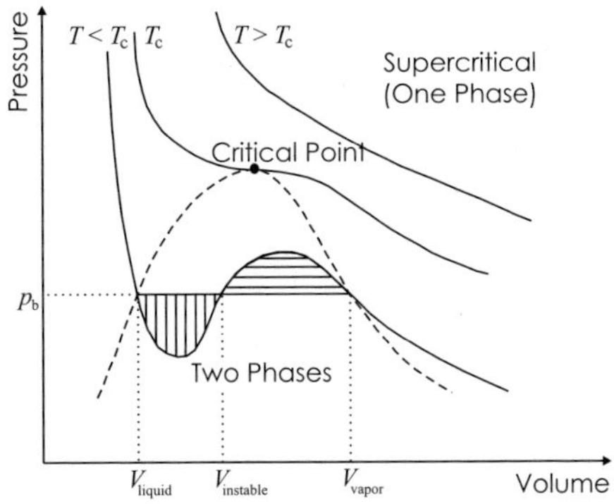

Above the critical point there exists only one phase. The location of the critical point can be calculated from the conditions

$$\left.\frac{\partial p}{\partial v}\right|_{T_c} = 0 \quad \text{and} \quad \left.\frac{\partial^2 p}{\partial v^2}\right|_{T_c} = 0 \tag{5.13}$$

which yield to

$$b = v_c/3 \quad \text{and} \quad a = 3\,p_c v_c^2 \quad \text{with} \quad p_c v_c = (3/8)RT_c\,. \tag{5.14}$$

Although the vdW equation shows all required qualitative features, it still does not have the accuracy needed for practical purposes in basin modeling. Many improvements have been proposed (Danesh, 1998; Pedersen et al., 1989). The best known are the Soave–Redlich–Kwong EOS (SRK)

$$\begin{aligned} p &= \frac{RT}{v-b} - \frac{a^2}{v(v+b)} \quad \text{or} \\ Z^3 &- Z^2 + (A - B - B^2)Z - AB = 0 \end{aligned} \tag{5.15}$$

and the Peng–Robinson EOS (PR):

$$\begin{aligned} p &= \frac{RT}{v-b} - \frac{a^2}{v(v+b)+b(v-b)} \quad \text{or} \\ Z^3 &- (1-B)Z^2 + (A - 2B - 3B^2)Z - AB + B^2 + B^3 = 0\,. \end{aligned} \tag{5.16}$$

Accurate methods for the derivation of the parameters a and b from experimental data are available with these EOS:

$$a = \alpha\, a_c\,, \qquad \alpha = \left[1 + m\left(1 - \sqrt{T/T_c}\right)\right]^2. \tag{5.17}$$

$$
\begin{aligned}
\text{SRK:} \quad & a_c = 0.42747\, R^2 T_c^2 / p_c \,, \qquad b = 0.08664\, R T_c / p_c \,, \\
& m = 0.480 + 1.574\,\omega - 0.176\,\omega^2 \,. \\
\text{PR:} \quad & a_c = 0.457235\, R^2 T_c^2 / p_c \,, \qquad b = 0.077796\, R T_c / p_c \,, \\
& m = 0.3796 + 1.485\,\omega - 0.1644\,\omega^2 + 0.01667\,\omega^3 \,.
\end{aligned}
\tag{5.18}
$$

In contrast to the vdW EOS which depends on two parameters a and b, the PR and SRK EOS implicitly depend on three parameters namely p_c, T_c, and the dimensionless acentric factor ω which is defined by

$$\omega = -\log_{10} p_b(T = 0.7\,T_c)/p_c - 1.0 \tag{5.19}$$

with $p_b(T)$ as the pressure on the vapor–pressure line. The acentric factor is zero for spherical molecules and related to their deviation from a spherical shape (Danesh, 1998). PR and SRK EOS are known to be quite accurate for $p < 100\,\mathrm{MPa}$ and $300\,\mathrm{K} < T < 500\,\mathrm{K}$. Densities of some light components, which are calculated with the SRK EOS and the ideal gas equation, are compared in Fig. 5.9. It can be seen, that the results differ enormously at high pressures and temperatures. A comparison of different EOS for methane only is shown in Fig. 5.10. The ideal gas equation is only accurate at low pressures and temperatures. As expected, SRK and PR EOS differ little. The modified Benedict–Webb–Rubin (MBWR) EOS (J.1) is exclusively adjusted to methane behavior (McCarty, 1974). It agrees well with the SRK EOS for temperatures, which are not too high.

5.4.1 Mixing Rules

The vdW equations can also be used for the description of multi–component phases with appropriate phase parameters. These phase parameters can be acquired through so called mixing rules which are often arithmetic or geometric averages of the component parameters. Formally, the approach is similar to lumping procedures described in (5.5) and (5.6). Usually, lumping refers to average component and mixing to average phase properties. However, there is no sharp boundary because the properties of lumped components are often identical with the phase properties of the corresponding mixture.

The mixing rule for co-volume b is usually

$$b = \sum_i z_i b_i \tag{5.20}$$

with b_i as the co-volumes of the components and z_i as molar fractions of the components in the phase.

A reasonable mixing rule for the attractive parameter a is a geometric average with corrective factors $k_{ij} \approx 0$ which take into account special intermolecular forces between components i and j. It has the form

$$a = \sum_{ij} z_i z_j a_{ij} \quad \text{with} \quad a_{ij} = \sqrt{a_i a_j}(1 - k_{ij}) \,. \tag{5.21}$$

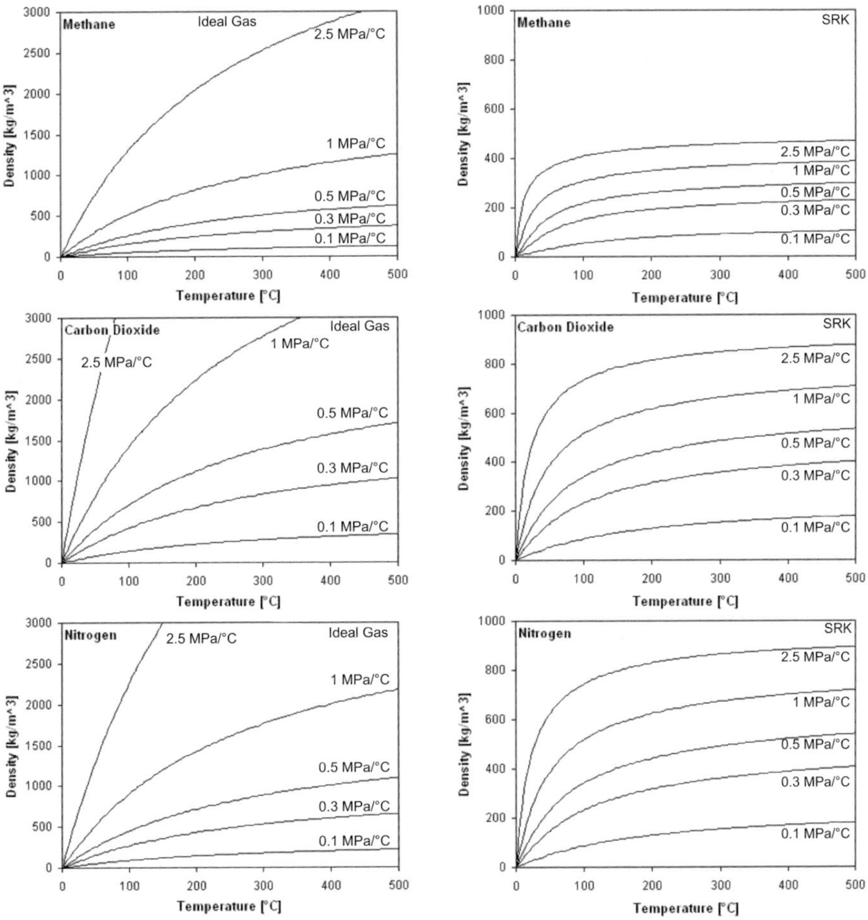

Fig. 5.9. Density $\rho = M_w p/(RT)$ of methane, nitrogen, and carbon dioxide according to (5.7) for different PT–paths with all of them starting at 0 MPa and 0 °C. The left column is calculated with the ideal gas equation (5.7) and the right column with the SRK EOS (5.15)

Many sets of these binary interaction parameters (BIP) k_{ij} are in use (Danesh, 1998; Reid et al., 1987, App. I). They are sometimes customized as tuning parameters for "fine–adjustment" of EOS. It is sometimes argued that this usage is problematic (Pedersen et al., 1989).

5.4.2 Phase Equilibrium

Gibbs' energy G represents the thermodynamic potential for systems with pressure and temperature as independent variables (Becker, 1985; Huang, 1987). The potential must be minimal for systems in equilibrium. This leads

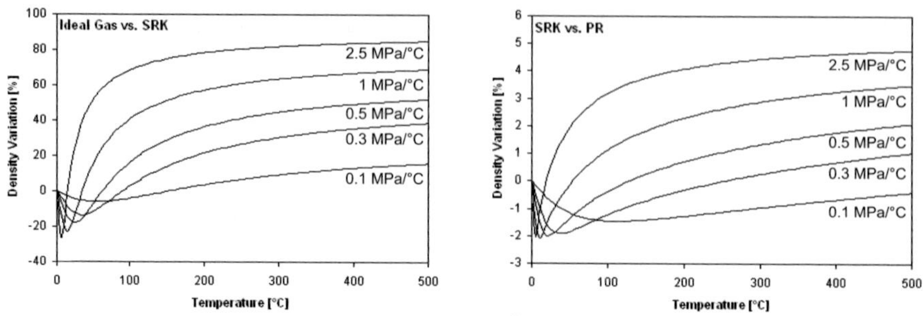

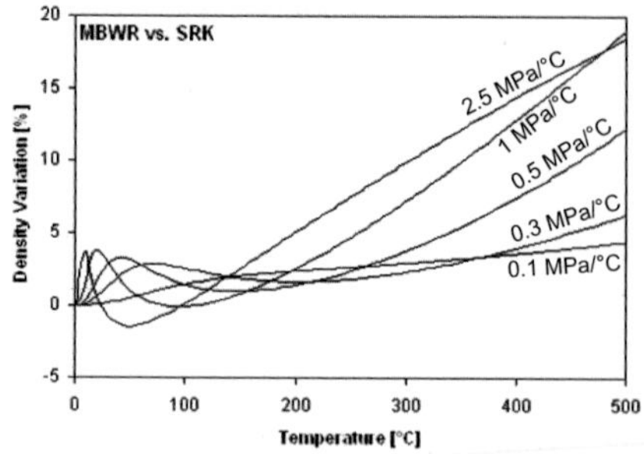

Fig. 5.10. Methane density variation between ideal gas equation (5.7) and SRK EOS (5.15) on the top left, between SRK and PR EOS (5.16) on the top right and between MBWR (J.1) and SRK EOS on the right for PT–paths as in Fig. 5.9

to the conclusion that the chemical potential μ of each component in each phase must be equal, namely

$$\mu_{l,i} = \mu_{v,i} \, . \tag{5.22}$$

The chemical potential change from $\mu_{0,i}$ to μ_i can be calculated for an ideal gas with an isothermal pressure change from p_0 to p as

$$\mu_i - \mu_{0,i} = RT \ln p/p_0 \tag{5.23}$$

in one phase. Similar to the definition of the compressibility factor Z in (5.8) a quantity labeled fugacity is commonly defined with reference to the behavior of an ideal gas implicitly with

$$\mu_i - \mu_{0,i} = RT \ln f_i/f_{0,i} \tag{5.24}$$

for non–ideal systems. The fugacity coefficient ϕ_i is defined as

$$\phi_i = \frac{f_i}{p z_i} \tag{5.25}$$

which must behave as an ideal gas for low pressures, so $\phi_i \to 1$ for $p \to 0$.It can be evaluated explicitly to

$$\begin{aligned} \ln \phi_i = {} & \frac{b_i}{b}(Z-1) - \ln(Z-B) \\ & - \frac{A}{(\delta_2 - \delta_1)B}\left[\left(\frac{2}{a}\sum_j z_j a_{ij}\right) - \frac{b_i}{b}\right] \ln \frac{Z + \delta_2 B}{Z + \delta_1 B} \end{aligned} \tag{5.26}$$

with $\delta_{1,2} = 1, 0$ for SRK and $\delta_{1,2} = 1 \pm \sqrt{2}$ for PR (Danesh, 1998).

5.5 Flash Calculations

The straight forward generalization of the material balance Equations (5.2) – (5.3) for a two–phase N–component system are

$$\sum_i x_i = 1, \qquad \sum_i y_i = 1, \qquad \sum_i z_i = 1 \tag{5.27}$$

and

$$n_l + n_v = 1, \qquad z_i = x_i n_l + y_i n_v \,. \tag{5.28}$$

With the definition (5.25) the equality of the chemical potentials in a two–phase system leads to

$$x_i \phi_{l,i} = y_i \phi_{v,i} \tag{5.29}$$

for given pressure and temperature.

Equations (5.27) – (5.29) with the fugacity coefficients as defined in (5.26), form a set of $2N + 2$ independent equations[6] with corresponding unknowns x_i, y_i, n_l, and n_v. The solution of this set of nonlinear equations is referred to as a "flash calculation" (Fig. 5.11).

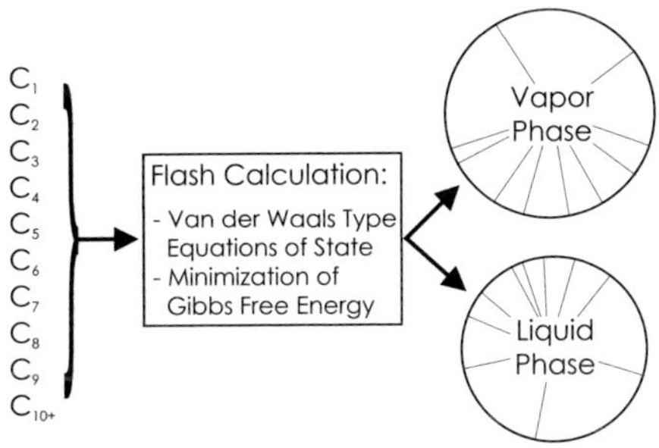

Fig. 5.11. Flash model which is based on thermodynamics

Equations (5.27) are linear in the unknowns and have a simple structure. Thus it is possible to solve these partly with the help of equilibrium ratios, which are defined as $K_i = y_i/x_i$. With knowledge of K_i and n_v it is then possible to calculate

[6] The condition $\sum_i z_i = 1$ does not count since it does not contain independent variables. But from $1 = \sum_i z_i = n_l \sum_i x_i + n_v \sum_i y_i = 1 + n_v \sum_i (y_i - x_i)$ it follows directly $\sum_i (y_i - x_i) = 0$. So one of the sums $\sum_i x_i = 1$ or $\sum_i y_i = 1$ is dependent and should also not be counted.

$$x_i = \frac{z_i}{1+(K_i-1)n_v}, \quad y_i = \frac{K_i z_i}{1+(K_i-1)n_v}, \quad \text{and} \quad n_l = 1 - n_v\,. \tag{5.30}$$

From

$$0 = \sum_i (y_i - x_i) = \sum_i \frac{(K_i-1)z_i}{1+(K_i-1)n_v} \tag{5.31}$$

however, it is possible to determine n_v numerically. The remaining K_i can now be treated iteratively. The complete algorithm is made up of the following sequence:

1 Estimate K_i
2 Calculate n_v, n_l and x_i, y_i with (5.31) and (5.30)
3 Calculate a_l, b_l and a_v, b_v with mixing rules
4 Calculate Z_l and Z_v with (5.15) or (5.16)
5 Calculate $\phi_{l,i}$ and $\phi_{v,i}$ with (5.26)
6 Calculate $K_i^{\text{new}} = \phi_{l,i}/\phi_{v,i}$
7 If $\sum_i (1 - K_i^{\text{new}}/K_i)^2 > \epsilon$ adjust $K_i \leftarrow K_i^{\text{new}}$ and go back to step 2

The criterion of convergence in step 7 is given by ϵ, which should be a small number such as $\epsilon = 10^{-12}$.

Good initial estimates of K_i are very important. Wilson (Danesh, 1998; Pedersen et al., 1989) proposed approximations of equilibrium ratios below $p < 3.5\,\text{MPa}$ with the formula

$$K_i = \frac{p_{ci}}{p} \exp\left[5.37\,(1+\omega_i)\left(1 - \frac{T_{ci}}{T}\right)\right] \tag{5.32}$$

which can be used as start values for flash calculations at even higher pressures.

A failure in convergence indicates a one–phase solution. Additionally, a stability analysis for the explicit evaluation of the number of phases can be performed (Danesh, 1998; Pedersen et al., 1989). However, these methods are very sophisticated and in themselves unstable. Ambiguous cases are rare in real case studies.

In case of a one–phase state, two alternative thermodynamic stable roots may exist, according to Fig. 5.8. The root with the smaller Gibbs energy is the searched one. The difference of the Gibbs energy between two roots Z_l and Z_v can be calculated as

$$\begin{aligned} \frac{G_v - G_l}{RT} &= Z_v - Z_l + \ln \frac{Z_l - B}{Z_v - B} \\ &\quad - \frac{A}{B(\delta_2 - \delta_1)} \ln\left(\frac{Z_l + \delta_1 B}{Z_l + \delta_2 B} \times \frac{Z_v + \delta_2 B}{Z_v + \delta_1 B}\right)\,. \end{aligned} \tag{5.33}$$

Due to limited data availability and system resources in basin modeling the number of components in flash calculations are commonly restricted to about

14 (Sec. 4.3.3). Thus, it is also necessary to use lumped pseudo components in flash calculations. Often they replace the heavier components. The C_{15+} pseudo–component, which lumps together all HCs of the fluid heavy end with more than or equal to 15 carbon atoms, is an example of this. Examples of some flash calculations are shown in Fig. 5.12 and Fig. 5.13.

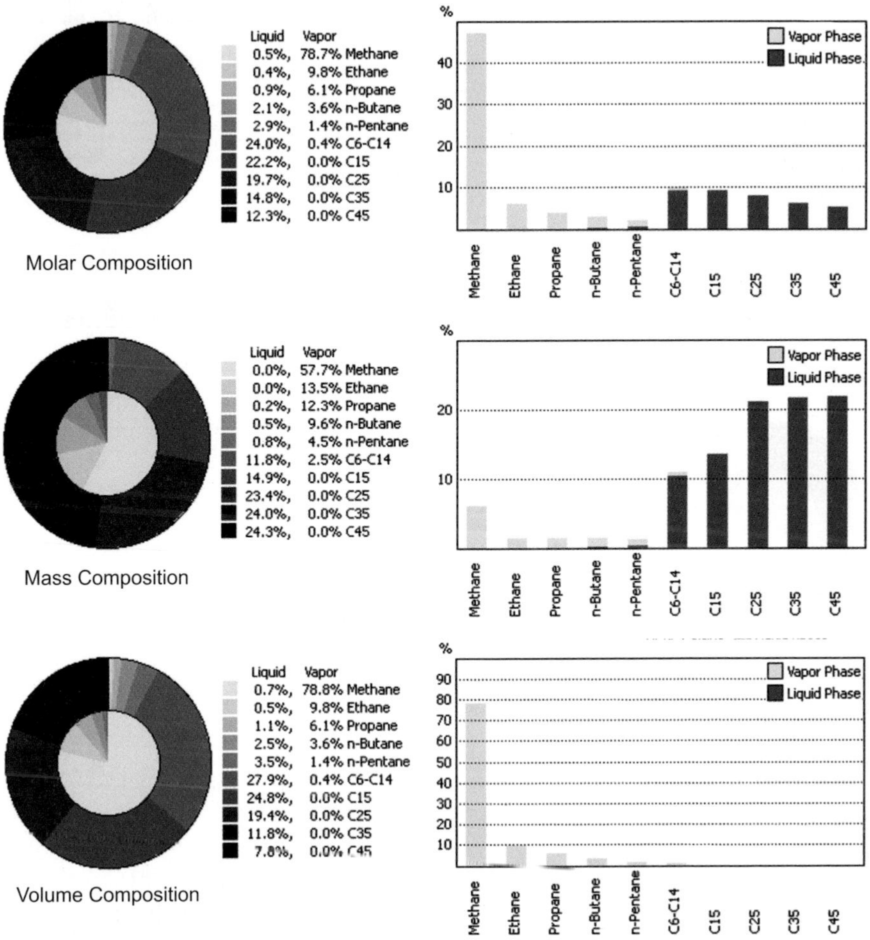

Fig. 5.12. Pie charts for phase composition and bar graphs for component distribution to phases. The overall composition is the same as the "Black Oil" of Fig. 5.14. It is listed in App. K. All figures are calculated with the SRK EOS for surface conditions of 0.1 MPa and 15 °C. The inner circle of each pie chart refers to vapor, the outer to liquid. The sum of all bars of each bar graph is 100%. All components are found in both phases, although only pentane and butane contribute with significant amounts to both phases

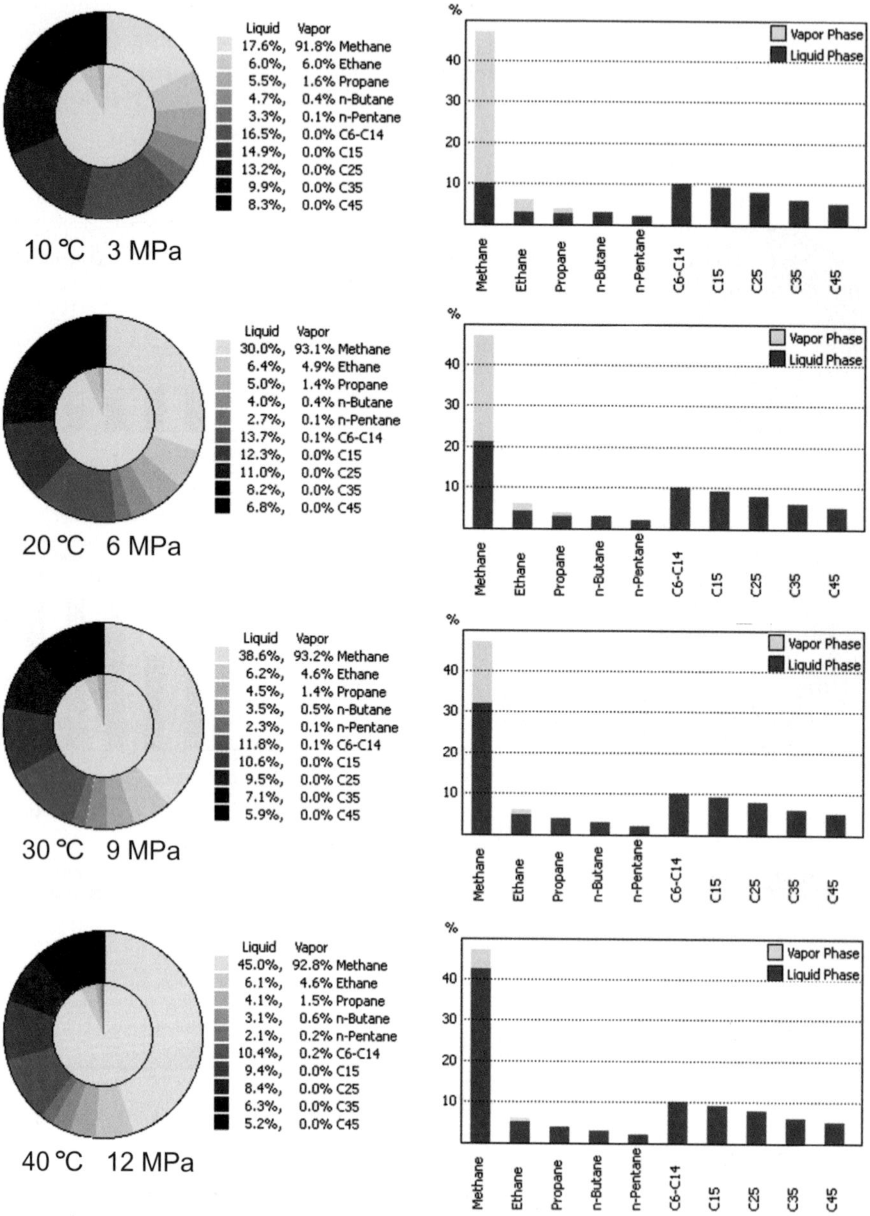

Fig. 5.13. Pie charts and bar graphs for the same composition as in Fig. 5.12. Only molar fractions are displayed. Light components are dissolved in the liquid phase with increasing pressure. The pressure temperature values correspond to a PT–path with 0.3 MPa/°C, which starts at 0°C. The composition at standard conditions is depicted in Fig. 5.12

The performance of flash calculations is very good when a relatively small number of components are involved. Flash calculations can even be explicitly performed during solution of differential equations for each grid cell and in each time step in basin modeling. By experience, the extra amount of time compared to a symmetrical black oil model is about 10% of total simulation time.[7] Phase diagrams, such as calculated for Fig. 5.14 from more than 10.000 flash calculations can be generated on a modern PC within a few seconds. However, flash calculations are also very helpful for the modeling of fluid flow with symmetrical black oil descriptions, because they can be used for the consistent construction of needed lookup tables (Sec. 5.3).

5.5.1 Classification of Petroleum

Petroleum is commonly classified as dry gas, wet gas, gas condensate, volatile oil, and black oil (Danesh, 1998). The main classification parameter is the gas oil ratio (GOR) (Table 5.1). Examples, which are constructed with flash calculations are shown in Fig. 5.14. The corresponding compositions and component properties are listed in App. K.

Class	GOR [m^3/m^3]	°API	Composition
Dry Gas	–	–	CH_4 + other light gas comp. only
Wet Gas	> 10 000	–	mainly CH_4 + other light gas comp.
Gas Condensate	570 ... 10 000	40 ... 60	< 12.5 mol% C_{7+}
Volatile Oil	310 ... 570	> 40	12.5 ... 20 mol% C_{7+}
Black Oil	< 310	< 45	> 20 mol% C_{7+}

Table 5.1. Classification of petroleum according to Danesh (1998). The API gravity refers to liquid after outgassing. Black oil is additionally classified by a relatively low bubble point pressure

5.5.2 PT–Paths

A spectrum of typical geological PT–paths is introduced in Fig. 5.2. Phase compositions can quickly be evaluated along PT–paths with flash calculations (Figs. 5.15, 5.16).

In typical geological environments the critical point can be reached. In this case strong variations of phase compositions might occur within certain pressure temperature intervals (Fig. 5.15).

---

[7] The number is highly dependent on the fraction of the grid cells that hold HCs inside.

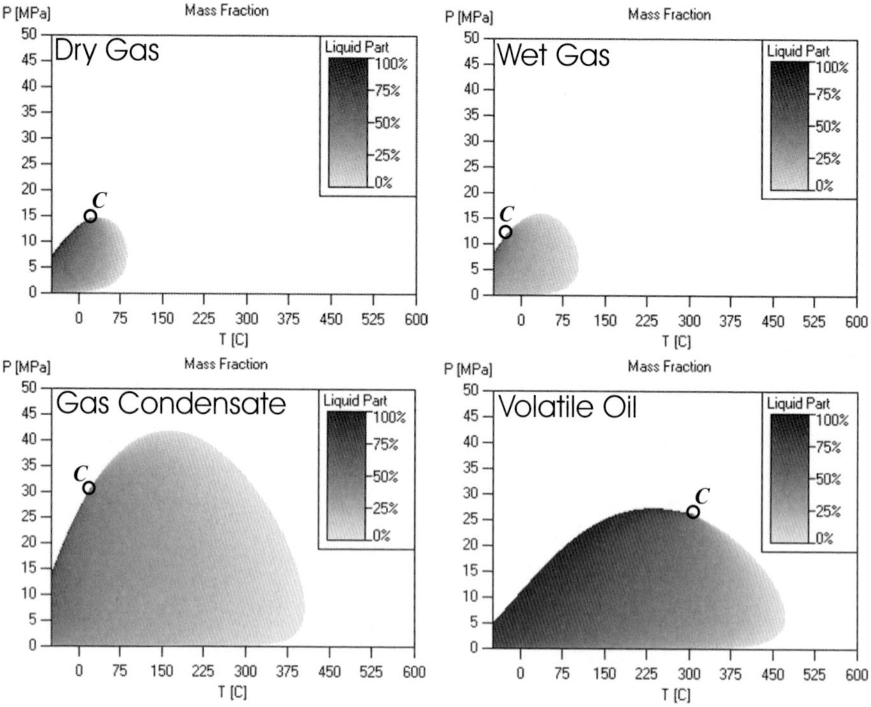

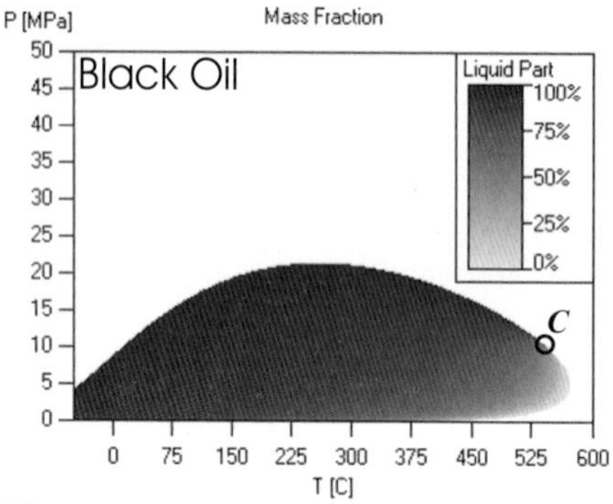

Example	GOR $[m^3/m^3]$	°API	C_{7+} [mol %]
Dry Gas	–	–	0.0
Wet Gas	22 452	(79)	≈ 1.5
Gas Condensate	1 997	52	≈ 4.0
Volatile Oil	473	48	≈ 13.0
Black Oil	110	32	≈ 36.0

Fig. 5.14. Five *PT* phase diagrams which are calculated with the SRK EOS. The critical point is marked with a "*C*". Compositions and component properties are listed in App. K

5.6 Property Prediction

5.6.1 Density

Density can directly be calculated from the results of a flash calculation. The molar volume v is known from (5.15) or (5.16) and the composition from (5.30). Hence, density can be directly calculated from

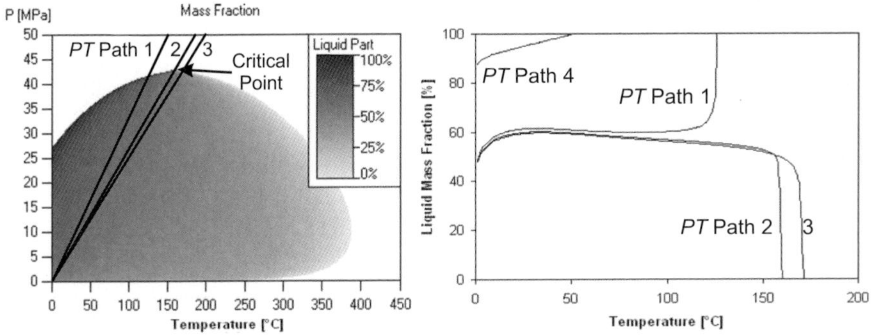

Fig. 5.15. Pressure–temperature phase diagram of a fluid with a high amount of light components on the left. Three PT–paths are marked. The liquid mass fraction along these paths is plotted on the right. Path 1 corresponds to a temperature gradient of 30 °C/km and 10 MPa/km, which is typical in geological systems. Path 3 has the same pressure gradient but 40 °C/km. The intermediate path 2 cuts almost exactly at the critical point. It can be seen, that the phase compositions vary strongly in the vicinity of the critical point. The highest curvature is found, as expected, on path 2 at the critical point. The three PT–paths 1–3 are passing the critical point quite close. The system, which is shown in Fig. 5.23, has a critical point far away from these PT–paths. For comparison, path 4 is extracted with the same *PT* conditions as path 1 from this system. The transition from the two–phase to the one–phase region is here rather smooth

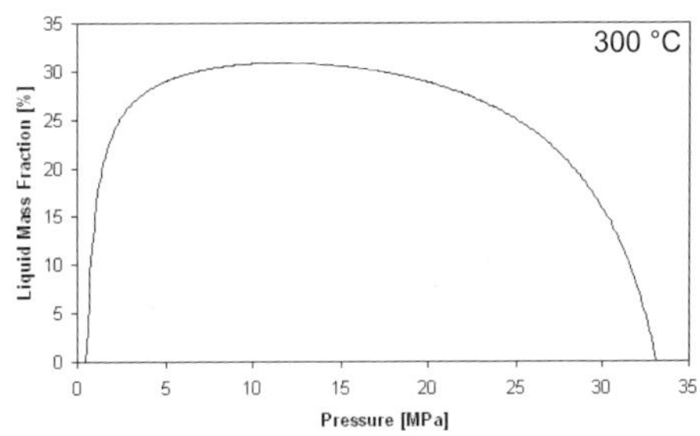

Fig. 5.16. Liquid mass fraction of the same fluid as in Fig. 5.15 at 300°C. The formation of liquid under isothermal pressure release is called "retrograde condensation" (Danesh, 1998)

$$\rho = \frac{1}{v} \sum_i M_i z_i \tag{5.34}$$

with M_i as the molar mass and z_i as molar fraction of component i. The molar volume v and the composition z_i must be replaced by v_l and x_i for liquids and by v_v, y_i for vapor.

Volume Shift

Although the SRK and PR EOS are quite accurate for the prediction of the number of phases and their composition, density calculations show some systematic deviations. Peneloux (Danesh, 1998; Pedersen et al., 1989) proposed to use a corrected molar volume v' instead of v. It can be calculated by a shift parameter c according to

$$v' = v - c\,. \tag{5.35}$$

For SRK, the shift factor c can be estimated from the Rackett compressibility factor[8] Z_{RA} with

$$c = 0.40768\,(0.29441 - Z_{RA})\,\frac{RT_c}{p_c}\,. \tag{5.36}$$

For PR, it is sometimes tabulated in the form of the shift factor $S_E = c/b$ according to Jhaveri and Youngeren (Danesh, 1998).

The mixing rule for the volume shift is arithmetic similar to co-volume mixing in (5.20).

It must be explicitly noted that volume shifts, according to Peneloux or Jhaveri and Youngeren for density corrections, do not affect the compositional results of PT–flash calculations (Danesh, 1998). Hence, volume shift values can be modified independently of compositional changes for density calibration. Volume shifts of lumped pseudo–components are often calibrated with their densities.

Liquid Density

The density of a liquid at standard conditions is often quantified with °API as the unit of choice in the petroleum industry. This emphasizes the fact that the oil price is mainly determined by the oil density, which represents the most common quality factor. The conversion from the specific gravity S_o to °API is presented as

$$°\mathrm{API} = \frac{141.5}{S_o} - 131.5 \tag{5.37}$$

with S_o in g/cm^3. For example, densities from 1000 ... 702 kg/m^3 are mapped to 10 ... 70 °API. It should be noted that this definition becomes difficult for density differences: a density uncertainty stemming from a measurement error or an estimated error in a model of, for example, 1% of an 800 kg/m^3 oil produces an relative error of almost 4% for the corresponding value of 45 °API. This makes the API gravity a very sensitive parameter. High measurement and simulation accuracy must be reached for common °API accuracy.

Alternative to the use of densities, which are directly extracted from EOS, empirical methods such as the API or Standing–Katz methods or the optimized EOS of Alani–Kennedy, can be employed if phase compositions are known (Pedersen et al., 1989; Danesh, 1998). These formulas often yield very accurate density values but they are usually not interpreted as EOS because

[8] The Rackett compressibility factor has been introduced for considerations of critical properties (Danesh, 1998). It is tabulated in App. I for may components.

they are strongly limited in their pressure and temperature ranges. Moreover, predetermined grouping and lumping of components restricts their usability. However, the predictions of empirical formulas can be compared to the results of the more general EOS and used for the calibration of phase properties.

API Method

The API density prediction is based on an available compositional analysis of the components up to C_6 with appropriate component densities and the measured density of C_{7+} at standard conditions. The density ρ_l at temperature T and pressure p can be evaluated according to the following formulas:

$$\rho_l = \rho_1 \frac{C(288.706\,\mathrm{K})}{C(T)}\,, \qquad \rho_1 = \frac{\sum_i z_i M_i}{\sum_i z_i M_i/\rho_{0,i}}\,, \tag{5.38}$$

with M_i the molecular weight of component i, z_i the molar fractions, and $\rho_{0,i}$ the dissolved component density at standard conditions according to

Component	N_2	CO_2	H_2S	C_1	C_2	C_3	i–C_4	n–C_4	i–C_5	n–C_5	C_6
Density [kg/m^3]	804	809	834	300	356	508	563	584	625	631	664

Furthermore it is

$$\begin{aligned} C &= A_1 + A_2 T'_r + A_3 T'^2_r + A_4 T'^3_r \quad \text{with} \\ A_i &= B_{i1} + B_{i2} p'_r + B_{i3} {p'_r}^2 + B_{i4} {p'_r}^3 + B_{i5} {p'_r}^4 \,. \end{aligned} \tag{5.39}$$

The coefficients B_{ij} are defined through

$i\backslash j$	1	2	3	4	5
1	1.6368	−0.04615	2.1138×10^{-3}	-0.7845×10^{-5}	-0.6923×10^{-6}
2	−1.9693	0.21874	-8.0028×10^{-3}	-8.2328×10^{-5}	5.2604×10^{-6}
3	2.4638	−0.36461	12.8763×10^{-3}	14.8059×10^{-5}	-8.6895×10^{-6}
4	−1.5841	0.25136	-11.3805×10^{-3}	9.5672×10^{-5}	2.1812×10^{-6}

and it is $T'_r = T/T'_c$, $p'_r = p/p'_c$ with $T'_c = \sum_i z_i T_{ci}$ and $p'_c = \sum_i z_i p_{ci}$.

Standing–Katz Method

Standing and Katz proposed another set of formulas for density calculations by known oil composition: The density of a C_{3+} mixture is

$$\rho_{C_{3+}} = \frac{\sum\limits_{i=C_3}^{C_{7+}} z_i M_i}{\sum\limits_{i=C_3}^{C_{7+}} z_i M_i/\rho_{0,i}} \tag{5.40}$$

with $\rho_{0,i}$ again as the dissolved component density at standard conditions. Just as for the API method, $\rho_{C_{7+}}$ is a measured density. Methane and ethane are treated separately in

$$\begin{aligned}\rho_{C_{2+}} &= \rho_{C_{3+}}(1 - 0.01386\, w_{C_2} - 0.000082\, w_{C_2}^2) \\ &\quad +0.379\, w_{C_2} + 0.0042\, w_{C_2}^2 \,, \\ \rho_{C_{1+}} &= \rho_{C_{2+}}(1 - 0.012\, w_{C_1} - 0.000158\, w_{C_1}^2) \\ &\quad +0.0133\, w_{C_1} + 0.00058\, w_{C_1}^2\end{aligned} \tag{5.41}$$

with w_{C_1} as weight percent of C_1 in the total phase and w_{C_2} as weight percent of C_2 in C_{2+}. The density $\rho_{C_{1+}}$ is the apparent oil density, which must be corrected for pressure and temperature so that

$$\rho_l = \rho_{C_{1+}} + \Delta\rho_p - \Delta\rho_T \tag{5.42}$$

with

$$\begin{aligned}\Delta\rho_p &= \left(0.167 + 16.181 \times 10^{-0.0425\, \rho_{C_{1+}}}\right) \frac{p}{1000} \\ &\quad - 0.01 \left(0.299 + 263 \times 10^{-0.0603\, \rho_{C_{1+}}}\right) \left(\frac{p}{1000}\right)^2 , \\ \Delta\rho_T &= \left[0.0133 + 152.4(\rho_{C_{1+}} + \Delta\rho_p)^{-2.45}\right] (T - 60) \\ &\quad - \left[8.1 \times 10^{-6} - 0.0622 \times 10^{-0.0764(\rho_{C_{1+}} + \Delta\rho_p)}\right] (T - 60)^2 .\end{aligned} \tag{5.43}$$

Herein p is given in psi, T in Fahrenheit, and the densities in lbm/ft^3. The formulas are set up for $40 < \rho_{C_{3+}} < 60$ lbm/ft^3, $w_{C_1} < 16$ and $w_{C_2} < 10$. For low concentrations, CO_2 can be taken into account with a specific gravity of 0.420.

Alani–Kennedy EOS

The Alani–Kennedy EOS is another alternative for density calculations when the liquid compositions are known. It is an EOS which has been optimized empirically for liquid density calculations only:

$$v^3 - \left(\frac{R(T + 460)}{p} + b\right) v^2 + \frac{av}{p} - \frac{ab}{p} = 0 \,, \tag{5.44}$$

with v as molar volume in ft^3/lbmol, T as temperature in Fahrenheit, p as pressure in psi, and $R = 10.7335$. The goal is to find the smallest root of (5.44). It is possible to calculate the parameters a and b by molar arithmetic mixing from the corresponding pure component values. These pure components are parametrized with parameters λ, n, m, and C by

$$a = \lambda\, e^{n/(T+460)} \quad \text{and} \quad b = m(T+460) + C \tag{5.45}$$

which are tabulated in (5.2). The C_{7+} component has exceptional parameters

$$\begin{aligned} a_{C_{7+}} &= \exp\Big[3.8405985 \times 10^{-3}\, M_{C_{7+}} - 9.5638281 \times 10^{-4}\, M_{C_{7+}}/S_{C_{7+}} \\ &\quad + 261.80818/(T+460) \\ &\quad + 7.3104464 \times 10^{-6}\, M_{C_{7+}}^2 + 10.753517\Big]\,, \\ b_{C_{7+}} &= 3.499274 \times 10^{-2}\, M_{C_{7+}} - 7.2725403\, S_{C_{7+}} \\ &\quad + 2.232395 \times 10^{-4}(T+460) \\ &\quad - 1.6322572 \times 10^{-2}\, M_{C_{7+}}/S_{C_{7+}} + 6.2256545 \end{aligned} \tag{5.46}$$

with $S_{C_{7+}}$ as the specific gravity of the C_{7+} component. Finally the density can be calculated easily from the molar volume, the composition, and the molecular weights of the components.

Component	λ	n	$m \times 10^4$	C
C_1 (70 – 300 °F)	9160.6413	61.893223	3.3162472	0.50874303
C_1 (301 – 460 °F)	147.47333	3247.4533	−14.072637	1.8326695
C_2 (100 – 249 °F)	46709.537	−404.48844	5.1520981	0.52239654
C_2 (250 – 460 °F)	17495.343	34.163551	2.8201736	0.62309877
C_3	20247.757	190.24420	2.1586448	0.90832519
i–C_4	32204.420	131.63171	3.3862284	1.1013834
n–C_4	33016.212	146.15445	2.9021257	1.1168144
C_5	37046.234	299.62630	2.1954785	1.4364289
C_6	52093.006	254.56097	3.6961858	1.5929406
H_2S	13200.0	0	17.900	0.3945
N_2	4300.0	2.293	4.490	0.3853
CO_2	8166.0	126.00	1.8180	0.3872

Table 5.2. Component parameters of Alani–Kennedy EOS

The different methods for calculation of liquid densities are compared in Fig. 5.17 for a black oil and a volatile oil. Their compositions and component properties are listed in App. K. The amount and molar weight of the C_{7+} fraction, which is needed as an input parameter for the API and the Standing–Katz methods, are calculated by molar average with the following component densities:

Component	C_{6-14}	C_{15}	C_{25}	C_{35}	C_{45}
Density [kg/m^3]	700	800	850	920	940

The direct density calculation from composition and molecular weight in many examples, deviates a lot from the results of the other methods. This can be corrected by calibrated volume shifts and molecular weights of the lumped pseudo components without affecting the composition. More about calibration can be found in Sec. 5.7.

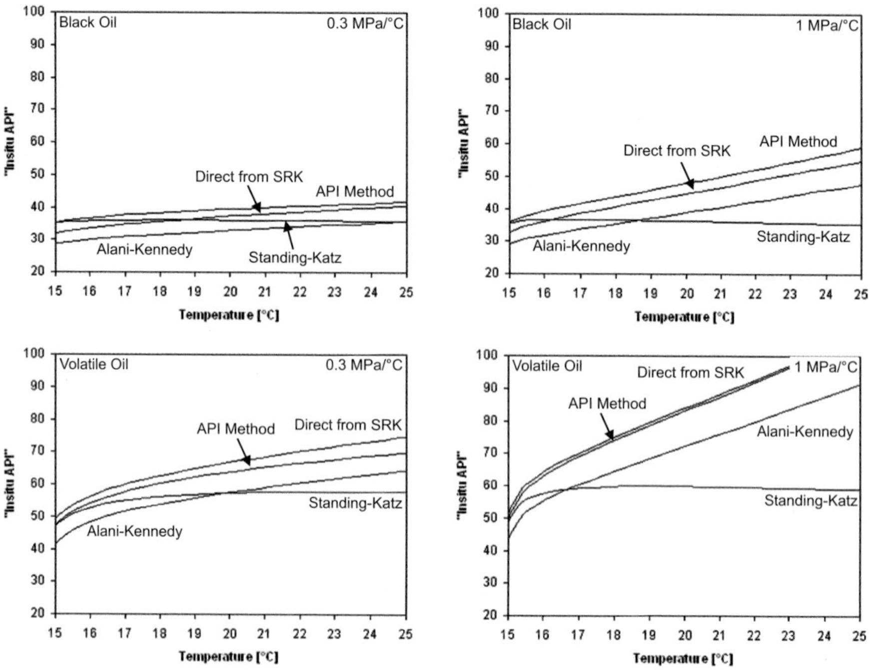

Fig. 5.17. Density of liquid phase calculated with different methods for the same black oil and volatile oil as shown in Fig. 5.14. Pressure and temperature follow different PT–paths with 0.3 MPa/°C and 1 MPa/°C. Both PT–paths start at normal conditions. The density is plotted in units of API and is therefore labeled "Insitu API". The composition of the liquid is calculated with the SRK EOS

API (at standard conditions)

	Volatile Oil	Black Oil
Direct from SRK	47.9	31.6
API Method	46.0	35.0
Standing–Katz	45.8	34.9
Alani–Kennedy	40.3	28.4

5.6.2 Bubble Point Pressure

The bubble point pressure p_b is also often called saturation pressure as can be easily seen from the discussion in Sec. 5.3. It can be read from diagrams such as shown in Fig. 5.23 or it can be calculated more accurately with special flash algorithms: At bubble point conditions it is

$$x_i = z_i \quad \text{and} \quad n_l = 1\,, \quad n_v = 0\,. \tag{5.47}$$

The relation

$$1 = \sum_i y_i = \sum_i x_i \frac{\phi_{l,i}}{\phi_{v,i}} \tag{5.48}$$

can be deduced from the equality of chemical potentials (5.29). Taking into account the definition (5.25) and the equality of fugacities $f_{v,i} = f_{l,i}$, Danesh (1998) proposed to iteratively solve for the bubble point pressure p_b according to

$$p_{b,n+1} = p_{b,n} \sum_i x_i \frac{\phi_{l,i}}{\phi_{v,i}} \, . \tag{5.49}$$

An algorithm for a bubble point flash can now be formulated as:

1 Calculate a_l, b_l with mixing rules
2 Estimate bubble point pressure p_b
3 Estimate K_i with (5.32)
4 Calculate $y_i = x_i K_i$ and a_v, b_v with mixing rules
5 Calculate Z_l and Z_v with (5.15) or (5.16)
6 Calculate $\phi_{l,i}$ and $\phi_{v,i}$ with (5.26)
7 Calculate $K_i^{\text{new}} = \phi_{l,i}/\phi_{v,i}$ and $p_b^{\text{new}} = \sum_i x_i K_i^{\text{new}}$
8 If $\sum_i (1 - K_i^{\text{new}}/K_i)^2 > \epsilon$ adjust $p_b \leftarrow p_b^{\text{new}}$, $K_i \leftarrow K_i^{\text{new}}$ and go back to step 4

The convergence criterion is, as in Sec. 5.5, again given by a small ϵ.

5.6.3 Gas Oil Ratio (GOR)

The GOR defines the gas to oil volume fraction of a produced fluid at standard conditions. It is often quantified in m^3/m^3 or in SCF/STB and it ranges up to 150.000 SCF/STB (Danesh, 1998). The GOR is an indicator of the amount of heavy components in the fluid. This is in accordance to the rule of thumb that mostly C_{6+} form the liquid phase at standard conditions. Inversely it can be interpreted as a measure of light components (especially methane) which are dissolved in the liquid under insitu conditions. Besides density it is the most important parameter for the classification of reservoir petroleum Sec. 5.5.1.

Outgassing along a typical PT path can be visualized very well with the GOR (Fig. 5.18). A black oil and a volatile oil both exsolve gas when reaching the bubble point from hotter and therefore deeper regions. These gases have a very high GOR. They are almost dry. The gas condensate behaves differently. Oil condensates when reaching the dew point.

5.6.4 Oil Formation Volume Factor B_o

Another important quantity is the oil formation volume factor B_o, which relates the subsurface insitu liquid volume (plus its dissolved vapor) to the

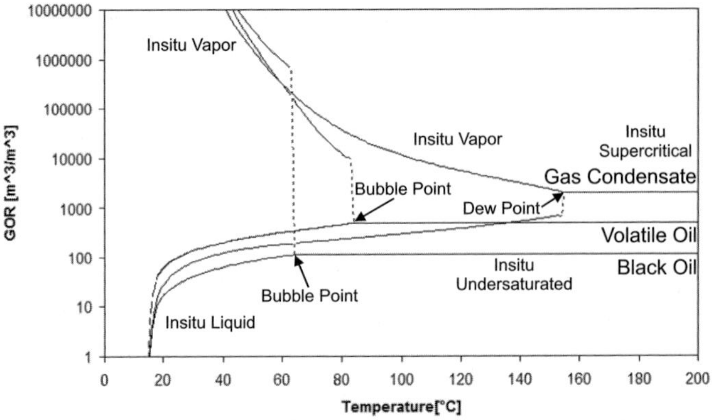

Fig. 5.18. GOR at standard surface conditions. It is calculated from separated phases at insitu conditions. The composition of the phases is calculated from the black oil, the volatile oil and the gas condensate of Fig. 5.14 with the SRK EOS along a PT–path of 0.3 MPa/°C, which ends at standard surface conditions. At the bubble or dew point each supercritical/undersaturated phase separates into liquid and vapor

liquid volume at surface conditions (Peaceman, 1977). It therefore defines the shrinkage of petroleum when it is produced. It can be evaluated to

$$B_o = \frac{W_l + W_v}{W_l} \frac{\rho_l}{\rho_{l,i}} \tag{5.50}$$

with liquid, vapor weights W_l, W_v, and liquid density ρ_l of a sample at standard conditions and liquid density $\rho_{l,i}$ at insitu conditions. It can empirically be determined with knowledge of liquid and vapor densities and the amount of vapor dissolved in the liquid under reservoir conditions. (Danesh, 1998). Obviously, the oil formation volume factor is tightly linked to the methane content in the liquid under insitu conditions. In modeling practice, two flash calculations, one for insitu and one for surface conditions must be performed for its calculation. It can be rewritten as

$$B_o = \left(1 + \frac{n_v \sum_k y_k M_k}{n_l \sum_k x_k M_k}\right) \frac{\rho_l}{\rho_{l,i}} \tag{5.51}$$

with x_k, y_k, n_l, and n_v at standard conditions. Two examples of B_o along two different PT–paths are shown in Fig. 5.19.

Both quantities, GOR and B_o are often referred to if considerations concerning light components especially methane are performed (di Primio and Horsfield, 2006). Methane content is often a key quantity in many geological processes e.g. the proper description of source rock kinetics, secondary cracking or biodegradation.

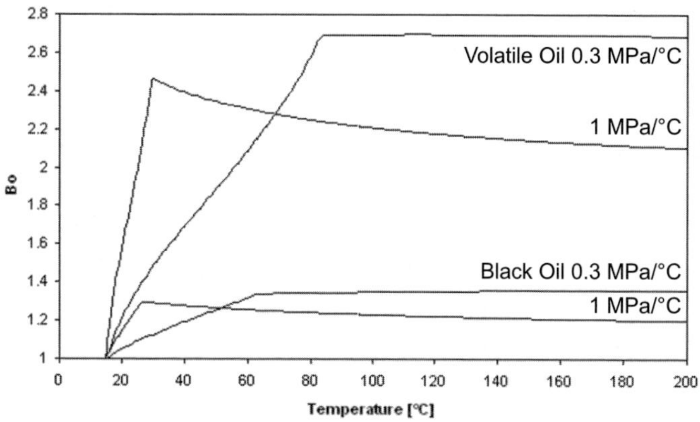

Fig. 5.19. Oil formation volume factor B_o calculated from the liquid phase of a black oil and of a volatile oil. The liquid phase is calculated with the SRK EOS along two different PT–paths which both end at standard surface conditions

5.6.5 Viscosity

Next to density, viscosity is an important indicator for phase property characterizations. It varies greatly between different oil types (Table 5.3). Due to the strong dependency of production recovery factors from viscosity it is of great interest.

	Viscosity in [cP] at		
Oil Type	50°C	100°C	150°C
very high viscosity	1000	25	2.5
high viscosity	100	2.5	0.5
medium viscosity	10	0.5	0.25
low viscosity	3	0.35	0.1
very low viscosity	1	0.25	0.05

Table 5.3. Revised viscosity values according to Ungerer et al. (1990)

Viscosity is often modeled as a quantity that is dependent only on pressure, temperature, density, and the amount of dissolved gas (Danesh, 1998). An overview about empirical correlations is given by Bergman and Sutton (2007). Models taking detailed compositional effects into account are found in (Pedersen et al., 1984; Pedersen and Fredenslund, 1987; Zéberg-Mikkelsen, 2001). Obviously, the effect of long–chained compounds has a significant influence on the viscosity. Hence a good characterization of the fluid heavy end is even more important for viscosity than for density prediction.

A recent comparison of actual models with measurement data can be found in Zéberg-Mikkelsen (2001). Advanced theories such as the friction–theory or the free–volume model match laboratory data best. But due to a lack of field data and unknown component parameters for the advanced theories, especially for many important heavy end compounds, it is common to follow more direct approaches such as the empirical Lohrenz–Bray–Clark (LBC) model (Lohrenz et al., 1964) or the corresponding states (CS) model (Pedersen et al., 1984; Pedersen and Fredenslund, 1987; Lindeloff et al., 2004) in practice.

Lohrenz–Bray–Clark (LBC) Model

The starting point of the LBC model are empirical formulas for the viscosity ν_0 of low–pressure pure component fluids:

$$\nu_0 = \begin{cases} 34 \times 10^{-5} (T/T_c)^{0.94} / \lambda & \text{for} \quad T/T_c \leq 1.5 \\ 17.78 \times 10^{-5} \left(4.58\,(T/T_c) - 1.67\right)^{5/8} / \lambda & \text{for} \quad T/T_c > 1.5 \end{cases} . \tag{5.52}$$

Herein, λ is called the viscosity reducing parameter. It has its origin in the kinetic theory of gases and is defined as

$$\lambda = T_c^{1/6}\, M^{-1/2}\, p_c^{-2/3} . \tag{5.53}$$

For higher pressures an empirical correlation is used. It has the form

$$\left[(\nu - \nu_0)\,\lambda + 10^{-4}\right]^{1/4} = a_0 + a_1\rho_r + a_2\rho_r^2 + a_3\rho_r^3 + a_4\rho_r^4 \tag{5.54}$$

with the reduced density $\rho_r = \rho/\rho_c$ and the coefficients

$$\begin{aligned} a_0 &= 0.1023, & a_1 &= 0.023364, & a_2 &= 0.058533, \\ a_3 &= -0.040758, & a_4 &= 0.0093324 . \end{aligned} \tag{5.55}$$

The model can be improved with an individual fit of the coefficients (5.55) to known viscosity data sets.

Multi–component systems are treated with Herning–Zipperer mixing rules. The low density viscosity is "mixed" according to

$$\nu_0 = \frac{\sum_i z_i \nu_{0i} \sqrt{M_i}}{\sum_i z_i \sqrt{M_i}} \tag{5.56}$$

and the viscosity reducing parameter following

$$\lambda = \left(\sum_i z_i T_{ci}\right)^{1/6} \left(\sum_i z_i M_i^{-1/2}\right)^{-1/2} \left(\sum_i z_i p_{ci}\right)^{-2/3} . \tag{5.57}$$

The mixing rule for the reduced density $\rho_r = \rho/\rho_c = v_c/v$ with v as the molar volume can be rewritten to a rule for the critical volume v_c. It is estimated with

$$v_c = \sum_i z_i v_{ci} \tag{5.58}$$

and for the C_{7+} according to

$$\begin{aligned} v_{c,C_{7+}} &= 1.3468 + 9.4404 \times 10^{-4} M_{C_{7+}} \\ &\quad -1.72651\, S_{C_{7+}} + 4.4083 \times 10^{-3} M_{C_{7+}} S_{C_{7+}} \end{aligned} \tag{5.59}$$

with S indicating the specific gravity. With these values and knowledge of the density ρ or the molar volume v, formula (5.54) can be used to estimate the viscosity of a mixture.

In the above formulas temperatures are in Kelvin, pressures in atmospheres, critical volumes in (5.59) in $\mathrm{m}^3/\mathrm{kmol}$ and viscosities in mPa s = cP.

Viscosities can be evaluated very fast due to the simple nature of the LBC–formulas. Models with lower performance are often not usable in fluid flow simulators. But it must be noted that the LBC–model is based on a polynomial of degree 16 as introduced in (5.54). Polynomials of such a high degree are known to easily become numerically unstable and therefore LBC based models must be evaluated with care.

Corresponding States (CS) Model

The principle of corresponding states is not only applied to viscosity predictions (Danesh, 1998). The central assumption is the observation that many properties behave similarly within the same "relative distance" from the critical point. This observation can be shown best with the vdW EOS in its "reduced" form

$$p_r = \frac{8\,T_r}{3\,v_r - 1} - \frac{3}{v_r^2}\,. \tag{5.60}$$

Here, "reduced quantities" are defined as $p_r = p/p_c$, $T_r = T/T_c$, and $v_r = v/v_c$. Equation (5.60) describes a universal behavior determined only by the relative distance to the critical point. It is only dependent on quantities which are occurring in reduced form.

Following the above scheme the CS model for viscosity prediction must basically consist of two parts: Firstly the viscosity behavior of a well known reference fluid must be quantified to a high degree of accuracy and secondly scaling procedures for mapping of this behavior to the fluid under investigation must be formulated.

In the model of Pedersen et al. (1984) methane is chosen as the reference fluid. Its viscosity can be calculated with formulas from Hanley et al. (1975, 1977):

$$\nu_{\mathrm{ref}}(\rho, T) = \nu_0 + \nu_1(T)\rho + \Delta\nu'(\rho, T)\,. \tag{5.61}$$

ν_0 is the dilute gas viscosity, which can be calculated with

$$\nu_0 = \sum_{i=1}^{9} GV_i T^{\frac{i-4}{3}} \tag{5.62}$$

and the coefficients

$$\begin{array}{ll} GV_1 = -2.090975 \times 10^5 & GV_2 = 2.647269 \times 10^5 \\ GV_3 = -1.472818 \times 10^5 & GV_4 = 4.716740 \times 10^4 \\ GV_5 = -9.491872 \times 10^3 & GV_6 = 1.219979 \times 10^3 \\ GV_7 = -9.627993 \times 10^1 & GV_8 = 4.274152 \\ GV_9 = -8.141531 \times 10^{-2} \, . & \end{array} \tag{5.63}$$

The term $\nu_1 \rho$ describes a low order density correction with

$$\nu_1 = 1.696985927 - 0.133372346 \left(1.4 - \ln \frac{T}{168.0}\right)^2 . \tag{5.64}$$

For methane of higher density the term $\Delta\nu'$ becomes more important. It is

$$\begin{aligned} \Delta\nu' = \exp &\left(j_1 + \frac{j_4}{T}\right) \times \\ &\left(\exp\left[\rho^{0.5}\left(j_2 + \frac{j_3}{T^{3/2}}\right) + \theta\rho^{0.5}\left(j_5 + \frac{j_6}{T} + \frac{j_7}{T^2}\right)\right] - 1.0\right) \end{aligned} \tag{5.65}$$

with

$$\theta = \frac{\rho - \rho_c}{\rho_c} \tag{5.66}$$

and the coefficients

$$\begin{array}{lll} j_1 = -10.35060586 & j_2 = 17.571599671 & j_3 = -3019.3918656 \\ j_4 = 188.73011594 & j_5 = 0.042903609488 & j_6 = 145.2902344 \\ j_7 = 6127.6818706 \, . & & \end{array} \tag{5.67}$$

Pedersen and Fredenslund (1987); Pedersen et al. (1989) extended (5.61) to temperatures below the freezing point of methane at $T_F = 91\,\mathrm{K}$ which corresponds to reduced temperatures below 0.4. Equation (5.61) becomes now

$$\nu_{\mathrm{ref}}(\rho, T) = \nu_0 + \nu_1(T)\rho + F_+ \Delta\nu'(\rho, T) + F_- \Delta\nu''(\rho, T) \tag{5.68}$$

with an additional term

$$\begin{aligned} \Delta\nu'' = \exp &\left(k_1 + \frac{k_4}{T}\right) \times \\ &\left(\exp\left[\rho^{0.5}\left(k_2 + \frac{k_3}{T^{3/2}}\right) + \theta\rho^{0.5}\left(k_5 + \frac{k_6}{T} + \frac{k_7}{T^2}\right)\right] - 1.0\right) , \end{aligned} \tag{5.69}$$

the coefficients

$$k_1 = -9.74602 \quad k_2 = 18.0834 \quad k_3 = -4126.66$$
$$k_4 = 44.6055 \quad k_5 = 0.976544 \quad k_6 = 81.8134 \tag{5.70}$$
$$k_7 = 15649.9$$

and the weight factors

$$F_\pm = \frac{1 \pm \tanh(T - T_F)}{2} \tag{5.71}$$

which ensure a continuous crossover between ν' and ν'' at the freezing temperature T_F. The viscosity is calculated in units of μP with the density in g/cm^3 and the temperature in K.

The viscosity can now be calculated as mentioned above by relative scaling to the critical point. The general formula has the form

$$\nu = \left(\frac{T_c}{T_{c,\mathrm{ref}}}\right)^{-1/6} \left(\frac{p_c}{p_{c,\mathrm{ref}}}\right)^{2/3} \left(\frac{M}{M_\mathrm{ref}}\right)^{1/2} \frac{\alpha}{\alpha_\mathrm{ref}} \nu_\mathrm{ref}(p_\mathrm{ref}, T_\mathrm{ref}) \,. \tag{5.72}$$

Obviously, the first three factors of (5.72) represent a scaling with the viscosity reducing parameter, which has already been introduced in (5.57) for the LBC model. Additionally corrective factors α and α_ref are introduced. These α–factors are analogously used for a correction of the reference pressure and temperature following

$$p_\mathrm{ref} = p \frac{p_{c,\mathrm{ref}}}{p_c} \frac{\alpha_\mathrm{ref}}{\alpha} \quad \text{and} \quad T_\mathrm{ref} = T \frac{T_{c,\mathrm{ref}}}{T_c} \frac{\alpha_\mathrm{ref}}{\alpha} \tag{5.73}$$

and can be calculated from

$$\begin{aligned} \alpha &= 1 + 7.378 \times 10^{-3} \rho_r^{1.847} M^{0.5173} \,, \\ \alpha_\mathrm{ref} &= 1 + 0.031 \rho_r^{1.847} \,. \end{aligned} \tag{5.74}$$

The reduced density ρ_r which is necessary for the calculation of the α–factors can obviously only be evaluated without the corrections itself:

$$\rho_r = \frac{\rho_\mathrm{ref}\left(p \dfrac{p_{c,\mathrm{ref}}}{p_c}, T \dfrac{T_{c,\mathrm{ref}}}{T_c}\right)}{\rho_{c,\mathrm{ref}}} \,. \tag{5.75}$$

The methane reference density without α–corrections in (5.75) or with α–corrections in (5.68) can be calculated with a modified Benedict–Webb–Rubin EOS proposed by McCarty (1974). For the sake of completeness this lengthy formula is listed in App. J.

Similar as in the LBC model the critical quantities T_c, p_c and the molecular weight M must initially be calculated with mixing rules from pure component values. The mixing of critical quantities follows a procedure similar to the Lee–Kesler rules (5.6):[9]

$$T_c = \frac{\sum_{ij} z_i z_j \sqrt{T_{ci} T_{cj}} V_{cij}}{\sum_{ij} z_i z_j V_{cij}} \quad \text{and} \quad p_c = \frac{8\,T_c}{\sum_{ij} z_i z_j V_{cij}} \tag{5.76}$$

with

$$V_{cij} = \left[\left(\frac{T_{ci}}{p_{ci}} \right)^{\frac{1}{3}} + \left(\frac{T_{cj}}{p_{cj}} \right)^{\frac{1}{3}} \right]^3 . \tag{5.77}$$

The molecular weight is calculated by

$$M = M_n + 1.304 \times 10^{-4} \left(M_w^{2.303} - M_n^{2.303} \right) \tag{5.78}$$

with the molecular weight M in g/mol and the molar average M_n and mass average M_w defined by

$$M_n = \sum_i z_i M_i \quad \text{and} \quad M_w = \frac{\sum_i z_i M_i^2}{\sum_i z_i M_i} . \tag{5.79}$$

CS for Heavy Oils

The CS model can be extended to heavy oils (Lindeloff et al., 2004). Instead using methane as the reference compound a correlation for heavy oils proposed by Rønningsen (1993) is used. It is for viscosity ν_0 under atmospheric pressure conditions

$$\log_{10} \nu_0 = -0.07995 - 0.01101 M' - \frac{371.8}{T} + \frac{6.125 M'}{T} \tag{5.80}$$

with

$$M' = M_n \quad \text{for} \quad \frac{M_w}{M_n} \leq 1.5 \quad \text{and} \quad M' = M_n \left(\frac{M_w}{1.5 M_n} \right)^{0.5} \quad \text{else.} \tag{5.81}$$

The first case describes a stabe oil and the second a live oil.[10] The exponent 0.5 and the coefficient 1.5 can be used as tuning parameters. Viscosity ν_0 is in mPa s = cP and temperature in K.

---

[9] Here it is assumed that all components have the same critical compressibility Z_c!

[10] A live oil contains dissolved gas that may be released at surface conditions (www.glossary.oilfield.slb.com). Correspondingly, a stable oil (dead oil) does at surface conditions not contain dissolved gas (anymore). It is in thermodynamic equilibrium.

The correlation is only valid for atmospheric pressure. A pressure correction of the form

$$\frac{1}{\nu_0}\frac{\partial \nu}{\partial p} = 0.008/\text{atm} \tag{5.82}$$

is assumed.

The heavy oil correlation is applied for methane reference temperatures $T_{\text{ref}} < 65\,\text{K}$. Below 75 K a smooth crossover from methane as the reference compound can be assumed. This can be modeled with a formula such as (5.71).

Other modifications and extensions of the CS theory are reported in the literature. For example $n-$decane is used as the reference fluid by Dexheimer et al. (2001).

Viscosities of two example oils according to LBC and CS theory are shown in Fig. 5.20. They are strongly dependent on composition and the type of components. A heavy end characterization might be necessary and a calibration of fluid data against the LBC coefficients (5.55) or against properties of lumped components can usually not be avoided. The curves in Fig. 5.20 are therefore only example curves.

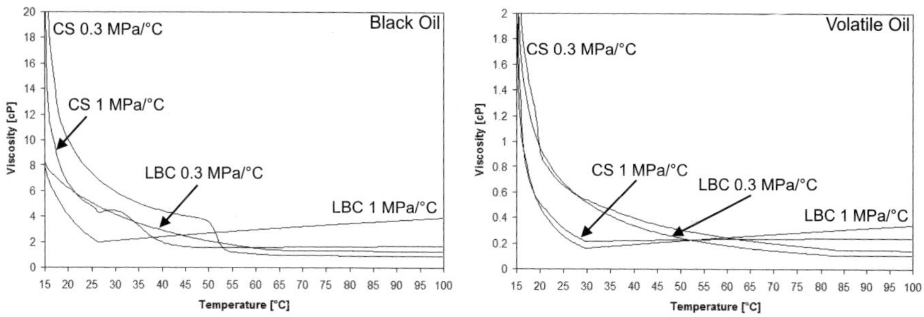

Fig. 5.20. Liquid phase viscosities according to LBC and CS theories. The black oil and the volatile oil, which are shown in Fig. 5.14 and which are more precisely defined in (App. K), are chosen as examples. Viscosities are calculated for two different PT–paths

5.6.6 Interfacial Tension (IFT)

A precise prediction of IFT is very sophisticated. It can only be performed if a detailed analysis of the phase composition and physical behavior is performed down to a molecular level. In basin modeling, IFT prediction is therefore restricted to basic approximations.

Interfacial tension is a property which is dependent on two phases. All combinations between different fluid phases must be treated explicitly. Obviously, the correct distribution of the components in the different phases is a prerequisite. This can be gained by flash calculations.

Liquid–Vapor Interfacial Tension

The liquid–vapor IFT $\gamma_{l,v}$ is usually determined with the parachor method

$$\left(\gamma_{l,v}\right)^{1/4} = P_\gamma(\rho_{n,l} - \rho_{n,v}) \tag{5.83}$$

with $\gamma_{l,v}$ in mN/m, molar densities $\rho_{n,l}$, $\rho_{n,v}$ in mol/cm^3, and "parachor" constant P_γ. Parachor values are tabulated for components and usually mixed based on molar fractions (Danesh, 1998). Equation (5.83) then becomes

$$\left(\gamma_{l,v}\right)^{1/4} = \rho_{n,l} \sum_i x_i P_{\gamma,i} - \rho_{n,v} \sum_i y_i P_{\gamma,i} \,. \tag{5.84}$$

Petroleum–Water Interfacial Tension

Usually petroleum–water IFT is listed in lookup–tables for simulation. IFT values are mainly used for determination of capillary pressure for migration and accumulation in basin modeling (Sec. 6.3.1). Very often rough estimates such as constant values of $\gamma_{p,w} = 42\,\text{mN/m}$ are used because uncertainties in IFT can be neglected compared to uncertainties of knowledge about pore throat radii.

However, it is found that pressure dependency of IFT is much weaker than temperature dependency. An improved empirical correlation which reflects this behavior is

$$\gamma_{p,w} = 111(\rho_w - \rho_p)^{1.024}(T/T_{p,c})^{-1.25} \tag{5.85}$$

with the densities in g/cm^3 and $\gamma_{p,w}$ in mN/m (Danesh, 1998, Fig. 5.21). The critical temperature $T_{p,c}$ of the petroleum phase must be known. Its calculation with mixing rules for critical parameters such as the Lee–Kesler–Average (5.6) are reported in the literature.

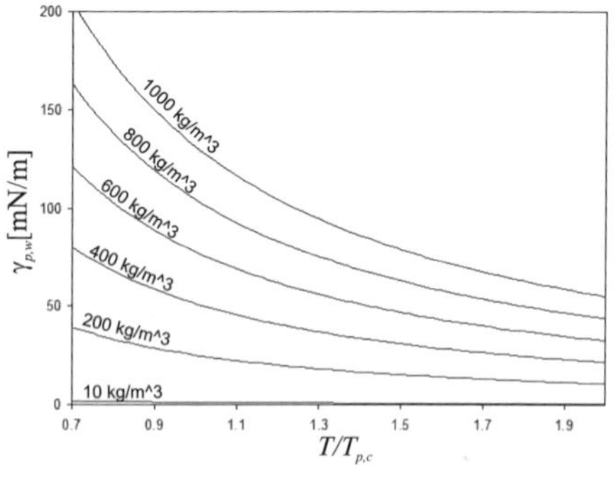

Fig. 5.21. Petroleum–water interfacial tension $\gamma_{p,w}$ according to (5.85) for different density contrasts $\rho_w - \rho_p$

5.7 Calibration of a Fluid Model

Besides overpressure and heat flow calibration (Secs. 2.4, 3.9) it is also common to calibrate against fluid data. The type of EOS, the choice of an appropriate set of binary interaction parameters and the method of grouping and lumping of pseudo components, especially for the heavy end of the fluids, are frequently calibrated against easily available fluid sample data. In addition to the compositional analysis, these are often basic properties such as density, viscosity, gas oil ratio (GOR) at surface conditions and bubble point pressure at surface temperature. Other properties, such as heat capacities, or more detailed results from fluid analysis, such as swelling tests, are usually not taken into account for flash model calibrations in basin modeling.

Flash calculations are sometimes used for compositional predictions but not for density calculations of liquids. Liquid densities at or near standard conditions can also be calculated with the API method, the Standing–Katz procedure or the optimized EOS of Alani–Kennedy (Sec. 5.6.1) in a post processing approach in case of known composition. These methods have a high degree of accuracy within a limited range of applicability. Evidently, they can only be used for the analysis of final simulation results and not for the entire simulation. Consistency between the more general EOS and the precise empirical formulas can be reached by a calibration of both methods against each other. Typically this can be done by proper selection of the pseudo–components and heavy end analysis (see 5.7.1 and 5.7.2 below). Obviously, this only makes sense if fluid sample data is not available.

In basin modeling the overall fluid model calibration is not a matter of fluid analysis only. The choice of a proper set of pseudo–components is not exclusively based on fluid characterization. The set of pseudo–components should be on the one hand as small as possible for high simulation performance and on the other hand sufficient to correctly model the PVT–behavior as well as the basic geochemical processes such as generation kinetics and source rock maturation. Secondary cracking or biodegradation are examples of component specific processes, which should also be taken into account for the choice of components.

Methane content is a key quantity in many geological processes. It strongly effects all easily accessible fluid parameters such as the gas oil ratio, the oil formation volume factor or the bubble point pressure. Due to its volatile nature, it is often used as a calibration parameter for fluid analysis. However, the methane content is usually calibrated against the source rock type and its associated kinetic (di Primio and Skeie, 2004; di Primio et al., 1998; di Primio and Horsfield, 2006). Fluid model calibration with the methane content therefore becomes obsolete with the quantification of a source rock kinetic.

In general it is assumed that the overall composition is not a matter of calibration in fluid analysis. The stoichiometry of geochemistry and mass conservation in basin modeling permit the calibration of parameters which change the composition of the simulation results.

Many kinetics are formulated for a predetermined set of pure and pseudo–components. They thus specify the components of the model. The number of the components is usually relatively small. Good quantitative information about the geochemistry of a source rock is often rare and a small set is necessary or at least very advantageous for a good simulation performance of a large basin model. Hence, the number of HC fluid components seldom exceeds 14.[11] However, there are still degrees of freedom for the calibration of the fluid model.

5.7.1 Calibration and Fluid Heavy End

The fluids heavy end is often relatively unspecified due to the limited number of pseudo–components given by typical kinetics. Hence, heavy end characterization parameters can be used as calibration parameters in basin modeling.

A rough approximation can be based on a fit of the molecular weight of the heaviest pseudo–component against density or API. The PR, the SRK EOS and the fugacities do not depend explicitly on molecular weights of components. Thus the composition of the phases does not change under the variation of molecular weights, whereas density varies linearly with its change. Such a fit is straight forward and easy to perform.

The lumped heavy end $z_{\mathrm{C_{n+}}}$ with its mass $M_{\mathrm{C_{n+}}}$ is usually known from modeling results or from sample analysis. From detailed analysis of real samples it has been found that heavy HCs are often distributed exponentially in single carbon number (SCN) groups $\mathrm{C_n}$ according to

$$z_{\mathrm{C_n}} \propto \exp(-F M_{\mathrm{C_n}}) \tag{5.86}$$

with F as a sample–dependent constant and $M_{\mathrm{C_n}}$ as the molecular weight of the lumped component $\mathrm{C_n}$ consisting of pure components with n carbon atoms (Danesh, 1998). Frequently, fluid PVT–property predictions can be improved by former expansion of heavy pseudo–components according to this scheme.[12]

Partly grouping and lumping these expanded components with regard to the improved predictions, leads to a new reduced set of pseudo–components, which is much smaller than the expanded one but more complex than the

[11] If component tracking is modeled the number of components is sometimes much higher. However, in such cases the number of components of the kinetic of each source rock is still limited to a small number. Because of this and the fact that source rock potentials can be compared better if the "assigned components" are physically the same, it turns out that in general the overall number of physically distinct components is rather small.

[12] With the assumption $M_{\mathrm{C_{n+1}}} - M_{\mathrm{C_n}} =$ const. it is possible to analytically calculate a good initial guess $F \approx 1/(M_{\mathrm{C_{n+}}} - M_{\mathrm{C_{n_0}}})$ from (5.86) for a heavy end distribution starting at $\mathrm{C_{n_0}}$. In practice it is conveniently assumed that $z_{\mathrm{C_n}} = 0$ for large n, e.g. $n > 45$.

original set. A workflow of this type is often integrated in the proper selection of the entire component set.

The predicted fluids in a basin model are sometimes "re-expanded" for higher accuracy of the final result. However, operations such as heavy end expansions, grouping or lumping during the simulation are usually not performed, due to a lack of computing performance.

More details about heavy end characterizations can be found in Pedersen et al. (1989) or Pedersen and Christensen (2007).

5.7.2 Tuning of Pseudo–Component Parameters

Sometimes systematic errors in the process of pseudo–component definition concerning fluid analysis appear. Lumping rules are often only based on rough semi–empirical formulas such as molar averages. The resulting pseudo–component parameters are not of the expected quality. They describe physical properties of a pure one pseudo–component fluid. But they are multi-component systems themselves and therefore more complex than a pure chemical species. It is both theoretically and experimentally difficult to determine the critical quantities of a pseudo–component. Furthermore, lumping heavy end components usually decreases the size of the coexistence area because the most heavy components are removed (Fig. 5.22). Hence, the resulting pseudo–component parameter sets are often only starting points for a calibration against real data. It must also be mentioned that EOS such as SRK or PR are themselves based on various assumptions and approximations. A calibration of component parameters individually to SRK or PR EOS enhance at least the predictive quality of the basin model. Therefore, tuning or calibration of pseudo–components can be very helpful in basin modeling.

It is easy to modify pseudo–component parameters so, that some quantities, such as the size of the coexistence area, fit better after lumping (Fig. 5.23). However, all quantities are non-linearly coupled and an improvement such as in (Fig. 5.23) could here only be reached for the price of making the GOR prediction worse.

The most important quantities for calibration are API, GOR, bubble point pressure p_b and oil formation volume factor B_o (Stainforth, 2004). They should at least be calibrated against all pseudo–component parameters namely critical temperature T_c, pressure p_c, acentric factor ω, and molecular weight M.

In principle this can be subdivided into two major steps: Firstly T_c, p_c, and ω can be fitted against GOR, p_b, and B_o as these quantities are independent of the molecular weights of the pseudo–components. Molecular weights enter equations of PVT–analysis only for the calculation of API and densities. As mentioned in the previous section, this relation is linear and it therefore follows that API can easily be fitted by variation of the molecular weights of the pseudo–components.

In practice both steps can be performed at once with a Markov chain Monte Carlo (MCMC) inversion algorithm. Flash calculations should there-

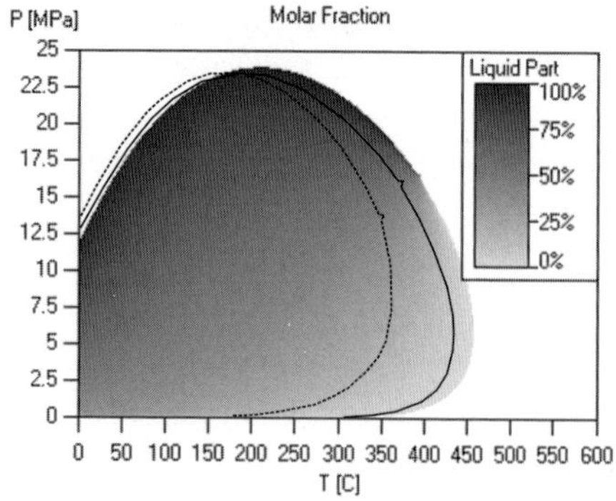

Fig. 5.22. Coexistence area of a fluid which consists of SCN components ranging from methane up to C_{45}. The solid line limits the coexistence area for the same fluid modeled with a lumped heavy end of the form C_{7-25} and C_{26-45} and the dashed line refers to further lumping with only one C_{7+} component

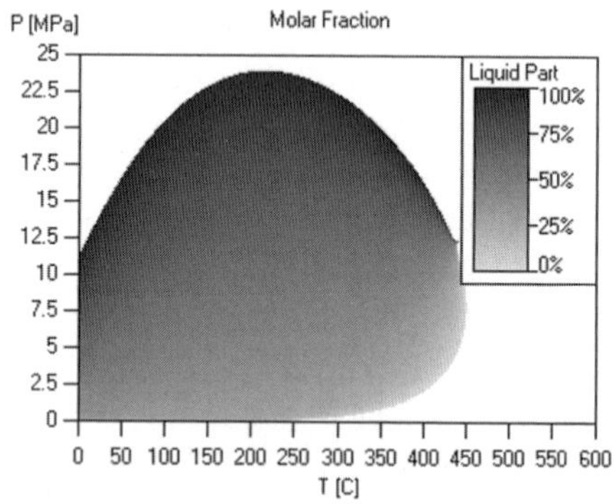

Fig. 5.23. Coexistence area of the same fluid as in Fig. 5.22 with one heavy pseudo–component C_{7+}. Here p_c and T_c of C_{7+} were shifted manually from $p_c = 2.14\,\text{MPa}$ and $T_c = 415.8\,°\text{C}$ to $p_c = 1.7\,\text{MPa}$ and $T_c = 500\,°\text{C}$ so that the size of the coexistence area again approaches the original size. The composition is listed in Table 5.4. Note that some deviations still exist. Especially the location of the critical point is different. Additionally, the fraction of liquid and vapor is altered. GOR changed from 212 to $165\,\text{m}^3/\text{m}^3$. API could be adjusted manually by variation of $M_{C_{7+}}$ from 193.3 to 247 g/mol. In summary, C_{7+} parameters were changed dramatically to achieve a calibration against a coexistence area size and API only

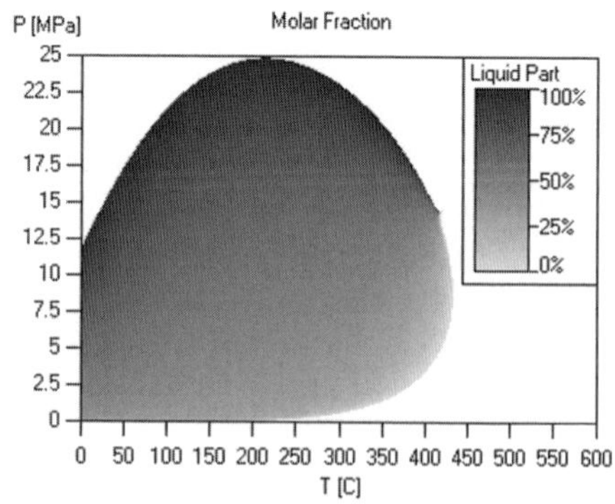

Fig. 5.24. Coexistence area of the same fluid as in Fig. 5.22 and Fig. 5.23 with one heavy pseudo–component C_{7+}. Here the pseudo–component parameters are fitted automatically with the MCMC against the results from the unlumped composition. The fitting procedure was repeated under manual change of T_c and p_c to achieve a coexistence area of almost the same size as for the original fluid in Fig. 5.22. A Bayesian objective provided a limit of change in respect to M, T_c, and p_c of C_{7+}. GOR and API are exactly matched with $M = 191\,\text{g/mol}$, $T_c = 492°\text{C}$ and $p_c = 2.07\,\text{MPa}$

	Molar Fraction [%]	M [g/mol]	T_c [°C]	p_c [MPa]	v_c [m^3/kmol]	Acentric Factor	Volume Shift [m^3/kmol]
Methane	50	16.043	−82.59	4.599	0.0986	0.0115	0.0007
Ethane	6	30.070	32.17	4.872	0.1455	0.0995	0.0028
Propane	5	44.096	96.68	4.248	0.2000	0.1523	0.0052
n–Butane	3	58.123	151.97	3.796	0.2550	0.2002	0.0080
n–Pentane	2	72.150	196.55	3.370	0.3130	0.2515	0.0122
C_6	2	84.000	236.85	3.271	0.3480	0.2510	0.0134
C_{7+}	32	247.000	500.00	1.700	0.7237	0.5295	0.0778

Table 5.4. Composition of an "Example–Oil". The corresponding phase diagram is shown in Fig. 5.23

fore be implemented efficiently because MCMC inversions are based on many model evaluations, typically 100.000 times and more.

A prototype for the objective χ^2_{PVT} of the MCMC inversion is given by (7.12) and becomes

$$\begin{aligned} \chi^2_{\mathrm{PVT}} = & \left(\frac{\mathrm{API} - \mathrm{API}_m[p_{ci}, T_{ci}, \omega_i, M_i]}{\Delta_{\mathrm{API}}}\right)^2 \\ & + \left(\frac{\mathrm{GOR} - \mathrm{GOR}_m[p_{ci}, T_{ci}, \omega_i]}{\Delta_{\mathrm{GOR}}}\right)^2 \\ & + \left(\frac{p_b - p_{bm}[p_{ci}, T_{ci}, \omega_i]}{\Delta_{p_b}}\right)^2 + \left(\frac{B_o - B_{om}[p_{ci}, T_{ci}, \omega_i]}{\Delta_{B_o}}\right)^2 . \end{aligned} \tag{5.87}$$

The index m points out that the related quantities are modeled with flash calculations and the square brackets indicate a dependency on the set of all pseudo–components, which are labeled here with the index i. The importance of each quantity in the inversion is controlled by its Δ factor. Decreasing Δ increases the importance in the same way as the reduction of a measurement error increases the importance of the related measurement. Minimization of (5.87) calibrates the fluid model. The MCMC algorithm is briefly outlined in Sec. 7.5.6.

An application of this algorithm to the lumped fluid is depicted in Fig. 5.22 with the calibration values taken from the original unlumped results quickly providing a perfect calibration. Unfortunately, the size of the coexistence area remains smaller than the one from the original data. Changing the critical pressure or temperature of the C_{7+} pseudo–component manually, before performing a new MCMC inversion, leads to a different calibrated end result with a differently coexistence area. The MCMC finds just one solution out of multiple possibilities. In such a case it is advantageous to add a Bayesian term to (5.87) to ensure that the MCMC algorithm does not alter the pseudo–component parameters too much. The objective becomes now

$$\phi_{\mathrm{PVT}} = \chi^2_{\mathrm{PVT}} + \phi_{T_c} + \phi_{p_c} + \phi_\omega + \phi_M \tag{5.88}$$

with e.g.

$$\phi_{T_c} = \sum_{i=1}^{N} \left(\frac{T_{ci} - T_{c0i}}{\Delta_T} \right)^2 , \tag{5.89}$$

N as the number of components, and T_{c0i} the critical value from which the calibration result should not deviate too much. The formulas for ϕ_{p_c}, ϕ_ω, and ϕ_M can be constructed analogously. Smaller the Δ factors are chosen as less variations of the related parameters are allowed. Iterative modification of the pseudo–component parameters and usage of the MCMC inversion quickly leads to the result shown in Fig. 5.24.

It must be noted that a calibration via the variation of molecular weight M, violates mass conservation. This is demonstrated with an extreme case: a lumped two-component system with API = 29, GOR = 300 m^3/m^3 and p_b = 11 MPa and pseudo–component parameters, as in Table 5.5, is calibrated against API = 30, GOR = 100 m^3/m^3, and p_b = 14 MPa. The molar fraction of both pseudo–components is kept constant. A calibration is easily possible by varying "Medium Oil" to "Medium Oil A". However, the variation is dramatic. The original "Medium Oil" has properties in the range of C_5 – C_8 on a SCN – scale, whereas the calibrated component parameters can be found in a wide region around C_{15}. The molecular weight is more than doubled. Alternatively, a calibration can also be performed under a constant mass fraction constraint for each component. Here this is unfortunately not possible via variation of only the heavy component. However, a calibration of both components yields quickly new parameters "Dry Gas B" and "Medium Oil B". The variation of the pseudo–component parameters is less drastic. Note the variation of the molar composition.

	Molar Fraction [%]	M [g/mol]	T_c [°C]	p_c [MPa]	Acentric Factor
Dry Gas	50.0	17.943	−75.730	4.850	0.02210
Medium Oil	50.0	101.141	238.770	3.459	0.26060
Medium Oil A	50.0	208.075	292.118	1.768	1.24167
Dry Gas B	34.7	37.498	−90.806	5.298	0.10547
Medium Oil B	65.3	112.313	207.621	2.790	1.07016

Table 5.5. Two component example

5.7.3 Tuning of the Binary Interaction Parameter (BIP)

It is rather unusual in basin modeling to perform a fine–tuning of the BIP because it is assumed that the modeling of large compositional variations under the strongly varying subsurface conditions of geological processes, are not significantly improved with fine–tuned parameters, which are often adapted to near surface conditions. Besides this, the usage of the BIP decreases the performance of flash calculations so that the overall simulator might slow down significantly. As for heavy end expansions, BIPs are sometimes tuned for higher quality of the final results.

5.8 Gas Hydrates

Clathrate hydrates are solid state phases which consist of water in "icy form" with other chemical species occupying the "ice–cavities". These other chemical species are often gases under normal conditions such as methane, carbon dioxide or hydrogen sulfide which are small enough to occupy even small sized molecular cavities. Additionally, these species appear frequently enough to form significant hydrate amounts in natural systems. It is therefore common to restrict the modeling to gas hydrates. Very often only methane hydrates are modeled. Methane is generated in each depth level and time scale of a basin model. It is produced in deeply buried gas prone source rocks, by secondary cracking in reservoir rocks, by biodegradation or by direct transformation of deposited organic matter in shallow sequences. It is also highly mobile in regards to primary and secondary migration with sufficient water solubility to allow it to be transported by groundwater (Sec. 5.2). Thus methane forms the most important hydrate deposits in natural systems.

The *PT* phase diagram for methane hydrate stability of a pure water methane system is depicted in Fig. 5.25. The phase boundaries are calculated from an empirical equation of the form

$$p = \exp(a - b/T) \tag{5.90}$$

with p in kPa (Sloan Jr., 1998). Parameters a and b and values for the quadruple points[13] are listed in Table 5.6. The influence of other chemical species, such as salt must be taken into account in more realistic systems. Impurities often act as hydrate inhibitors.

The formation of gas hydrates has some implications on bulk lithology properties such as a change of thermal conductivity (Fig. 3.8). However, gas hydrate layers are usually thin compared to the extensions of a basin model and for that reason it can be assumed that their overall impact is small. An exception is a drastic reduction of permeability because pore throats are choked by hydrates. A layer which contains hydrates usually acts as a seal for water flow and petroleum migration.

[13] Four phases coexist at a quadruple point.

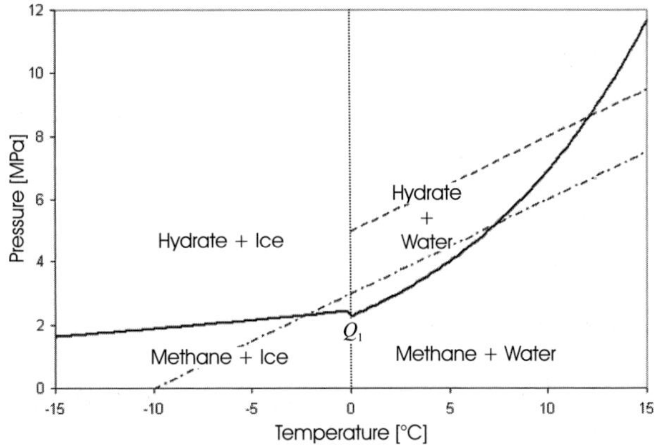

Fig. 5.25. Methane water phase diagram according to (5.90). Methane is meant to be in the vapor and water in the liquid phase. The quadruple point Q_1 is located at −0.2°C and 2.563 MPa (Table 5.6). The dashed line indicates a temperature pressure PT–path. It starts at 5 MPa, which is equivalent to a water depth of about 500 m, and continues with a linear temperature gradient of 30 K/km and hydrostatic conditions of 10 MPa/km. The dash–dotted line depicts a permafrost example with the same temperature and pressure gradients. For both PT–path examples it is possible that hydrates form in a limited depth range above the phase separation line (Gas Hydrate Stability Zone – GHSZ). Note that according to (5.90) the phase boundary between hydrate and water and the boundary between hydrate and ice are fitted empirically and independently. This leads to a small artificial discontinuity at the quadruple point

	Q_1		Q_2	
	T[°C]	p[MPa]	T[°C]	p[MPa]
CH_4	−0.2	2.563	–	
CO_2	0	1.256	9.9	4.499
H_2S	−0.3	0.093	29.6	2.293

Table 5.6. Quadruple points and parameters for (5.90) from Sloan Jr. (1998). Note, that the formula for CO_2 is limited to 9.9 °C due to a second quadruple point Q_2

Methane CH_4			
$-25\,°C < T < -0.2\,°C$		$-0.2\,°C < T < 25\,°C$	
a	b[K]	a	b[K]
14.717	1886.79	38.980	8533.80
Carbon Dioxide CO_2			
$-25\,°C < T < 0\,°C$		$0\,°C < T < 9.9\,°C$	
a	b[K]	a	b[K]
18.594	3161.41	44.580	10246.28
Hydrogen Sulfide H_2S			
$-25\,°C < T < -0.3\,°C$		$-0.3\,°C < T < 25\,°C$	
a	b[K]	a	b[K]
16.560	3270.41	34.828	8266.10

Summary: Water, liquid HCs and vapor HCs are the fluid phases, which occur most frequently in sedimentary basins.

HCs are sometimes only present in one supercritical or undersaturated phase with methane bearing the only HC component which is found in non–negligible amounts in water. H_2O does not dissolve in petroleum and separates strongly from the other phases. Therefore, the prediction of phase amounts, compositions and properties is reduced to a consideration of the HCs only.

Some basic aspects of phase separation are demonstrated with the highly simplified fixed phase and the symmetrical black oil model. These models are only valid in limited pressure and temperature intervals.

Basin wide predictions can only be performed with a more precise specification of the pressure–volume–temperature (PVT) relationships of the fluids. These relationships are called equations of state (EOS). The most well known EOS are the Soave–Redlich–Kwong and the Peng–Robinson EOS. Phase amounts and compositions can be calculated by minimization of thermodynamical potentials, namely the Gibbs free energy, in multi-compositional resolution. The resulting algorithm is called a flash calculation and is well known for its accurate results.

The properties of phases can be further studied based on compositional information and the consideration of empirical correlations. The focus is put on density and API predictions, bubble point pressure, gas oil ratio (GOR), the oil formation volume factor (B_o) and the viscosity. However, viscosity predictions are very difficult.

Overall fluid compositions are not well known in basin models. Sparsely available source rock data, limited geochemical component resolution and many uncertainties about the details of the geological processes, involved in generation and expulsion, do not allow high compositional resolutions. Besides this, memory restrictions in computer simulations also limit the amount of processable components. Hence components are often lumped crudely together. A maximum number of 14 pseudo components is seldom exceeded. However, pseudo components are specified by parameters which have degrees of freedom for calibration. A procedure for simultaneous calibration of pseudo component parameters against known properties of fluid samples is therefore possible. The predictive quality of the fluid analysis can be improved significantly when fluid sample data is available.

Finally, the conditions for formation of gas hydrates (Gas Hydrate Stability Zone – GHSZ) are outlined.

References

R. Battino. *Methane*, volume 27/28 of *Solubility data series*. Pergamon Press., 1987.

R. Becker. *Theorie der Wärme*. Springer–Verlag Berlin Heidelberg, 3. Auflage, 1985.

D. F. Bergman and R. P. Sutton. A consistent and accurate dead–oil–viscosity method. *SPE 110194*, 2007.

A. Danesh. *PVT and Phase Behaviour of Petroleum Reservoir Fluids*. Number 47 in Developments in petroleum science. Elsevier, 1998.

D. Dexheimer, C. M. Jackson, and M. A. Barrufet. A modification of Pedersen's model for saturated crude oil viscosities using standard black oil PVT data. *Fluid Phase Equilibria*, 183–184:247–257, 2001.

R. di Primio. *Private communication*, 2008.

R. di Primio and B. Horsfield. From petroleum–type organofacies to hydrocarbon phase prediction. *AAPG Bulletin*, 90:1031–1058, 2006.

R. di Primio and J. E. Skeie. Development of a compositional kinetic model for hydrocarbon generation and phase equilibria modelling: A case study from Snorre field, Norwegian North Sea. In J. M. Cubitt, W. A. England, and S. Larter, editors, *Understanding Petroleum Reservoirs: Towards an Integrated Reservoir Engineering*, Special Publication, pages 157–174. Geological Society of London, 2004.

R. di Primio, V. Dieckmann, and N. Mills. PVT and phase behaviour analysis in petroleum exploration. *Organic Geochemistry*, 29:207–222, 1998.

J. Haas. An empirical equation with tables of smoothed solubilities of methane in water and aqueous sodium chloride solutions up to 25 weight percent, 360 °C and 138 MPa. Open–file rep 78–1004, USGS, 1978.

H. J. M. Hanley, R. D. McCarty, and W. M. Haynes. Equations for the viscosity and thermal conductivity coefficients of methane. *Cryogenics*, pages 413–417, 1975.

H. J. M. Hanley, W. M. Haynes, and R. D. McCarty. The viscosity and thermal conductivity coefficients for dense gaseous and liquid methane. *J. Phys. Chem. Ref. Data*, 6:597–609, 1977.

K. Huang. *Statistical Mechanics*. John Wiley & Sons, second edition, 1987.

N. Lindeloff, K. S. Pedersen, H. P. Rønningsen, and J. Milter. The corresponding states viscosity model applied to heavy oil systems. *Journal of Canadian Petroleum Technology*, 43:47–53, 2004.

J. Lohrenz, B. G. Bray, and C. R. Clark. Calculating viscosities of reservoir fluids from their compositions. *Journal of Petroleum Technology*, pages 1171–1176, 1964.

W. D. McCain Jr. *The Properties of Petroleum Fluids*. Pennwell Books, second edition, 1990.

R. D. McCarty. A modified Benedict–Webb–Rubin equation of state for methane using recent experimental data. *Cryogenics*, pages 276–280, 1974.

J. S. Nelson and E. C. Simmons. Diffusion of methane and ethane through the reservoir cap rock: Implications for the timing and duration of catagenesis. *AAPG Bulletin*, 79:1064–1074, 1995.

D. W. Peaceman. *Fundamentals of Numerical Reservoir Simulation*. Number 6 in Developments in petroleum science. Elsevier, 1977.

K. S. Pedersen and P. L. Christensen. *Phase Behavior of Petroleum Reservoir Fluids.* CRC Taylor & Francis, 2007.

K. S. Pedersen and A. Fredenslund. An improved corresponding states model for the prediction of oil and gas viscosities and thermal conductivities. *Chemical Engineering Science*, 42:182–186, 1987.

K. S. Pedersen, A. Fredenslund, P. L. Christensen, and P. Thomassen. Viscosity of crude oils. *Chemical Engineering Science*, 39:1011–1016, 1984.

K. S. Pedersen, Aa. Fredenslund, and P. Thomassen. *Properties of Oils and Natural Gases*, volume 5 of *Contributions in Petroleum Geology & Engineering.* Gulf Publishing Company, 1989.

R. C. Reid, J. M. Prausnitz, and B. E. Poling. *The Properties of Gases and Liquids.* McGraw–Hill Book Company, 4th edition, 1987.

H. P. Rønningsen. Prediction of viscosity and surface tension of north sea petroleum fluids by using the average molar weight. *Energy and Fuels*, 7: 565–573, 1993.

J. L. Sengers and A. H. M. Levelt. Diederick Korteweg, pioneer of criticality. *Physics Today*, pages 47–53, 2002.

E. D. Sloan Jr. *Clathrate Hydrates of Natural Gases.* Marcel Dekker Inc., second edition, 1998.

J. G. Stainforth. New insights into reservoir filling and mixing processes. In J. M. Cubitt, W. A. England, and S. Larter, editors, *Understanding Petroleum Reservoirs: Towards an Integrated Reservoir Engineering*, Special Publication, pages 115–132. Geological Society of London, 2004.

P. Ungerer, J. Burrus, B. Doligez, P. Y. Chenet, and F. Bessis. Basin evaluation by integrated two–dimensional modeling of heat transfer, fluid flow, hydrocarbon gerneration and migration. *AAPG Bulletin*, 74:309–335, 1990.

C. K. Zéberg-Mikkelsen. *Viscosity Study of Hydrocarbon Fluids at Reservoir Conditions.* PhD thesis, Technical University of Denmark, Lyngby, Denmark, 2001.

6

Migration and Accumulation

6.1 Introduction

The processes of petroleum migration are still under discussion and not very well understood. Reservoir engineering and production modeling, which are usually based on Darcy type separate phase flow and mass conservation, are successfully applied to model petroleum flow, at least in reservoirs (Peaceman, 1977; Aziz and Settari, 1979; Barenblatt et al., 1990; Dake, 2001). Engineering success and the persuasiveness of the approach justify a transfer of the methodology from reservoirs to petroleum systems and from timescales of years to millions of years. The resulting differential equations constitute a consistent and complex description for modeling migration in porous media. The argumentation is supported by the fact that Darcy's law has already been applied successfully to model water flow and compaction in sedimentary basins on geological timescales (Chap. 2). The straight–forward extension of the single phase water flow model to include petroleum phases yields the most comprehensive and consistent formulation of multi–phase Darcy flow in one set of coupled differential equations. The separation of water flow for the calculation of compaction as demonstrated in Chap. 2 is an accurate approximation.

The chapter starts with a short introduction to the geological aspects of migration Sec. 6.2. Fundamental aspects of Darcy flow based migration modeling are described in Sec. 6.3. Therein, a general description of all the driving forces and of the transport parameters is given without a detailed introduction to the mathematics of fluid flow. The complete set of coupled differential equations for fluid flow are formulated in Sections 6.3.3—6.3.5.

Petroleum transport via diffusion is assumed to be of lesser importance for migration. However, it is shortly outlined in Sec. 6.4.

Darcy flow based differential equations are often too complex to be solved in acceptable times. As a consequence, model resolutions are usually very poor. To overcome these difficulties, methods with higher performance are presented in the following sections.

T. Hantschel, A.I. Kauerauf, *Fundamentals of Basin and Petroleum Systems Modeling*, DOI 10.1007/978-3-540-72318-9_6,
© Springer-Verlag Berlin Heidelberg 2009

Firstly, flowpath based reservoir analysis is introduced in Sec. 6.5. A hybrid method, which merges the advantages of this approach and Darcy flow, is presented afterwards in Sec. 6.6. Even faster but more approximative is pure flowpath modeling (Sec. 6.7). Alternatively, overall migration modeling can be performed with an invasion percolation technique, which is described in detail in Sec. 6.8. The different methods are discussed and compared in Sec. 6.9.

A detailed analysis and understanding of a petroleum system relies on comprehensive bookkeeping of all the petroleum amounts involved in different geological processes. Generation, expulsion, cracking, migration losses, basin outflow, and accumulation, to name only the most important processes, must be tracked in multi–component resolution for all layers and facies. Rules for bookkeeping are finally discussed in Sec. 6.10.

6.2 Geological Background

The aim of this section is merely to summarize some basic knowledge about the processes of petroleum migration.

It is known that petroleum amounts which are transported via secondary migration are very small, at least on average over space and geological time. Source rocks generate petroleum on a basin–wide length scale over time intervals of millions of years. Flow rates based on average expulsion amounts have values in the range of $8 \times 10^{-15} \ldots 8 \times 10^{-14} \mathrm{m}^3/\mathrm{m}^2/\mathrm{s}$ (England et al., 1987). However, it is also known that migration pathways focus and that migration flow is pulsed in time, see e.g. Haines jumps (Wilkinson, 1984; England et al., 1987; Carruthers, 1998; Sylta, 2004; Dembicki Jr. and Anderson, 1989; Catalan et al., 1992). This implies that localized peak flow rates of moving petroleum might be much higher.

The overall direction of migration is known to be vertical due to buoyancy. Highly permeable features such as reservoir rocks, or on a smaller scale faults, can act as conduits of petroleum. The resulting transport paths follow these conduits in an upward direction (Sec. 6.5). Structures of low permeability and high capillary pressure can be invaded under sufficient pressurization of the petroleum phases, due to buoyancy forces associated with accumulations formed in traps (Dembicki Jr. and Anderson, 1989; Catalan et al., 1992; England et al., 1987).

Residual amounts of hydrocarbons (HC) are immobile and "lost" along migration paths.

Density and compositional changes of petroleum phases during migration are important. Symmetrically with decreasing pressure, the liquid phase loses light components and the vapor phase, heavy components. The density varies more with compositional changes than with immediate pressure and temperature changes (England et al., 1987).

Diffusion effects are negligible (England et al., 1987).

This picture of migration seems unique and consistent. However, in a more detailed view, many questions arise. For a first impression a few shall be listed here:

It is often assumed that petroleum migrates in disconnected stringers or filaments (Berg, 1975; Tissot and Welte, 1984; Sylta, 2004; Schowalter, 1979). What is their size and form? Do these stringers cover a lot of microscopic pores? Obviously, there is at least no connection in terms of pressure from source rock to reservoir. It is commonly assumed that on a macroscopic scale stringer sizes are rather small. Otherwise overpressuring would lead to break through processes and the petroleum would be able to pass barriers almost everywhere. The size must be known more accurately for a detailed flow description.

Do flow pulses occur on a macroscopic or on a microscopic scale in space? Such a question can only be answered if more is known about the stringers. How long do the flow pulses last? What is the velocity of a moving stringer? In the case of fast moving macroscopic stringers, viscosity and internal displacement might play a significant role. Obviously, pores might be filled which would not be invaded by a slow, almost static flow.

Is the flow pattern rather homogeneous in space or does it show a fractal appearance on a macroscopic scale?[1] This question is of great importance, as migration losses can be better estimated if "the petroleum traversed pore volume" is known. In a fractal petroleum migration pattern, only parts of space are traversed.

As special effects in a more continuous picture, saturation "shocks" or discontinuities occur on a macroscopic level. Immiscible fluid displacement often occurs at the surface of a sharp saturation boundary (Barenblatt et al., 1990).[2] This is well known in production and seems to be contradictory to fractal flow patterns. However, production induced flow rates are much higher than the flow rates during migration. For that reason such effects might not appear during migration. On the other hand, the local flow rate belonging to one moving stringer is definitely higher than average migration rates. Hence a stringer might have sharp boundaries while moving through the rock matrix.

Observations originating from laboratory experiments and core samples support the fractal view. Does such fractal behavior originate from macroscopic large scale variations of rock properties or does it evolve through upscaling of self–similar fractal structures coming from random pore size variations?

These are some basic questions. Surely, the list can be further extended and many fundamental questions concerning migration are still open.

---

[1] Fractal means self–similar under a given magnification. More precise definitions are formulated in the literature, e.g. in Stüwe (2007).

[2] The saturation is only rapidly varying and appears to be discontinuous on a macroscopic level.

Finally it must be noted that in basin modeling migration is considered as the flow or movement of HCs in the free pore space. It is not principally distinguished into primary migration inside and secondary or tertiary migration outside of source rocks.

6.3 Multi–Phase Darcy Flow

Separate flow of non–mixing phases is commonly assumed to be the dominant transport mechanism for secondary migration. The driving forces for fluid flow are pressure potential differences. The fluid pressure potential is just the pressure reduced by the pressure of a static fluid column with a corresponding vertical pressure gradient, which is necessary to balance the weight of the column itself (Fig. 2.2).[3] Exceeding amounts of pressure cause flow. The potential u_p for any phase p is thus defined as

$$u_p = p - \rho_p g z \tag{6.1}$$

with ρ_p the density of fluid phase p, g the gravitational acceleration, and z the depth.[4]

Darcy's law states that a potential difference causes a flow according to

$$v_p = \mu_p \frac{\Delta u_p}{\Delta l} . \tag{6.2}$$

Herein v_p is the velocity of flow of phase p and μ_p its mobility. The symbol Δu_p indicates a potential difference over a distance Δl in space. The flow direction is from high to low pressure potential. Timing of the fluid flow is included and quantified via the introduction of flow velocities.

The mobility in multi–phase fluid systems is usually split into three factors according to

$$\mu_p = \frac{k k_{rp}}{\nu_p} . \tag{6.3}$$

The relative permeability k_{rp} is introduced additionally in comparison with the single–phase formulas of Sec. 2.2.3. The numerator kk_{rp} is called the effective permeability and k the absolute or intrinsic permeability. Models and approximations for the estimation of absolute permeability are already discussed in Sec. 2.2.3.

Effective permeability and viscosity can vary drastically with temperature, pressure, saturation, and porosity. Hence, flow and migration velocities might also show variations over several orders of magnitude (Fig. 6.1).

---

[3] Plus a depth independent shift for zero level adjustment.

[4] For the indication of different petroleum phases an "oil–gas" instead of a "liquid–vapor" notation is used in this chapter. The authors prefer the latter but the "oil–gas" notation is standard in the literature.

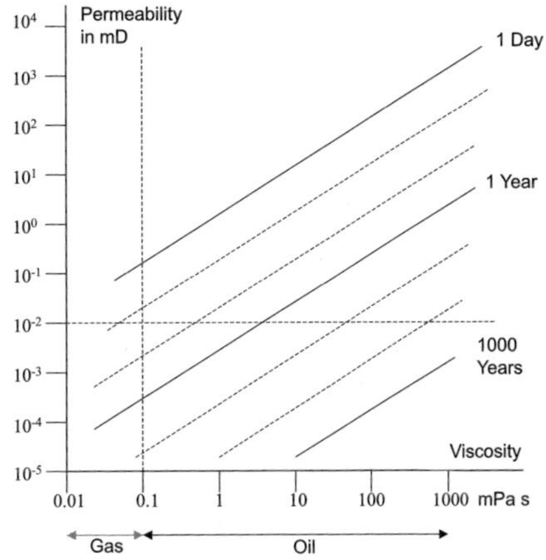

Fig. 6.1. Petroleum flow velocity isolines. Assuming a fixed viscosity and a varying permeability or vice versa the time for the fluid to travel a given distance for a fixed pressure difference varies here between one day and thousands of years

The petroleum saturation defines the fraction of pore space which is used for flow (England et al., 1987). It is thus possible to roughly approximate the relative permeability k_{rp} with the petroleum saturation S_p by $k_{rp} = S_p$ based on the strongly simplified tube bundle model for fluid flow from Sec. 2.2.3. More generally, the fraction of tubes available for fluid flow is proportional to saturation. Tube radii are assumed to be fixed and randomly distributed within certain limits. At low petroleum saturations, the permeability is determined by the probability of randomly drawing an arbitrary tube with radius r. It is $k \propto \langle r \rangle^2$ with $\langle r \rangle$ as the expectation value of drawing a value r. At full saturation the sum of all tubes is used for flow. The permeabililty is proportional to the slice plane area of the tubes in the rock and thus $k \propto \langle r^2 \rangle$. Generally $\langle r^2 \rangle > \langle r \rangle^2$. Intermediate saturations are described by a continuous crossover. However, such a simplified tube model does not take into account some important effects. Actual path lengths might decrease with increasing saturation. Pathways which emerge due to flow branching and flow through partially filled tubes are also not considered.

As saturation increases, the permeability also increases in real porous media. Relative permeability dependencies of a more realistic form are shown in Fig. 6.2. Here, water is considered to be a wetting phase with a low contact angle between the rock and the water and petroleum to be generally non-wetting, which implies that the grain surface is covered by at least a thin layer of water. The three saturation end points, namely critical oil S_{oc}, critical gas S_{gc} and connate water saturation S_{wc}, are threshold values. They distinguish between initial saturations, which must be overcome to allow flow, and residual saturations, which are immobile. Very important is the critical oil saturation value of a source rock. It controls expulsion because it defines a threshold saturation which has to be overcome before oil starts moving and is expelled.

Generally, critical saturation values are lithology dependent and must be distinguished according to the surrounding phases Yuen et al. (2008). Some approximations are usually made in basin modeling. Critical gas saturations S_{gc} are usually assumed to be negligibly small, allowing every small gas bubble to be mobile. Critical oil saturations S_{oc} in sandstones, which roughly range from 0.1...10%, are much smaller than in shales, with values from about 0.5...50.0%. Specific values depend on proper upscaling. Therefore the typical flow channel width, its density and the microscopic saturation distribution along the channels must, in principle, be correctly estimated and rescaled to the common gridcell dimensions of the basin model.

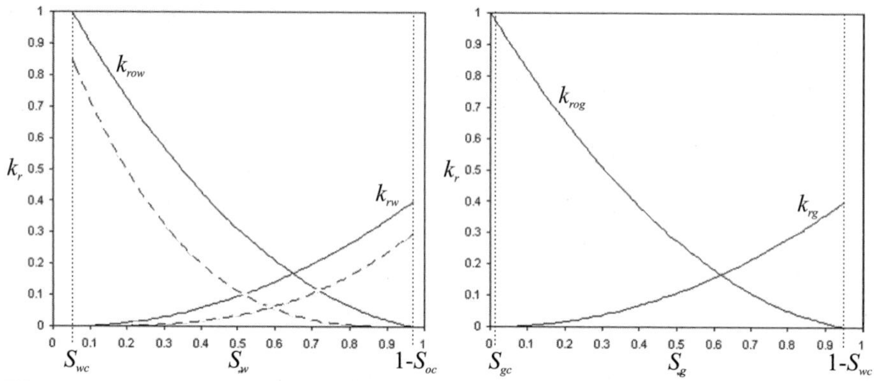

Fig. 6.2. Relative permeability curves for $S_{wc} = 5\%$, $S_{oc} = 3\%$, and $S_{gc} = 1\%$ according to Table 6.1. The dashed curves are according to (6.6). Note that the k_{rog}–curve starts at $S_g = 0$ and that $k_{rg} = 0$ for $S_g < 1\%$

	$S_e = 0$	$S_e = 0.5$	$S_e = 1$	Quadratic Fit
k_{rw}, k_{rg}	0.0	0.1	0.4	$0.4\,S_e^2$
k_{row}, k_{rog}	1.0	0.3	0.0	$1 - 1.8\,S_e + 0.8\,S_e^2$

Table 6.1. Supporting points and quadratic fit of relative permeability curves which are shown in Fig. 6.2

Relative permeability curves of water and gas are assumed to be a function of water and gas saturation S_w and S_g respectively and the relative permeability of oil depends on both water and gas saturation in the most common relative permeability models (Aziz and Settari, 1979). Hence it is

$$k_{rw} = f(S_w), \qquad k_{rg} = f(S_g) \tag{6.4}$$

and

$$k_{ro} = k_{row}\, k_{rog}, \qquad k_{row} = f(S_w), \qquad k_{rog} = f(S_g)\,. \tag{6.5}$$

The flow of one phase is treated here as if the other phases are part of the solid rock matrix. This assumption is not valid if fluid phases interact during flow. Other relations than (6.4) and (6.5) can be found in Aziz and Settari (1979) amongst others.

The relative permeability of any fluid is zero below its critical saturation, it becomes immobile. Saturations are for often rescaled into normalized or effective saturations S_e which map the saturation interval between the connate and the critical saturation to an interval of $0\ldots1$ as $S_{we} = (S_w - S_{wc})/(1 - S_{wc} - S_{oc})$ for k_{rw} and k_{row}, $S_{goe} = S_g/(1 - S_{wc})$ for k_{rog} and $S_{ge} = (S_g - S_{gc})/(1 - S_{wc} - S_{gc})$ for k_{rg}. Due to the lack of precise data in basin modeling it is common to approximate the overall shape of relative permeability curves with a universal form, which is lithology and phase property independent. Such general shapes are sometimes modeled with quadratic functions, which are based for example on three points as defined in Table 6.1. This is a crude approximation but it is justified by huge uncertainties in absolute permeabilities, which are often only known to one order of magnitude in accuracy. Small relative permeability uncertainties are of comparatively no consequence, at least in basin modeling.

The data base is exceptional for sandstones. Empirical formulas exist such as

$$k_{rw} = 0.3\, S_{we}^3 \quad \text{and} \quad k_{row} = 0.85\,(1 - S_{we})^3 \tag{6.6}$$

from Ringrose and Corbett (1994). Other frequently used formulas are the Brooks and Corey equations

$$k_{rw} = S_{we}^{(2+3\lambda)/\lambda} \quad \text{and} \quad k_{row} = k_{row}^0 (1 - S_{we})^2 \left(1 - S_{we}^{(2+\lambda)/\lambda}\right) \tag{6.7}$$

with a constant k_{row}^0 and a parameter λ which describes a "sorting" of the rock. It can vary between 0 and ∞. A small value indicates a poorly sorted (inhomogeneous) rock (Sylta, 2002a; Ataie-Ashtiani et al., 2002). The relative permeabilities in (6.6) and (6.7) become the same for $\lambda \to \infty$, $k_{rwo}^0 = 0.85$, and an additional pre-factor of 0.3 in the Corey equation for k_{rw}.

Equation 6.2 is only a basic formulation of Darcy's law. Some details have not been mentioned yet. The comprehensive formulation is

$$\mathbf{v}_p = -\boldsymbol{\mu}_p \cdot \boldsymbol{\nabla} u_p \tag{6.8}$$

with the mobility tensor $\boldsymbol{\mu}_p = \mathbf{k} k_{rp}/\nu_p$. This law is formulated in terms of vectors. The driving force $-\boldsymbol{\nabla} u_p$ is a gradient, which points in the direction of the steepest decrease of the potential field u_p. It is multiplied with a tensor $\boldsymbol{\mu}_p \propto \mathbf{k}$ which describes the anisotropy of the rock permeability so that the resulting flow velocity $\mathbf{v}_p$ is not necessarily pointing in the same direction as $-\boldsymbol{\nabla} u_p$, (compare with 8.2, 8.3).

The gradient $-\boldsymbol{\nabla} u_p$ is a mathematical formulation of pressure potential differences over infinitesimally small distances in spatial directions. This point is of a technical nature and ensures that spatially varying potentials are treated correctly in three dimensions.

Darcy's law is a typical friction law with a friction force proportional to the velocity. Fluids which obey this law are called "Newtonian fluids". Note that for mechanics without friction, acceleration is proportional to the driving force. In other cases (e.g. if flow velocities are very high) other quantities might be related to friction. The kinetic energy of the displaced fluid, which is proportional to v^2, might quantify the energy loss from friction. Other relationships are possible for such "non–Newtonian" fluids.

It is sometimes argued that viscous resistance is so small that it can generally be neglected. This implies that flow–velocity analysis is abandoned, which might be reasonable for many geological situations, especially if long geological timescales are considered. However, time–control of migrating petroleum is lost. Instead, generation rates become the only time controlling factor in the petroleum systems model. The topic is discussed in more detail in Sec. 6.8.

The origin of the driving forces for petroleum, which have not been mentioned yet, is the topic of the following sections.

6.3.1 Capillary Pressure

Interfacial tension occurs at the interface between two adjacent immiscible phases. It rises due to differently sized attraction forces between molecules within one phase and across a boundary to molecules within another phase. The corresponding effect in porous media is named capillary pressure which indicates an additional fluid pressure due to geometry and contact forces. It is commonly described by the term p_{co} for oil–water and p_{cg} for oil–gas systems and can be calculated for an ideal capillary tube of radius r as

$$p_c = \frac{2\gamma}{r}\cos\theta \tag{6.9}$$

with γ as the interfacial tension and θ the contact angle (Fig. 6.3). In porous media the capillary pressure depends on the pore throat radii r, which are pure rock properties, interfacial tensions, which are pure fluid properties, and contact angles for the specific combination of both.

Capillary pressure is usually measured for mercury–air systems in laboratory experiments. It can be transformed to water–petroleum systems by

$$p_{c\mathrm{Pet}} = p_{c\mathrm{Hg}} \frac{\gamma_{\mathrm{Pet}}\cos\theta_{\mathrm{Pet}}}{\gamma_{\mathrm{Hg}}\cos\theta_{\mathrm{Hg}}} \tag{6.10}$$

according to (6.9). The interfacial tension of mercury is $\gamma_{\mathrm{Hg}} = 471\,\mathrm{mN/m}$ and the contact angle for a mercury–air system is $\theta_{\mathrm{Hg}} = 140°$. Assuming an ideal water wet petroleum–water system with $\theta_{\mathrm{Pet}} = 0°$ the above formula reduces to

$$p_{c\mathrm{Pet}} = p_{c\mathrm{Hg}} \frac{\gamma_{\mathrm{Pet}}}{360.8\,\mathrm{mN/m}}\,. \tag{6.11}$$

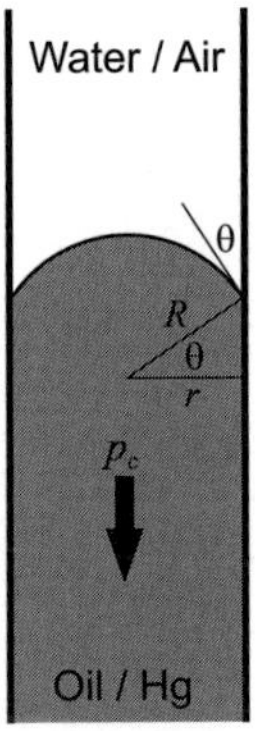

Fig. 6.3. Pressure in a capillary

Capillary pressure is saturation dependent for macroscopic rock samples because rock contains many pores with varying sizes, which can be invaded at different pressures (Fig. 6.4). Commonly, a capillary pressure threshold $p_{\rm int}$ has to be exceeded to start petroleum saturation. It corresponds to the largest pore throat at the petroleum water contact. Higher pressures are necessary to increase petroleum saturation. The entry pressure p_{ce} has to be overcome to saturate the rock up to the residual petroleum saturation, which corresponds to the first connected petroleum path through the sample, and which is not removable anymore due to hysteresis. Residual and critical petroleum saturation are commonly approximated with the same value in basin modeling. In the following text they are not distinguished anymore. With further pressure increase smaller throats are overcome, smaller pores are filled and a higher saturation is achieved.

Idealized forms of capillary pressure curves are used in basin modeling. Hysteresis, as depicted in Fig. 6.4, is usually not taken into account. The only exception are immobile losses.

A porous medium can be interpreted as a bundle of capillary tubes. According to (2.44) $r \propto \sqrt{k/\phi}$ with permeability k for the radius r of the tubes. This can be inserted into (6.9). The resulting equation can be rearranged and according to Fig. 6.4 interpreted as a function of saturation only which is called the Leverett J–function (Barenblatt et al., 1990)

$$J(S_w) = \frac{p_c(S_w)\sqrt{k/\phi}}{\gamma \cos\theta} \,. \tag{6.12}$$

This dimensionless function allows measurement results obtained for the same rock type with different fluids at different porosities and permeabilities to be compared. The values should lie approximately on one saturation dependent $J(S_w)$ – curve. For example, Ringrose and Corbett (1994) proposed $J(S_w) \sim S_{we}^{-2/3}$ for sandstones.

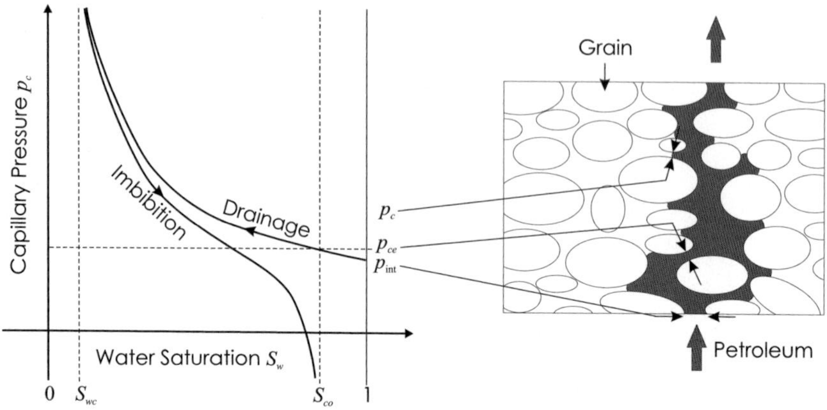

Fig. 6.4. Illustration of drainage and imbibition curves similar to Aziz and Settari (1979), Schowalter (1979), and Wilkinson (1986) on the left. The capillary pressure p_c refers to the minimal pore throat radius which is actually reached by petroleum within the rock on the right. Note that the x–axis on the left side shows water saturation. The nomenclature "drainage" and "imbibition" refers to water drainage and imbibition

Due to lack of data, only simplified functions of $p_c(S_w)$ are used in modeling practice. The simplest functional form is

$$p_c(S_w) = p_{ce} + p_0(1 - S_w) \tag{6.13}$$

with e.g. $p_0 = 1\,\text{MPa}$. However, such a simple model is unrealistic for high petroleum saturation. Capillary pressure rises drastically when water is drained and the connate water saturation S_{wc} is approached.

As with permeability, more accurate formulas for sandstones such as

$$p_{co} = p_{oe} S_{we}^{-1/\lambda} \tag{6.14}$$

with p_{oe} as the oil entry pressure are available (Sylta, 2002a). The sorting factor λ is the same as in (6.7). The previously mentioned case of $\lambda \to \infty$ yields a constant capillary pressure of form $p_{co} = p_{oe}$. This simple model is used in invasion percolation, where the capillary pressure curve is sampled by just one value, the entry pressure (Fig. 6.4). It is not sufficient for Darcy flow models where capillary pressure must increase continuously with petroleum saturation S_p (Fig. 6.6).

The most important parameter for all commonly used capillary pressure models is the capillary entry pressure p_{ce}. It is usually given by mercury–air values p_{cHG} in lithological data bases or empirical equations and can be converted to oil–water and gas–water values with interfacial tension values of oil–water γ_o and gas–water γ_g. The interfacial tension values of oil–water and gas–water are temperature and pressure dependent and described in Chap. 5. In the following discussion, the interfacial tension values of $\gamma_o = 42\,\text{mN/m}$

and $\gamma_g = 72\,\mathrm{mN/m}$ are used to compare reference data for capillary entry pressures, which are reported for different fluid systems. According to equation (6.11), this yields a general relationship between oil–water, gas–water and mercury–air capillary entry pressures as

$$p_{cOil} = p_{cHG}/8.6\,, \qquad p_{cGas} = p_{cHG}/5.0. \tag{6.15}$$

Due to the dependency of capillary pressure on pore throat radii, it is also strongly dependent on the compaction state, which is usually described by porosity or permeability (App. A). It can quite accurately be estimated for many lithologies with the following models.

Permeability dependent models

Some authors proposed general exponential relationships between capillary entry pressures and permeabilities, which can be used for all lithotypes of the following type:

$$p_{ce} = a(k/k_0)^b\,. \tag{6.16}$$

Ibrahim et al. (1970) in Ingram and Urai (1999) derived $a = 0.548\,\mathrm{MPa}$, $b = -0.33$, and $k_0 = 1\,\mathrm{mD}$ from calculations of column heights. Hildenbrand et al. (2004) performed gas break through experiments on pelitic rocks with N_2, CO_2, and CH_4, which yield $a = 0.0741\,\mathrm{MPa}$, $b = -0.24$, $k_0 = 1\,\mathrm{mD}$ for methane–water, which is equivalent to $a = 0.37\,\mathrm{MPa}$ for mercury–air according to (6.16).

Porosity dependent models

The combination of the above relationship $p_{ce}(k)$ with a linear relationship between $\log k$ and ϕ yields

$$p_{ce} = a\,10^{-b\phi} \tag{6.17}$$

for capillary entry pressure dependency on porosity. The parameters a and b are lithology dependent as the permeability versus porosity curves are also lithotype specific functions. Typical mercury–air values for various lithologies calculated with the Hildenbrand model from permeability curves of App. A are also tabulated in the appendix.

The advantage of the porosity dependent curves is, that they can be fitted for each rock type independently of permeability calibrations. The proposed equation (6.16) from Hildenbrand and Ibrahim for a typical shale with a piecewise linear permeability curve such as in Fig. 2.16 yields capillary entry pressure curves as shown in Fig. 6.5.a. Application in basin modeling shows, that the Hildenbrand model works well for low porosity shales and good experiences are made with the Ibrahim model for high porosity shales. Hantschel and Waples (2007) proposed a linear interpolation between the capillary entry pressure calculated for 1 % porosity from the Hildenbrand model, the capillary entry pressure calculated for 25 % porosity from the Ibrahim model and zero capillary entry pressure for the depositional porosity (Fig. 6.5.a).

This construction of piecewise linear capillary entry pressure versus porosity curves can generally be used for other lithotypes. The corresponding capillary entry pressure values for 1 % and 25 % porosities are tabulated in App. A and shown in Fig. 6.5.b. The conversion of mercury–air into oil–water and gas–water values using (6.15) yields the curves of Fig. 6.5.c.

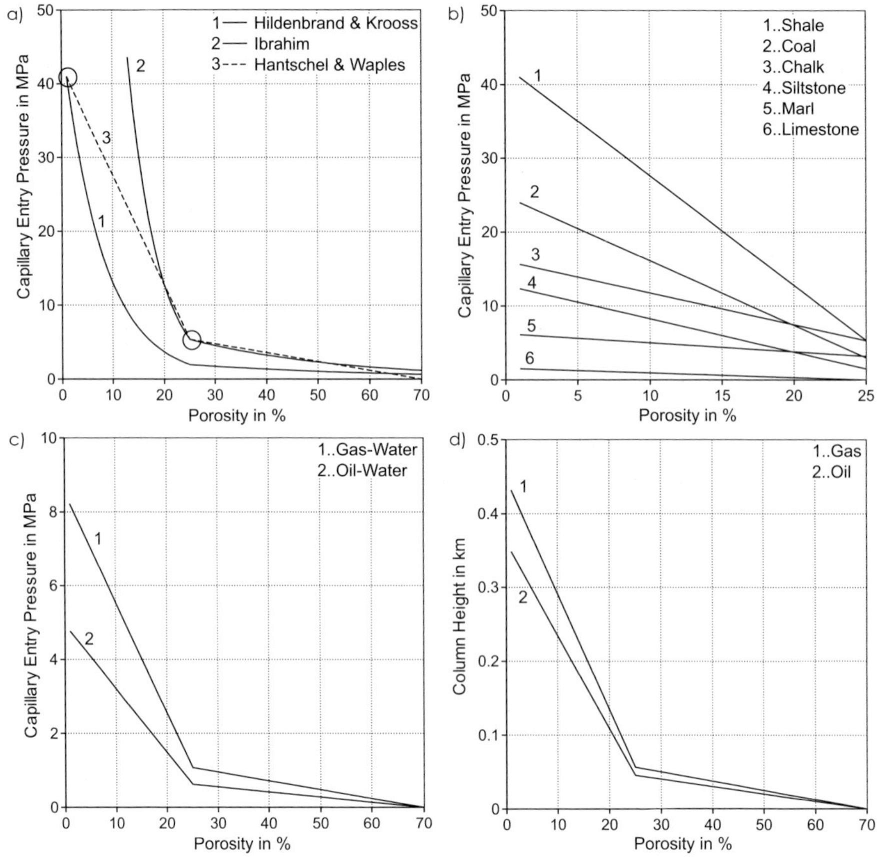

Fig. 6.5. Capillary entry pressure curves

It must further be noted that basin scale entry pressure values are often calculated from hand specimen values divided by an upscaling factor which lowers the entry pressure. An upscaling factor of 2.56 is proposed for clastic rocks and carbonates.[5] Upscaling of the piecewise linear capillary entry

[5] Assuming an upscaling factor for the permeability of 50 (Sec. 2.2.3), the Hildenbrand parameter b=-0.24 yields an upscaling factor for capillary entry pressure of $50^{-0.24} = 1/2.56$.

pressure curve of typical shale yields maximum column heights as shown in Fig. 6.5.d for oil and gas densities of $\rho_o = 600\,\mathrm{kg/m^3}$ and $\rho_g = 100\,\mathrm{kg/m^3}$.

Note, that all basin modeling input capillary pressure values are usually given as mercury–air reference values and that the conversion into oil–water and gas–water values with the corresponding maximum column heights are calculated during the simulation with the actual reservoir porosities and the temperature and pressure controlled oil and gas interfacial tension and density values.

6.3.2 Pressure at Phase Boundaries

In the absence of other forces, petroleum migrates from high to low capillary pressure regions. Obstacles such as cap rocks with high capillary pressures inhibit migration. For example, the higher the capillary pressure contrast between a reservoir and a seal, the higher the buoyant pressure of an accumulation must be to allow fluid flow into and through the seal.

Another force arises through water flow, more precisely through its pressure potential u_w, which has been discussed in Chap. 2. In general, any phase might be driven by forces originating at the boundary to other phases.

Without capillary pressure changes from water to pretroleum the pressure would be continuous at phase boundaries. A pressure jump of the height of the capillary pressure must be taken into account. With (6.1) this yields

$$u_o = u_w + (\rho_w - \rho_o)gz + p_{co} \tag{6.18}$$

and

$$u_g = u_w + (\rho_w - \rho_g)gz + p_{co} + p_{cg} \; . \tag{6.19}$$

Petroleum buoyancy in water is directly identified by inserting the last two equations into Darcy's law (6.2) and looking at the terms with the density contrasts. Buoyant forces are strong enough to cause the rapid migration of oil through sandstones. Migration times can often be neglected on geological timescales and replaced by instantaneous movements as done in the hybrid and flowpath methods (Sec. 6.5). The behavior is quite different for water which experiences no buoyant forces in the surrounding medium and no capillary pressure thresholds.

Due to the fact that the overall amount of petroleum in a basin is very small compared to the amount of water, it is often assumed that neither the overall water flow nor compaction is distorted very much by the HCs and that the methods presented in Chap. 2 do not need to be or need to be only slightly corrected. Effects such as fluid expansion (Sec. 2.3.2) are usually incorporated in the water flow equations but not implicitly coupled to migration equations. In principle, a correction based on iterative solutions of water and petroleum flow equations for the improvement of the coupling between both flow equations can be easily performed.

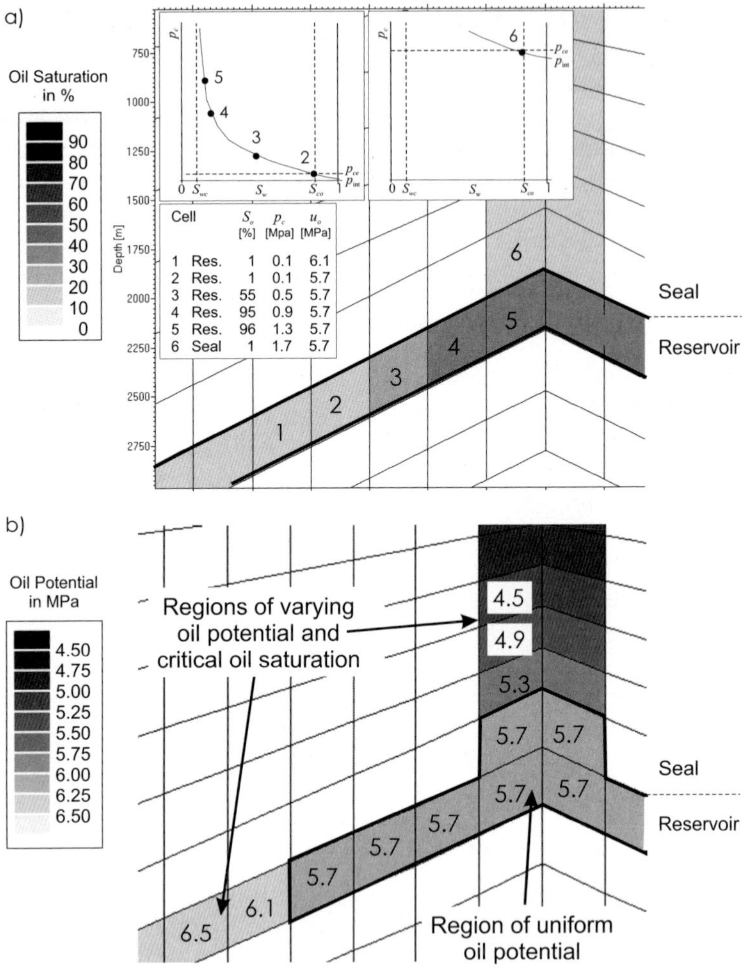

Fig. 6.6. Reservoir in static equilibrium

A principle scheme of an accumulation in static equilibrium below a seal is shown in Fig. 6.6. The oil potential at base seal (cell 6) is equal to the oil potential at the top reservoir (cell 5). Thus, the two corresponding capillary pressures differ only by the amount of buoyancy pressure between the two adjacent cells. The capillary pressure in the base of the seal is equal to the capillary entry pressure, while the same capillary pressure at the top reservoir location is related to a very high saturation value. The oil potential is constant in the entire accumulation, which means that the capillary entry pressure decreases linearly with depth in the reservoir (cells 5 — 2). The buoyancy decreases by the same amount.

There is no migration within the oil accumulation and from the reservoir into the seal as there is no gradient of the oil potential. Note that the capillary pressure difference between the base reservoir and base seal cells (2 and 6) differ by the difference of the capillary entry pressure of the reservoir and sealing lithology and that it is equal to the buoyancy of the accumulation. This result is used as the fundamental sealing rule in flowpath and invasion percolation methods.

The petroleum potential below the accumulation increases with depth. Due to a missing pressure communication between the immobile petroleum droplets, buoyancy is not varying with depth and the capillary pressure is equal to the capillary entry pressure for all locations saturated with the critical oil saturation. Although there is an oil potential gradient here, no migration occurs, as the relative permeability for critical oil saturation is equal to zero.

In summary, the petroleum system around a petroleum accumulation in static equilibrium consists of two types of domains: migration regions with critical oil saturations and accumulation regions with constant oil potentials. Due to the fact, that source rock expulsion rates are usually very small, it can often be assumed, that even an accumulation under continuous feeding with a break through on top is well approximated by a static equilibrium. Other accumulation examples under static equilibrium are explained in Fig. 6.7. However, Darcy flow equations are also capable to model dynamic behavior.

An example model is shown in Fig. 6.8. It consists mainly of two sandstone layers and a deeply buried source rock. Expelled petroleum is moving upwards. The permeability is highly anisotropic in the shales and thus the petroleum is moving in an intermediate direction between vertically upward and the dipping angle of the layers. Leaking accumulations are found in the sandstone structures beside the impermeable fault. The capillary curve is still below 50 % saturation, reaching pressure values which allow break through from sandstone into shale below 2000 m depth. The resolution of the grid is too rough for a precise estimate of the column height. Some vectors indicate small amounts from locations where flow traversed in paleo times and residual amounts were captured. With ongoing compaction and reduced porosity, residual saturation values are exceeded and small petroleum amounts started to move again.

It can finally be summarized that Darcy type migration modeling can be interpreted as a balance of all external forces, such as capillary pressure, buoyancy, and water pressure at the phase boundary, with a viscous resistance force, where, from the viewpoint of a balanced Newtonian force, each force has a counterpart of the same strength but with opposite direction.

6.3.3 Three Phase Flow Formulation without Phase Changes

A starting point for the formulation of separate multiphase flow equations are mass balances. Mainly three immiscible fluid phases, namely water, oil, and gas are found in a basin. The mass fraction for any phase p can be calculated from

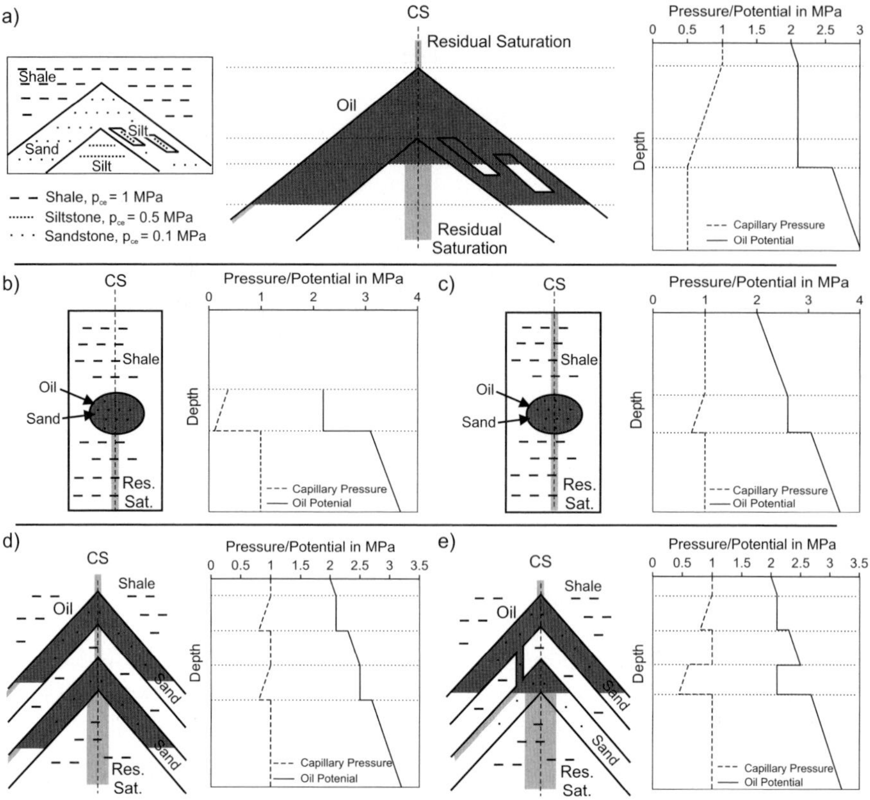

Fig. 6.7. Capillary pressure and oil potential for accumulations under static equilibrium: (a) With siltstone facies in and below a sandstone reservoir. Note, that the oil–water contact in all silt facies have the same depth. (b) With a sandstone lens in a shale layer at the time when the sandstone has been entirely filled. (c) At the time, when the sandstone pressure is increased until break through. Note, that the accumulation saturation from the first entire filling to break–through increases corresponding to the capillary pressure versus saturation curve. (d) With two disconnected reservoirs. (e) With two connected reservoirs. The pressure and potential curves are shown for the marked cross–sections (CS) through the accumulations

$$m_p = \phi \rho_p S_p \tag{6.20}$$

with ρ_p as the density, S_p the saturation, and ϕ as the porosity of the rock. Correspondingly, mass fluxes $\dot{\mathbf{m}}_i$ are given by

$$\dot{\mathbf{m}}_p = \rho_p \mathbf{v}_p \ . \tag{6.21}$$

The velocities $\mathbf{v}_p$ can be calculated with Darcy's law according to (6.8). The mass balance for each phase p is

$$\frac{\partial m_p}{\partial t} + \boldsymbol{\nabla} \cdot \dot{\mathbf{m}}_p = q_p \ . \tag{6.22}$$

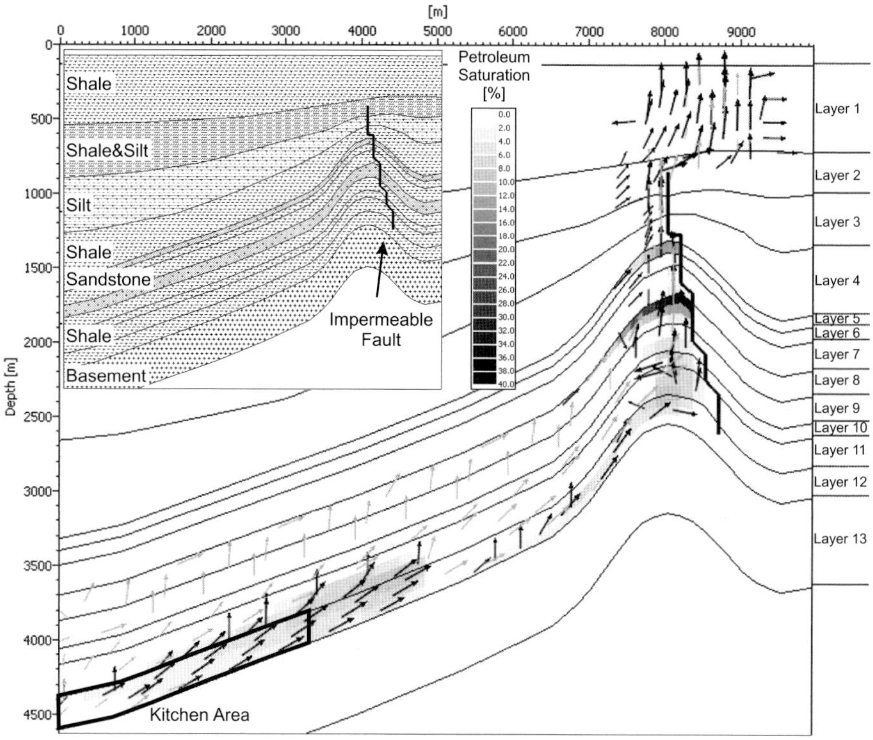

Fig. 6.8. Two dimensional example model with reservoir equilibrium as outlined in Fig. 6.6. The black vectors indicate oil and the light grey vectors gas flow. Layers 5 and 8 are sandstone reservoirs

A source term q_p which describes the generation or loss of phase p (e.g. by chemical reactions) is added here for the sake of completeness.

Furthermore, compaction is described by $d\phi/dt = -C\, d\sigma_z'/dt$ with σ_z' as the vertical component of the effective stress and C as the matrix compressibility. For simplicity, the equations are here, according to (2.3), only noted for a coordinate system which moves with the solid rock. The compaction term becomes

$$\frac{\partial \phi}{\partial t} = -C \frac{\partial (u_l - u_w)}{\partial t} \tag{6.23}$$

with $u_l = p_l - p_h$ as the depositional or overburden load potential (Chap. 2). A comprehensive treatment in a spatially fixed coordinate system can e.g. be found in Luo and Vasseur (1992). Additional factors of form $1/(1-\phi)$ such as in (2.10), which are also known from Yükler et al. (1979), enter the equations (App. B).

Equations 6.20, 6.21, and 6.23 can be inserted in (6.22) for the construction of final three phase flow equations:

$$\begin{aligned}
\frac{\rho_w\phi}{1-\phi}\frac{\partial S_w}{\partial t} &- \nabla\cdot\rho_w\boldsymbol{\mu}_w\cdot\nabla u_w - \frac{\rho_w S_w C}{1-\phi}\frac{\partial(u_l-u_w)}{\partial t} = q_w \,, \\
\frac{\rho_o\phi}{1-\phi}\frac{\partial S_o}{\partial t} &- \nabla\cdot\rho_o\boldsymbol{\mu}_o\cdot\nabla u_o - \frac{\rho_o S_o C}{1-\phi}\frac{\partial(u_l-u_w)}{\partial t} = q_o \,, \\
\frac{\rho_g\phi}{1-\phi}\frac{\partial S_g}{\partial t} &- \nabla\cdot\rho_g\boldsymbol{\mu}_g\cdot\nabla u_g - \frac{\rho_g S_g C}{1-\phi}\frac{\partial(u_l-u_w)}{\partial t} = q_g \,,
\end{aligned} \tag{6.24}$$

$$S_w + S_o + S_g = 1 \,. \tag{6.25}$$

Equations (6.18) and (6.19) can be rearranged to

$$u_o - u_w = p_{co}(S_w) + (\rho_w - \rho_o)gz \,, \tag{6.26}$$

$$u_g - u_o = p_{cg}(S_o) + (\rho_o - \rho_g)gz \,. \tag{6.27}$$

They close the system. The unknowns are the three pressure potentials and the three saturations. For simplicity, it is here assumed that the densities are only slowly varying over time. Derivatives concerning $\partial\rho_i/\partial t$ could easily be added to (6.24).

The saturation dependency of capillary pressure is noted for clarification. However, the oil saturation dependency of p_{cg} only describes the case where gas is surrounded by oil. The gas capillary pressure p_{cg} is similar to the oil capillary pressure p_{co} which is also dependent on water saturation in the case of $S_o = 0$. Generally, p_{co} and p_{cg} depend on all saturations S_w, S_o, and S_g.

There are two principal methods, with many variations, for solving such complicated sets of equations. First of all, they can be solved directly after implicit gridding in time (Sec. 8.4). This scheme is generally called implicit and usually involves a huge effort for the construction and inversion of the corresponding matrices. The equation system is solved simultaneously for many variables. Such methods are therefore also called "simultaneous solution" methods (Aziz and Settari, 1979). The "classical" implicit approach is based on a simultaneous solution for u_o, S_w, and S_g.

Alternatively, it is possible to solve such equation systems partially explicit. Some values (e.g. the saturation values) are kept fixed for a small time step and the system is solved implicitly, e.g. for pressures only. Afterwards, HC saturations are updated explicitly according to the calculated pressure gradients and Darcy's flow law. The "classical" explicit approach is based on an implicit solution for the water pressure only and an explicit treatment of all other unknowns. Such an approach is also called "implicit pressure and explicit saturation" (IMPES) (Aziz and Settari, 1979). The whole procedure must be repeated multiple times because it works only for small time steps.

However, iterations are always necessary for both approaches as there are non–linear dependencies within the equations. Densities depend non–linearly on pressures, capillary pressures and relative permeabilities depend non–linearly on saturations, and so on.

6.3.4 Multicomponent Flow Equations with Phase Changes

Dissolution and exsolution of components in and out of phases was not considered in the previous section. It can be incorporated explicitly with flash calculations before or after each time step (Chap. 5). For an implicit treatment, it is necessary to switch to a multicomponent formulation with N–components.

The mass m_{ip} denotes the mass of the component i dissolved in the phase p, while m_i and m_p are the total component and phase masses at a given location. Thus the conditions

$$\sum_{i=1}^{N} m_{iw} = m_w, \quad \sum_{i=1}^{N} m_{io} = m_o, \quad \sum_{i=1}^{N} m_{ig} = m_g \tag{6.28}$$

and

$$m_{iw} + m_{io} + m_{ig} = m_i \quad \text{for} \quad i = 1, \ldots, N \tag{6.29}$$

must be fulfilled. The mass fractions

$$C_{ip} = \frac{m_{ip}}{m_p} \tag{6.30}$$

are PVT–functions and can be calculated under consideration of thermodynamic laws or via lookup tables (Chap. 5). Besides this,

$$\sum_{i=1}^{N} C_{ip} = 1 \ . \tag{6.31}$$

The mass of each component inside the rock can now be calculated according to

$$m_i = \phi(C_{iw}\rho_w S_w + C_{io}\rho_o S_o + C_{ig}\rho_g S_g) \tag{6.32}$$

with ϕ as the porosity of the rock and ρ_p the density of phase p. Correspondingly, mass fluxes $\dot{\mathbf{m}}_i$ are given by

$$\dot{\mathbf{m}}_i = C_{iw}\rho_w \mathbf{v}_w + C_{io}\rho_o \mathbf{v}_o + C_{ig}\rho_g \mathbf{v}_g \tag{6.33}$$

The phase velocities $\mathbf{v}_p$ obey Darcy's law (6.8). The local formulation of mass balance for each component i is given by

$$\frac{\partial m_i}{\partial t} + \boldsymbol{\nabla} \cdot \dot{\mathbf{m}}_i = q_i \tag{6.34}$$

with newly introduced component sources q_i.

Insertion of (6.32), (6.33), and (6.23) into (6.34) with consideration of App. B yields

$$
\begin{aligned}
&\frac{\phi}{1-\phi}\left(C_{iw}\rho_w\frac{\partial S_w}{\partial t}+C_{io}\rho_o\frac{\partial S_o}{\partial t}+C_{ig}\rho_g\frac{\partial S_g}{\partial t}\right)\\
&-(\rho_w S_w C_{iw}+\rho_o S_o C_{io}+\rho_g S_g C_{ig})\frac{C}{1-\phi}\frac{\partial(u_l-u_w)}{\partial t}\\
&-\nabla\cdot C_{iw}\rho_w\boldsymbol{\mu}_w\cdot\nabla u_w-\nabla\cdot C_{io}\rho_o\boldsymbol{\mu}_o\cdot\nabla u_o\\
&-\nabla\cdot C_{ig}\rho_g\boldsymbol{\mu}_g\cdot\nabla u_g \qquad = q_i
\end{aligned}
\tag{6.35}
$$

for all $i = 1, \ldots, N$. The equation system consists of n differential equations for the $3N + 6$ unknowns C_{ip}, S_p, and p_p. The closing equations are again (6.25), (6.26), and (6.27). Additionally, the $2n$ PVT–functions defining (6.30) must be specified and (6.28) must be fulfilled. Again it is here assumed that the densities are not or only slowly varying over time.

An implicit solution is very difficult to process as a simultaneous solution with so many unknowns yields huge matrices, which cannot be inverted without serious problems.

The application of the classical explicit method to the multicomponent model is relatively simple as the equations are implicitly solved for the water pressure only. Afterwards all of the above multicomponent equations are sequentially evaluated for an incremental mass change of each component.

Finally, it must be noted that the densities and the PVT–functions (6.30) are strongly temperature dependent. Temperature is also explicitly time dependent as can be seen in Chap. 3. Hence, in a correct formulation, partial derivatives of ρ_p and C_{ip} must also be taken into account and added to (6.35). However, in the classical explicit approach only water pressure is calculated implicitly. Water density does not vary much. Additionally, water does generally not contain much HC in solution and vice versa HC phases do not absorb much H_2O so these derivatives can be neglected for water. Other phases are treated explicitly and thus correctly, when small enough time steps are chosen. Therefore the potential error is expected to be small.

6.3.5 Black Oil Model

Two dissolution models are widely used in reservoir engineering and basin modeling. These are the black oil model and the multicomponent model.

The black oil model is an approximation which consists of three components only, namely H_2O, and two petroleum components, one lighter gas and one heavier oil component (Sec. 5.3). In the following, the water, oil, and gas components are denoted with 1, 2 and 3 respectively. The water component consists of 100% H_2O and the oil component is dissolved in the oil phase only. The light gas component can be dissolved in both petroleum phases, dependent on temperature and pressure. Thus, only one independent PVT–function is found in the model. It is the ratio of the gas component in the oil phase, named here as x with $C_{3o} = x$. Most of the C_{ip} are zero, except

$$C_{1w} = 1, \quad C_{2o} = 1 - x, \quad C_{3o} = x, \quad C_{3g} = 1\,. \tag{6.36}$$

The mass balance equations for the three components now yield

$$\begin{aligned}
\frac{\rho_w \phi}{1-\phi}\frac{\partial S_w}{\partial t} - \boldsymbol{\nabla}\cdot\rho_w \boldsymbol{\mu}_w \cdot \boldsymbol{\nabla} u_w - \frac{\rho_w C S_w}{1-\phi}\frac{\partial (u_l - u_w)}{\partial t} &= q_1 \\
(1-x)\frac{\rho_o \phi}{1-\phi}\frac{\partial S_o}{\partial t} - \boldsymbol{\nabla}\cdot(1-x)\rho_o \boldsymbol{\mu}_o \cdot \boldsymbol{\nabla} u_o - (1-x)\frac{\rho_o C S_o}{1-\phi}\frac{\partial (u_l - u_w)}{\partial t} &= q_2 \\
\frac{\rho_g \phi}{1-\phi}\frac{\partial S_g}{\partial t} + x\frac{\rho_o \phi}{1-\phi}\frac{\partial S_o}{\partial t} - \boldsymbol{\nabla}\cdot\rho_g \boldsymbol{\mu}_g \cdot \boldsymbol{\nabla} u_g - \boldsymbol{\nabla}\cdot x\rho_o \boldsymbol{\mu}_o \cdot \boldsymbol{\nabla} u_o & \\
- \frac{C(\rho_g S_g + x\rho_o S_o)}{1-\phi}\frac{\partial (u_l - u_w)}{\partial t} &= q_3
\end{aligned} \tag{6.37}$$

instead of (6.24).

The above formulation is very similar to the three phase flow formulation without phase changes. In principle, explicit and implicit solution methods can be applied in the same way. Obviously, the incorporation of oil dissolution in the gas phase, as described for the symmetrical black oil model, is straight forward (Sec. 5.3).

6.4 Diffusion

Petroleum transport as a diffusion process in aqueous solutions, in molecular or micellar form, is of lesser importance for migration. Exceptions are very poor source rocks, which are not able to build up sufficiently large amounts of fluid for separate phase flow, and diffusion of gas through dense unfractured seals, where leakage is prohibited. Another application of diffusion flow is transport of HC components in accumulations. For example, diffusion may act against gravitational separation and thus components are transported from the core of an accumulation to its oil–water contact, where biodegradation might occur. In basin modeling it is commonly assumed that lateral diffusion inside of accumulations over geological times is strong enough to achieve a homogeneous mixing of petroleum from different oil types with individual compositions. However, the grid resolution of typical basin models does usually not allow spatial effects within accumulations to be modeled.

Diffusion models for secondary petroleum migration are only related to the transport of methane in aqueous solution. The dissolution of methane in water is PVT controlled and described in Sec. 5.2. The resulting concentration gradients cause a diffusion flux, which can be formulated with Fick's law applied to porous media and a effective rock diffusion coefficient D according to

$$\mathbf{J} = -D\boldsymbol{\nabla} c \tag{6.38}$$

for a methane concentration c (Krooss, 1992; Krooss et al., 1992a,b).

Diffusion coefficients for different temperatures were measured by Schlömer and Krooss (1997, 2004). Values between $0.018 \times 10^{-10}\text{m}^2/\text{s} = 56.8 \times 10^{-6}\text{km}^2/\text{My}$ and $4.46 \times 10^{-10}\text{m}^2/\text{s} = 0.014 \times \text{km}^2/\text{My}$ were found.

It is commonly assumed that the temperature dependency of a diffusion coefficient follows an Arrhenius law according to

$$D(T) = D_0 \exp\left(-E_A/RT\right) \tag{6.39}$$

with D_0 as a frequency type pre-factor, activation energy E_A, and the universal gas constant R. Activation energies of about $16 \ldots 50\,\text{kJ/mol}$ are reported (Krooss, 1992).

A diffusion equation of form

$$\frac{\partial c}{\partial t} = D\nabla^2 c \tag{6.40}$$

can be derived from (6.38) with consideration of mass conservation. Following Krooss et al. (1992a), an example solution of the diffusion equation for methane diffusion through a cap rock is presented in App. L.

The transport via diffusion is regarded as a two step process comprising the dissolution of methane in water and diffusion within the water phase. Once the methane is dissolved in water, transport via water flow also becomes relevant, especially when high rates of aquifer flow exist.

6.5 Reservoirs

Accumulations of hydrocarbons are often found in reservoir rocks with high porosities, which can accommodate large amounts of hydrocarbons. High porosities correlate commonly with high permeabilities. Reservoirs with such properties are called carriers. Inside them HCs can migrate long distances in short periods of time before they finally accumulate in traps.

The most accurate formulation of migration physics leads to a complicated set of coupled non–linear partial differential equations (Sec. 6.3), which can only be solved with a huge amount of resources (Aziz and Settari, 1979; Øye, 1999). This effort is commonly taken into account in production related reservoir modeling. However, these solutions are limited to fixed geometries and to time scales of less than a few years. In contrast, geological timescales and non–rigid geometries caused by compaction are studied in basin modeling. It is practically impossible to directly solve the differential equations with varying geometries. However, long geological timescales in basin modeling permit an alternative approach which needs only a fraction of the resources needed for a direct solution:

Due to the high mobility of the HCs in the carriers and the density contrast with surrounding pore water, the resulting force governing migration is buoyancy (Hubbert, 1953; Sylta, 1993; Lehner et al., 1987). Thus HCs move primarily upwards in carriers until they reach a barrier. A barrier is a region with high capillary pressure and low permeability such as a sealing rock. Still driven by buoyancy, the HCs migrate just below the seal following the interface to the highest point of the carrier where they are trapped and accumulate (Fig. 6.9). The migration can be almost lateral over long distances and happens almost instantaneously compared to other geological processes such as deposition and compaction or transient temperature compensations.

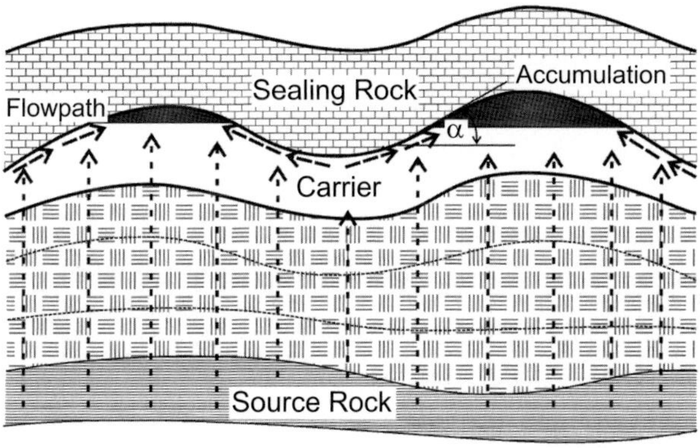

Fig. 6.9. Section view of the general scheme for reservoir analysis

In total, HC migration in carriers can be modeled by construction and analysis of flowpaths, drainage areas and the calculation of volumetrics for accumulation. This is almost a geometric procedure and thus can be accomplished much more quickly than the direct solution of coupled partial differential equations.[6]

Reservoir analysis is based on flowpath evaluation. It is therefore also known as "ray tracing" (Hantschel et al., 2000) and is the topic of Sec. 6.5.1. Flowpath analysis can be used for the construction of drainage areas which collect all the flow of a region in one trap. Drainage area decomposition of carriers and properties of drainage areas are treated in Sec. 6.5.2. Volumetrics for accumulation analysis as one major step of reservoir analysis is the subject of Sec. 6.5.3. The treatment of faults in this picture is described in Sec. 6.5.4. Later, in Sec. 6.5.6 non–ideal reservoirs are discussed.

The terminology "flowpath modeling" is in this volume used for basin wide migration and is not restricted to reservoirs only Sec. 6.7.

[6] The solution performance of the Darcy flow problem can be enhanced by the formulation of two–dimensional Darcy flow based equations for the reservoir (Lehner et al., 1987).

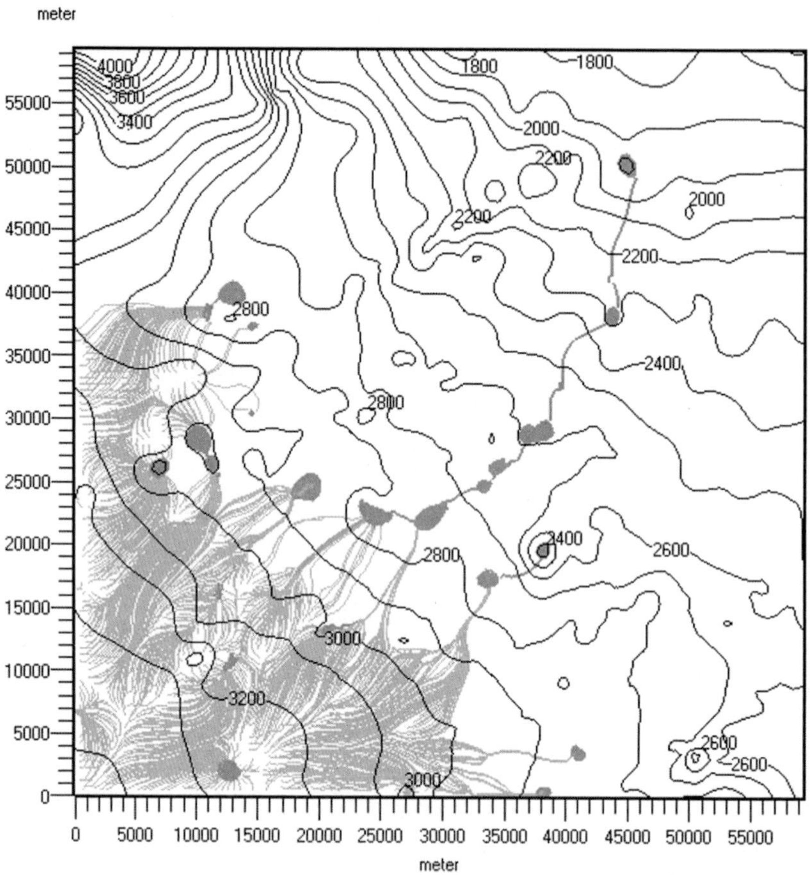

Fig. 6.10. Map view with an example of reservoir analysis. Isolines depict the depth of the sealing interface. Flowpaths and accumulated amounts of liquid HCs are displayed in grey. In the bottom left corner, below the region of 3000 meter depth, a kitchen area expels HCs into the carrier. They migrate as far as the top right of the map. The map has 120 × 120 gridpoints and open borders. Some flowpaths are filtered for better visualization

6.5.1 Flowpath Analysis

According to Darcy's law the velocity $\mathbf{v}$ of HC flow is

$$\mathbf{v} = -\frac{\mathbf{k}k_r}{\nu} \cdot \boldsymbol{\nabla} u \tag{6.41}$$

with permeability tensor $\mathbf{k}$, relative permeability k_r, viscosity ν, and overpressure gradient $\boldsymbol{\nabla} u$. With buoyancy as the driving force $|\boldsymbol{\nabla} u| = (\rho_w - \rho_p)g$. Herein $\rho_{w,p}$ are the densities of water and petroleum and $g = 9.80665\,\mathrm{m/s^2}$. A low estimation of the velocity in carriers with $kk_r \approx 1\mathrm{mD} \approx 10^{-15}\mathrm{m}^2$,

$\nu \approx 10^{-3}$Pa s, $\rho_w - \rho_p \approx 600\,\mathrm{kg/m^3}$ and $1\,\mathrm{My} \approx 3 \times 10^{13}$s yields

$$v \approx 6\,\frac{\mathrm{nm}}{\mathrm{s}} \approx 180\,\frac{\mathrm{km}}{\mathrm{My}} \tag{6.42}$$

for upward movement (Fig. 6.1). The velocity is reduced if migration follows a seal with a dipping angle α by a factor $\sin\alpha$ (Fig. 6.9). With an angle of about ten degrees the estimated velocity is

$$v \approx 30\frac{\mathrm{km}}{\mathrm{My}} \tag{6.43}$$

and with one degree the velocity becomes $v \approx 3\,\mathrm{km/My}$. This is a conservative estimate. Velocities of up to 1000 km/My have been reported in the literature (Sylta, 2004). Generally, HCs have the capability to migrate long distances through carriers on geological timescales.

This conclusion is the basis for migration modeling in carriers. It implies that migration in carriers continues until a trap is reached. Only HCs which have not entered the carrier during the last geological event may not have reached a trap.[7,8]

HCs entering a carrier move straight upwards until they reach a seal and then follow the steepest direction upwards below the seal. Flowpaths help to visualize the migration just below the seal (Fig. 6.10). They end at the local heights of the carrier indicating the end of migration. Thus flowpaths can easily be constructed on a map of the reservoir seal interface. Migration modeling with flowpaths has an "implicit" high resolution because flowpaths can be constructed from interpolations of mapped gridpoints (Fig. 6.14). Flowpaths passing between two neighboring gridpoints can end in different traps (Fig. 6.11) and therefore even low–resolution maps can show a complicated migration pattern.

Residual amounts of petroleum, which indicate that petroleum traversed or is actually traversing, are usually only found in the zone below the sealing interface but not in the bulk reservoir rock (Dembicki Jr. and Anderson, 1989; Schowalter, 1979). This confirms the flowpath concept with migration directly below the seal. Flowpaths are constructed from purely geometric analysis and indicate only the direction of flow. A quantification of immobile losses in this picture is postponed until Sec. 6.5.6.

Several flowpaths starting at great distances from each other can end at one local height. Thus all the HCs reaching this trap must be summed up before it is possible to perform an accumulation analysis (Sec. 6.5.2). Finally, for the determination of column heights, volumetrics must be performed in each trap (Sec. 6.5.3). Hence reservoir analysis is a three step process which consists of "flowpath analysis", "drainage area analysis", and volumetrics for trapping or alternatively "accumulation analysis".

[7] Except for small amounts of immobile HCs

[8] In streamline modeling timing is taken into account on each streamline (Datta-Gupta et al., 2001). This approach is necessary on short production timescales.

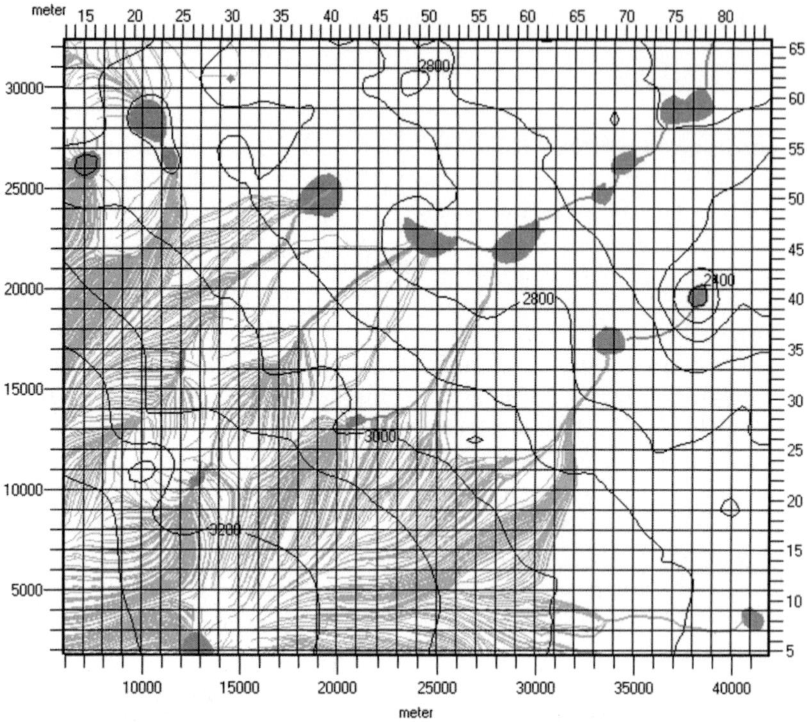

Fig. 6.11. Zoomed cutout of Fig. 6.10 with a mesh which contains every second gridline. Flowpaths and accumulations have a very high "implicit" resolution caused by the smooth interpolation of gridded maps

6.5.2 Drainage Area Analysis

A carrier can be subdivided into drainage areas or fetch areas. All HCs entering the carrier in the same drainage area migrate to the same trap. HCs entering the carrier at other locations migrate to other traps. Each trap belongs to one drainage area and vice versa. Any point of the carrier belongs to exactly one drainage area. The subdivision into drainage areas is a domain decomposition of the carrier. Drainage areas can be calculated by construction of "possible" flowpaths for each point of a carrier (Fig. 6.12).

Migration becomes a purely geometric problem of HC distribution in traps when the drainage areas are known. Prior to anything else it is then necessary to know the maximum trap capacity of a structure. A trap can be filled until the accumulated HCs reach the border of the drainage area. The contact point at the border is called the "spill point" and the maximum possible filling volume is called the pore volume of the "closure" (Fig. 6.13).

The shape of an accumulation can be constructed using the following scheme. HCs move to the highest point which has not been occupied yet.

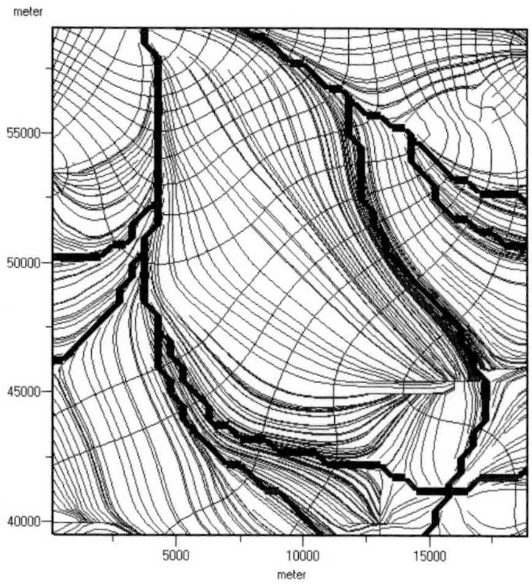

Fig. 6.12. Zoomed cutout of the same map as in Fig. 6.10 with highlighted drainage area borders (thick lines) and "possible flowpaths" perpendicular to some depth isolines. Flowpaths originating at the border of a drainage area might follow the border over wide distances until they bend off towards the interior

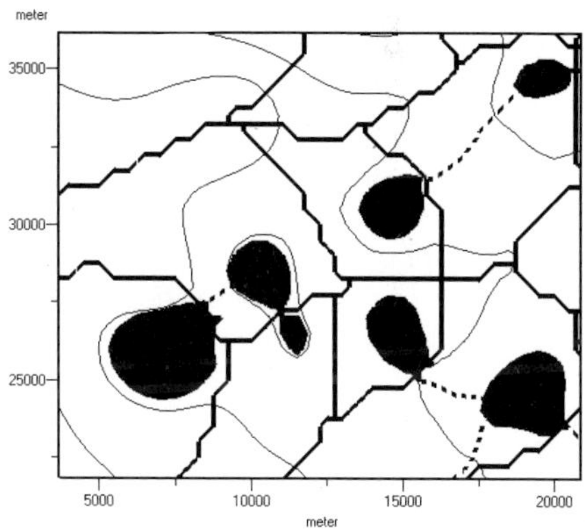

Fig. 6.13. Zoomed cutout of the same map as in Fig. 6.10 with drainage area borders, closures, thin depth isolines, and dotted spill paths. Spill points are located at the beginning of the spill paths. This is the location where the closure touches the drainage area border

Thus, the meniscus of the accumulation is flat and horizontal. The spill point must be the highest point on the border of the drainage area. It is easy to construct the shape of the closure by cutting the reservoir with a horizontal plane through the spill point. The volume between this meniscus plane and the reservoir seal interface is the closure volume.

The entire algorithm of the drainage area, spill point, and closure volume calculation is in practice complicated because the construction of flowpaths is based on the calculation of depth gradients. A gradient is dependent on the local shape of the map and therefore strongly dependent on the interpolation method and on more than one grid point (Fig. 6.14). It is therefore very sensitive to gridding uncertainties at the border of drainage areas because it is supposed to be zero or at least perpendicular to the border. Small variations in interpolation might shift the drainage are border.

It might be argued that twisted cells, as in Fig. 6.14, are rather rare and therefore not representative for depth maps. However, due to the fact that flowpaths are perpendicular to depth isolines[9], it can be assumed that twisted cells, such as in Fig. 6.14 with almost the same value at more than one corner point, occur more frequently at the borders of the drainage areas. The accurate location of these borders is obviously very important for the determination of the spill point.

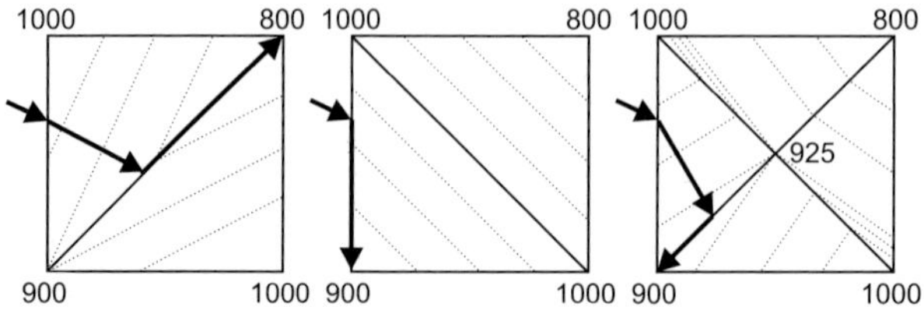

Fig. 6.14. Three different examples of interpolation between four gridpoints. The rectangles are cells in a map view with given depth at the corner points. In the two left examples the values are interpolated linearly within two triangles and in the right example within four triangles. The center point on the right is calculated with an arithmetic average. Depth isolines are dotted. An example of a flowpath entering from left is visualized with arrows. The flowpath is perpendicular to the depth isolines. In the first two examples different flowpaths based on locally different gradients are emerging. It can easily be verified that flowpaths entering the cell on the left side from any position always end–up at the top right corner, whereas the other examples show a drainage area border from the top left to bottom right. The first two examples represent the same scheme differing only in the orientation of the decomposition triangles. The right example does not have a comparable inherent orientation and yields a drainage area border consistent with expectations. Hence, the right interpolation scheme is superior

Fortunately, the exact location of the border usually does not contribute significantly to volume calculations because the closure is often far inside the drainage area. Besides the fact that fetched amounts of HCs depend on the size of the drainage area itself,[10] there is one further problem. The depth of the spill point is the depth of the highest point at the drainage area border. It strongly influences the closure volume (Fig. 6.15) and therefore well interpreted and gridded maps are needed to allow good volumetrics.

Locations of HC injection into a reservoir should be the unique starting points of the flowpaths. This implies two different technical approaches for handling drainage areas. If the injection of HCs is performed at grid point locations then drainage areas can be described by areas of grid points with the border lines in between. Alternatively, if HCs are injected according to a cell picture in the center between the grid points then drainage areas should be treated as collections of cells with the drainage area borders on the grid

[9] Only if the map is displayed with an equal horizontal to vertical aspect ratio

[10] Below one drainage area, source rock properties and expulsion amounts do not usually vary very much.

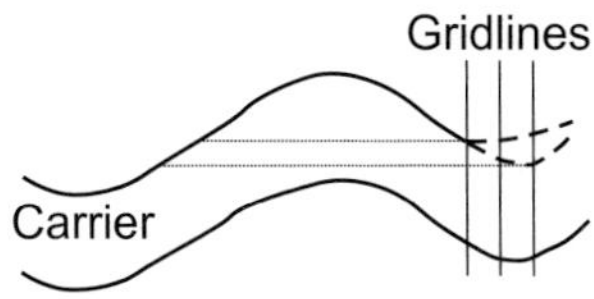

Fig. 6.15. Illustration of how slight variations of the spill point depth severely alter the trap capacity. Two possible scenarios of spill point depth are indicated by dashed carrier seal interface lines on the left. The closure volume varies between the dotted lines. Note that the horizontal axis is often stretched by a factor of about 10 in typical basin models and that the volume must be interpreted three dimensionally. Large volumetric errors can arise from poor gridding or low resolution

lines. The first approach is easier in practical workflows because everything is handled on the same grid, e.g. data export and import. The second is more natural because non–unique flow directions seem to be less frequent (Fig. 6.16). However, the best results can be achieved if flexible drainage area borders are allowed on sub–gridding level (Fig. 6.17).

Fig. 6.16. Non–unique flowpath direction for HCs reaching the seal from below. The left case is a section view with petroleum injected at a grid location and the right case is a map view of a cell similar to the left case of Fig. 6.14, with injection at the cell center. In practice, the flowpath is often simply assumed to follow the steepest gradient

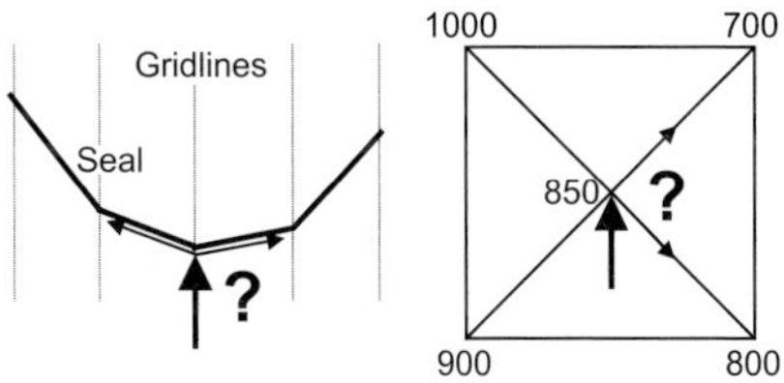

Fig. 6.17. An example of a gridding problem at a spill point location. In this example the closure is allowed to be extended half a grid distance behind the displayed drainage area border. The values between the gridlines are smoothly interpolated. The closure should not only be restricted to the gridcells, which belong completely to the drainage area. This ensures the best possible calculation of the spill point depth and therefore the best closure amount and trap capacity estimation

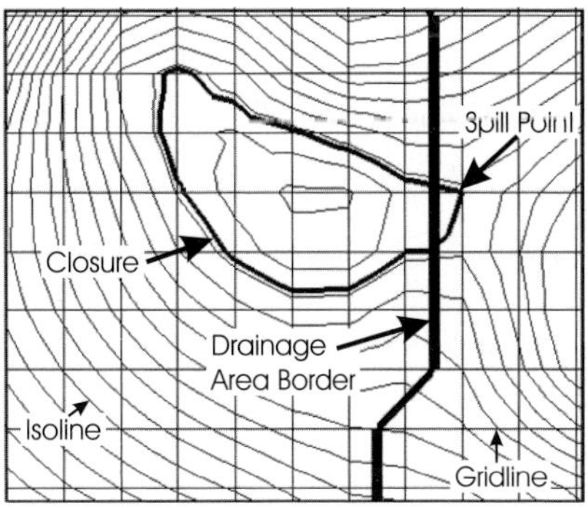

Absolutely flat horizontal areas are particularly problematic where flowpaths are not defined and in cases where the location of the drainage area border is undefined (Fig. 6.18). For simplicity it is better not to separate such areas. The volumetrics is not affected anyway. A similar problem occurs if a given drainage area has multiple spill points at the same depth. Nevertheless even such unrealistic cases can be handled consistently. Such problems are academic, as absolutely flat areas or different spill points at exactly the same depth do not exist in real case studies.

Fig. 6.18. Schematic view of two cases with a flat reservoir seal interface. Obviously, the flat area belongs to the drainage area on the left because all HCs coming from below will sooner or later migrate to the left. In the right case the drainage area border and the spill point might be located at any location on the flat area

If a trap is filled up to the spill point, it starts spilling excess amounts of HCs over the border of the drainage area into a neighboring drainage area. All these spilled HCs follow one flowpath, the "spill path". Spill paths can become main migration pathways because huge amounts of HCs may migrate along only a few paths (Fig. 6.10).

Drainage area analysis can become more sophisticated if drainage areas merge: This happens if two areas spill into each other and the total amount of HCs entering these areas exceeds the summed capacity of both. This is often the case when excessive amounts of HCs do not fit into the trap of the area into which they are spilled and then they are spilled back into the area where they originally came from. In such cases both areas merge. A new closure and a new spill point for the merged area have to be calculated (Fig. 6.19). With the continuous generation of HCs over geological time the merging of already merged areas may occur (Fig. 6.20).

The performance of drainage area analysis can be improved if small drainage areas or drainage areas with overall negligible closure volumes are directly merged with their larger neighbors (Fig. 6.21). In such cases flowpaths sometimes seem to end inside a merged drainage area but outside of the trap itself (right example of Fig. 6.21). However, the calculation of volumetrics and HC distribution is not affected.

6.5.3 Accumulation Analysis

HC accumulations are found as interconnected zones of high porosity and permeability. Therefore a HC pressure arises. It can be determined by cal-

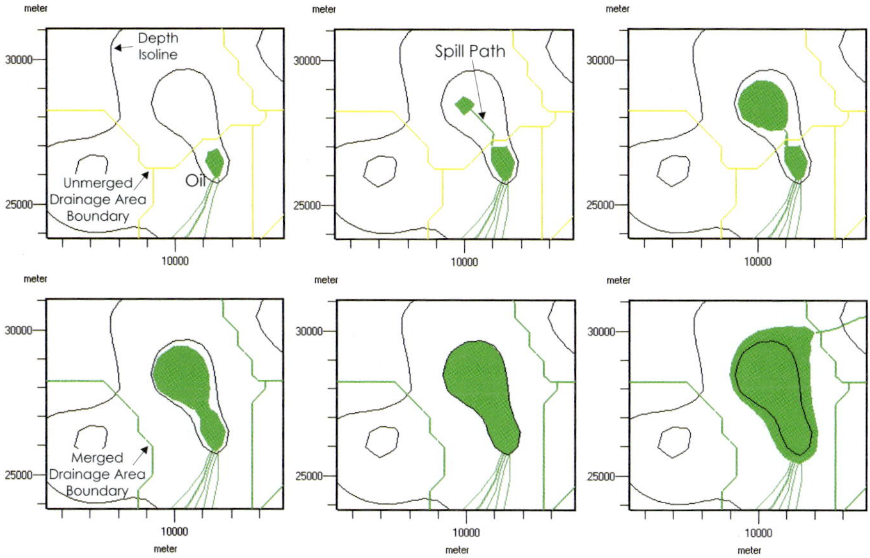

Fig. 6.19. Continuous trapping of HCs with filling, spilling, and merging of accumulations and drainage areas followed again by spilling in a cutout region from Fig. 6.10

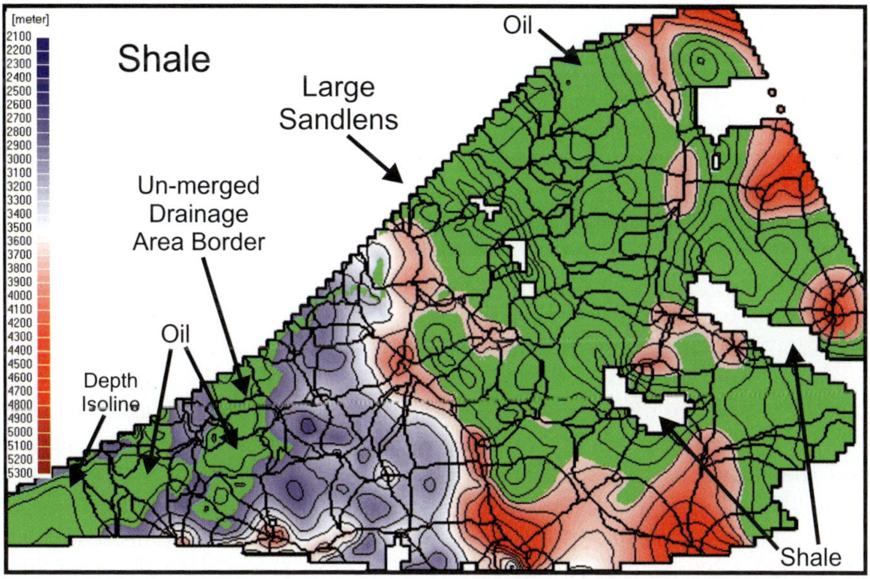

Fig. 6.20. Map view of a reservoir layer with some shale lenses inside. It is surrounded by low permeability shales which implies closed boundaries. Oil is accumulated into big bodies covering multiple structural drainage areas

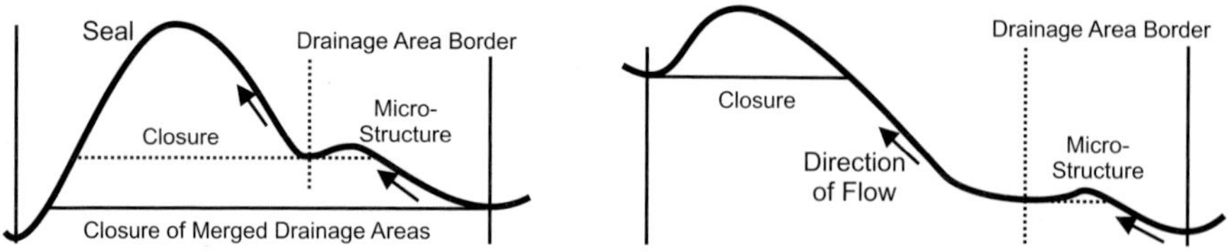

Fig. 6.21. Two schematic examples of drainage areas in section views with micro–structures which are merged before accumulation analysis. The trap capacity rises enormously in the left example

culation of the column height which acts against a seal. When it exceeds a certain limit, a break through occurs (Fig. 6.22). The limiting pressure is given by the capillary pressure of the sealing rock. The meniscus (or column) height which balances the column pressure with the sealing capillary pressure and which fits the remaining volume into the trap must be found.

Fig. 6.22. Pressure acting on top of an accumulation against a seal. The column pressure p_{col} can be evaluated from the density contrast $\Delta\rho = \rho_w - \rho_p$ of water and petroleum and column height h via $p_{\mathrm{col}} = \Delta\rho g h$ due to pressure communication in the accumulation

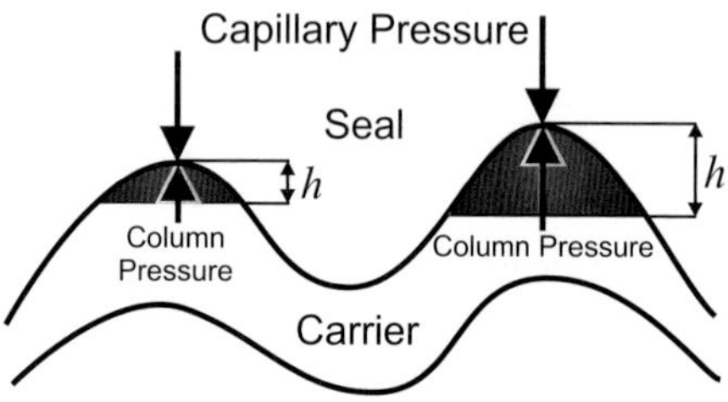

For all the considerations which were made up to now, perfect component mixing in an accumulation is assumed. Accumulations are studied over geological timescales in basin modeling. So this assumption is reasonable for accumulations with universal pressure contact but must not be valid in every case. Processes such as compositional grading are not taken into account because in basin modeling the spatial model resolution is usually not high enough. Accumulations often have a size of only a few gridcells.[11]

Very often HCs occur in two phases, vapor and liquid (Chap. 5). The migration of these phases can be treated separately. Both will reach the trap on geological timescales. In the trap they will interact: the vapor is lighter and therefore it moves to the top of the accumulation displacing the liquid (Fig. 6.23). Liquid can only occupy space which is not occupied by vapor. In extreme cases the liquid and vapor drainage areas must be distinguished. Liquid areas have merged, while at the same time the vapor still "sees" unmerged areas (Fig. 6.24).

[11] An exception are local grid refinements (LGR) around accumulations such as described in Sec. 8.9

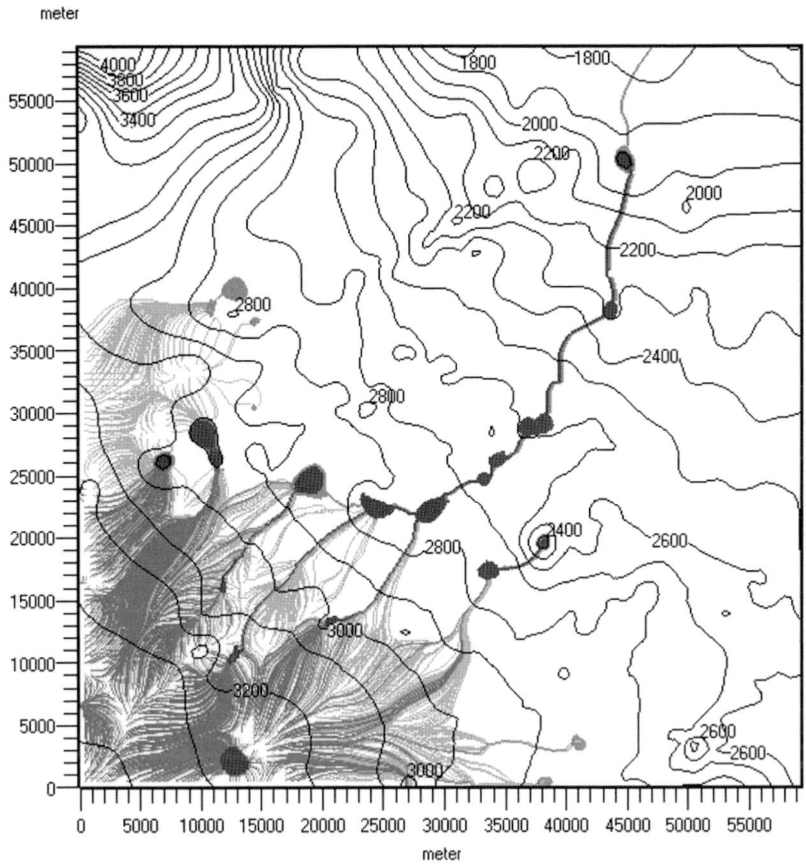

Fig. 6.23. Vapor (dark grey) displacing liquid (light grey) in comparison with Fig. 6.10: vapor entered the carrier from the kitchen area below 3200 meter depth from the bottom left corner of the map. The vapor displaced most of the liquid

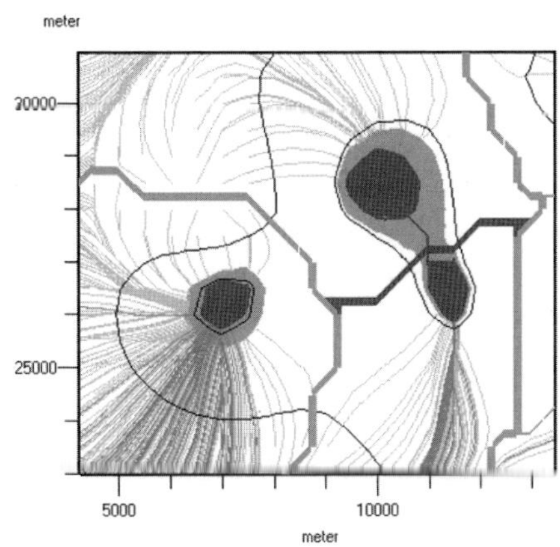

Fig. 6.24. Close up view of a scenario with less gas than in Fig. 6.23. Two vapor accumulations (dark grey) are located above one liquid accumulation (light grey) in the structure on the right side. Here vapor accumulates along different drainage areas than liquid. The "liquid" drainage area has merged, the "vapor" areas have not

Column pressure calculations for the determination of break throughs must take into account the different densities of vapor and liquid in each phase and the continuity of pressure at the liquid–vapor interface (Fig. 6.25, Watts 1987, Dake 2001).

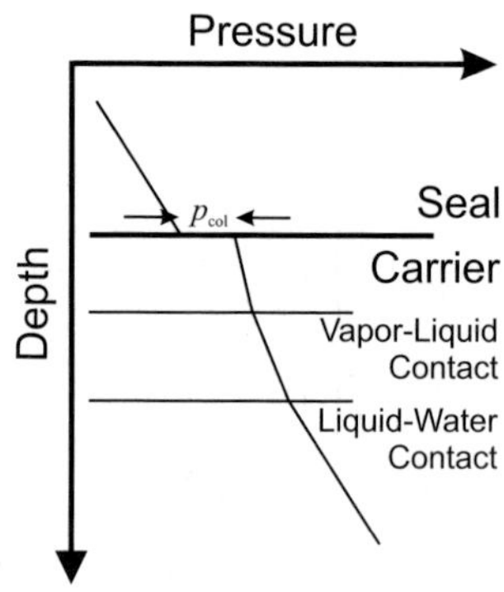

Fig. 6.25. Schematic figure of pressure build–up in an accumulation containing liquid and vapor. The pressure gradient in the vapor is steeper than in the liquid HCs, which again is steeper than in the water

Component mixing between the two phases and redistribution of components to the phases after fluid analysis or recalculation of phase densities in the accumulations are possible but usually not necessary because carriers often lie within a limited pressure–temperature interval. Variations in phase composition or density are thus assumed to be rather small. Huge effects of outgassing inside of reservoirs are expected to appear more during ongoing reservoir uplift than during migration within the reservoir. An exception is long distance lateral migration, e.g. due to spilling (Gussow, 1954). However, outgassing over geological timescales is automatically treated by the hybrid method and flowpath modeling (Secs. 6.6, 6.7). Flash calculations can easily be performed before and after each reservoir analysis or Darcy migration time step within each gridcell which contains hydrocarbons.

6.5.4 Faults and Small Scale Features

Faults are often interpreted as almost perfect HC "conduits" or barriers, open or closed faults respectively (Chapman, 1983). Both cases can be combined with reservoir analysis. The reservoir is usually a horizontally oriented layer of smaller thickness. By contrast faults are typically vertically oriented without any thickness and thus modeled as vertical planes inserted into the reservoir. They can be visualized by lines on maps (Fig. 6.26). Open or closed faults are easy to combine with reservoir analysis.

Open faults act as the endpoints of flowpaths. They are vertical HC conduits. HCs reaching open faults are transported out of the reservoir and are therefore no longer the subject of reservoir analysis.

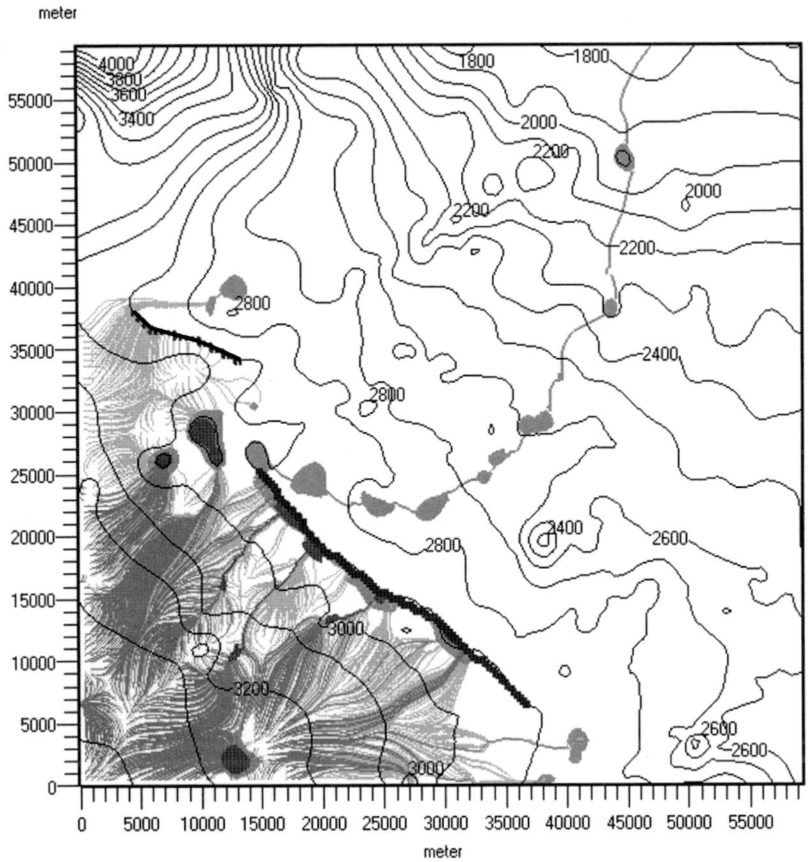

Fig. 6.26. A map with one impermeable and one permeable fault. Vapor is colored dark and liquid light grey. The impermeable fault acts as a barrier whereas the HCs are leaving the carrier at the permeable fault

Closed faults act as barriers which cannot be crossed by flowpaths. Drainage area and accumulation analysis are automatically adjusted if flowpaths have been correctly calculated.

Very often faults are assumed to have a finite fault capillary pressure (FCP) (Sec. 2.7, Yielding et al. 1997; Clarke et al. 2005a,b, 2006). Due to the fact that column pressure is highest at the top of a petroleum column, it is possible to model faults with lower FCP than the seal capillary pressure by simply replacing the seal capillary pressure by the FCP at the location where the fault crosses the seal (Figs. 6.27, 6.28). Break through amounts at fault locations are assumed to migrate into the fault following its structure upwards.

In a grid which consists of hexahedron type cells such as described in Chap. 8 with the gridpoints at the corners, the sealing capillary pressure value of a gridpoint in the reservoir layer is finally selected out of a maximum of 15 values, namely three possible surrounding juxtaposition cells, four overlaying top seal cells, four possible in–reservoir fault walls and four possible reservoir to seal horizontal fault cells.

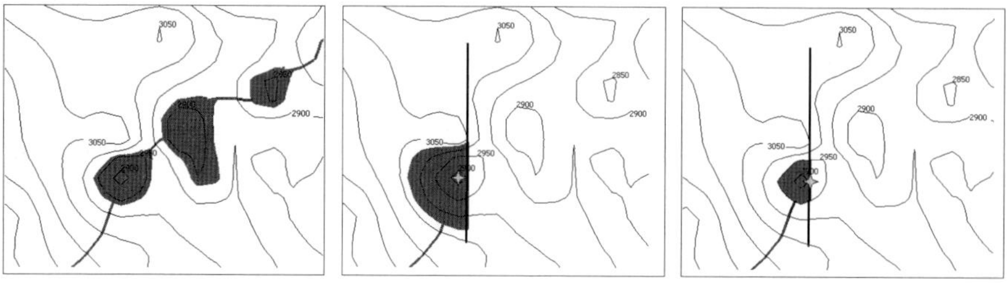

Fig. 6.27. A map view of a reservoir with accumulations but without any faults or break through (left), with a high FCP fault (center) and a low FCP fault (right). As indicated by the "star" a break through appears on top of the structure in the center and on the highest point of the fault at the reservoir seal interface in the scenario to the right

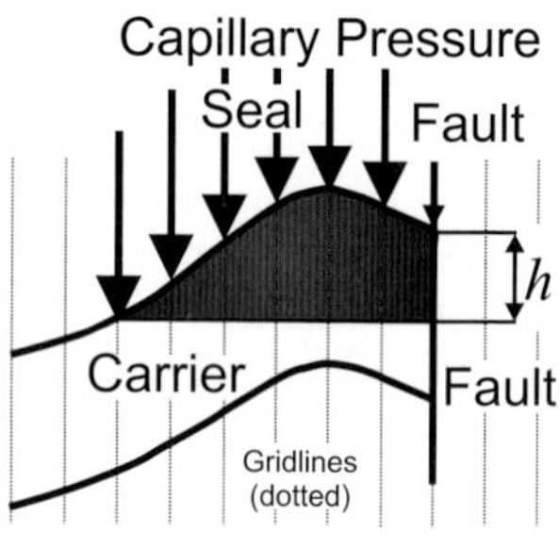

Fig. 6.28. Oil column of height h in equilibrium with fault capillary pressure (FCP). Break through amounts follow the fault upwards. The column pressure can be calculated as in Fig. 6.22

Lithology variations which affect migration or accumulation such as porosity variations, fracturing or facies changes, can easily be modeled with additional porosity maps, capillary pressure maps of the seal or in complicated geometries, with artificial faults which limit the horizontal extension of the reservoir (Fig. 6.34). It is often more challenging to create the maps than modeling the effects.

6.5.5 Overpressure and Waterflow

Pressure gradients twist flowpaths and deform HC–water contact planes (Hubbert, 1953; Hindle, 1997). Reservoir rocks with high permeabilities have the ability to balance overpressures directly with outflow of water.[12] Thus only very small pressure gradients are found in reservoir rocks and therefore water flow can only significantly affect the direction of flowpaths in regions of relatively slow migration, e.g. in regions of low reservoir seal tilting. Due to the small size of the pressure gradients and a lack of high precision data describing the geometry of the reservoir, it is often impossible to determine these locations and the pressure gradient to the precision which is necessary for accurate flowpath bending in basin modeling. However, it is usually assumed that the effect is small and does not contribute significantly to the overall picture of migration and accumulation.

Additionally, continuous aquifer flow might occur. It can originate on length–scales which are even larger than the arbitrary extensions of a basin model (Hubbert, 1953; Ingebritsen and Sanford, 1998; Freeze and Cherry, 1979). The necessary overpressure data is usually not available but it is generally assumed that it can be approximated by a hydraulic head $h_w = p/g\rho_w$ which follows onshore topography (Chap. 2). The dipping angle γ of the hydraulic head is in this case equivalent to the tilting angle of the surface. Hence, topography driven flow occurs in basins with water flow originating in high (on–shore) mountains.

The calculation of bending of the lateral petroleum flow direction below a seal is demonstrated in App. M. For example, the angle ψ between the x–direction and the projection of the petroleum flow direction into the horizontal xy–surface is given by

$$\tan\psi = \frac{\rho_w - \rho_p}{\rho_w} \frac{\cos\alpha \sin\alpha}{\tan\gamma} \tag{6.44}$$

for dipping angles γ of the hydraulic head in the x–direction and α of the seal in the y–direction (Fig. 6.29).

6.5.6 Non–Ideal Reservoirs

Until now, migration losses have been neglected. Two different types of losses are distinguished. Petroleum becomes immobile below the critical saturation. It forms small droplets which are stuck. This phenomenon is described with a relative permeability of exactly zero below the critical oil saturation S_{oc}. Besides immobile losses, all petroleum which is not found within accumulations is commonly named "losses of the petroleum system" (PS losses). Some amounts are lost at the basin boundaries, e.g. on top, or are mobile and might reach an accumulation later but they are currently not available for production.

However, compared to losses during primary migration or in other regions of low permeability, it can be assumed that losses in reservoirs are mostly small so reservoir analysis without losses is a good approximation in basin modeling (Dembicki Jr. and Anderson, 1989).

[12] At least on geological timescales

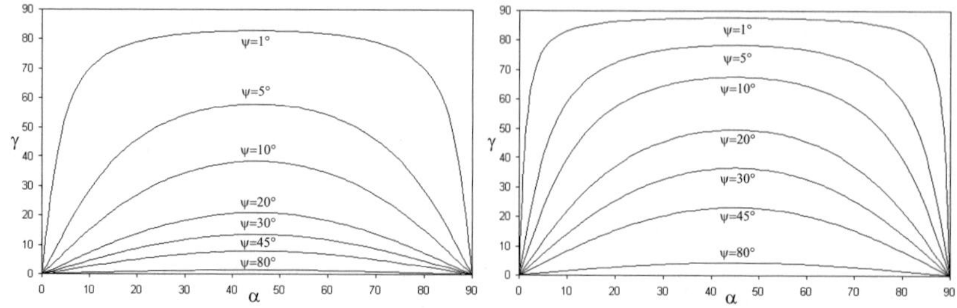

Fig. 6.29. Bending of a flowpath for an oil with $\rho = 750\,\mathrm{kg/m^3}$ on the left and a gas with $\rho = 150\,\mathrm{kg/m^3}$ on the right according to (6.44). The water density is here $\rho_w = 1040\,\mathrm{kg/m^3}$. As expected, the same bending angle ψ at the same tilting α needs less "lateral overpressure" γ for the oil . The curves are symmetric around $\alpha = 45°$. A high tilting α of the seal implies a moderate lateral flow component in cases without lateral overpressure. Thus only a moderate overpressure is necessary to bend the lateral flow component significantly

However, losses in reservoirs can be treated via an extra processing step before accumulation analysis. For an estimation of both immobile losses and PS losses, it is necessary to determine the height of the migrating oil stringer below the reservoir seal interface. Let Q be a volume of petroleum which is transported per time unit along a flowpath of thickness H and width w (Fig. 6.30). It has the velocity $v = Q/w/T$ which can also be calculated by (6.41). Hence it becomes

$$T = \frac{Q\nu}{wkk_r \Delta\rho g \sin\alpha} \tag{6.45}$$

for isotropic permeability and α as the dipping angle of the seal. This formula can only be used for rough estimates of thickness H because the relative permeability k_r is not known. It is mainly determined by the oil saturation of the migrating oil which is not known very well (Fig. 6.2). Very often it is assumed that the saturation is rather high and therefore the influence of k_r is small.

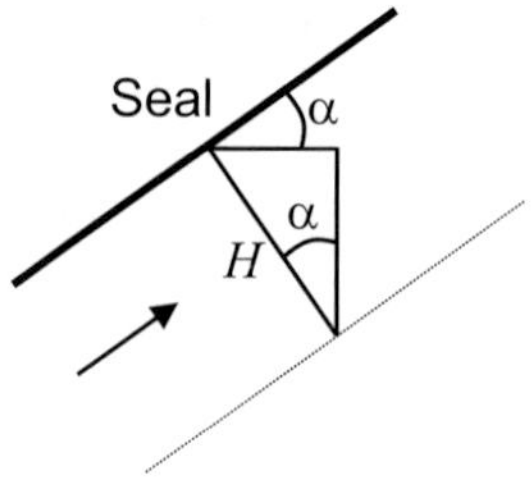

Fig. 6.30. Stringer flow of thickness H below seal

Sylta (1991, 2002a) derived formulas for a better estimation of thickness H. The volume flow along a flowpath is given by

$$Q = w \int_0^H v(h)dh = w \int_0^H \frac{k}{\nu} \Delta\rho g \sin\alpha \, k_r(S_w(h))dh \qquad (6.46)$$

with a saturation dependent relative permeability, which is again dependent on the distance h below the seal within the moving stringer. Based on an equilibrium of buoyancy with capillary pressure according to Fig. 6.4, which can be formulated in equations such as (6.14), it is possible to calculate the saturation within the moving oil stringer by inversion of

$$p_{co}(S_w) = \Delta\rho g(H - h)\cos\alpha + p_{oe} \qquad (6.47)$$

for distance h. The result can be inserted into (6.46) and numerically solved for total stringer thickness H. It was shown that migration often takes place under low average saturations. This implies that k_r becomes small and hence the thickness H large. Finally immobile losses, which increase with thickness H, are significantly higher than initially expected and PS losses are rather small due to the low average saturation (Sylta, 2002a).

In (6.46) and (6.47) it is not considered that the flow rate Q changes due to saturation losses along the flow path. For example, this can technically be taken into account by gridding along a flowpath and stepwise reduction of Q.

The idea of adjusting saturation and permeability according to a given flow amount can be formulated more generally for reservoirs. Reservoirs are often rather small structures compared to the size of the basin. They are in flow balance with their environment (Fig. 6.31), which is equivalent to an approximation without any time derivative in the Darcy flow equations.[13] Thus the Darcy flow equations for oil flow can be reduced from an initial value problem such as (6.24) to a boundary value problem of the form

$$\mathbf{v}_o = -\frac{\mathbf{k}k_{ro}(S_w)}{\nu_o} \cdot \nabla u_o, \quad \nabla \cdot \mathbf{v}_o = 0, \quad u_o - u_w = p_{co}(S_w) + \Delta\rho g z \quad (6.48)$$

with five unknowns u_o, S_w, and the three components of $\mathbf{v}_o$. Here, for simplicity, migration of gas is not taken into account. Oil inflow velocities $\mathbf{v}_\upsilon$ at the bottom of the reservoir are known from source rock expulsion rates. Top capillary pressure and permeability at areas of leakage must be estimated by taking into account leakage flow rates. In principle, sophisticated boundary conditions must be formulated or evaluated by repeated iterative reservoir analysis. However, capillary pressure curves and permeabilities of seals are usually not known very well. Rough approximations are therefore justified.

Additionally, a "homogeneous" solution without inflow (and therefore without leakage) but with a finite oil saturation, must also be constructed

[13] Compaction is also neglected here because it occurs on longer timescales than petroleum flow in reservoirs.

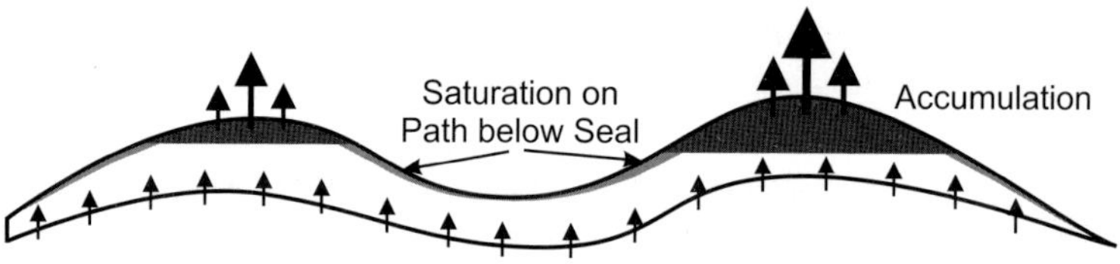

Fig. 6.31. A reservoir with overall flow balance. Petroleum saturations are adjusted in such a way that flow rates, capillary pressure and buoyancy are kept locally in balance. Thus total inflow and leakage is also in balance

and added in unsaturated regions for the modeling of preserved accumulations from previous geological events.

An exact numerical solution of these equations is still very complex. High resolution reservoir models must be constructed. Thus a high effort is necessary for small improvements in accuracy compared to a more simple flowpath analysis.

6.6 Hybrid Method

Multi–phase flow can be described by coupled nonlinear partial differential equations (Sec. 6.3.3). Due to irregular geometries varying through time combined with wide ranges of parameters and long geological timescales in basin modeling, it is practically impossible to solve these equations. Based on flowpath analysis, an approximate solution in high permeability reservoir regions can be constructed. On the other hand the flow in low permeability regions is slow, so it can be calculated based on an explicit and smooth evolution of the saturation pattern through time. Putting these two approaches together yields the hybrid method of flow modeling (Hantschel et al., 2000).

The flow velocity $\mathbf{v}$ can be calculated with Darcy's law (6.8) or (6.41). It can easily be estimated that it becomes very small in low permeability regions. For example, with $k = 10^{-4}$mD and conditions such as in Sec. 6.5.1 it becomes smaller than $20\,\mathrm{m/My}$. Isolines of flow velocity taking permeability and viscosity variations into account are depicted in Fig. 6.1.

In low permeability regions and during one geological event, HCs might only move a total of a few gridcells in basin modeling. Thus, it is possible to explicitly solve the differential equations by calculating velocities from pressure gradients and then updating HC saturations according to this velocity field.

Numerical stability is achieved by introducing small migration time steps. According to the Courant–Friedrichs–Lewy criterion, explicit solutions of differential equations are stable if the time steps are small enough (Press et al., 2002). The size of the time steps scales inversely with the velocity of flow. Thus, in regions of low velocity, this approach is feasible whereas in highly permeable reservoir regions the number of time steps explodes so that an explicit solution becomes impossible.

A domain decomposition in low and high permeability regions is the basis for the hybrid method of basin wide flow modeling. A special challenge of the hybrid method are break throughs, which are treated in Sec. 6.6.2, and fault flow, which is described in Sec. 6.6.3.

6.6.1 Domain Decomposition

A domain decomposition for the hybrid method is a spatial disjunction of the model into low and high permeability regions. Highly permeable regions are reservoir rocks which act as carriers and containers of accumulations. Obviously, these accumulations are objects of special interest. In most cases only a few layers in a basin model are reservoirs. A much larger part is occupied by low permeability rocks. Usually, the source rock belongs to this part (Fig. 6.32). Hence the domain decomposition is a cut–out of reservoirs. One must be careful, because the permeability limit which defines if a gridcell belongs to a reservoir or not depends on the grid distances and the size of the time steps. In practice is more relevant to ensure that the cut–out does not become disjunkt into small pieces, e.g. when the permeability limit is exceeded in some celles during subsidence. This can be achieve by application of the cut–out condition on a fixed instead of the insitu porosity. A value of 10^{-2} mD for a porosity of 30% is found as a good default value, which works well in many basins. A model with flow in a layer in which the permeability jumps between regions with low permeability and reservoirs is shown in Fig. 6.33.

Reservoirs can have arbitrary outlines due to sometimes complicated facies distributions. In such cases a technical complication arises concerning the flow boundaries in the reservoir. The borders of the low permeability regions act as barriers, which create the possibility of stratigraphic traps, whereas the model boundaries are usually treated as open, which indicates a continuity of the facies and allows outflow into neighboring regions. Thus, a tool for reservoir analysis must be able to allow closed and open boundary conditions in one reservoir (Fig. 6.34).

One main advantage of reservoir analysis is rapid processing. For example, maps of 200×200 cells can be processed within seconds on modern PCs. Detailed geometric information on small scales, which is necessary for accurate flow modeling, can be retained. Thus it is possible to use a higher grid resolution for the reservoir analysis than for migration in the regions with low permeability (Fig. 6.32). The price to be paid for this multigrid technique are grid transformations, which are more elaborate than initially appears. Besides the reservoir cut–out itself, saturation values in full multicomponent resolution must be transformed between the fine and the coarse grid (Fig. 6.35).

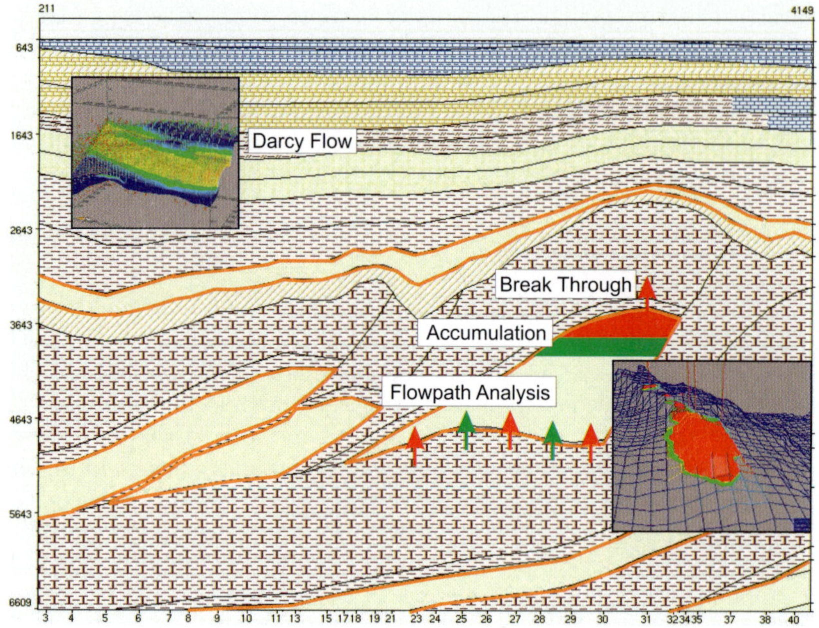

Fig. 6.32. The principle scheme of domain decomposition into reservoirs and low permeability regions for petroleum flow analysis with the hybrid method. In the cut-out on the right the finite element grid is displayed. The accumulation is calculated on a finer grid

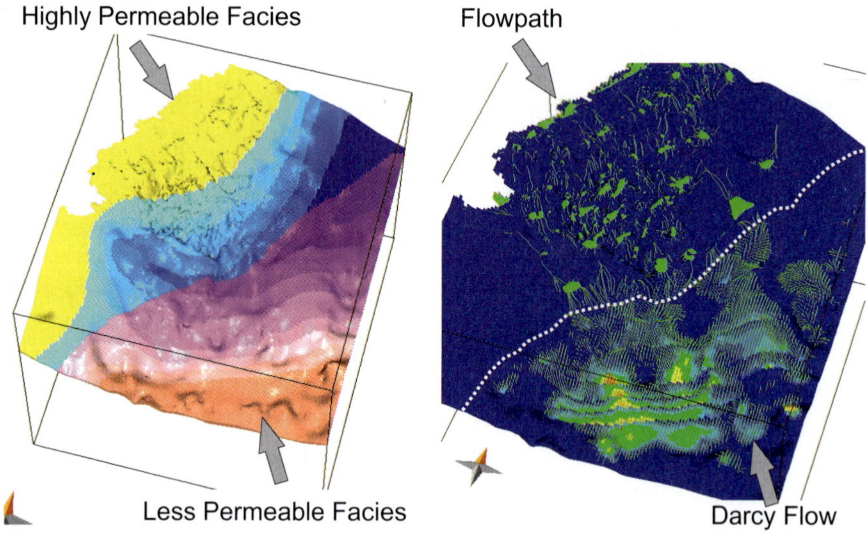

Fig. 6.33. Flow in a layer with facies change from low to high permeability. Darcy flow is indicated by vectors in the low permeability region, whereas flowpaths and accumulation bodies represent the reservoir analysis

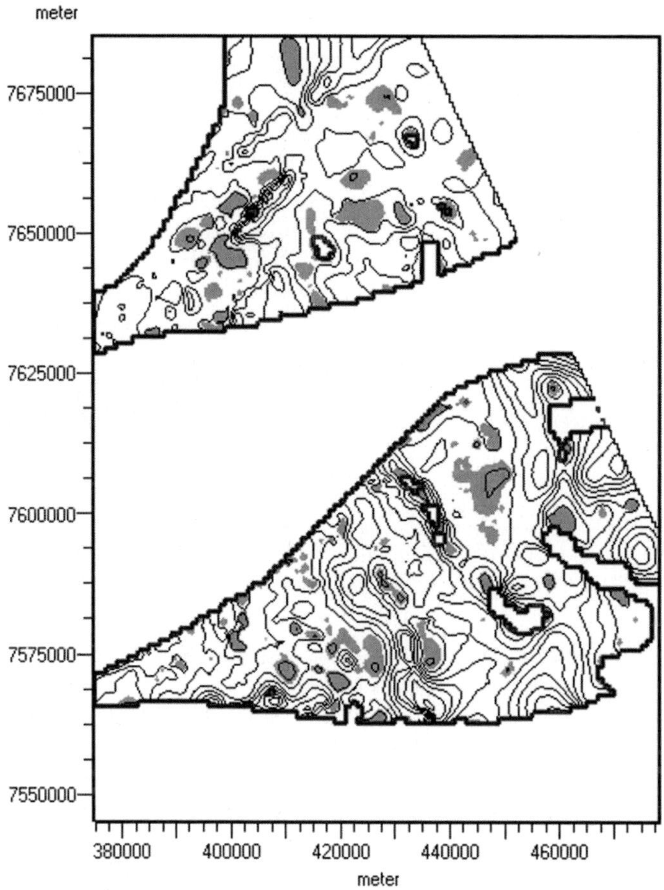

Fig. 6.34. A map view of a reservoir from a study of the Campos basin, offshore Brazil. Isolines indicate the depth in 200 meter intervals from 2000 to 5000 meters. Oil accumulations are colored grey. The thickly gridded outlines mark barriers to facies with low permeability. The two reservoir "lenses" are here open for HC migration to the sides, compared with a different scenario shown in Fig. 6.20

6.6.2 Break Through

Break through and leaking of accumulations are of special interest. This effect is caused by a column pressure acting against the reservoir seal (Fig. 6.22). If the pressure becomes high enough, a break through occurs. The pressure is highest at the highest point of the accumulation. The break through appears at the point where the contrast between column pressure and capillary pressure is the highest.[14] Exeeding amounts are transported into the seal. Advandced

[14] In general this need not be the highest point if there are capillary pressure variations in the seal.

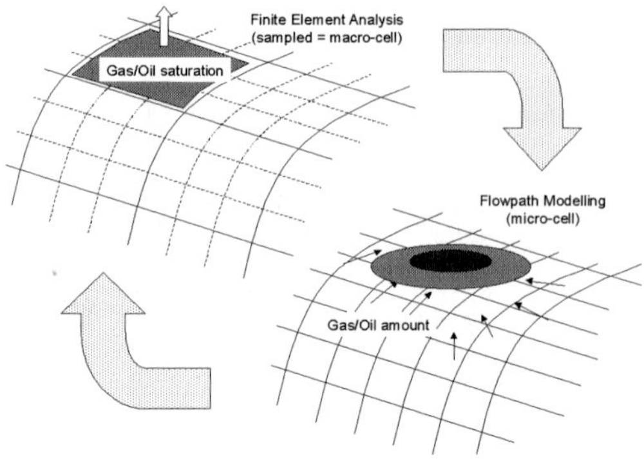

Fig. 6.35. Multigrid technique for integrated flow analysis, figure taken from Hantschel et al. (2000)

break through models take dynamic leaking into account. The sealing strength decreases during leakage down to 35% of its original value Vassenden et al. (2003).

Break throughs can be problematic concerning timing. Break through amounts which cross the interface between both domains cannot easily be split into small portions which are moved in many small time steps, as necessary for the framework of explicitly treated Darcy flow. Time control has already been dismissed for reservoir analysis on the highly permeable side of the interface. The use of simplified break through models is quite common for both spatial and timing reasons.

The simplest model is a purely vertical break through (Fig. 6.36). Grid cell after grid cell is filled with residual saturation upwardly in a vertical direction according to an average flow direction based on buoyancy (Sec. 6.5). This continues until the break through amount of petroleum is completely distributed. If the model's top surface is reached, the excess amount is assumed to leave the basin. A special case occurs if another reservoir is reached. This reservoir is then assumed to accommodate the remaining petroleum. Alternatively, a barrier in the form of a highly sealing lithology such as salt might be reached. In this case the procedure is stopped and the excess amount is redistributed into the initial reservoir (reinjection). Lateral migration is assumed not to occur.

Such a model might be improved by taking into account permeability or capillary pressure gradients for the determination of the flow direction. Of course, this is only consistent if lateral flow is allowed. Cell by cell is now saturated in decreasing order of capillary entry pressure with an upward tendency instead of an exclusively upward flow. Such models are usually called "Invasion Percolation" (IP) models (Wilkinson and Willemsen, 1983; Meakin,

1991; Carruthers, 1998; Carruthers and Ringrose, 1998). A more detailed discussion is postponed to Sec. 6.8.

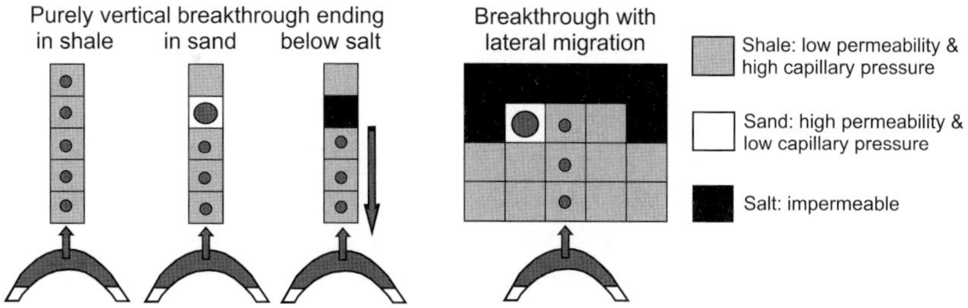

Fig. 6.36. A schematic section view of a break through from a reservoir with either simplified vertical flow or invasion percolation (IP). The arrow pointing down indicates a case of reservoir reinjection

Some additional complications must be taken into account:

In cases of reinjection an additional reservoir analysis must be performed. This is necessary because column heights would have changed if even a small amount of petroleum were transported through the seal. Obviously, this reservoir analysis must be performed artificially with ideal seals.

If cells which already contain petroleum are percolated, this petroleum should be added to the excess amounts for consistency. In rare cases of non–residual amounts in the cells, this might have a significant influence on the migration pattern.

A break through is commonly assumed to appear on a scale which is smaller than the grid resolution used for the solution of the Darcy flow equations (Fig. 6.45). Even whole accumulations may not be resolved properly on the Darcy flow grid if a higher resolution for the reservoir analysis was chosen. A localized break thought path could become difficult to model in a Darcy flow picture. The effect is commonly attributed by residual saturations.

However, it has to be mentioned that local and small scale break throughs are unlikely and that accumulations may leak over a wide range. In an accumulation with one break through point, all migration amounts that originate in a possibly huge source rock volume and are continuously feeding the accumulation from a laterally wide drainage area focus on one break through migration path. Such a huge amount cannot be transported through a small localized break through area on top of the accumulation. A "wide" leakage area might be more appropriate (Sylta, 2005, 2004, 2002b; Vassenden et al., 2003). However, a percolation based break through method can easily be adapted to this situation. Grid cells on the break through path are now filled with full saturation instead of residual saturation. The disadvantage of too low spatial resolution in the Darcy flow region does not exist in this picture anymore and a modeling of focused break through paths with residual saturations in huge grid cells is not necessary anymore. Both break through types, wide area leaking or focused break through paths, can thus be properly modeled.

A two dimensional example model with two leaking accumulations is shown in Fig. 6.40. The results of the hybrid and Darcy runs are almost the same (Fig. 6.8). A corresponding invasion percolation run is displayed in Fig. 6.48.

6.6.3 Fault Flow

A single fault cell is said to be a locally open fault when the capillary pressure is lower then the value of the adjacent cell and otherwise named locally closed. Closed faults act as seals and open faults as conduits, e.g. as additional flow avenues.

This approach allows migration within faults (Fig. 6.37). Following the literature, such cases might be questionable (Allan, 1989; Knipe, 1997). Much literature is related to production with much higher flow rates. A fault which is sealing in production might be open for migration over geological time scales of million of years. However, avoiding migration within faults can in any case easily be achieved with high capillary pressure values.

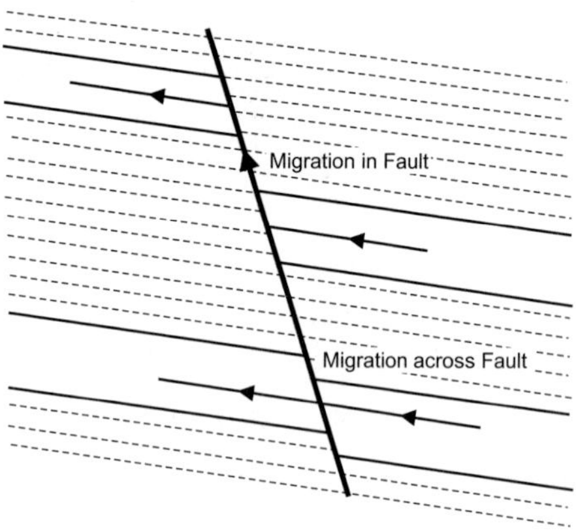

Fig. 6.37. Section view of migration in a fault at the top and across a fault at the bottom according to Chapman (1983)

Reservoir analysis with faults is discussed in Sec. 6.5.4. An example with ideally open and closed faults is shown in Fig. 6.26. Generally, fault capillary pressures (FCP) are compared with capillary pressure values of the seal (Sec. 6.5.4). Excess break through amounts are injected into the fault wall (Fig. 6.38).

The Darcy flow method does not allow zero volume elements in flow equations. Thus fault flow must be modeled with thin locally refined volume cells. A realization of this approach is exhausting and yields long computing times. Alternatively, migration in the fault is performed with the same percolation methods as for break through. It is controlled by capillary pressure only. The

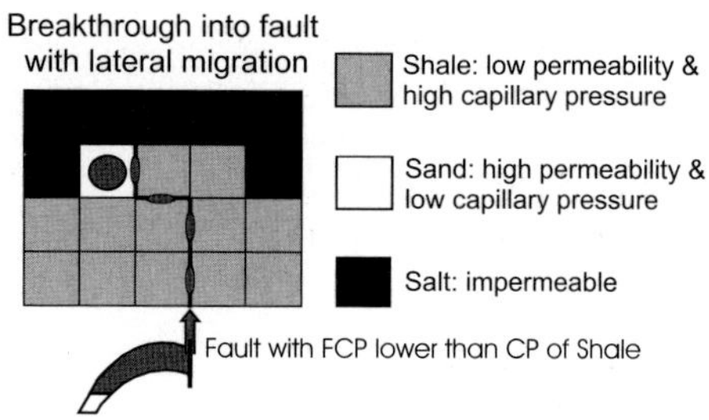

Fig. 6.38. Section view of a break through from a reservoir into a fault system. In comparison with Fig. 6.36 this case only makes sense if lateral flow is allowed

only difference is that the fault cells cannot contain significant amounts of petroleum.

Fault inflow does not necessarily originate from reservoirs only. It might occur also in pure Darcy flow regions (Fig. 6.39). This is especially important when permeable faults cross source rocks and significant amounts are transported away along these faults. For that reason an additional "petroleum drainage" or "fault inflow" function is added and called after each Darcy step.

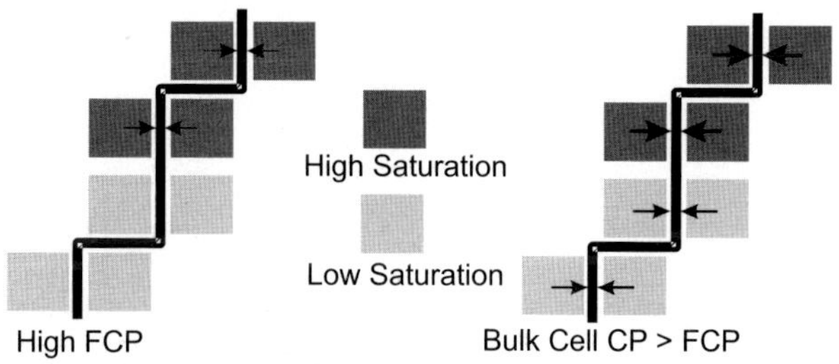

Fig. 6.39. Schematic view of flow from bulk cells into faults. According to capillary drainage and imbibition curves (Figs. 6.4, 6.6), only small amounts from highly saturated cells can migrate into a fault in the case of high FCP (left side). In the case of generally lower FCP than bulk cell capillary pressure, the fault might quickly drain the neighboring cells of petroleum (right side). Note that due to the negligible bulk volume of a fault, an inflow is only possible if there is a corresponding outflow at some other location

Additionally, fault inflow might also occur from break through paths. However, in practice only one generalized break through and fault flow method is used for all cases.

A break through and fault flow routine as described in this and the previous section is an invasion percolation technique. Invasion percolation is capillary–driven flow without any timing, which results in high flow velocities. The approach is especially meaningful for highly permeable cells with shorter flow times than the Darcy time steps, or for small amounts which are quickly redistributed over short distances. Invasion percolation is discussed in more detail in Sec. 6.8.

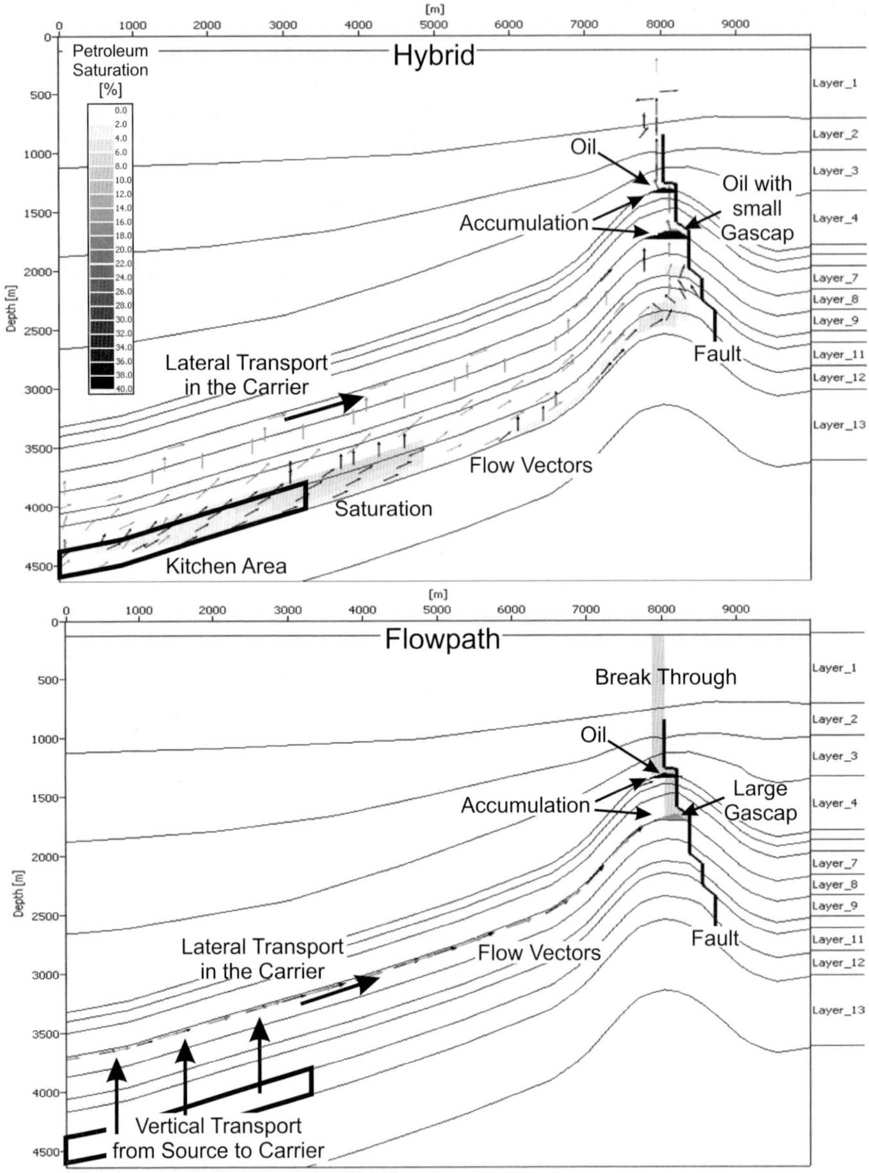

Fig. 6.40. Hybrid and flowpath results of the same model as in Fig. 6.8. The hybrid model has the same break through path as the flowpath model

6.7 Flowpath Modeling

Very often it can be assumed that HC migration through layers of low permeability is almost vertical in an upward direction. Typical vertical migration distances are at least one order of magnitude smaller than lateral distances in basin modeling. In such cases migration may happen quickly on geological timescales. The basin wide migration pattern can therefore become an almost pure reservoir analysis. The HCs are generated in the source rock and after expulsion they are directly "injected" into the next reservoir above. Losses can roughly be estimated as proportional to the thickness of the rocks which are passed through (Fig. 6.41). Break through and fault flow can be treated in a similar way as in the hybrid method (Fig. 6.42). Models of this type are called "flowpath models". They are usually not referred to as hybrid, although they are based on a domain decomposition and contain elements of invasion percolation.

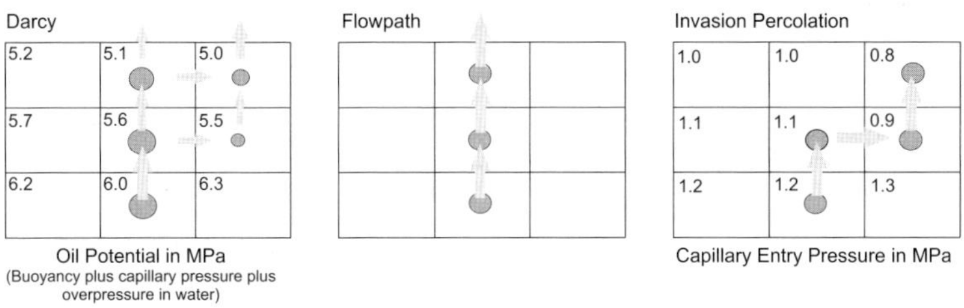

Fig. 6.41. Section view for comparison of Darcy flow, flowpath modeling, and invasion percoalation migration schemes in low permeable regions

The advantage of flowpath modeling is that processing is extremely fast. Small migration time steps are completely avoided. The price paid is an approximation which disregards timing and lateral migration in low permeability regions almost completely.

The same example model, which was processed with Darcy flow or with the hybrid method, is for comparison also simulated as a pure flowpath model and shown in Fig. 6.40. Migration pathways, size, and location of the accumulations are almost the same as in the Darcy or hybrid runs. The composition of the lower accumulation is very different. It contains much more gas from late expulsion of the source. These gas amounts have not yet reached the structure in the Darcy and hybrid models.

Flowpath models are very advantageous if the overall basin model is "simple". In general: heatflow can very often be modeled one–dimensionally in the vertical direction if lateral effects are small. If overpressures do not occur the pressure is hydrostatic and can also be calculated one–dimensionally. In such cases all submodels are very efficient and the complete simulation performance is very high. Very often an speedup of more than one order of magnitude can

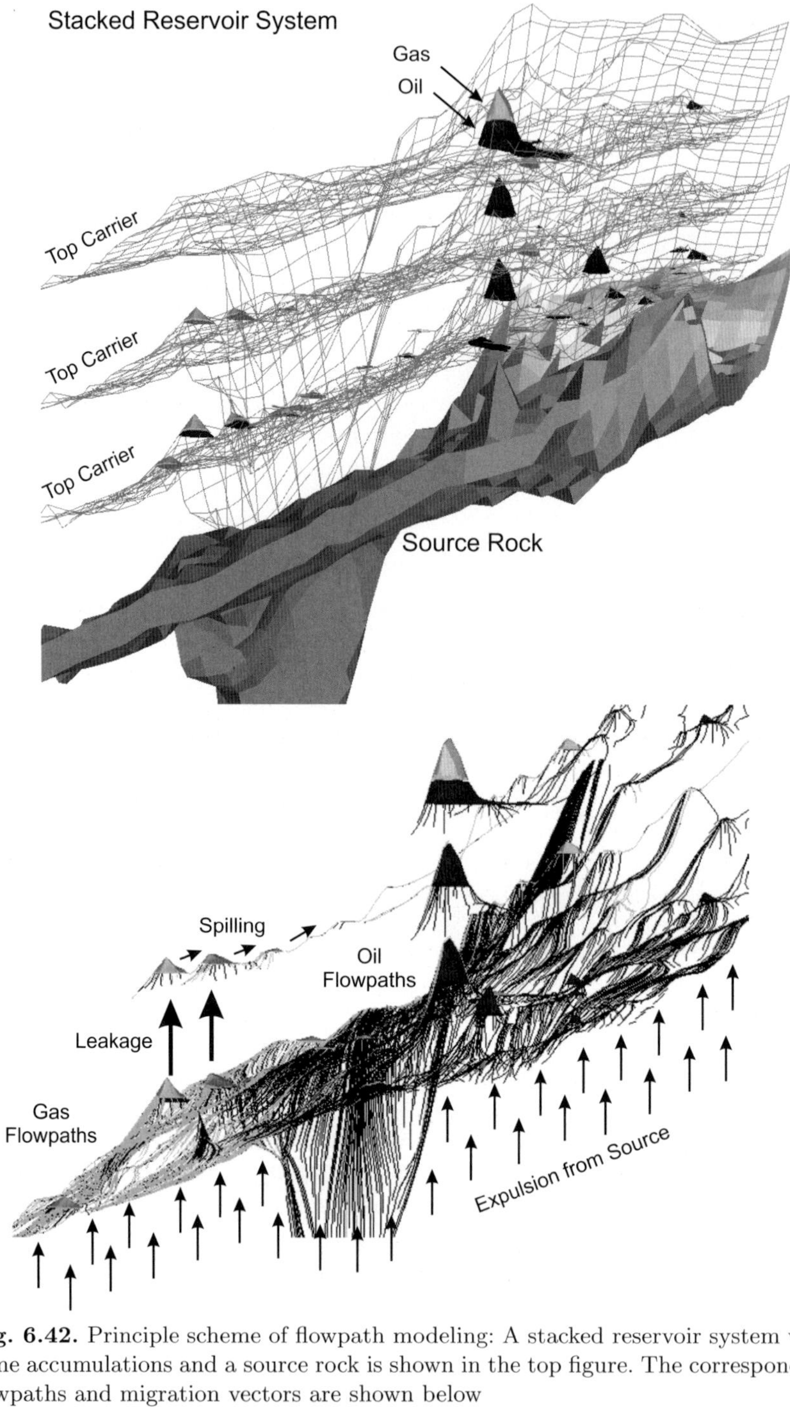

Fig. 6.42. Principle scheme of flowpath modeling: A stacked reservoir system with some accumulations and a source rock is shown in the top figure. The corresponding flowpaths and migration vectors are shown below

be reached compared to hybrid models. Simulation runs can be performed in an hour instead of a day.

It is even possible to easily integrate simple schemes of source rock downward expulsion into the methodology just by splitting up the expelled amounts of HCs by simple formulas into two parts. One is moved vertically upwards and the other downwards to the next reservoir layer.

Due to their high efficiency, flowpath models are usually processed as special scenarios, as first–guess models or as crude approximations, even if the assumptions are poorly fulfilled. However, basic constraints such as an overall HC balance are kept and a good overview of fundamental issues such as the generated amount of HCs or reservoir capacities with consistent distributions of accumulations can be acquired in a fast and efficient manner, unrivaled by other methodologies. The approach is very often justified by good results but care needs to be taken.

6.8 Invasion Percolation

Fluid flow in porous media is assumed to be best and most consistently described by Darcy flow equations (Sec. 6.3), although it was shown in Sec. 6.2 that a lot of unknowns concerning petroleum migration still exist. It is further widely assumed that the application of Darcy flow for migration modeling might only be crude due to the low grid resolution coming from limited computer resources, although the method is physically correct on the macroscopic length scales of rough gridding. Darcy flow is based on microscopic averaging and upscaling of "smooth and continuous" flow to macroscopic length scales. One of the biggest disadvantages is the resulting low spatial resolution, which constricts modeling of migration channels. In its formulation, as in Sec. 6.3, it is also doubtful if it can be used for the modeling of migration based on disconnected stringers with sharp boundaries, fractal distribution, and the possibility to perform almost instantaneous spatial jumps.

Although often yielding good results it must be concluded that Darcy flow equations might sometimes not be applicable for migration. Some important technical drawbacks of pure Darcy flow modeling can be overcome with the hybrid approach (Sec. 6.6). Flowpath modeling is also a well accepted alternative (Sec. 6.7). Unfortunately these improvements are restricted to reservoirs only. Invasion percolation can be interpreted as an extension of such approaches.

Invasion percolation has already been introduced in Sec. 6.6.2 for the modeling of break through processes, which are often believed to be difficult to model on a "rough Darcy grid". The technique is commonly used for low permeability seals. Therefore it should also be possible to apply invasion percolation for migration in rocks with low permeability. Consequently, Darcy flow in hybrid models can in some cases be completely replaced by invasion percolation. This allows processing of higher spatial resolutions. The price is cruder approximations, especially in regard to the timing of migration.

Timing is not of importance in reservoirs (Sec. 6.5). Reservoir analysis is based on buoyancy driven flow and can be modeled from purely geometrical analysis. Thus it should also be possible to use the percolation approach in reservoirs.

Finally, invasion percolation can be applied as one migration method for the whole model. Domain decompositions which are necessary for hybrid or flowpath modeling can be avoided. Migration patterns that have been created without domain decomposition and which have been modeled in only one technique are easier to present and to understand. Additionally, there is the benefit that one can easily include complex and detailed geometries such as vertical sandstone channels, which pose the greatest challenge in the hybrid approach (Fig. 6.43).

On the other hand some benefits such as the extremely high processing speed of flowpath modeling (Sec. 6.7), especially for multi–phase models with displacement, or the implicitly high resolution of the flowpaths traversing the grid in slightly skewed directions, are lost (Fig. 6.11).

A more detailed discussion of the physics of migration with special consideration of the applicability and restrictions of the invasion percolation method is presented in 6.8.1. Percolation on microscopic length scales is introduced in Sec. 6.8.2. Upscaling of a microscopic percolation technique to a method usable in basin modeling is treated in Sec. 6.8.3. The complete technique is described in Sects. 6.8.4 and 6.8.5. Gridding, small scale property variations and the incorporation of high resolution data are the subjects of Sec. 6.8.6.

6.8.1 Physical Background

The physical background which leads to the idea of developing the rule–based invasion percolation migration method shall now be summarized.[15]

Three forces are commonly assumed to determine petroleum flow (Hubbert, 1953). These are buoyancy, which originates from gravity and the density contrast between petroleum and the surrounding water, capillary pressure, which is due to interfacial tension between water and petroleum, and friction of the moving fluid, which is usually described by viscosity and mobility (Sec. 6.3). As mentioned in the previous section, average flow velocities are very small in secondary migration. This leads to the assumption that viscous forces, which are proportional to velocity according to Darcy's law, might be negligible. Prior to a further analysis, viscous forces are therefore compared with forces originating from interfacial tensions.

Pressure drops due to viscous forces over a distance R have, according to Wilkinson (1984), the magnitude

$$\Delta p_{\mathrm{visc}} \sim \frac{\nu v R}{k} , \tag{6.49}$$

[15] The term invasion percolation was originally defined in a slightly different context (Wilkinson and Willemsen, 1983; Meakin, 1991) but it is also used in basin modeling.

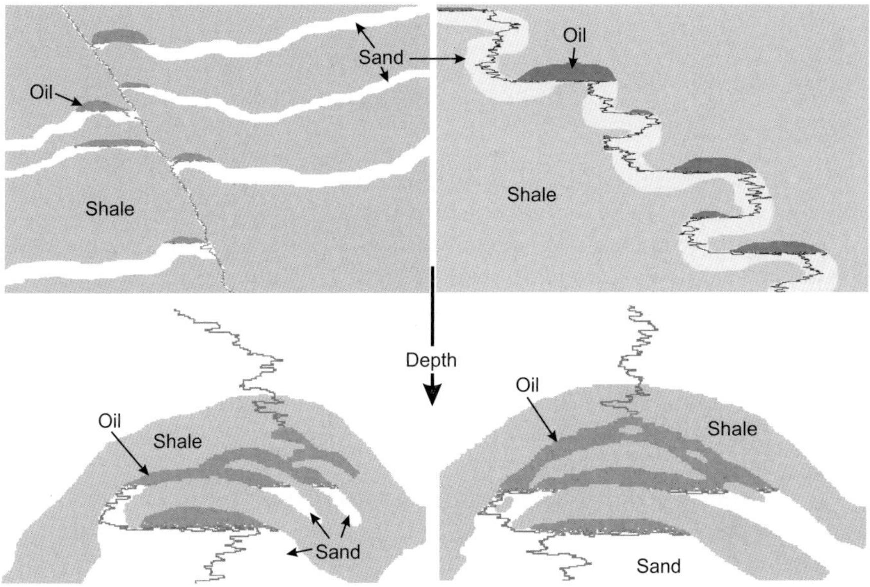

Fig. 6.43. Geometries which cannot easily be modeled with layer based domain decomposition. The sand layers are overthrusted in the example on the top left. Such cases are difficult to handle in layer based domain decomposition of hybrid and flowpath models. All sand layer parts on the left and on the right side of the fault must be treated separately. This causes an enormous effort in big models with many overthrusted layers. For example, the determination of the correct order of processing of these layer fragments is not trivial. The example on the top right is even worse for hybrid modeling. The sand object can principally not be separated into horizontally aligned layer parts, which are a prerequisite for a map based flowpath analysis. Accumulations which span over branching sand layers or sand layers with large shale lenses, are shown on the bottom. Such geometries are obviously also problematic in hybrid models. All examples in this figure are calculated with invasion percolation

with velocity v, viscosity ν, and permeability k (6.41). The magnitude of capillary pressure differences is given by

$$\Delta p_{\text{int}} \sim \frac{\gamma}{R} \tag{6.50}$$

with γ being the interfacial tension (Sec. 5.6.6). With this choice the size of R is of the order of a minimal pore throat radius. Comparison of both quantities yields

$$\frac{\Delta p_{\text{visc}}}{\Delta p_{\text{int}}} = \frac{C}{K} \tag{6.51}$$

with the capillary number

$$C = \frac{\nu v}{\gamma} \tag{6.52}$$

and the dimensionless geometrical factor

$$K = \frac{k}{R^2} , \tag{6.53}$$

(Wilkinson, 1984). Both factors can be estimated.

The capillary number is very small and has a value of $C \sim 10^{-6}$ in reservoir flow for production (Barenblatt et al., 1990). An limit of $C < 10^{-10}$ can be estimated by taking into account the average velocity of a gravity driven flow derived from a Darcy flow model of secondary migration (England et al., 1987).

The estimation of the factor K is more complicated. Evaluation of (2.44) yields $K \sim \phi S/24 \sim 10^{-3}$ for tortuosity $\tau = \sqrt{3}$ and $\phi S \sim 0.05 \times 0.5 = 0.025$ in agreement with Wilkinson (1984).

This estimate can be confirmed with some values based on experience. Example values are grain sizes of 0.1 mm for sandstone with estimated $R \sim 1\,\mu\text{m}$ and a permeability of $10\,\text{mD} \sim 10^{-15}\text{m}^2$ or $1\mu\text{m}$ grain sizes for clay with estimated $R \sim 10\,\text{nm}$ and permeability $10^{-4}\text{mD} \sim 10^{-19}\text{m}^2$ (England et al., 1987; Wood, 1990). The examples yield a value of $K \sim 10^{-3} \ldots 10^{-2}$.

Finally with $C < 10^{-10}$ it follows from (6.51)

$$\frac{\Delta p_{\text{visc}}}{\Delta p_{\text{int}}} < 10^{-6} . \tag{6.54}$$

A possible conclusion would be that, at least to a certain degree of accuracy, viscous effects can be neglected. This conclusion can be criticized. A macroscopic estimate of velocity is applied to a microscopic scale of varying capillary pressures. Macroscopic capillary variations such as the Hobson type

$$\Delta p_{\text{int}} \sim \gamma \left(\frac{1}{R_t} - \frac{1}{R_b} \right) \tag{6.55}$$

are a better choice for the comparison (Berg, 1975). Herein R_t is the smallest throat radius, which is currently filled with petroleum and usually found on top of the stringer, and R_b is the smallest throat belonging to the same stringer which is simultaneously drained from petroleum and usually found at the bottom of the stringer (Fig. 6.44). The values R_t and R_b can be assumed to be of the same order of magnitude. Otherwise significant overpressuring must occur within the stringer. Hence $1/R_t - 1/R_b$ is smaller than $1/R_t$ or $1/R_b$. The geometric factor now has the form

$$K = \frac{k}{R_e} \left(\frac{1}{R_t} - \frac{1}{R_b} \right) . \tag{6.56}$$

Note that in this picture, according to (6.49), R_e describes the extension of a stringer moving through the rock matrix. In the case of macroscopic stringers

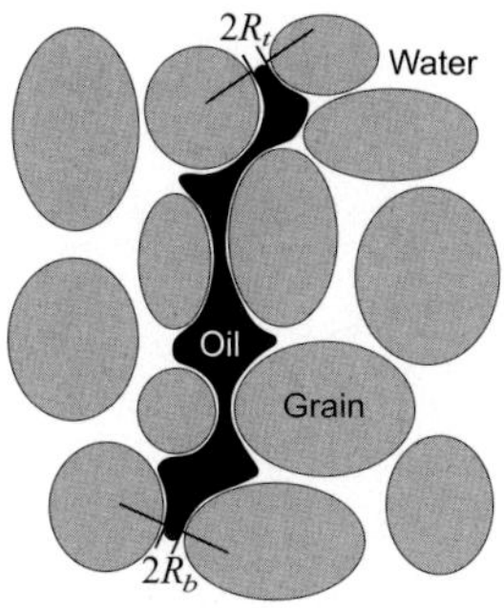

Fig. 6.44. Schematic diagram of a stringer according to Berg (1975); Tissot and Welte (1984)

$R_e \gg R$. Finally, K decreases significantly and so the small estimate (6.54) rises drastically.

Pathway focusing is an additional effect disturbing the estimate (6.54). It is argued that a typical accumulation is filled with $10^6\text{m}^3/\text{My}$ (Sylta, 2004, 2002b). This corresponds to a flow rate $v = 10^{-14}\text{m}^3/\text{m}^2/\text{s}$ for a drainage area of $A \sim 3\,\text{km}^2$ belonging to the accumulation. In the case of a break through on top of the accumulation, this flow amount must be transported through the break through area. Otherwise the column height would increase leading to higher pressures, which would finally create additional break through paths. If the break through occurs only at "weak" points in the seal over small distances with, for example, less than 30 m in diameter, this leads to a break through area, which is about 3000 times smaller than the drainage area (Fig. 6.45). Hence the flow rate is about 3000 times higher and thus the estimation (6.54) must be enhanced by this factor.[16]

Flow pulsing is difficult to estimate. Permeabilities and capillary pressures were introduced with consideration of flow in tubes (Sec. 6.3). Hence a starting point for the study of flow pulses could be the rate of fluid penetration into a thin lengthy capillary, which has been well researched by Washburn (1921). It is a dynamic process which depends on time and penetration length. A further consideration of flow snap–off (e.g. such as described in Vassenden et al. 2003) with periodical or continuous supply of HCs from below, suggests the occurrence of flow pulses. However, flow pulsing indicates time intervals with high flow and time intervals with low flow rates. Correspondingly, the flow velocity during a flow pulse must be higher than the time averaged flow velocity. An enhanced flow velocity increases the capillary number (6.52) and thus the viscosity numerator in (6.54).

In summary, a more detailed view of geometry and stringer sizes, the incorporation of pathway focusing and the consideration of flow pulsing leads

[16] It is even argued that such high flow rates cannot occur in break through processes and therefore leaking must occur on wider areas (Sylta, 2004, 2002b).

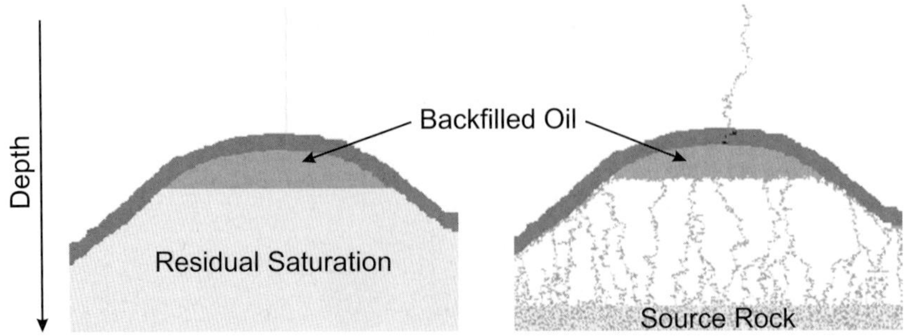

Fig. 6.45. Schematic section views of an accumulation with break through, which is fed over the whole drainage area from a source rock at the bottom of the picture. Random capillary entry pressure heterogeneities are present in the right but not in the left. Dark grey indicates backfilled oil and light grey residual amounts. All the oil coming from below has to pass a small break through path. Note, that the column height is slightly lower in the right figure due to capillary entry pressure variations in the seal

to the conclusion that the limiting factor in (6.54) rises dramatically. Viscous effects should generally not be neglected. Even low velocity viscous flow probably explains the existence of flow pulses and migration in disconnected stringers:

On migration pathways, microaccumulations form below barriers with enhanced capillary pressure. The saturation within a microaccumulation rises until the saturation dependent capillary pressure is high enough to overcome the barrier (compare with Fig. 6.6) All capillary pressure gradients are evened out at the barrier. The resulting forces acting on the petroleum are only viscous friction and buoyancy. Hence, the minimum bulk velocity of continuous petroleum outflow can be estimated with Darcy's law. It is

$$v \sim \frac{k}{\nu} \Delta\rho g > 10^{-13} \frac{\mathrm{m}}{\mathrm{s}} \tag{6.57}$$

with $\Delta\rho = 300\,\mathrm{kg/m^3}$, $\nu = 3 \times 10^{-3}\mathrm{Pa\ s}$, and $k \geq 10^{-19}\mathrm{m}^2 = 10^{-4}\mathrm{mD}$.[17] The estimated velocity v is faster than source rock expulsion velocities, which can be estimated with a maximum of $8 \times 10^{-14}\mathrm{m/s}$ (Sec. 6.2, England et al. 1987, Carruthers 1998). Hence snap–off must occur because feeding of the microaccumulation is slower than its outflow. Finally, petroleum migrates in small disconnected blobs which are the stringers.

It must be noted, that the migration velocity of such a stringer is not given by (6.57). Porosity ϕ must be taken into account. The actual velocity of movement would thus become $v_a = v/\phi$. Additionally, the continuous pen-

[17] Velocity reduction due to a (possible) small relative permeability is assumed to balance approximately with velocity raise due to upscaling and anisotropy effects (Sec. 2.2.3).

etration of petroleum into pores which are not or are only partially saturated must be considered (Washburn, 1921). As argued previously, stringers might therefore move in pulses with strongly varying velocity. However, knowledge of the details of stringer migration are fortunately not necessary for rough estimates such as in (6.57).

Darcy's law joins timing with viscous effects. Even pure capillary–driven flow must take into account viscous effects (Washburn, 1921). A model based on entirely static capillary effects neither explains dynamic effects such as snap–off nor gives hints about migration velocities (Meakin et al., 2000).

A stringer follows a migration path by combination of buoyancy–driven movement in an upward direction and following of the smallest capillary resistance given by the widest pore throats. During movement, some petroleum is lost as immobile microscopic droplets and truncated parts. It shrinks until it finally disappears or becomes trapped below a barrier of small throats. A barrier of small throats might occur due to the laws of probability in an environment with randomly distributed pore throats, if the stringer is very small, or due to a macroscopic variation of the lithology. The trapped stringer stays there until a new stringer from below reaches it. The newly arriving stringer is usually bigger, since its losses are balanced with collected droplets from its predecessor. It did not follow saturated dead ends and moved on the backbone of the migration pattern. Both stringers merge and the movement might continue, depending on the maximum throat width on top of the merged stringer and its height, which determines its buoyancy pressure. Alternatively a microaccumulation may arise. If a big capillary threshold has to be overcome, stringers can continue to merge and a visible accumulation might form. The principle scheme is depicted in Fig. 6.46.

Moving stringers do not dissipate since the petroleum at their inner and bottom part may even move faster than on top, which enforces cohesion. In the inner part the saturation is higher and therefore the permeability increases. Extra upward forces act due to interfacial tension at the bottom.[18] In total the petroleum in the bottom and inner part is "pushing" against the slower moving top. Sharp stringer boundaries evolve and inner stringer convection might arise eventually.

Finally, an overall picture of migration might be a percolation of stringers. The similar behavior of moving droplets, connected strings of blobs, snap–off and disconnected fingers are reported from experiments (Frette et al., 1992; Meakin et al., 1992, 2000; Catalan et al., 1992; Vassenden et al., 2003).

The overall estimation of migration velocity is given by an expulsion controlled average velocity of a first front of stringers. Oil expelled into existing pathways moves faster and arrives earlier in traps than oil expelled into un-

[18] This is expressed by the saturation dependency of relative permeabilities denoted by e.g. the Buckeley–Leverett function (Barenblatt et al., 1990). In production it leads to well defined petroleum water boundaries.

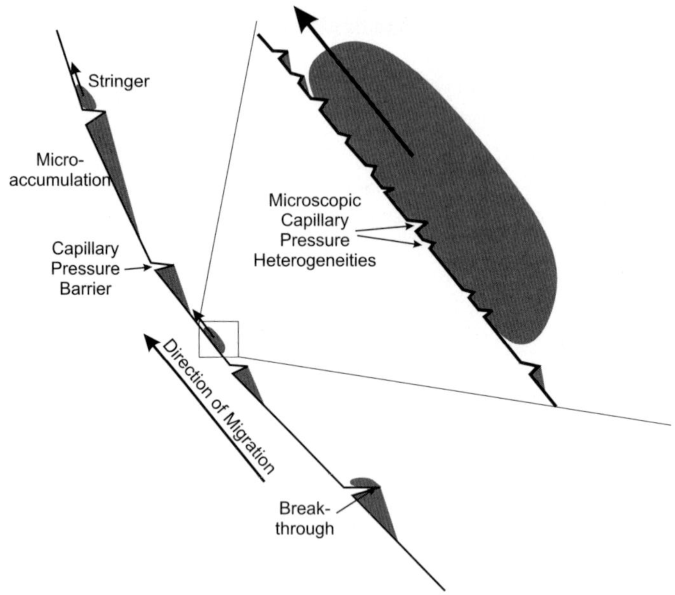

Fig. 6.46. Scheme of stringer migration along a migration pathway. Migration channels exist due to inherent properties of the rock matrix such as fractures or faults or they evolve due to microaccumulations, which "planate" or "smooth" the pathways for the stringers. Microscopic heterogeneities can be passed automotively by stringers

saturated pathways. However, the average flow rate remains limited by the expulsion amounts.

It has been argued that migration can be described as a percolation of stringers through a network of throats. In the following a new migration algorithm is derived from an enhancement and modification of existing microscopic percolation models.

6.8.2 Percolation on Microscopic Length Scales

In a simplified view, the space between the sediment grains can be divided into pores and throats connecting the pores. Only the case of water as the wetting phase is considered here, so the entire mineral surface is covered with a thin layer of water. The migration of petroleum is mainly controlled by the throats representing the smallest structures which must be traversed. A throat of size R can only be entered if a capillary pressure barrier has been overcome (6.50).

Pore and especially throat sizes are randomly distributed within some limits in a natural rock. Ordinary percolation theory mainly deals with the subject of finding paths through a network of random sized pores and throats assuming that a fluid under pressure p can overcome barriers of capillary pressure $p_{\text{int}} < p$. Regular grids are studied analytically and with computer simulations (Fig. 6.47). The gridcells are either called sites, if they indicate pores or bonds, if they are related to pore throats (Stauffer and Aharony, 1994; Nickel and Wilkinson, 1983). The main results are critical pressure values p_{crit} describing the threshold for the creation of a path through a sample and that the probability P of finding such a percolation path scales according to

$$P \sim (p - p_{\text{crit}})^{\beta} \tag{6.58}$$

for $|p - p_{\text{crit}}| \ll 1$ with an exponent $\beta = 0.41$, which can be determined from simulations (Winter, 1987; Stauffer and Aharony, 1994). The critical pressure can be interpreted as the entry pressure p_{ce}, which is necessary to achieve a significant macroscopic fluid penetration.

The number N of percolated sites scales analogously as

$$N \sim L^{D} \tag{6.59}$$

with L as the grid or sample size and D as the exponent of the fractal percolation dimension. The fraction of percolated sites is commonly interpreted as saturation. Thus saturation is also expected to follow this scaling law.

The exponents are usually dependent on the grid dimension only and not on the specific choice of the grid, which is why they are called universal (Wilkinson and Willemsen, 1983). A value of $D \approx 2.5$ has been calculated theoretically (Stauffer and Aharony, 1994) and proved experimentally (Hirsch and Thompson, 1995).

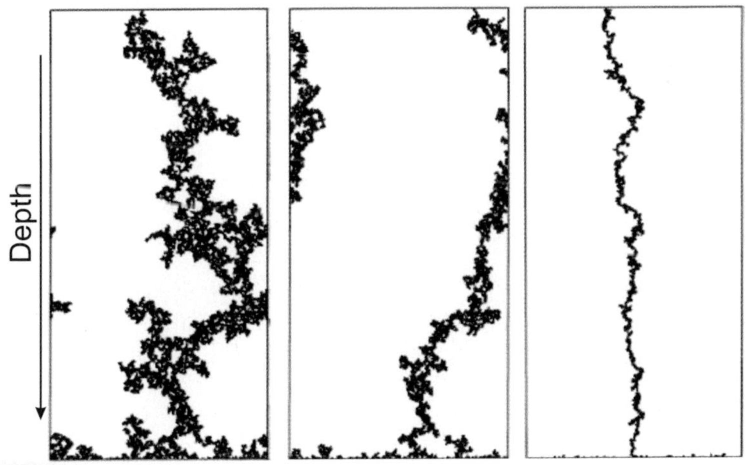

Fig. 6.47. Microscopic invasion percolation patterns for three different values of density contrast which increase from left to right. The pictures are from Meakin et al. (2000). Note the periodic boundary conditions

In invasion percolation, percolating paths are typically created starting from one given boundary, and describe a process of invasion and displacement. This alone does not yield significant differences to the "original" percolation method (Wilkinson and Willemsen, 1983; Wilkinson, 1984). Very often water trapping processes are also simulated. Water trapping might occur if a water cluster is completely surrounded by petroleum, with the consequence that sites belonging to this cluster cannot be invaded anymore (Wilkinson and Willemsen, 1983). Controversially, it is argued that the water can escape through the thin wetting layer covering the grains, giving rise to the assumption that trapping phenomena do not exist or that the trapping probability is so small that the effect can be neglected (Wilkinson, 1986; Carruthers, 1998; Frette et al., 1992).

Extensions to the method take correlated disorder or buoyancy into account (Meakin, 1991; Meakin et al., 1992). Experiments related to buoyancy–driven invasion percolation have also been successfully performed (Frette et al., 1992; Meakin et al., 2000; Hirsch and Thompson, 1995).

6.8.3 Upscaling of Microscopic Percolation

Basin models are so large that migration cannot be modeled based on percolation within microscopic pores. Percolation must be simulated with macroscopic sites. The degree to which pore throat distribution and macroscopic capillary pressure variations affect percolation with macroscopic sites has yet to be investigated. A fundamental question is weather upscaling also yields fractal patterns, or weather it instead yields more continuous saturation patterns as commonly assumed for the Darcy flow equations. Fractal means self–similar under a given magnification. This definition includes the possibility of fractal patterns with macroscopic extension.

The relationship of buoyant to capillary forces is the most important parameter for the characterization of invasion percolation processes. The dimensionless Bond number is defined as

$$\mathrm{B} = \frac{\Delta\rho g R^2}{\gamma} \tag{6.60}$$

(Wilkinson, 1984, 1986). With $\Delta\rho = 300\,\mathrm{kg/m^3}$, $R = 0.01 \ldots 1\,\mu\mathrm{m}$ as in Sec. 6.8.1 and interfacial tension $\gamma \sim 30\,\mathrm{mN/m}$ (Danesh, 1998) it can be estimated to have a rather small value of

$$\mathrm{B} \sim 10^{-11} \ldots 10^{-7}\,. \tag{6.61}$$

Wilkinson introduced a length scale ξ above which the characteristic behavior of the system alters. It scales as $\xi \sim R/B^{0.47}$ with $R/B^{0.47} \approx 1 \ldots 2\,\mathrm{mm}$. For a "stabilizing gradient" system, which means injection at the top and not at the bottom of the system, percolative character is lost above this length scale (Wilkinson, 1986). In the case of a "destabilizing gradient" with injection at the bottom, investigated here, it was shown that the saturation pattern is characterized by a "directed random walk" of small "blobs" (Meakin et al., 1992; Frette et al., 1992).

The maximum horizontal deviation w of the random walk scales according to

$$w \sim h^{1/2} \qquad \text{or} \qquad \frac{w}{h} \sim \frac{1}{\sqrt{h}} \tag{6.62}$$

with h as the vertical height. Upscaling from a length scale of $\xi = 1\,\text{mm}$ to $10\,\text{m}$ yields a reduction for w/h of about $\sqrt{10^{-3}/10} = 0.01$. Blobs which perform random walks with $w/h = 1$ on a scale of ξ show horizontal deviations of about $0.1\,\text{m}$ in sites of $10\,\text{m}$ in size. Hence, it is not necessary to consider pore throat variations when studying capillary pressure variations.

In basin modeling invasion percolation cells are huge compared to a microscopic scale and small compared to the basin scale. They are even small compared to "Darcy flow" cells. The same argument which has just been used for upscaling from pore to invasion percolation size can now be used again. Due to the small size of an invasion percolation site, which is usually above seismic resolution, it does not contain macroscopic variations relevant for basin modeling. It is therefore a realistic assumption to treat a site as rather smooth, homogeneous, and non–fractal in capillary pressure variation. Hence an invasion percolation site has properties similar to a pore in a microscopic percolation model. For example, its capillary pressure can be discretized by just one value. This behavior can even be verified for sandstones: non–fractal behavior corresponds to the Corey equations according to Timlin et al. (1999) with $\lambda \to \infty$. Application of this limit to (6.14) yields a saturation independent capillary pressure as already demonstrated in Sec. 6.3.1. Finally, the capillary pressure curve Fig. 6.4 can be described by one entry pressure value and two saturation values, a residual start– and a connate water end–saturation for each invasion percolation site.

When dealing with macroscopic invasion percolation, the capillary pressure should vary randomly between sites. This variation should reflect the impact of random heterogeneity of capillary pressure variation on length scales above the site size. The range of capillary pressure variation, which defines the range of these random variations, must be specified. Again, formula (6.62) yields some hints. It can roughly be estimated down to which length scale ξ' noise coming from heterogeneities must be considered. If $w \approx 1\,\text{m}$ is the cut–off threshold of the influence of noise in a $10\,\text{m}$ spacing grid, which indicates an almost straight line crossing the grid, then it is $w/h = 0.1$. The corresponding length scale ξ' for $w/h = 1$ becomes $\xi' = 0.1^2\,10\,\text{m} = 0.1\,\text{m}$. Roughly estimated, capillary heterogeneity variations shall be considered down to a resolution of 1/100th of the grid spacing. Variations due to smaller structures do not play a significant role. They do not affect the macroscopic appearance of migration pathways.

However, site saturations must be upscaled properly based on the fractal dimensionality of the microscopic pattern of the blobs. Scaling of the saturation S with the fractal dimension $D \approx 2.5$, provides the behavior $S \sim L^{2.5}/L^3 = L^{-0.5}$, with L as a typical edge length of the considered volume. Extrapolating from experimentally measured saturations of well cores with a diameter of 0.1 m to grid site sizes of 10 m; yields a reduction factor of $100^{-0.5} = 0.1$ for site saturation (Meakin et al., 2000; Hirsch and Thompson, 1995; Stauffer and Aharony, 1994). Hence, residual saturations are expected to be very small on a layer scale (Hirsch and Thompson, 1995). On the other hand micro–accumulations due to macroscopic heterogeneities may also form on all length scales. This must be remembered for the estimation of average residual saturations.

An upper limit of the stringer size h_{max} can be estimated by the height necessary to overcome the percolation pressure p_{crit} as

$$h_{\mathrm{max}} = \frac{p_{\mathrm{crit}}}{\Delta\rho g} \tag{6.63}$$

in a microscopic view. The percolation pressure in a cubic lattice with uniformly distributed capillary pressures between 0 MPa and 1 MPa is $p_{\mathrm{crit}} = 0.3116\,\mathrm{MPa}$ (Stauffer and Aharony, 1994; Wilkinson and Willemsen, 1983).[19] A more realistic distribution would be between 1 MPa and 1.1 MPa. It must be taken into account that filling a new pore with petroleum at the top of the stringer causes another at the bottom to be drained of petroleum. Hence, only a pressure difference such as described by (6.55) needs to be overcome by buoyancy (Fig. 6.44). Big pores with up to 1 MPa are filled because the stringer is already located within the rock and spans multiple pores. So only 31.16% of the remaining 0.1 MPa variation must be overcome and with $\Delta\rho = 300\,\mathrm{kg/m^3}$ the maximum stringer height becomes $h_{\mathrm{max}} \approx 10\,\mathrm{m}$. Analogously, an estimated variation of capillary pressures of 1 ... 2 kPa in sandstone leads to a stringer height of 10 cm (compare with Berg 1975, Schowalter 1979).

The phenomenon that pore distributions are usually not uniform, is assumed to be unimportant (Meakin et al., 2000). Nevertheless the estimates for the stringer heights are crude. Stringers of such size are obviously connected to a lot of pore throats. They have a macroscopic size. An entrapment based on a statistical distribution of small pore throats becomes highly improbable because in reality, layers are non–random heterogeneous over such macroscopic distances. Variations of capillary pressure in the form of heterogeneities might occur in all sizes and on all possible length scales above grain size. For example, fractures occurring on or below a macroscopic length scale must also be considered. Finally, it can be estimated that a realistic stringer height is often smaller.

It is important that stringers are smaller than the high resolution grid sites of the invasion percolation models for basin modeling. Because of this, a site describes a small ensemble of migrating stringers. Pressure snap–off occurs within a site. Sites are often the smallest objects which are modeled in basin

[19] The value refers to site percolation. It is $p_{\mathrm{crit}} = 0.2488\,\mathrm{MPa}$ for bond percolation.

modeling. For that reason snap–off is modeled to occur at site height because stringer heights are directly below the site height in shales and down to one order of magnitude below it in sandstones. This introduces a vertical gridding error with a small error for the calculation of overpressuring. However, it does not influence the saturation significantly because the saturation of the macroscopic sites is chosen independently of pressure.

In the case of macroscopic capillary pressure variation, accumulations can form below capillary pressure barriers. This can be mapped by backfilling sites. These sites have "full saturation" and only a residual amount of water. Almost all pores are filled and the sites are pressure connected. The value of "full saturation" can be specified by accounting for the amount of pores which can be reached. In practice, drainage and imbibition curves are considered (Fig. 6.4). They often show a plateau (Schowalter, 1979). For such cases, residual and connate saturations are chosen as limiting values of this plateau.

Finally, upscaling can be summarized by comparison of the macroscopic invasion percolation processes to well established microscopic approaches: on the macroscopic scale, migration can be treated as a random walk of stringers. The stringers are not connected during migration. Hence the macroscopic picture is equivalent to a microscopic picture of invasion percolation without buoyancy but with a bias of preferred vertical migration direction. The stringers are pressure–connected during backfilling. Here the macroscopic model is equivalent to a microscopic description of invasion percolation with buoyancy. The method uses two values of saturation, namely the residual petroleum saturation associated with exceeding the capillary entry pressure, and a full saturation associated with backfilling. Random capillary pressure variations are based on macroscopic heterogeneities down to a scale of 1/100th of the grid spacing.

A big advantage of an invasion percolation approach is that capillary entry pressures can easily be biased by overpressure in the water. Aquifer flow, which causes a petroleum water contact area deformation at the bottom of an accumulation, can easily be taken into account (Fig. 6.52). The biased entry pressure is called the threshold pressure.

6.8.4 One Phase Invasion Percolation

A one phase invasion percolation algorithm for migration can now be formulated.

The space is subdivided into grid sites with a higher resolution than finite elements for Darcy flow analysis. A threshold pressure value is assigned to each site. The value is determined randomly from a distribution which describes the capillary pressure variations due to heterogeneities of the flow unit it belongs to plus the overpressure in the water.[20]

[20] A flow unit is defined here as a region of space with similar flow properties. In its definition it has originally been limited to reservoir rocks (Stolz and Graves, 2005).

Migration starts in the source rock. Migration sites are saturated with residual saturation following a path of decreasing capillary pressure resistance with a preferential direction upwards. The first filling of a site with residual petroleum saturation is called invasion.

If no path of decreasing or constant capillary pressure is found, backfilling begins. A column pressure due to backfilling is taken into account and a break through is sought. If a break through point is found, migration continues at this point, otherwise the last site according to the opposite order of invasion is investigated. If a neighboring site with lower threshold pressure exists, migration continues in this direction, otherwise the site is backfilled and the algorithm continues as above.

Three non–trivial examples are shown in Figs. 6.49 and 6.51. The first one on the left side of Fig. 6.49 shows a rather complicated example of capillary equilibrium in sand and silt. The liquid–water contact of the upper accumulation in the silt is in equilibrium with the contact height in the intermediate sand. The difference of these contact heights is the same as the column height of the lower accumulation.

The second example on the right side of Fig. 6.49 demonstrates that it is possible to correctly model completely filled and overpressured sand lenses with invasion percolation. Although the column height in the sandlens is less than the one of the accumulation below it, a break through occurs. The reason for this behavior can be easily understood: The saturation throughout the sand lens reaches the point of high saturation with complete filling. According to the drainage curve Fig. 6.4 pressure rises in the lens until it balances with the shale (see also Fig. 6.6). Finally, a break through occurs.

The third example is about expulsion, with the overall primary migration downward due to a vertical overpressure variation (Fig. 6.51).

The migration object evolving from one expulsion point is called a stringer path. During the whole procedure a mass and volume balance must be kept. The algorithm stops when feeding amounts at the injection point of the stringer path have been distributed.

The biggest disadvantage of the algorithm is that the time necessary for the migration of the petroleum from the source rock expulsion point to the top of the stringer path is not taken into account. Such an estimation would be difficult because the velocity of stringers moving along the stringer paths is difficult to calculate. Darcy flow methods can only be used for the estimation of flow velocities in pressure connected flow regions. Displacement of water and complicated intra–stringer flow patterns can in principle only be evaluated at resolutions which must be higher than the size of the stringers. This is obviously not possible and one must therefore rely on crude average flow velocity estimates such as given by (6.57). Usually, invasion percolation is performed without time control and based on the assumption that migration happens on a faster timescale than generation and expulsion.

It must be noted that different stringer paths evolving from different expulsion points might merge if a site is invaded from both. The result is a

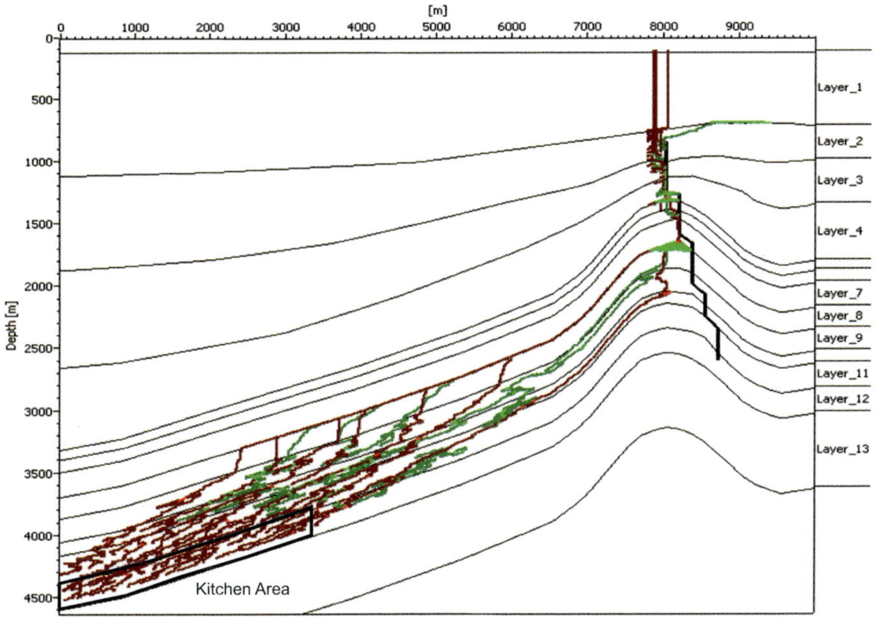

Fig. 6.48. Invasion percolation results of the same model as in Figs. 6.8 and 6.40. Light red and green cells are backfilled whereas the dark cells contain only residual amounts of petroleum. The same color scheme is used in Figs. 6.49 and 6.50

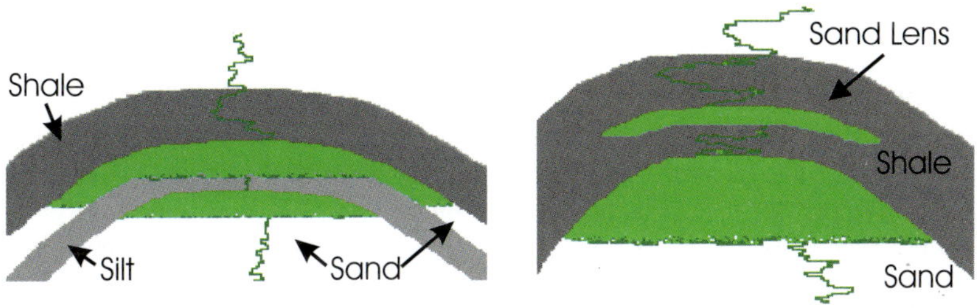

Fig. 6.49. A percolation example with accumulations in a shale – sand – silt – sand layering on the left and an example of a sand lens which is completely filled with oil

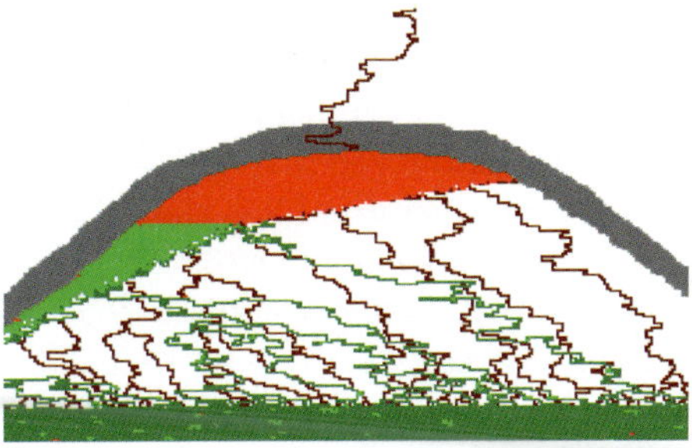

Fig. 6.50. An example of tilted accumulations under constant lateral overpressure in a reservoir below a seal (grey)

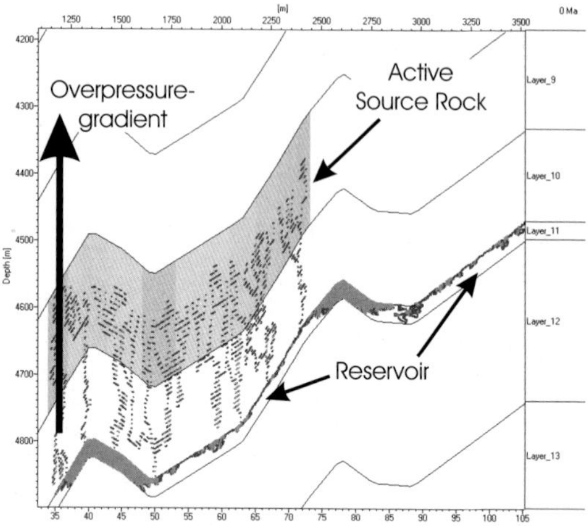

Fig. 6.51. An example of downward expulsion due to overpressure rising in the shales above the reservoir

stringer path with two feeding points. This behavior models the process of pathway focusing (Fig. 6.45). Obviously, the order of invasion and backfilling is dependent on the order of processing different stringers. This order is not predetermined by the method itself and must be chosen in a geologically meaningful manner.

6.8.5 Two Phase Migration with Displacement

Petroleum often occurs in the two phases liquid and vapor. To a first–order approximation, liquid and vapor migration can be treated almost independently. Each of the phases has its own density contrast and its own pathways. Hence, sites can be traversed by liquid and vapor without interaction. Only sites with backfilling must be treated in a special way. Vapor usually displaces liquid whereas liquid cannot enter fully vapor–saturated sites.

An additional complication is pressure build–up in accumulations which contain a gas cap with connection to an oil body. The column pressure on top of the oil must be added to the gas pressure. A simple example of a two phase accumulation under lateral overpressure conditions is shown in Fig. 6.50. Tilting of petroleum water contact areas under lateral water pressure variation is demonstrated. Vapor buoyancy is higher than liquid buoyancy and therefore the vapor–water contact is tilted less than the liquid–water contact. The liquid–vapor contact is not tilted due to the assumption that lateral overpressure does not occur within the accumulation. The petroleum is in static equilibrium and hence it can be assumed that it does not move. Note that a constant lateral water pressure is applied here. This is a first order approximation because water flow and overpressure must be calculated under consideration of layer and accumulation geometry. The latter is an obstacle for the water flow. However, the accumulation forms on top of the reservoir and it can be assumed that the water flow and the overpressure pattern are not disturbed very much.

It is known that compositional changes during migration give rise to important effects during migration. Especially the changing pressure and temperature conditions during vertical migration affect these processes (England et al., 1987). Phase properties can only be properly determined if the composition is known (Chap. 5). It is not possible to take compositional changes into account in each percolation cell, due to computer memory restrictions. Computing resources are not sufficient for advanced fluid analysis on each individual tiny site either. However, this problem can be overcome by additional approximations. For example, phase compositions can be modeled with simple symmetrical black oil models. But this approach is not very sophisticated because symmetrical black oil models only work well in restricted pressure and temperature intervals (Sec. 5.3).

In an alternative approach phase properties are calculated only after long time steps. After each geological event, all hydrocarbon amounts are transformed to the rough finite element grid. On this grid scale fluid analysis can be performed. Afterwards, the high resolution percolation sites can be updated accordingly. During the percolation steps themselves the phase compositions remain fixed. Migration will be limited in range if source rock expulsion is small in one event step, especially if the event step is short. Hence the error in phase composition will also be small. This method allows an iterative refinement with decreasing time step length, which can be tested for convergence.

Invasion percolation results are shown in Fig. 6.48 for the same example as modeled with Darcy flow (Fig. 6.8), the hybrid method and flowpath modeling (Fig. 6.40). The results are almost the same as for the Darcy and hybrid calculations. Even the migration pathways from the source to the carrier are tilted with similar angles as in the Darcy and hybrid cases. Most of the petroleum does not enter the carrier vertically above the source and it can be seen that the deeply buried source is only generating gas. However, due to micro–accumulations and further compaction, which also drives some oil migration from the residual saturations, it is found that some gas is stuck below and some oil is still reaching the carrier layer 8. Hence, the compositions and phases which are found in the lower accumulation of layer 8 are similar to the hybrid and not to the flowpath results.

6.8.6 Discretization of Space and Property Assignment

There is a choice between completely regular gridding or a gridding adapted to the geological structures of interest, i.e. the flow units.

Regular grids have the advantage of easier algorithms, faster implementation and higher performance in execution. On the other hand, a regular gridding might be insufficient for the modeling of small scale structures, such as thin layers. Even a high resolution model with one billion sites and a

$1000 \times 1000 \times 1000$ grid resolution has, for a maximum depth of $10,000$ meters, a vertical grid resolution of 10 meters, which is too poor for modeling migration in thin reservoir rocks. Thus non–regular grids are recommended (Fig. 6.55). This is consistent with arbitrary percolation theory where many results are independent of the type of the grid (Stauffer and Aharony, 1994).

A disadvantage of non regular grids is related to the shape of the petroleum water contact below an accumulation. Due to the irregularity of the grid the meniscus is not sampled horizontally but instead follows the structure of the layer. In general, this problem can only be overcome with high resolution models (Fig. 6.52). In models with aquifer flow and lateral pressure gradients, the contact is not horizontal, so even regularly gridded models have this problem.[21]

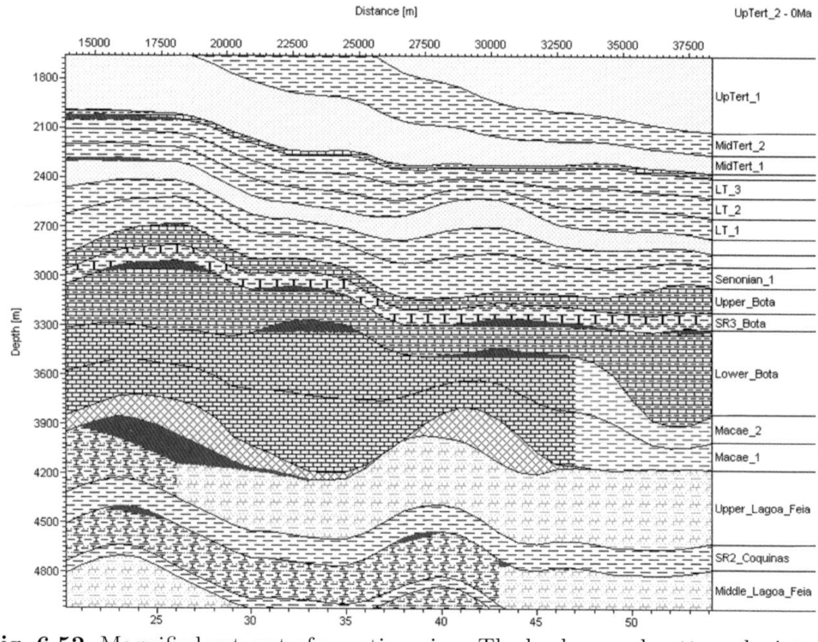

Fig. 6.52. Magnified cut–out of a section view. The background pattern depicts the facies. Backfilled irregular sites are marked with regular grey rectangles. Some layers contain water flow and overpressure. Tilted and deformed oil water contacts are visible. Sampling of the column height is difficult to observe due to high resolution

Another problem concerning gridding in general, is long distance migration in dipped reservoirs. Gridding directions usually do not follow the dipping

[21] Regularly gridded models have the additional problem, that the top surface of an accumulation does not arbitrarily follow seal dipping. Hence a smooth oil–water contact is sometimes achieved for the price of a poorly modeled seal interface.

directions. Migrating petroleum "sees" only the neighboring sites and moves to the highest neighbor site if capillary pressure variations are small. It therefore follows the grid direction instead of the direction of the steepest ascent.

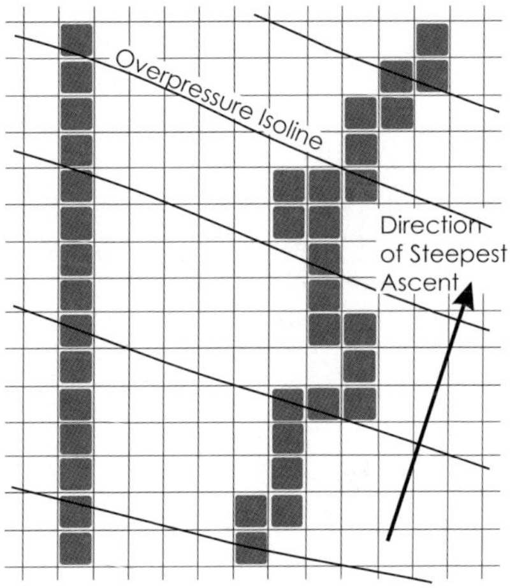

Fig. 6.53. Section view of two schematic invasion percolation paths in a reservoir rock. No capillary pressure variation is assigned on the left side. The capillary entry pressure is modified by overpressure in the water. Obviously, invaded sites follow the grid in an upward direction. On the right side a capillary pressure variation in the form of random noise is assigned to each site. Hence, jumps to the left or right might occur and in average the path approximately follows the direction of the steepest ascent

The problem does not exist in cases of significant capillary pressure variation. Here the mean migration direction follows the direction of steepest ascent, at the price of smearing out the migration path (Figs. 6.53, 6.54). Therefore a proper amount of heterogeneity is often needed in invasion percolation models.

Fig. 6.54. Invasion percolation pathways below an accumulation according to Fig. 6.53. The figures are calculated without and with heterogeneities, respectively

Faults are sometimes of special importance for migration. They are two dimensional on a basin scale and can be modeled by surface pieces adjacent to the "volume sites" of the rock matrix. However, in a microscopic picture they are fully three–dimensional and the fluid flow through them can be modeled with percolation methods in the same way as through rock. Therefore, fault surfaces can easily be integrated in an invasion percolation algorithm simply by treating them as arbitrary sites whose volume, consistent with their surface description, is zero (Figs. 6.55, 6.56). However, they act as conduits or barriers which means that backfilling might occur. Backfilled amounts are negligible but the pressure rise due to pressure connection is the same as in an ordinary accumulation in a sandstone below a seal. It must be noted that the modeling of faults as surfaces is not possible in a pure Darcy flow framework where all cells must have a finite volume and surfaces appear only as boundaries.

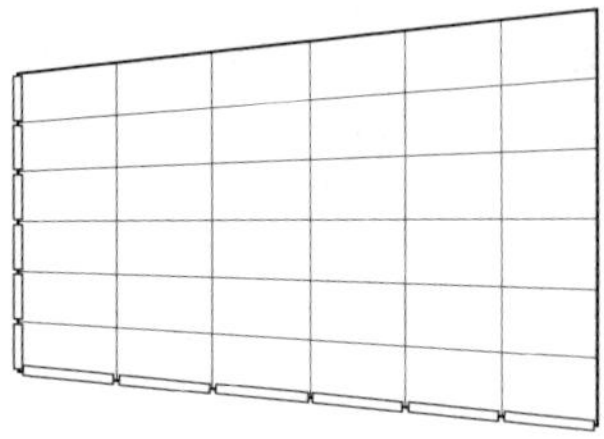

Fig. 6.55. Regular 6 × 6 subdivision of a two dimensional non–regular finite element with fault sites at the element boundary as implemented in PetroMod®. A finite element grid is shown in Fig. 8.6

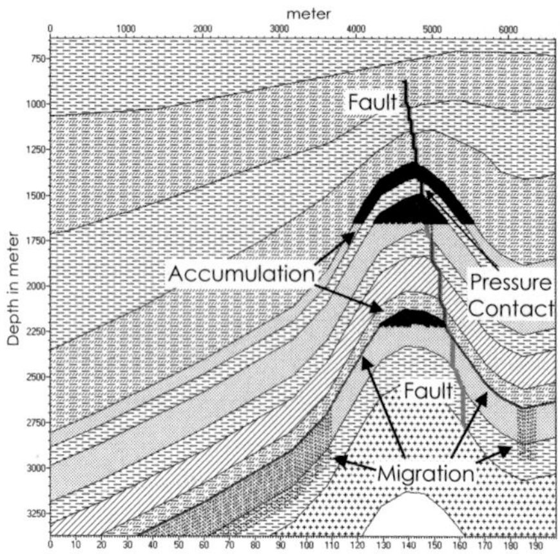

Fig. 6.56. Section view of an example with migration. Migration is colored dark grey, migration through the fault light grey and accumulated petroleum black. The two upper accumulations are in pressure contact along the fault. Due to this pressure contact both upper accumulations have the same oil water contact height. The pressure at top of the fault originates from the column down to the oil water contact of both upper accumulations

Invasion percolation can be performed on grids with a higher resolution than temperature calculations or Darcy flow. It is therefore possible to directly incorporate high resolution data such as seismic facies in the population of the capillary pressure field. Seismic inversion also provides porosity and clay content, which are good indicators for the size of capillary entry pressure.

An example of direct incorporation of seismic data for invasion percolation into a basin model is shown in Fig. 6.57. Seismic velocities have been converted to porosities and permeabilities and subsequently into capillary entry pressures. Finally, facies maps have been refined on the site resolution scale.[22] Calculated gas chimneys from seismic attribute analysis agree with leaking accumulations.

Three basic pitfalls must be recognized when seismic data is used. Firstly, seismic data is usually only available for the present day and not for paleo times when migration actually took place. This problem can be theoretically overcome by backstripping the seismic data according to basin evolution. However, an enhancement porosity and therefore an reduction of capillary pressure due to decompaction must be taken into account. This problem can be handled with a well defined basin model.

Secondly, interpreted seismic data is needed because flow units must be assigned. The invasion percolation method needs an overall underlying knowledge of the rock types for construction of flow units. Up to now this cannot be done automatically. Again, this problem can be handled with a well defined basin model. Facies distributions from basin models can be used for construction of flow units for migration. However, direct incorporation of seismic data is not as easy as it seems at first glance.

As a last pitfall it must be mentioned that care needs to be taken with a "direct" usage of seismic data. Seismic data contains noise which is not related to the variation of rock properties but comes from the restrictions of the measurement setup and the physical processes of sound propagation. For example, noise is created from microphones or the dissipation of sonic waves. This contribution of noise to the measured signal must be clearly distinguished from the "pure" signal due to the rock specific distribution of pores, throats and macroscopic heterogeneities. Obviously, this is very difficult. Direct usage of seismic data must be performed very carefully, otherwise there is a danger of miscalculating basic quantities and producing uncertain results. Simple deterministic mapping of noisy seismic data to throat distributions or capillary pressure heterogeneities may be unreliable.

6.8.7 Anisotropy

Migration based on invasion percolation, as formulated up to now, does not incorporate effects of anisotropy. Anisotropy is described in the Darcy equations (Sec. 6.3) with a permeability tensor. Capillary pressure is usually defined as a scalar quantity for the description of porous media. It has a direction independent nature, which is stated explicitly with the word "pressure". Hence the introduction of a horizontal and a vertical capillary pressure component apparently makes no sense. The only simple possibilities for introducing anisotropy are given by site size anisotropy and by anisotropic variation of the capillary pressure distribution.

[22] The regional model of 80×100 km has gridcells of 1 km in length. The area of interest is refined with a resolution of 300 m (see LGR in Sec. 8.9). An invasion percolation grid with $5 \times 5 \times 5$ sampling according to Fig. 6.55 yields 60 m resolution. The inverted seismic cube has a resolution of 30 m.

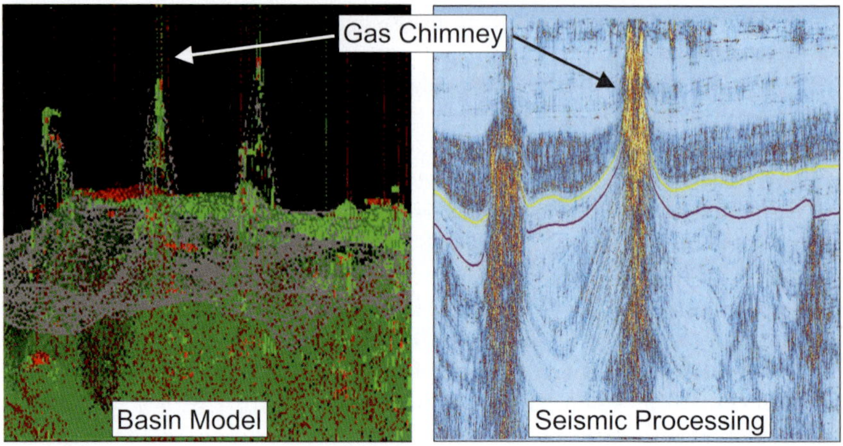

Fig. 6.57. Chimneys which are modeled with IP on the left. Similar structures are observed in processed seismic on the right. Pictures are courtesy of MAERSK

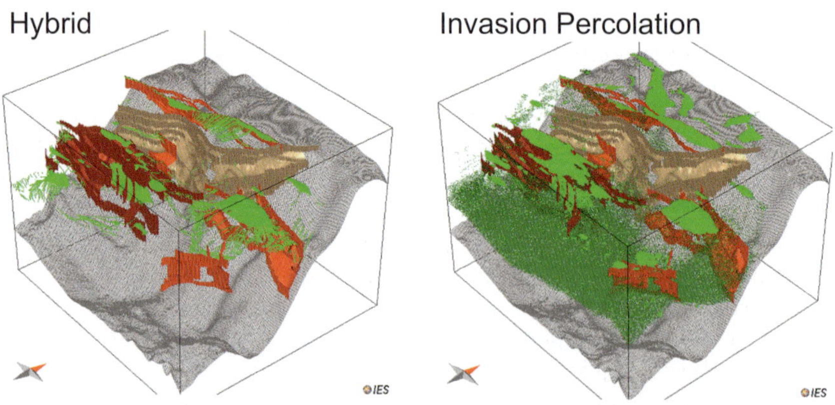

Fig. 6.58. Comparative study of Shengli basin, China

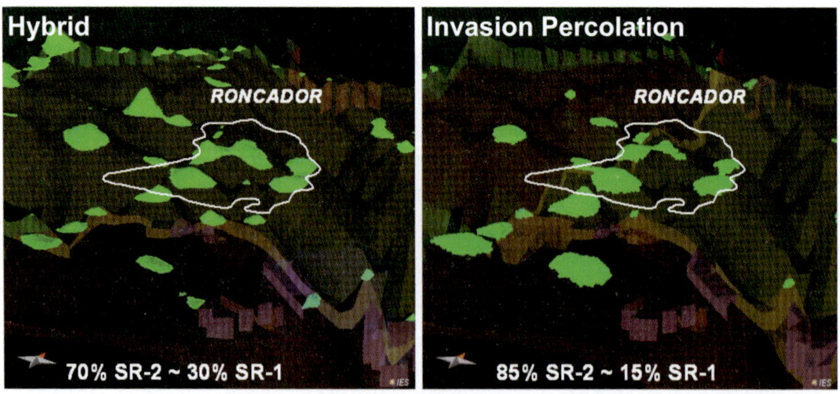

Fig. 6.59. Roncador field in the Campos basin, Brazil (Bartha, 2007)

The requirements to construct a meaningful anisotropic capillary pressure field are rather high. Without incorporation of heterogeneities migration will be completely vertical or horizontal depending on the buoyancy force, site dimensions and capillary pressure variation. As in Fig. 6.53, heterogeneities must be incorporated to achieve a smooth crossover from these extreme cases. But even backfilling based on a very small enhancement of capillary pressure due to low heterogeneities might lead to a relatively drastic buoyant pressure rise in the case of not very thin sites. Strong vertical capillary variations might be overcome and migration will follow upwards. Therefore, even small variations of capillary pressure in a lateral direction might destroy an anisotropy of preferred horizontal migration pathways.

This problem can be overcome technically by introducing anisotropy as a vertical heterogeneity variation at a sub–site resolution. This corresponds to a picture with "sub–layers" crossing the sites. Micro–accumulations build up until preferred sub–layers with reduced threshold in a lateral direction are found. Anisotropy can then be interpreted as a site averaged quantity defining a capillary pressure heterogeneity level, which can be overcome in a lateral direction by migration (Fig. 6.60). A continuous crossover from non–anisotropic to very anisotropic can be achieved easily by increasing this anisotropy value. This method only works consistently if heterogeneity (e.g. in the form of random noise) is assigned in the model. Both, the sub–site and the global heterogeneity variation are directly and quantitatively related. An anisotropy much smaller than the global heterogeneity does not affect the migration picture and an anisotropy much bigger than the global heterogeneity has the same effect as completely ignoring it in the lateral direction. Therefore, this is a consistent approach and, as expected, the number of micro–accumulations at the site scale will decrease with increasing anisotropy. Obviously, the determination of residual saturation values and general upscaling procedures are affected and become quite complicated in this picture.

By visual inspection the authors found 10% capillary pressure heterogeneity variations with a value of 10% for an anisotropic threshold level as an appropriate default for most lithologies and typical gridcell sizes in basin modeling.

6.9 Discussion

The Darcy flow model is a well established and physically consistent formulation of a transport problem with separate phases considering buoyancy, overpressure, capillary, and viscous forces. It also incorporates water flow and compaction in the most comprehensive formulation. The basic idea is a summation and balance of all the forces acting on the fluids.

Multi–phase Darcy flow models are too complex and too inefficient for basin modeling. Alternative methods are therefore presented in this chapter. They rely exclusively on the assumption that overall migration timing can

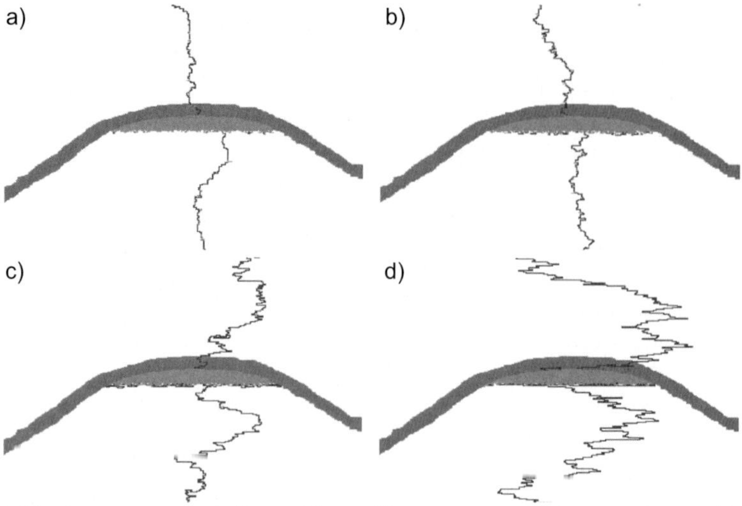

Fig. 6.60. Invasion percolation with capillary pressure heterogeneity variations of 10% and different anisotropy threshold levels: a) 0%, b) 5%, c) 10%, d) 15%

be neglected. An application of this approximation in reservoirs, for fault flow and for break through processes yields the hybrid method. Flowpath modeling is a further approximation with even better processing performance and with the main advantages of the hybrid approach, but it is rather vague regarding the overall timing and lateral migration in low permeability regions. Both, flowpath and invasion percolation methods delegate the whole time control to HC generation and expulsion. Migration is treated as if it occurs instantaneously on geological timescales.

Hybrid and flowpath models are characterized by reservoir analysis. Accumulation bodies with well defined volumes and column heights are calculated. Two phase effects, such as the displacement of liquid by vapor in accumulations, are taken very precisely into account. The spatial resolution can usually be chosen to be significantly higher than for heat analysis or compaction calculations of the overall basin model. High spatial resolution allows high precision volumetrics and accurate petroleum–water contact height predictions. Hence, correct calculations of column pressures from accumulations are very easy. A reservoir analysis can be carried out very quickly on modern computers. Data uncertainties can be tested by interactive risking of different migration and accumulation scenarios on most PCs.

The biggest problem of hybrid and flowpath models are complex geometries such as vertical sandstone channels or permeable faults, which connect several reservoirs (Fig. 6.43). The concept of reservoir analysis as discussed in this chapter is map based and thus essentially a two dimensional concept. Hybrid and flowpath models are limited in complexity of geometry. They work best when the geometry follows a layer cake topology.

However, without loss of generality the flowpath concept could be extended into the third dimension. Drainage areas become drainage volumes and flowpaths, which are located at the reservoir seal interface, would become complete three dimensional pathways. This approach ensures a more "natural" domain decomposition, which enables a more complex geometry. However, an implementation in the form of a computer program would cause some difficulties. Standard data formats such as maps have to be replaced by more proprietary 3D – formats. Three dimensional data sets, which are obviously less common than simple maps, must be populated. Increasing data amounts and reduced performance confine the options of interactive handling. It is an open question whether the implementation of a full 3D flowpath concept would be worth the large amount of effort required.

Besides the complex geometry, some smaller problems exist concerning hybrid and flowpath modeling. The automatic construction of a dataset by domain decomposition is often an expensive task. A large amount of data must be collected for a reservoir sub–model. Among the main components are the mapped reservoir, capillary pressures, porosities, faults, and the component resolved amounts of HCs entering the reservoir. Grid transformations require extra efforts in modeling. Fortunately, a cut out from a fully populated basin model can be automated. A disadvantage comes with parallelization. Reservoir analysis itself is to some degree parallelizable (Bücker et al., 2008). The main challenge is found in parallelization of the domain decomposition and the collection of injection amounts.

Other small problems rise due to effects such as HC loss in the carriers, which are only processable with extra efforts or the impact of water flow within a carrier, which is easy to consider for flowpath bending but difficult to integrate into an appropriate prediction of contact area deformation.

Invasion percolation offers an alternative migration method, which is based on an interpretation of migration as a movement of separated stringers. It can be performed on a higher resolution grid than commonly used in basin modeling for Darcy flow or temperature analysis. Physically, the method does not differ significantly from flowpath based migration. However, its technical implementation is completely different. High resolution seismic data can be directly incorporated and a domain decomposition is not a prerequisite anymore. It is a big advantage of the technique that it is possible to model HCs percolating through a whole basin model. Complicated geometries with strongly varying migration properties can be easily modeled (Fig. 6.43). Effects of laminated, crossbedded, and pervasively faulted strata can be taken into account to a high degree of accuracy (Ringrose and Corbett, 1994). Small scale structures such as faults can be integrated into the algorithm in a "natural" and "intuitive" way. Principally, migration losses and hence migration efficiency can be calculated to a high degree of accuracy (Luo et al., 2007, 2008). Column heights in accumulations are calculated quite accurately even under aquifer flow conditions.

A simple two dimensional example model for comparison of hybrid, flowpath, and invasion percolation is shown in Fig. 6.61. Accumulated amounts are quite similar here. The invasion percolation and flowpath models are almost the same. The hybrid model differs slightly, in that the structure on the top left is not fed by a break through from below. Besides this all accumulations in all examples have the same break through behavior. On the right, in layers 5—8, some migration activity over wider regions can be seen in the hybrid and the IP models. This flow originates from additional break throughs in previous events, which give rise to residual saturations or microaccumulations and further transport of small HC amounts with ongoing compaction. This process is completely ignored by the flowpath model.

The composition of the big accumulation within Layer 5 is also depicted in Fig. 6.61. It is very similar in all three models. A major difference in PVT–behavior can be found in the accumulation in Layer 9 beside the fault on the right. The accumulation consists mainly of vapor in the hybrid model, whereas it is almost entirely liquid in the flowpath and IP results. A closer inspection shows that the composition is very similar in all three models. During the last event the geometry changed significantly. Flash calculations are performed more often in the hybrid than in the flowpath and IP models. Thus the phases of the hybrid are already updated whereas the "faster migrating" flowpath and IP models show a systematic error due to time steps being too long. However, pressure and temperature are not far from the critical point. The density difference between vapor and liquid is relatively small and the error is not very big.

A Darcy flow model is also calculated for comparison (Fig. 6.62). The principal migration scenario including the location of the accumulations is very similar. The biggest differences are found for saturation values inside the accumulations. Due to the low grid resolution and an improper capillary pressure curve (compare Fig. 6.4), it is not possible to model saturation patterns such as in the hybrid case, which approach almost 100 % petroleum saturation in the center of some accumulations.

Another important difference in the Darcy model in Fig. 6.62 is the saturation in Layer 9 on the right side, which is not located in a structure or stratigraphic trap. Layer 9 is a highly permeable sandstone. The amounts on the right are stuck due to convergence problems of the explicitly treated Darcy flow. A reduction in the length of the migration time steps reduces this artifact. However, the calculation time of the Darcy flow model is already three times as much as for the hybrid model. The difference in simulation time increases even more for three dimensional models.

Finally it must be noted that the composition within the accumulations of the Darcy flow model differs significantly from the other results. The saturation never exceeds 40 % inside the gridcells. The corresponding capillary pressure is so high that liquid cannot enter these cells. The vapor displaces all the liquid and finally most of the liquid leaves the model at the sides. Only in the Darcy model do all accumulations contain vapor. This is an error which is inherited from low gridding and improper capillary pressure curves.

Invasion percolation demands the analysis of multiple migration scenarios if, due to the random assignment of capillary pressure values, several widely distinct saturation patterns evolve. The resulting scenarios are not deterministic and cannot be reproduced (at least not in detail). Thus, in theory, one specific migration pattern is not representative because major migration paths might be dependent on small scale variations. Further analyses, such as sensitivity studies with multiple scenarios, are necessary. A risking procedure should be integrated into the migration analysis (Fig. 6.63). However, this is not a major drawback, as risking should be performed in basin modeling anyway (Chap. 7).

High resolution seismic data combined with advanced flow unit interpretation enables sophisticated percolation analysis on a high resolution scale. The data, however, must be available and interpreted. Fully populated three dimensional models must be constructed. Possible pitfalls associated with slightly dipped seals or with the interpretation of noise in seismic data complicate the workflow (Sec. 6.8.6).

Flowpath models might be a better alternative in cases of incomplete data sets or simple geometries such as "layer cake" structures. Generated amounts are vertically transported into the reservoir rocks. The implicit high resolution of the flowpath calculation (Sec. 6.5.1), petroleum water contact height analysis without vertical gridding problems and the high performance of the calculation, especially for multi phase migration, leads often to almost the same results with a simpler and faster workflow.

As mentioned before, timing is a big disadvantage of invasion percolation as well as for pure flowpath modeling. The time the hydrocarbons need to migrate from the source rock to the structure is generally not taken into account by the method. An example is shown in Fig. 6.58. Colored planes indicate faults which are partially open for migration. Similar structures are charged in both models. In the IP model, petroleum is transported to the reservoirs more quickly and in larger amounts Another example with severely affected timing is shown in Fig. 6.59. In the Campos basin, salt windows might open at a certain time allowing HCs to pass. Charging and accumulated amounts in the hybrid and in the invasion percolation model are almost the same for the Roncador field. But the two source rocks charge the structures differently in each model and there are huge migration differences in sub salt layers which are not visible in the figure. Elaborated mass balances such as Table 6.2 are here very advantageous. Important information can often be extracted from balance tables more efficiently than by visual interpretation of migration and accumulation scenarios. Mass balances are discussed in general in Sec. 6.10.

Additionally, it should not be forgotten that break through and leakage flow rates are usually so high that the basic assumption of low flow rates for invasion percolation is violated. Darcy flow is argued to be the better alternative here (Sylta, 2004, 2002b). However, this problem could principally

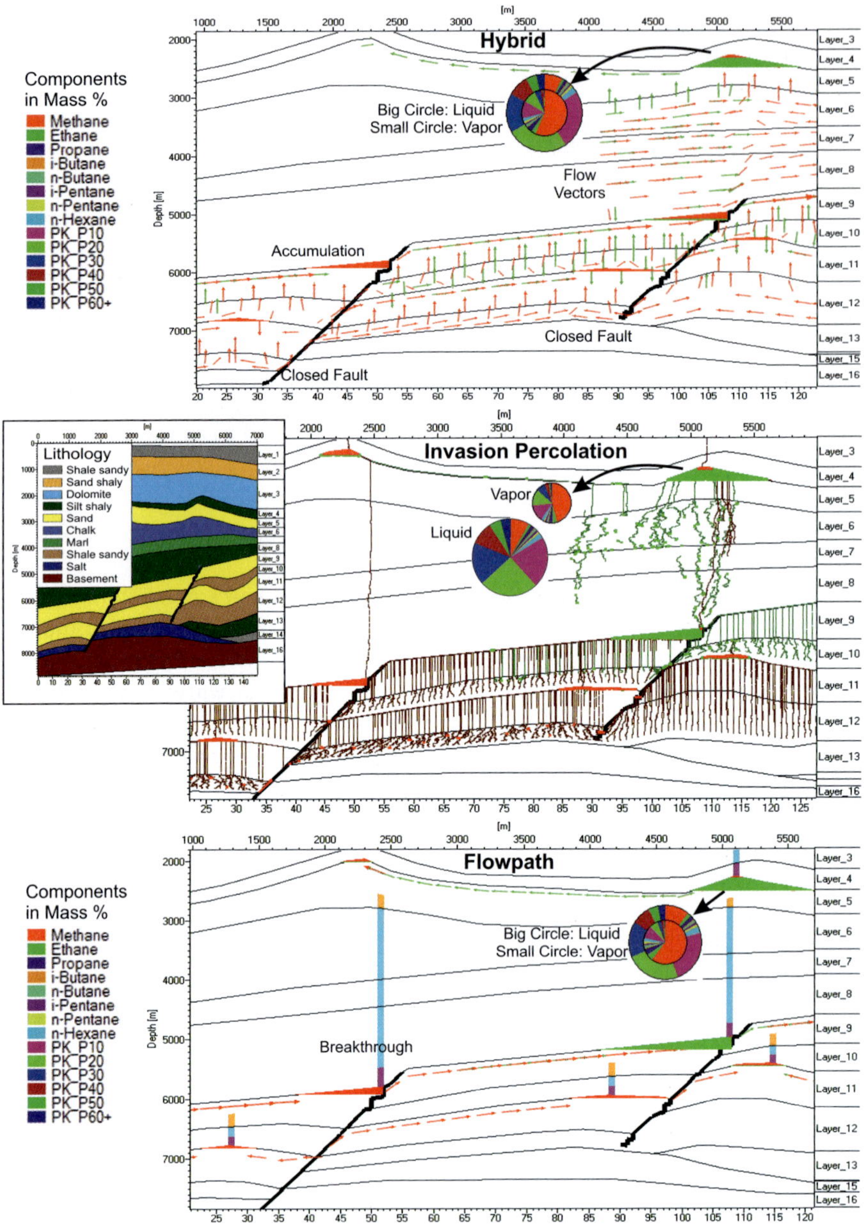

Fig. 6.61. Hybrid, flowpath, and IP runs of a simplified North Sea example model. Layer 10 and 12 are source rocks. Break through pathways are not displayed in the hybrid plot. The resolution of this figure is unfortunately too poor to clearly distinguish between full and residually saturated IP cells. Flowpaths in the reservoir layers of the hybrid and the flowpath model are graphically converted here to flow vectors for clearer presentation

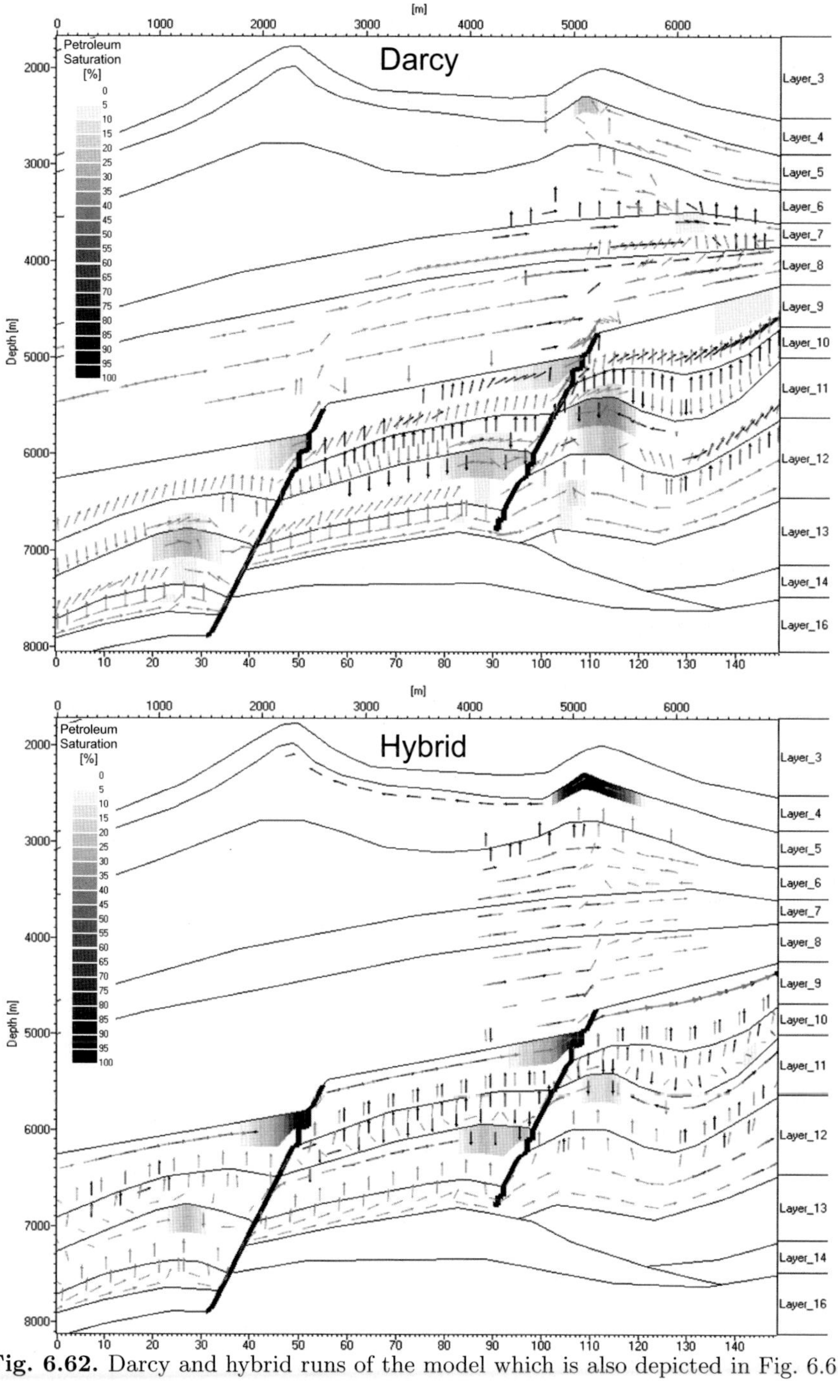

Fig. 6.62. Darcy and hybrid runs of the model which is also depicted in Fig. 6.61

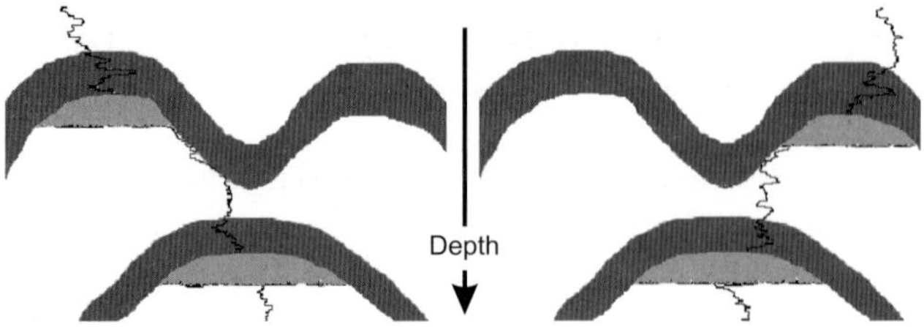

Fig. 6.63. Invasion percolation scenarios with two different realizations of heterogeneity. The top left structure is filled in the left example and the top right structure in the right example

Petroleum System	Hybrid	IP
Generated	60.0	60.0
Expelled	54.0	54.0
HC in Reservoirs	2.0	2.0
Losses	52.0	52.0

Losses in Detail	Hybrid	IP
Migration	29.5	0.5
Secondary Cracking	3.0	0.0
Outflow Top	7.0	2.5
Outflow Sides	12.5	49.0

Table 6.2. The masses of the petroleum Systems in Gtons for the two versions of the Campos model which are shown in Fig. 6.59. The main differences are due to different migration scenarios in the shale between source and salt. In the IP model most of the mass is transported to the border and lost at the basin sides whereas petroleum moves slowly in the hybrid case. The mass balance is additionally affected by secondary cracking in the hybrid model. The petroleum is kept long enough in hot regions just above the source rock. Sub salt migration timing is important in this study (Bartha, 2007)

be solved: break through flow rates could be estimated and additional break through points could be calculated if necessary.

Explicitly treated Darcy flow and invasion percolation are not so different from their technical approach. Both are grid cell (site) based and both modify the saturation in the grid cells according to physical conditions in a range containing neighboring cells. The central differences are that invasion percolation usually works on a higher spatial grid resolution but at a much lower resolution in time. The explicit Darcy method performs many small migration time steps to achieve a picture of continuously moving saturation under time control, which is neglected in invasion percolation. Saturation is modeled with continuous values in Darcy flow whereas discrete fillings, namely residually and fully saturated, are used in invasion percolation. The implications of the saturation discretization are expected to be rather small, because the smearing out of the saturation values of invasion percolation in the low resolution Darcy cells also yields (almost) continuous pictures. However, a big difference is found in column height calculations, which cannot be performed accurately within a low resolution Darcy grid.

All approaches have advantages and disadvantages. Neither Darcy flow, hybrid approach, flowpath modeling nor invasion percolation is superior in general. All known migration models are based on a number of assumptions, simplifications and approximations. There is no modeling method which fully covers the whole spectrum of all migration related effects to a high degree of accuracy. However, depending on the specific geological environment and the availability of data in a concrete case study, one of the methods is often more suitable than the others. Better modeling results can be achieved just by selection of the most appropriate method. It must also be mentioned that the comparison of different methods for the same study yields at least an idea of result variations and uncertainties.

6.10 Mass Balances

An accurate analysis of a petroleum system is equivalent to the detailed quantification of all the petroleum involved in each geological process. Petroleum amounts must be therefore separately quantified and tracked for the processes of deposition, generation, cracking, adsorption, phase separation, dissolution, migration, and accumulation. The large number of processes, layers, phases, and components involved, make this topic a challenge. A systematic approach is necessary.

6.10.1 Fundamental Laws of Mass Conservation

A sediment with kerogen within it has at time t a potential to generate a mass $M_P(t)$ of petroleum. The potential might increase by additional deposition of sediment with mass generation potential $M_{\text{Sed}}(t)$ and decrease via erosion by $M_{\text{Ero}}(t)$. The amount of kerogen available for generation before the start of generation is $M_P(t) = M_{\text{Sed}}(t) - M_{\text{Ero}}(t)$. Generated HC mass is quantified by $M_G(t)$. Hence, the potential through time is finally given by

$$M_P(t) = M_{\text{Sed}}(t) - M_{\text{Ero}}(t) - M_G(t) \, . \tag{6.64}$$

The mass amounts in balance equations such as (6.64) are cumulative over and uniquely dependent on time t. Amount differences $\Delta M(t,t') = M(t) - M(t')$ for any time interval, which starts at time t' and ends at time t, can be calculated just as differences between the values at the two different time points. Hence $\Delta M_P(t,t')$ becomes

$$\Delta M_P(t,t') = \Delta M_{\text{Sed}}(t,t') - \Delta M_{\text{Ero}}(t,t') - \Delta M_G(t,t') \, . \tag{6.65}$$

It is not called cumulative anymore. Equation 6.65 has the same structure for mass differences ΔM as its cumulative counterpart (6.64) for total masses M. Thus all the following cumulative balance equations can also be interpreted as non–cumulative. The time dependence is, for the simplicity of the expressions, not noted anymore in this section.

HC mass amounts M_{HC}, which are found in a sediment, are given by the sum of liquid and vapor petroleum $M_{L,V}$, HCs which are dissolved in water M_W and HCs which are adsorbed M_{Ad}:

$$M_{\mathrm{HC}} = M_L + M_V + M_W + M_{\mathrm{Ad}} \,. \tag{6.66}$$

Generated mass and current HC mass differ by secondary cracked and by migrated amounts only. The amount of the secondary cracking product coke with mass M_C can be calculated as the difference of HC mass M_D which is destroyed by cracking and newly generated (transformed) mass M_T by $M_C = M_D - M_T$. Migrated masses are usually distinguished by in– and outflow $M_{\mathrm{In/Out}}$ across the border of the object under consideration. Hence the continuity equation can be written as

$$M_{\mathrm{HC}} + M_C + M_{\mathrm{Out}} = M_G + M_{\mathrm{In}} \tag{6.67}$$

or

$$M_L + M_V + M_W + M_{\mathrm{Ad}} + M_D + M_{\mathrm{Out}} = M_G + M_{\mathrm{In}} + M_T \,. \tag{6.68}$$

The last equation contains on both sides only terms with positive values, typical for bookkeeping. The left side defines where the HCs are found and the right from where they come (assets and liabilities).

The last two equations (6.67) and (6.68) are obviously valid for any geological object, especially any layer or facies at each time. For example, there is usually no generation in a reservoir r, hence it is $M_{G,r} = 0$. The equations are also separately valid for each HC component. Hence each term is usually labeled with additional indices such as $M_{V,l,i}$, which indicate the mass of component i found in the vapor phase inside of layer l. The index i is skipped in this section. It makes the equations rather ugly and it is not necessary for the argumentation. Sums of all components yield total masses and formulas with skipped component indices can also be interpreted as total mass balances.

Tracking should be performed for each finite element cell in modeling practice. Any geological object in a basin model, which is important for HC balance considerations (e.g. any layer l), can be constructed by sums over its cells. It is therefore possible to calculate any balance for any geological object in post processing steps.

A difference formulation such as (6.65) between two succeeding time steps with infinitesimal duration and infinitesimally small volumes for (6.67) yields a continuity equation such as (6.34) for mass balance. A rearrangement of (6.67) in difference form such as (6.65) and division by $\Delta t = t - t'$ yields

$$\frac{\Delta M_{\mathrm{HC}}}{\Delta t} + \frac{\Delta(M_{\mathrm{Out}} - M_{\mathrm{In}})}{\Delta t} = \frac{\Delta(M_G - M_C)}{\Delta t} \tag{6.69}$$

and makes the relationship with (6.34) more obvious. Existing HCs constitute the transient term and in– and outflow the flow term. Generation is formulated with source terms on the right side of the Darcy flow equations.

Equation 6.68 contains the important quantities which are necessary for an understanding and a detailed analysis of petroleum systems whereas (6.69) demonstrates the relationship to the flow equations.

Losses from the border of the whole model also need to be tracked to ensure the completeness of the bookkeeping. The masses which leave the model are usually quantified and collected under the term model outflow M_{MO}. Sometimes it is additionally distinguished between outflow at the top and outflow at the sides with $M_{\mathrm{MO},l} = M_{\mathrm{MO\ top},l} + M_{\mathrm{MO\ sides},l}$. Any outflow from a geological object is an inflow into another geological object except the amounts which leave the basin. Hence it is

$$\sum_l \left(M_{\mathrm{Out},l} - M_{\mathrm{In},l} \right) = \sum_l M_{\mathrm{MO},l} \,. \tag{6.70}$$

The summation index l must cover all geological objects with in– and outflow, e.g. all layers and faults.

Conducting faults are obviously of special interest for migration. In basin modeling resolution they have a two-dimensional character which implies that all quantities which are found in equations (6.64) — (6.68) except $M_{\mathrm{In/Out}}$ are zero. Additionally

$$\Delta M_{\mathrm{Flow},F} = \sum_{l(F)} \Delta M_{\mathrm{Flow},l} = 0 \tag{6.71}$$

with the sum over all layers l (and other faults) which touch the fault F.

Biodegradation is for convenience often included in the secondary cracking scheme. It usually cannot be mixed up with secondary cracking because it is found only in shallow sediments at temperatures too cold for cracking. Similarly, petroleum dissolved in hydrate phases is usually added to the water balance. However, an extension of the balance scheme (6.68) with a separate incorporation of these masses is straight forward.

Separate tracking of M_{Out} and M_{In} is not recommended in modeling as demonstrated in Fig. 6.64. Due to gridding artifacts neither M_{Out} nor M_{In} can be determined accurately. Only the difference $\Delta M_{\mathrm{Flow}} = M_{\mathrm{Out}} - M_{\mathrm{In}}$ is reliable. A similar problem rises with flowpath modeling and invasion percolation. Petroleum which traverses a facies in one event is not tracked as in– and outflow. For example, in flowpath models, the petroleum is transported vertically from source to reservoir without considering vertical migration pathways in detail (Sec. 6.7). An in– and outflow tracking is elaborate and time consuming especially in models with complex facies distributions and sublayering. Due to random capillary pressure variations and high resolution cells, similar problems as in Fig. 6.64 arise in invasion percolation. A stringer path might enter and leave a facies multiple times over short geological distances. This leads to large and meaningless in– and outflow amounts. Again only the difference is useful.

All amounts should be tracked in units of mass. Insitu volume is inappropriate because it changes with pressure and temperature (Chap. 5). For an

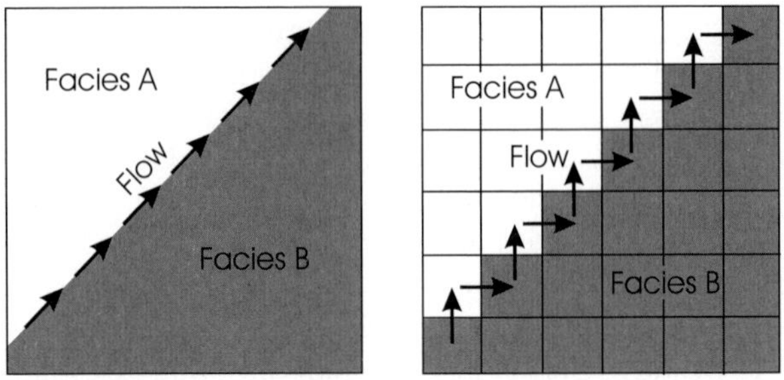

Fig. 6.64. Schematic map view of flow along a facies boundary on the left. A correctly modeled flow may follow the gridlines in a model on the right. Only the average flow direction is the same as on the left side. Each facies has multiple in– and outflows. Obviously, the vanishing difference of in– and outflow is correctly conserved. Models with many time steps such as explicitly treated Darcy flow models might erroneously cause enormous amounts of in– and outflow

impression of volumetrics it is sometimes advisable to transform all masses with one fixed density to volumes. Such a transformation conserves all balance and conservation laws. Nevertheless, vapor volumes are sometimes transformed separately with another density. Conservation laws such as (6.67) are not valid in such a case and must be treated with caution.

6.10.2 The Petroleum System

All HC masses which are involved in all the important geological processes of petroleum geology in a sedimentary basin, are tracked when all the quantities of (6.68) the generation potentials and the basin outflow are known all over the basin. It is now possible to evaluate all characteristic numbers for petroleum systems analysis. However, it is first necessary to introduce the expulsion masses M_E and HC losses M_{Loss} of a petroleum system.

In basin modeling expulsion masses of a source s are defined by

$$M_{E,s} = M_{G,s} - M_{C,s} - M_{\mathrm{HC},s} = \Delta M_{\mathrm{Flow,s}} \; . \qquad (6.72)$$

Classically, expulsion is defined slightly differently in geology. Expulsion is only related to the petroleum which is generated in the expelling source rock. The balance according to (6.72) adds masses of hydrocarbons which were generated in another source rock, migrated into the considered source and again left it.[23] Classically defined expulsion amounts can only be calculated with high effort in basin models with migration and multiple source rocks. The situation is shown in Fig. 6.65. First of all, each component must be tracked according to the source rock where it was generated. But source rock tracking alone is not sufficient because free pore space in source B might be occupied

[23] According to (6.72) even "non–source" rocks might expell if $\Delta M_{\mathrm{Flow},l} > 0$.

by petroleum of source A. An interaction of such a type causes an additional amount of expulsion from source B, which would not occur if source A did not exist. This additional amount is usually not be taken into account in the definition of classical expulsion. It is defined as a property of the considered source alone. Furthermore, classical expulsion is only defined for the first time when petroleum leaves the source. Inflow back into its generating source rock (e.g. after downward expulsion) followed again by outflow must also not taken into account. This causes additional complications according to Fig. 6.64. Finally, it must be stated that classical expulsion can not be directly calculated from or within a basin model with secondary migration.[24] However, it can easily be calculated from special simulation runs which model primary migration only without performing any further migration. These extra simulations can in both cases be performed very quickly if temperature and pressure results are reused from runs with migration instead of recalculating them every time (Sec. 1.3).

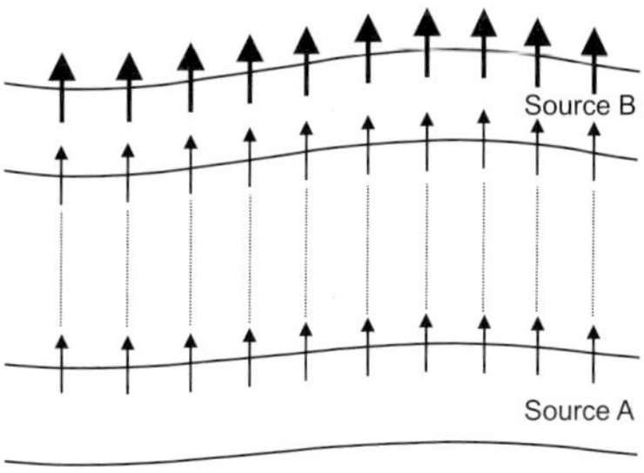

Fig. 6.65. Stacked source system. Expelled amounts of the deeper source A migrate into source B and mix with petroleum inside of source B. Free pore space in source B is occupied by petroleum which is generated in source A. Expulsion amounts of source B are therefore difficult to determine

HC losses of a petroleum system (PS) are now defined by the expelled masses minus the amount found in the reservoirs as

$$M_{\text{Loss,PS}} = \sum_{s(\text{PS})} M_{E,s} - \sum_{r(\text{PS})} M_{\text{HC},r} \,. \tag{6.73}$$

The first sum is over all sources s and the second over all reservoirs r belonging to the petroleum system. It does not matter, if a classically defined expulsion amount or the definition of (6.72) is used in (6.73). The flow balance is invariant because inflow into a source rock is counted negatively according to (6.72).

HC losses can alternatively be evaluated with

$$M_{\text{Loss,PS}} = \sum_{l(\text{PS})} M_{\text{MO},l} + \sum_{l'(\text{PS})} M_{C,l'} + \sum_{l''(\text{PS})} M_{\text{HC},l''} \,. \tag{6.74}$$

[24] Classical expulsion is often not present as an overlay for visualization in model viewers. Source rock leakage, which is defined as the outflow at the upside or downside of a source rock, is commonly available as a compensation.

The first sum represents the flow out of the basin. It is over all the layers of the petroleum system including the sources and the reservoirs. The second sum describes the amount of losses by cracking. The index l' runs over all layers exclusive of all source rocks. The last sum describes the amount of HCs in the petroleum system which is neither located within the source nor is found within a reservoir. The index l'' therefore runs over all layers exclusive of all source rocks and reservoirs.

Finally, a petroleum system is commonly characterized by the masses found in its reservoirs

$$M_{\mathrm{HC,PS}} = \sum_{r(\mathrm{PS})} M_{\mathrm{HC,r}} \,, \tag{6.75}$$

the losses $M_{\mathrm{Loss,PS}}$, which can be calculated from (6.73) or (6.74), the expelled masses

$$M_{\mathrm{E,PS}} = \sum_{s(\mathrm{PS})} M_{\mathrm{E,s}} \,, \tag{6.76}$$

and its generation potential given by

$$M_{\mathrm{P,PS}} = \sum_{s(\mathrm{PS})} M_{\mathrm{P,s}} \,. \tag{6.77}$$

Again, the sum over $s(\mathrm{PS})$ indicates a sum over all sources and $r(\mathrm{PS})$ over all reservoirs of the petroleum system.

A requirement for the evaluation of characteristic numbers of a petroleum system is knowledge of its encompassing elements (Sec. 1.4). This is not a trivial problem, neither in reality nor in models with multiple source rocks and a complicated geometry. Two supplementary methods are commonly used. Firstly, flowpaths, drainage areas, flow vectors and cells with residual saturations, which indicate that petroleum has passed through them, are tracked just by visualization with model viewers. This is a very fast but also rather crude approach and often does not reveal all affiliations. Compaction and varying geometry makes the mapping of flowpaths over time a sophisticated task. Secondly, source rock tracking is applied. Components which are generated in one source are separated during the whole simulation from identical components generated within other source rocks. This method allows a close tracking but unfortunately a lot of computer memory is needed if many source rocks are present in the model. Another disadvantage is the rather low spatial resolution, especially if the source rocks are wide. However, this method has the additional benefit, that petroleum can be accurately associated with its origin, even if it is mixed from different sources. The best results are achieved with a combination of both methods.

6.10.3 Reservoir Structures and Accumulations

Individual structures in reservoirs and accumulations are another important class of geological objects for which mass balances are commonly formulated.

Outflow out of a single structure is often subdivided into spilling, fault flow and leakage or break through

$$M_{\text{Out}} = M_{\text{Spill}} + M_F + M_{\text{BT}} \,. \tag{6.78}$$

Usually, only one of these three outflow types exists. Generation and absorption usually do also not exist in reservoirs and hence (6.68) becomes

$$M_L + M_V + M_W + M_C + M_F + M_{\text{Spill}} + M_{\text{BT}} = M_{\text{In}} \,. \tag{6.79}$$

Sometimes inflow is also separated into spilled amounts coming from other accumulations and amounts caught over the whole drainage area from below $M_{\text{in}} = M_{\text{SpillIn}} + M_{\text{DA}}$.

A severe problem arises with compaction and tectonic movements. Structures do not permanently exist over the whole lifecycle of a basin. They mostly develop ages after deposition and they may move laterally with an overall tectonic shift or vanish during further compaction. The tracking of such variable geological objects over several geological events is sometimes impossible and technically very difficult to automate. Similar problems arise with accumulations which merge over several traps with proceeding source rock expulsion (Fig. 6.19). Generally, it is only possible to formulate non–cumulative balances such as (6.65), e.g. for one geological event only.

A tracking algorithm for traps and accumulations can be based on a present day drainage area subdivision of a reservoir. Each structure of a previous event has a highest point which is often located in its center. If this point is found in the present day drainage area it is added to its time track. This rule has several consequences. Firstly, each structure found in paleo time is related uniquely to a present day structure. This implies that in sums over multiple structures no paleo time structure is counted twice and that no structure is skipped if a sum over all structures is performed. Secondly, no paleo time structure might be found. In such a case, a gap is found in time extractions. Finally, more than one paleo time structure may be related to one present day structure. In such a case all the amounts of liquid, vapor, break through etc. must be summed up over this paleo set of structures Fig. 6.66. An exception are spilled amounts (Fig. 6.67). It must be checked if they spill out of the tracked set to which they belong. Internal spills obviously need not be counted. The summed spill amount of the set outflow is called net spill.

The total paleo outflow M_{Out} according to (6.78) might not be provided by one term alone. For example, one of the related paleo structures might have a break through and another might spill.

The system of summing a group of many substructures has a big advantage. It can be easily extended to a group of present day structures or drainage areas which can be summed together with the same rules. A group of present day structures is associated with a group of structures at each paleo event, which is just the union of all sets belonging to each drainage area. This allows an almost continuous crossover from one present day structure to a larger

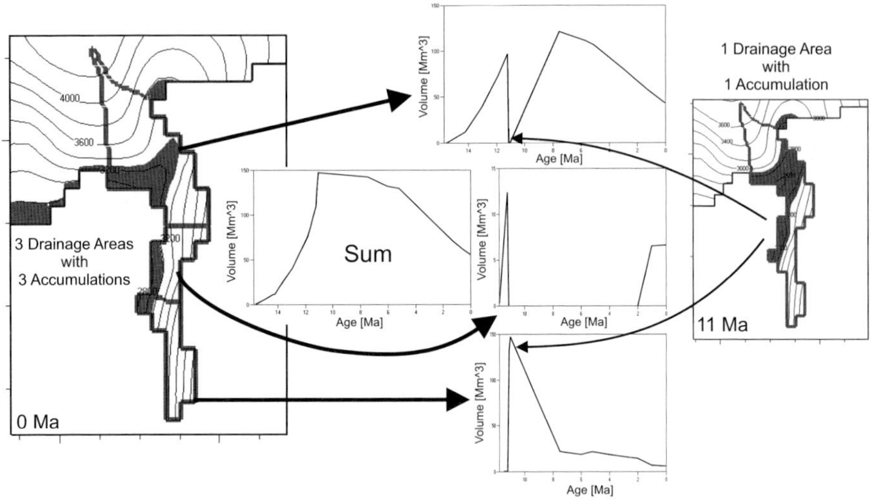

Fig. 6.66. Map view of tracked accumulations. Three accumulations are found at the present day but only one at 11 Ma. This leads to discontinuities in the tracking of accumulated volumes of the individual accumulations at 11 Ma. However, the sum over all drainage areas does not show discontinuities

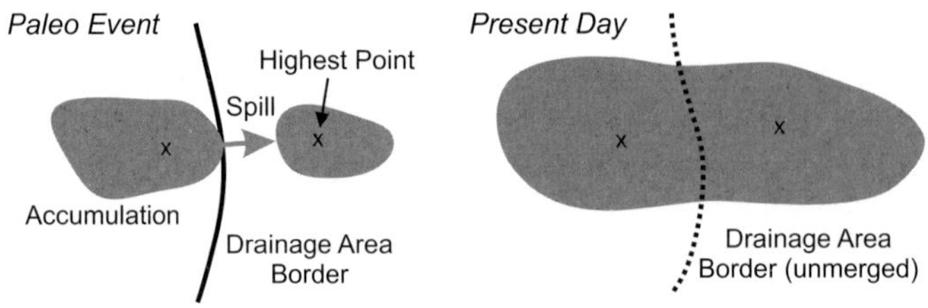

Fig. 6.67. A schematic map view of a reservoir with accumulations at a paleo event and at present day. Both small accumulations belong to the history of the big accumulation at present day. The spilled amounts at paleo time should not be accounted for when spilled amounts are tracked through time

field (e.g. a main structure and its satellites) and the full reservoir. Obviously, it can be expected that wrongly tracked paleo structures become relatively less important if the number of present day drainage areas increases.[25] Erroneously tracked paleo structures might only appear at the border of the tracked field, which becomes less important with the increasing size of the full area.

[25] The areas should not be disjunct.

Summary: The most comprehensive formulation of fluid flow in porous media is given by Darcy flow. It is based on a balance of all forces which act on the fluid. These include capillary pressures which rise due to interfacial tensions in rock pores and throats, friction which is described by petroleum viscosity and rock permeability, buoyancy of the petroleum in water, and additional external forces, e.g. due to aquifer flow. The differential equations with appropriate boundary conditions can be formulated. They allow for petroleum mass conservation. Practically, Darcy flow can only be modeled on a coarse grid resolution.

However, migration in carriers and reservoirs can be approximated very accurately with flowpath (ray tracing) methods. Due to high permeability, migration occurs instantaneously on geological timescales. Hence, migration and accumulation can be modeled geometrically including the flowpaths, drainage areas, and volumetrics of the structures.

In rocks with low permeability petroleum migrates slowly, which allows for a solution of Darcy flow based differential equations. Further more, a domain decomposition of a sedimentary basin into high and low permeability rocks can be performed. Highly permeable regions can be modeled with flowpath based reservoir analysis and low permeability regions with Darcy flow. This approach is called the hybrid method.

Vertical migration between source rock and reservoir occurs very quickly on geological timescales when the vertical distance is short. Darcy based flow can often be approximated by direct injection of expelled petroleum into the reservoirs above the source rocks. Reservoir analyses can be performed with the hybrid method. This highly efficient method is called flowpath modeling. Time control of migration is completely determined by expulsion. Obviously, flowpath modeling is not applicable everywhere.

Darcy flow can be criticized because it is based on continuous flow and macroscopic length scales. It might not be suitable when migration is interpreted as a non–uniform movement of microscopic disjunct petroleum stringers along preferred migration pathways. Upscaling considerations give rise to an alternative invasion percolation method, which is in many physical aspects very similar to flowpath modeling because time control of the migrating petroleum is again completely neglected. Domain decompositions can be avoided and migration can be performed in complex geometries on high resolution grids. Seismic data can be incorporated without upscaling.

Each of these methods have advantages and disadvantages. Darcy flow is the most comprehensive method but practically unusable. The hybrid method is very accurate and can be well applied in practice. Flowpath modeling is the most efficient method and often a very good approach when data is sparse. Invasion percolation works best on complex geometries such as multi–faulted regions and when high resolution seismic data is incorporated.

A systematic assembly of mass tracking in all geological processes of petroleum generation and migration can be incorporated.

References

U. S. Allan. Model for hydrocarbon migration and entrapment within faulted structures. *AAPG Bulletin*, 73:803–811, 1989.

B. Ataie-Ashtiani, S. M. Hassanizadeh, and M. A. Celia. Effects of heterogeneities on capillary pressure–saturation–relative permeability relationships. *Contaminant Hydrology*, 56:175–192, 2002.

K. Aziz and A. Settari. *Petroleum Reservoir Simulation*. Elsevier, 1979.

G. I. Barenblatt, V. M. Entov, and V. M. Ryzhik. *Theory of Fluid Flows Through Natural Rocks*, volume 3 of *Theory and Applications of Transport in Porous Media*. Kluwer Academic Publishers, 1990.

A. Bartha. Migration methods used in petrolem systems modeling – comparison of hybrid and invasion percolation: Case study, campos basin, brazilian offshore. Presentation at the IES Usermeeting, 2007.

R. R. Berg. Capillary pressures in stratigraphic traps. *AAPG Bulletin*, 59: 939–956, 1975.

H. M. Bücker, A. I. Kauerauf, and A. Rasch. A smooth transition from serial to parallel processing in the industrial petroleum system modeling package petromod. *Computers & Geosciences*, 34:1473–1479, 2008.

D. J. Carruthers. *Transport Modelling of Secondary Oil Migration Using Gradient–Driven Invasion Percolation Techniques*. PhD thesis, Heriot–Watt University, Edinburgh, Scotland, UK, 1998.

D. J. Carruthers and P. Ringrose. Secondary oil migration: oil–rock contact volumes, flow behaviour and rates. In J. Parnell, editor, *Dating and Duration of Fluid Flow and Fluid–Rock Interaction*, volume 144, pages 205–220. Geological Society of London, Special Publication, 1998.

L. Catalan, F. Xiaowen, I. Chatzis, and A. L. Dullien. An experimental study of secondary oil migration. *AAPG Bulletin*, 76:638–650, 1992.

R. E. Chapman. *Petroleum Geology*. Number 16 in Developments in Petroleum Science. Elsevier, 1983.

S. M. Clarke, S. D. Burley, and G. D. Williams. A three–dimensional approach to fault seal analysis: fault–block juxtaposition & argillaceous smear modelling. *Basin Research*, pages 269–288, 2005a.

S. M. Clarke, S. D. Burley, and G. D. Williams. Dynamic fault seal analysis and flow pathway modelling in three–dimensional basin models. In A. G. Doré and B. A. Vining, editors, *Petroleum Geology: North–West Europe and Global Perspectives—Proceedings of the 6th Petroleum Geology Conference*, pages 1275–1288. Geological Society of London, 2005b.

S. M. Clarke, S. D. Burley, G. D. Williams, A. J. Richards, D. J. Meredith, and S. S. Egan. Integrated four–dimensional modelling of sedimentary basin architecture and hydrocarbon migration. In S. J. H. Buiterand and G. Schreurs, editors, *Analogue and Numerical Modelling of Crustal–Scale Processes*, volume 253, pages 185–211. Geological Society of London, Special Publication, 2006.

L. P. Dake. *The Practice of Reservoir Engineering.* Number 36 in Developments in Petroleum Science. Elsevier, revised edition, 2001.

A. Danesh. *PVT and Phase Behaviour of Petroleum Reservoir Fluids.* Number 47 in Developments in petroleum science. Elsevier, 1998.

A. Datta-Gupta, K. N. Kulkarni, S. Yoon, and D. W. Vasco. Streamlines, ray tracing and production tomography: Generalization to compressible flow. *Petroleum Geoscience*, 7:75–86, 2001.

H. Dembicki Jr. and M. J. Anderson. Secondary migration of oil: Experiments supporting efficient movement of separate, buoyant oil phase along limited conduits. *AAPG Bulletin*, 73:1018–1021, 1989.

W. A. England, A. S. MacKenzie, D. M. Mann, and T. M. Quigley. The movement and entrapment of petroleum fluids in the subsurface. *Journal of the Geological Society, London*, 144:327–347, 1987.

R. A. Freeze and J. A. Cherry. *Groundwater.* Prentice Hall, 1979.

V. Frette, J. Feder, T. Jøssang, and P. Meakin. Buoyancy driven fluid migration in porous media. *Phys. Rev. Lett.*, 68:3164–3167, 1992.

W. C. Gussow. Differential entrapment of oil and gas: a fundamental principle. *AAPG Bulletin*, 5:816–853, 1954.

T. Hantschel and D. Waples. Personal communication, 2007.

T. Hantschel, A. I. Kauerauf, and B. Wygrala. Finite element analysis and ray tracing modeling of petroleum migration. *Marine and Petroleum Geology*, (17):815–820, 2000.

A. Hildenbrand, S. Schlömer, B. M. Krooss, and R. Littke. Gas breakthrough experiments on peltic rocks: comparative study with N_2, CO_2 and CH_4. *Geofluids*, 4:61–80, 2004.

A. D. Hindle. Petroleum migration pathways and charge concentration: A three–dimensional model. *AAPG Bulletin*, 81(9):1451–1481, 1997.

L. M. Hirsch and A. H. Thompson. Minimum saturations and buoyancy in secondary migration. *AAPG Bulletin*, pages 696–710, 1995.

M. K. Hubbert. Entrapment of petroleum under hydrodynamic conditions. *AAPG Bulletin*, 37(8):1954–2026, 1953.

M. A. Ibrahim, M. R. Tek, and D. L. Katz. Threshold pressure in gas storage. *Pipeline Research Committee American Gas Association at the University of Michigan, Michigan*, page 309 pp., 1970.

S. E. Ingebritsen and W. E. Sanford. *Groundwater in Geologic Processes.* Cambridge Univerity Press, 1998.

G. M. Ingram and J. L. Urai. Top–seal leakage through faults and fractures: the role of mudrock properties. Number 158 in Special Publications, pages 125–135. Geological Society, London, 1999.

R. J. Knipe. Juxtaposition and seal diagrams to help analyze fault seals in hydrocarbon reservoirs. *AAPG Bulletin*, 81:187–195, 1997.

B. Krooss. Diffusive loss of hydrocarbons through cap rocks. *Erdoel, Erdgas und Kohle*, 45:387–396, 1992.

B. Krooss, D. Leythaeuser, and R. G. Schaefer. The quantification of diffusive hydrocarbon losses through cap rocks of natural gas reservoirs - reevaluation. *AAPG Bulletin*, 76:403–406, 1992a.

B. Krooss, D. Leythaeuser, and R. G.Schaefer. The quantification of diffusive hydrocarbon losses through cap rocks of natural gas reservoirs - reevaluation. reply. *AAPG Bulletin*, 76:1842–1846, 1992b.

F. K. Lehner, D. Marsal, L. Hermans, and A. van Kuyk. A model of secondary hydrocarbon migration as a buoyancy–driven separate phase flow. In B. Doligez, editor, *Migration of Hydrocarbons in Sedimentary Basins*. Institut Français du Pétrole, Technip, 1987.

X. Luo and G. Vasseur. Contributions of compaction and aquathermal pressuring to geopressure and the influence of environmental conditions. *AAPG Bulletin*, 76(10):1550–1559, 1992.

X. R. Luo, B. Zhou, S. X. Zhao, F. Q. Zhang, and G. Vasseur. Quantitative estimates of oil losses during migration, part I: the saturation of pathways in carrier beds. *Journal of Petroleum Geology*, 30(4):375–387, 2007.

X. R. Luo, J. Z. Yan, B. Zhou, P. Hou, W. Wang, and G. Vasseur. Quantitative estimates of oil losses during migration, part II: measurement of the residual oil saturation in migration pathways. *Journal of Petroleum Geology*, 31(1): 179–190, 2008.

P. Meakin. Invasion percolation on substrates with correlated disorder. *Physica A*, 173:305–324, 1991.

P. Meakin, J. Feder, V. Frette, and T. Jøssang. Invasion percolation in a destabilizing gradient. *Phys. Rev. A*, 46:3357–3368, 1992.

P. Meakin, G. Wagner, A. Vedvik, H. Amundsen, J. Feder, and T. Jøssang. *Marine and Petroleum Geology*, 17:777–795, 2000.

B. Nickel and D. Wilkinson. Invasion percolation on the Cayley tree: Exact solution of a modified percolation model. *Phys. Rev. Lett.*, 51:71–74, 1983.

G. Å. Øye. *An Object–Oriented Parallel Implementation of Local Grid Refinement and Domain Decomposition in a Simulator for Secondary Oil Migration.* PhD thesis, University of Bergen, 1999.

D. W. Peaceman. *Fundamentals of Numerical Reservoir Simulation.* Number 6 in Developments in petroleum science. Elsevier, 1977.

W. H. Press, S. A. Teukolsky, W. T. Vetterling, and B. P. Flannery. *Numerical Recipes in C++.* Cambridge University Press, second edition, 2002.

P. S. Ringrose and P. W. M. Corbett. Controls on two–phase fluid flow in heterogeneous sandstones. In J. Parnell, editor, *Geofluids: Origin, Migration and Evolution of Fluids in Sedimentary Basins*, volume 78 of *Special Publication*, pages 141–150. Geological Society of London, 1994.

S. Schlömer and B. M. Krooss. Molecular transport of methane, ethane and nitrogen and the influence of diffusion on the chemical and isotopic composition of natural gas accumulations. *Geofluids*, 4:81–108, 2004.

S. Schlömer and B. M. Krooss. Experimental characterisation of the hydrocarbon sealing efficiency of cap rocks. *Marine and Petroleum Geology*, 14: 565–580, 1997.

T. T. Schowalter. Mechanics of secondary hydrocarbon migration and entrapment. *AAPG Bulletin*, 63:723–760, 1979.

D. Stauffer and A. Aharony. *Introduction to Percolation Theory.* Taylor & Francis, revised second edition, 1994.

A.-K. Stolz and R. M. Graves. Choosing the best integrated model for reservoir simulation. In *AAPG International Conference and Exhibition, Paris, France*, 2005.

K. Stüwe. *Geodynamics of the Lithosphere.* Springer, 2nd edition, 2007.

Ø. Sylta. Quantifying secondary migration efficiencies. *Geofluids*, (2):285–298, 2002a.

Ø. Sylta. Modeling techniques for hydrocarbon migration. In *EAGE 64'th Conference & Exhibition, Florence*, 2002b.

Ø. Sylta. *Hydrocarbon Migration Modelling and Exploration Risk.* PhD thesis, Norwegian University of Science and Technology, 2004.

Ø. Sylta. On the dynamics of capillary gas trapping: implications for the charging and leakage of gas reservoirs. In A. G. Doré and B. A. Vining, editors, *Petroleum Geology: North–West Europe and Global Perspectives — Proceedings of the 6th Petroleum Geology Conference*, pages 625–631. Petroleum Geology Conferences Ltd., Geological Society, London, 2005.

Ø. Sylta. Modelling of secondary migration and entrapment of a multicomponent hydrocarbon mixture using equation of state and ray-tracing modelling techniques. *Geological Society London*, (59):111–122, 1991.

Ø. Sylta. New techniques and their applications in the analysis of secondary migration. Basin Modelling: Advances and Applications, pages 385–398. Norwegian Petroleum Society (NPF), Special Publication No. 3, Elsevier, 1993.

D. J. Timlin, L. R. Ahuja, Ya. Pachepsky, R. D. Williams, D. Gimenez, and W. Rawls. Use of brooks–corey parameters to improve estimates of saturated conductivity from effective porosity. *Soil Sci. Soc. Am. J.*, 63:1086–1092, 1999.

B. P. Tissot and D. H. Welte. *Petroleum Formation and Occurrence.* Springer–Verlag, Berlin, second edition, 1984.

F. Vassenden, Ø. Sylta, and C. Zwach. Secondary migration in a 2D visual laboratory model. In *Proceedings of conference "Faults and Top Seals", Montpellier, France.* EAGE, 2003.

E. W. Washburn. The dynamics of capillary flow. *Phys. Rev.*, 17:273–283, 1921.

N. L. Watts. Theoretical aspects of cap–rock and fault seals for single– and two–phase hydrocarbon columns. *Marine and Petroleum Geology*, 4:274–307, 1987.

D. Wilkinson. Percolation model of immiscible displacement in the presence of buoyancy forces. *Phys. Rev. A*, 30:520–531, 1984.

D. Wilkinson. Percolation effects in immiscible displacement. *Phys. Rev. A*, 34:1380–1391, 1986.

D. Wilkinson and J. F Willemsen. Invasion percolation: A new form of percolation theory. *J. Phys. A: Math. Gen.*, 16:3365–3376, 1983.

A. Winter. Percolative aspects of hydrocarbon migration. In B. Doligez, editor, *Migration of Hydrocarbons in Sedimentary Basins.* Institut Français du Pétrole, Technip, 1987.

D. M. Wood. *Soil Behaviour and Critical State Soil Mechanics.* Cambridge University Press, 1990.

G. Yielding, B. Freeman, and D. T. Needham. Quantitative Fault Seal Prediction. *AAPG Bulletin*, 81(6):897–917, 1997.

B. Yuen, A. Siu, S. Shenawi, N. Bukhamseen, S. Lyngra, and A. Al-Turki. A new three–phase oil relative permeability simulation model tuned by experimental data. *International Petrolem Technology Conference IPTC 12227*, 2008.

M. A. Yükler, C. Cornford, and D. Welte. Simulation of geologic, hydrodynamic, and thermodynamic development of a sediment basin – a quantitative approach. In U. von Rad, W. B. F. Ryan, and al., editors, *Initial Reports of the Deep Sea Drilling Project*, pages 761–771, 1979.

7

Risk Analysis

7.1 Introduction

In previous chapters two assumptions were made about data needed for successful simulation runs. It was first proposed that necessary data is completely available and second that it is good quality. So it was implicitly concluded that each model is unique. In practice, this is usually not the case. Data sets have gaps and the data values often have wide error bars. These uncertainties lead to the following three types of questions:

1. What is the impact of uncertainties in the input data on the model?

What is the chance or the probability of having special scenarios? How large is the risk or the probability of failure? Is the simulation result stable or does a slight variation of some input parameters cause a completely different result? How sensitive is the relationship between a given parameter variation and the resulting model variation or how do the error bars of the input data map to the error bars of the results?

2. What are the important dependencies in our model?

Not every uncertainty of an input parameter has an impact on each uncertainty of the simulation result values. Which parameter influences which result? How strong are the different influences? Do they have a special form? Studying these questions is especially necessary for the understanding of the model and the processes it contains. Understanding is again necessary if conclusions are to be drawn, which go beyond a plain collection of results.

3. Which set of input data leads to agreement when considering additional comparison data?

Very often additional calibration data are available which cannot be used directly for the modeling but can be compared to simulation results. Is it possible to reduce the uncertainty in the input data by excluding models related to simulation results which are not matching the calibration data? In the literature, procedures treating this problem are often listed under the keywords "inversion" or "calibration".

T. Hantschel, A.I. Kauerauf, *Fundamentals of Basin and Petroleum Systems Modeling*, DOI 10.1007/978-3-540-72318-9_7,
© Springer-Verlag Berlin Heidelberg 2009

This chapter deals with these three topics under the headings "Risking", "Understanding", and "Calibration".

The classical approach of tackling such problems would be to perform scenario runs. Starting with a first best guess model, which is commonly named "reference model" or "master run" (Fig. 7.1), model parameters are modified manually and scenario runs are performed according to the knowledge, speculations, expectations, and understanding of the modeler (Fig. 7.2).

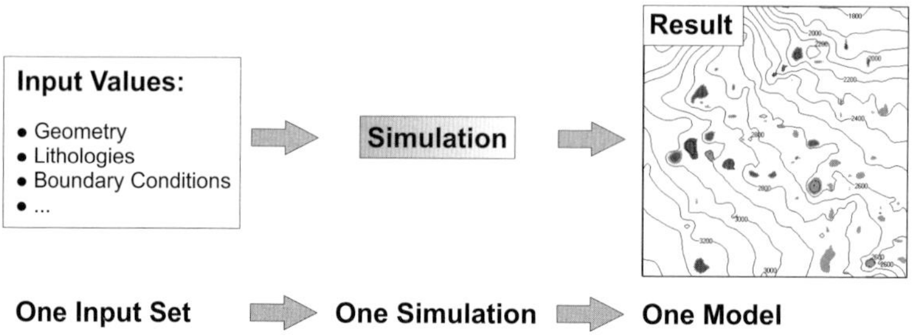

Fig. 7.1. Result of one – deterministic – 3D simulation. It is implicitly assumed, that uncertainties in input data do not exist

An example, with an uncertain temperature history caused by unknowns in heat flow and thermal conductivity could have this typical form: A high heat flow scenario and a low heat flow scenario are simulated, whereas other uncertain input parameters such as thermal conductivities are held at fixed values. The high heat flow scenario is found to be realistic by looking for example at the calibration data or through the experience of the modeler. Next, a high thermal conductivity scenario and a low conductivity scenario are modeled with a fixed high heat flow. High thermal conductivity matches the calibration data best. So, a high heat flow combined with high thermal conductivity is found to be the most realistic scenario.

Two main problems arise with such an approach:

The procedure of variation and selection of the input parameters is not systematic. There is a possibility of overseeing other realistic scenarios e.g.: in this example, low heat flow with low thermal conductivity. With a higher amount of uncertain parameters such mistakes can become normal.

The choice of "probable" scenarios is not very well quantified: A sensitivity analysis of how precise the heat flow has to be known to match the result is not performed, a quantification of the reduction of uncertainty is missing and the repercussions of the variation of other parameters, simultaneously with the heat flow, is omitted completely.

The reliability of risk results gained by scenario runs is primarily dependent on the knowledge of the involved modelers. Scenarios are usually not

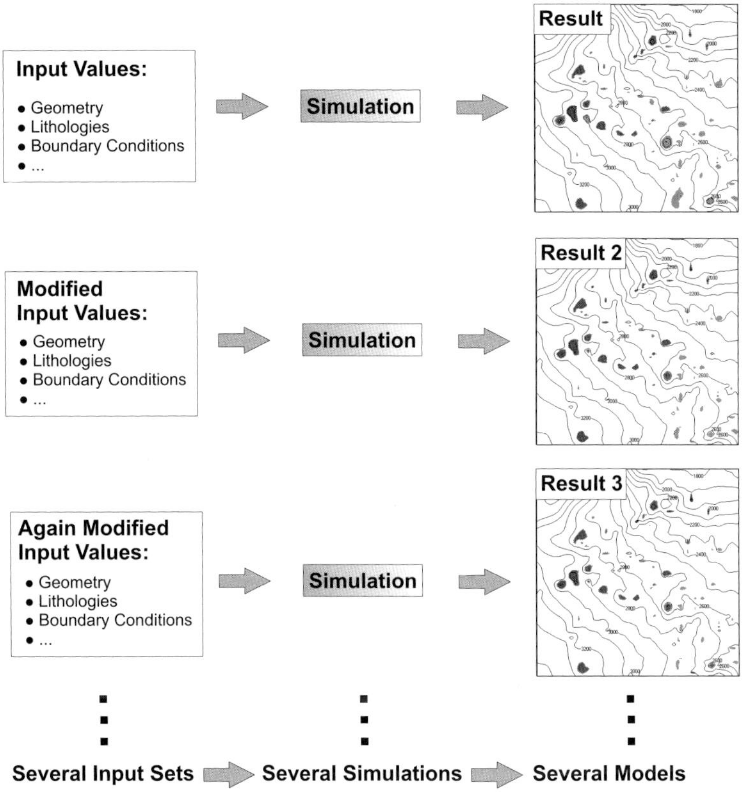

Fig. 7.2. Approach with "Scenario Runs". The first run is usually called the "Master Run" or "Reference Model"

performed systematically and the discussion of the results is qualitative but not quantitative.

The main goal of this chapter is to describe a systematic approach to deal with these issues. A more concrete formulation of the tasks involved with the three topics are:

- **Risking:** Calculation of probabilities, confidence intervals and error bars.
- **Understanding:** Calculation and analysis of correlations.
- **Calibration:** Calculation of the probability of how good a model fits calibration data and search for the best fitting model.

All topics contain words such as "probability" or "correlation" which are related to the language of stochastics and statistics. It is possible to treat all three topics simultaneously with a stochastic method such as a "Monte Carlo Simulation". This has the big advantage that expensive and time consuming simulation runs can be reused for the analysis of three distinct topics.

An introduction into probabilistic methods of applied basin modeling can be found in Thomsen (1998). Other approaches are usually less general and restricted in applicability or assumptions. Nevertheless, the efficiency can be significantly raised by studying limited problems or tasks with different methods. Highly specialized methods of inversion, for interpolation and extrapolation of simulation results are discussed in later sections.

7.2 Monte Carlo Simulation

The starting point for a Monte Carlo simulation is a reference model and a list of uncertainties belonging to the data. The reference model is based on a parameter set within the limits of these uncertainties, which typically represents a best first guess. Additionally, a quantification of the uncertainties must be known. The most precise quantification is a probability distribution (Fig. 7.3) which defines the probability of a data value to be exact.[1]

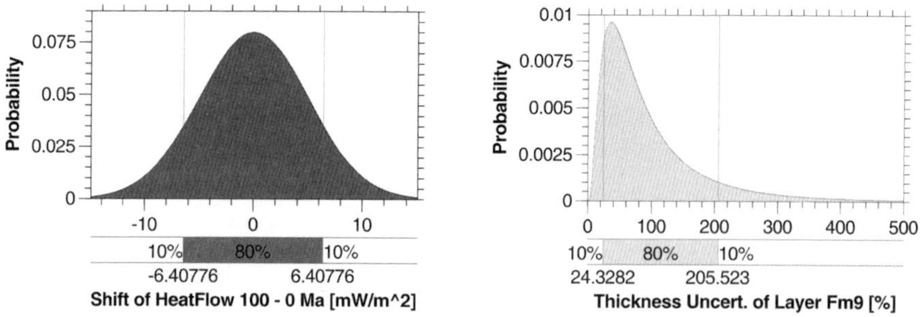

Fig. 7.3. Examples of normally and log. normally distributed uncertainties

Very often the distribution is not known but only some more general statements about the type and size of the uncertainties. It is usually not difficult and also not critical to construct a distribution from this knowledge. This is discussed in Sec. 7.2.1 and typical examples are demonstrated.

One important point, which must be mentioned, is that the uncertain model parameters should be independent. In Sec. 7.2.3 this is discussed in more detail.

With this setup the "Monte Carlo Workflow" is straightforward: A set of random numbers according to the distributions is drawn and a simulation run with this parameter set is performed. This procedure is repeated while the results are collected (Fig. 7.4). Output parameters are collected, visualized, and analyzed with statistical tools such as histograms.

[1] It more precisely defines a probability density with probabilities of values to be within certain intervals.

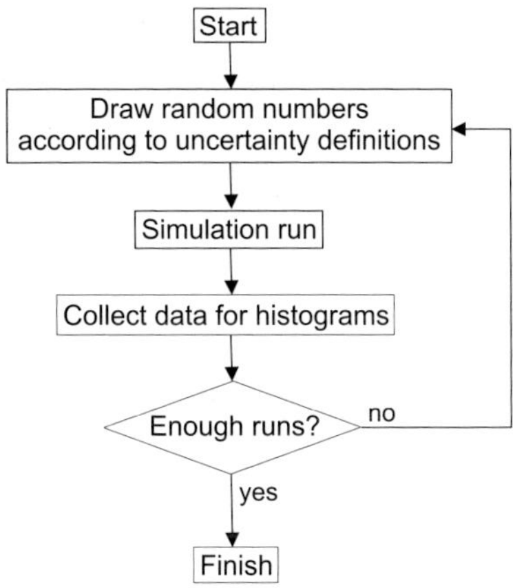

Fig. 7.4. Flow chart for Monte Carlo simulation runs

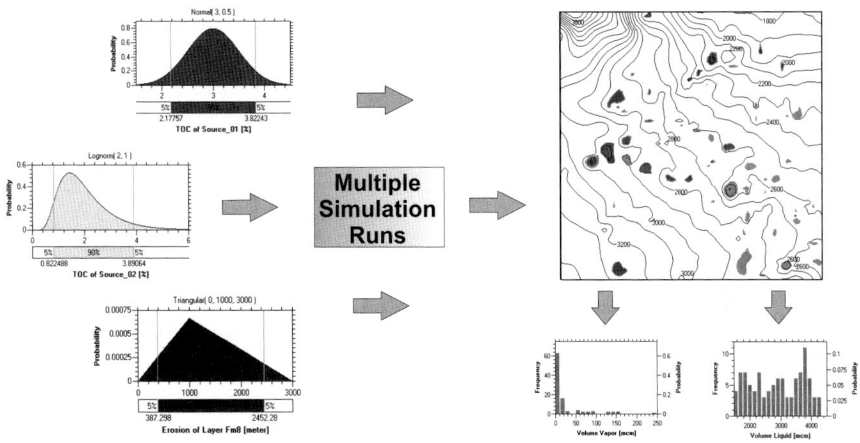

Fig. 7.5. Monte Carlo simulation with histograms of accumulated petroleum

It will now be shown that the main topics “risking”, “understanding”, and “calibration” can be solved with the Monte Carlo simulation approach:

Risking

Confidence intervals related to risking can directly be read off from result histograms. They define the probability to find a result within a given interval. E.g. it is possible to formulate statements such as *“With 80% probability the accumulated liquid petroleum amount is between 1623 and 1628 million cubic meters”* (Fig. 7.6).

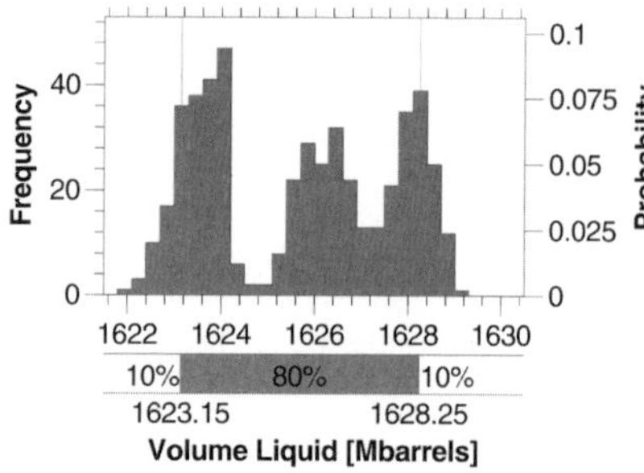

Fig. 7.6. Histogram of liquid accumulations: "With 80% probability the accumulated liquid petroleum amount is between 1623 and 1628 million cubic meters"

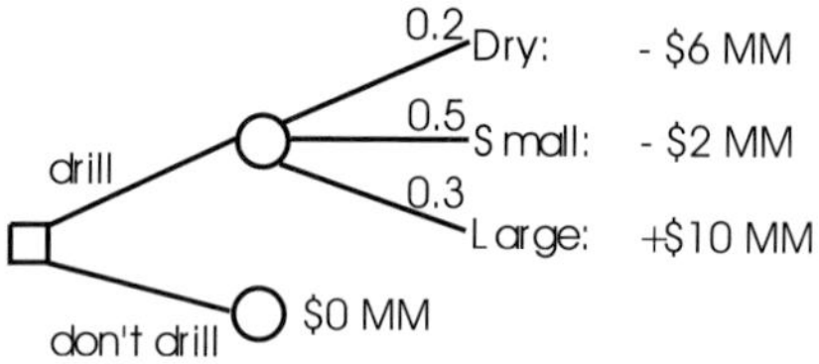

Fig. 7.7. Decision analysis with tree: The expected value EV for gains or losses of the "drill branch" is the average $EV(\text{drill}) = 0.2 \times (-6) + 0.5 \times (-2) + 0.3 \times 10 = \$\,0.8\,\text{MM}$

Other valuable characteristics of a histogram are the modus, which defines the location of the most probable result, or the average. The concept of calculating expectation values, such as the average, is extremely important in economics: For example complex decision procedures in companies are often analyzed with decision trees (Fig. 7.7). These trees are based on the evident statistical law, that the optimal decision strategy is found by following the branches with the highest expectation values, which can be calculated from averages.

A measure for the width of a histogram is the standard deviation. This quantity can be set in relation to the less precisely defined error bar. Together, average and standard deviation are often used as "value with error bar" (Fig. 7.8). Big standard deviations of resulting histograms indicate high uncertainties and give rise to the conclusion that the master run is not representative and therefore not probable.

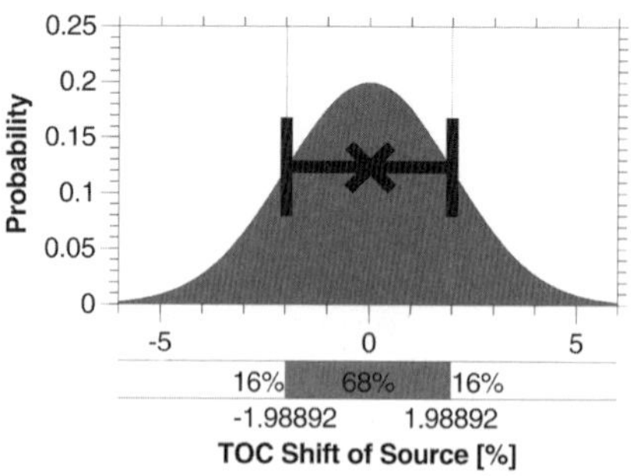

Fig. 7.8. Gauss distribution with mean $\mu = 0$ and standard deviation $\sigma = 2$. The standard deviation can be interpreted as the size of an error bar. About 68% of numbers drawn from this distribution will be inside the range of the error bar

The analysis of the result widths as functions of the uncertainty widths is called “sensitivity analysis”. More precisely, it can be represented by the relation between standard deviations of uncertainty and result parameters. A result is highly sensitive/unsensitive to an uncertainty if its error bar is large/small compared to the error bar of the uncertainty parameter. Therefore, sensitivity analysis can be a guiding tool for the understanding of a system.

Understanding

A very important problem arises with the question of where future efforts concerning the reduction of uncertainties should be spent. A reduction of uncertainties can be achieved by further data acquisition, which can be very costly. Sensitivity analysis directly leads to the parameters, which are of importance (Fig. 7.9). So, an expensive collection of unnecessary data could be avoided.

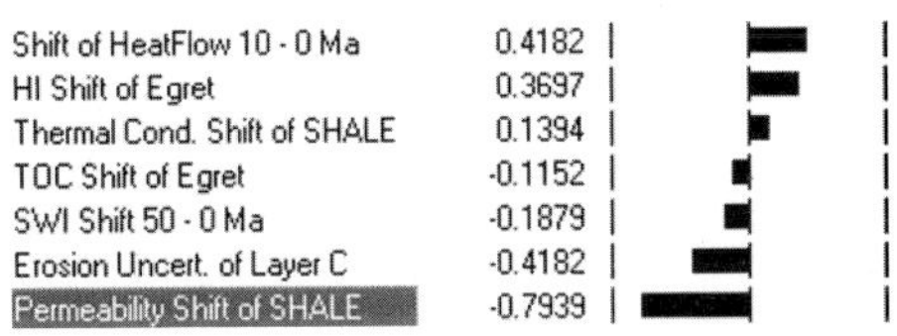

Fig. 7.9. Tornado diagram depicting the influence of some uncertainties on the porosity at a defined location in a well. Spearman rank order correlation coefficients (Press et al., 2002) are plotted as bars. As expected, the permeability shift (highlighted) (anti–)correlates mostly with the porosity

Understanding can be improved by searching for dependencies, e.g. via cross plots (Fig. 7.10). Correlations can be visualized and with the help of correlation coefficients quantified. In case of strong correlations, it is possible to interpolate the results and for forecasting purposes state formulas of dependency. For expensive simulation runs this is very valuable. Generalizations of such techniques are discussed in Sec. 7.5.

Calibration

It is obvious that calibration could be performed with Monte Carlo simulations in a simple way by just looking at the model that best fits the calibration data. The investigation of uncertainty space is performed by sampling the uncertainties according to their probability distributions. Random combinations of parameters are used for the Monte Carlo models. This method ensures a global sampling of the space of uncertainty. The risk of missing regions with good calibration, becomes small with a high number of Monte Carlo runs.

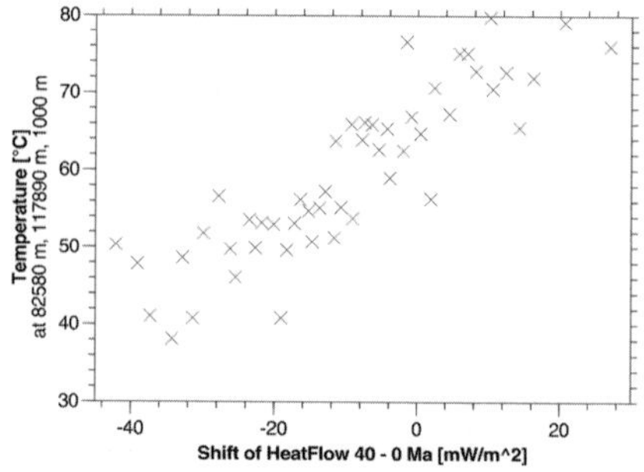

Fig. 7.10. Cross plot of temperature against heat flow shift. A correlation is visible and a linear interpolation might be performed

The Monte Carlo method is not intelligent in a way that it searches for models with good calibration. Search algorithms would be more efficient but obviously cannot be performed simultaneously in combination with risking and understanding. Additionally, these algorithms often search in local regions of space so it could be that they end up with an erroneous calibration. Therefore, a global investigation of the uncertainty space such as with a Monte Carlo analysis has to be performed as a first step before beginning with a search algorithm. Global stability and the prevention of extra simulation runs are often of higher importance compared to high–quality calibration. However, in Sec. 7.5 more sophisticated calibration methods, which combine the advantages of both approaches are discussed.

7.2.1 Uncertainty Distributions

Uncertainty distributions must be specified for Monte Carlo simulations. The properties of some well known distributions are now discussed with regard to their usage in Monte Carlo simulations.

Normal Distribution

The normal or Gauss distribution

$$p(x) = \frac{1}{\sigma\sqrt{2\pi}} \exp\left[\frac{-(x-\mu)^2}{2\sigma^2}\right] \tag{7.1}$$

with mean μ and standard deviation σ is the most widely used distribution in science (Fig. 7.3). Assume that a quantity X is measured independently N times with the values $x_1 \ldots x_N$. Following the central limit theorem of statistics, the average

$$x = \frac{1}{N}\sum_{i=1}^{N} x_i \tag{7.2}$$

is Gauss distributed for $N \to \infty$ with[2]

[2] In practice $N > 7$ is enough for high numerical accuracy.

$$\mu = \frac{1}{N}\sum_{i=1}^{N}\mu_i \qquad \text{and} \qquad \sigma^2 = \frac{1}{N}\sum_{i=1}^{N}\sigma_i^2 \,. \tag{7.3}$$

Here, μ_i and σ_i are the means and the standard deviations of the probability distributions for each measurement i. They are often the same for all i. Parameters which are used in large scale basin models are often provided as upscaled averages of higher resolution data or averages of multiple measurements. Assuming independency, it is often possible to assign a Gauss distributed uncertainty to such a parameter.

Logarithmic Normal Distribution

This distribution is also called lognormal or lognorm distribution and has the form

$$p(x) = \frac{1}{\sigma x\sqrt{2\pi}} \exp\left[-\frac{(\ln x - \mu)^2}{2\sigma^2}\right] \quad \text{for} \quad x > 0 \qquad \text{and} \qquad p(x) = 0 \quad \text{else} \,, \tag{7.4}$$

compare with Fig. 7.3. It has similar properties to the normal distribution. If a quantity Y is normally distributed then $X = \exp Y$ is lognormally distributed. The central limit theorem for the arithmetic average of some Y_i becomes a geometric average for the related X_i namely

$$x = \prod_{i=1}^{N} x_i^{1/N} \,. \tag{7.5}$$

The equations for μ and σ stay the same as in (7.3) but it should be remembered that μ and σ are not the mean and standard deviation of the lognormal distribution, they are only the mean and standard deviation of the related normal distribution.

A lognormal distribution is of special interest to "scale quantities", which by definition cannot be negative. The logarithm of a scale quantity can be calculated every time and the distribution is zero for negative values. Many physical quantities especially material properties such as thermal conductivities are limited to positive values. And the calculation of averages e.g. upscaling, is often performed with geometrical averaging (Chap. 8). The lognorm distribution can be a proper choice for the description of uncertainties related to such quantities.

Uniform Distribution

The uniform distribution (Fig. 7.11) is defined by

$$p(x) = \frac{1}{b-a} \quad \text{for} \quad a \le x \le b \qquad \text{and} \qquad p(x) = 0 \quad \text{else} \tag{7.6}$$

with $a < b$. The uniform distribution is a good choice if nothing except some limiting statements can be made about the form of the uncertainty. The two discontinuities of the distribution are often the subject of criticism: They have sharp edges and are therefore argued to be in contradiction to the assumption of ignorance about the "tails" of the distribution. Additionally, it often seems unreasonable that the central inner parts of an uncertainty have the same probabilistic weight as the more outer parts.

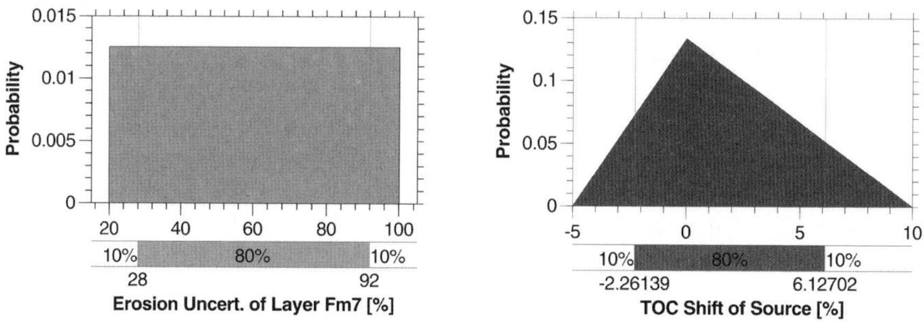

Fig. 7.11. Examples of uniform and triangular distributions

Triangular Distribution

The triangular distribution (Fig. 7.11) does not have the principal problems, which come with the uniform distribution. It is given by

$$p(x) = \begin{cases} \dfrac{2(x-a)}{(c-a)(b-a)} & \text{for} \quad a < x \leq b, \\ \dfrac{2(c-x)}{(c-a)(c-b)} & \text{for} \quad b < x < c, \\ 0 & \text{else} \end{cases} \tag{7.7}$$

with $a < b < c$. The distribution is zero at a and c and its median is located at b. A triangular uncertainty distribution can be directly constructed when the uncertainty limits and also the most probable value are known. It can also be used as an "easy to use" approximation to normal and lognorm distributions (Lerche, 1997; Thomsen, 1998).

Other distributions, such as exponential or beta distributions, are more sophisticated alternatives to uncertainty descriptions (Figs. 7.12, 4.5, Rinne 1997). They are only used under special circumstances.

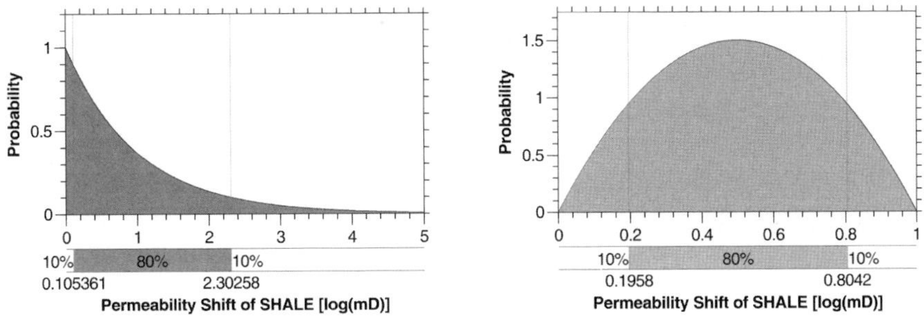

Fig. 7.12. Examples of exponential and beta distributed uncertainties

Nominal Distributions

Sometimes it is necessary to assign uncertainties to discrete parameters (Fig. 7.13). In the most general case these parameters are without order relation, which signifies that there are no "less than" or "bigger than" defined between them. Then they are called nominal parameters. Typical examples are lithologies or kinetic type assignments as lithologies and kinetics are usually specified by a large number of parameters. Therefore uncertain nominal parameters often imply strong result variations.

An interpretation of results derived from nominal uncertainties can be difficult especially if nominal and continuous uncertainties are mixed in the same sequence of risk runs. This should therefore be avoided.

Substit. of Lithology SANDSTONE_04

ID	Name	Prob. [%]	0 % 100 %
108	SANDSTONE_04	70.00	
42	SAND&SILT	15.00	
43	SAND&SHALE	15.00	

Fig. 7.13. Discrete distributed lithologies

7.2.2 Derived Uncertainty Parameters

An uncertainty is described as a distribution of one number, e.g. the thermal conductivity of shale. But very often it is associated with more than only one number. For example a heat flow uncertainty is related to the complete basal heat flow, which is space and time dependent and cannot be described with one number only.

However, the basal heat flow of the master run can be shifted, tilted, twisted, etc.. A restriction to special forms of variation which can be described with one number only allows the assignment of an uncertainty to this "derived parameter".

It is well known from mathematics that arbitrary variations can often be decomposed into infinite series of orthogonal functions which would yield infinite uncertainty parameters. In practice, one is thus restricted to the most important variations, which are often defined by the first terms of such series.

The simplest form of basal heat flow variations is a value shift in a defined time interval (Fig. 7.14). Therefore, it is possible to assign an uncertainty distribution to a shift of the whole basal heat flow.

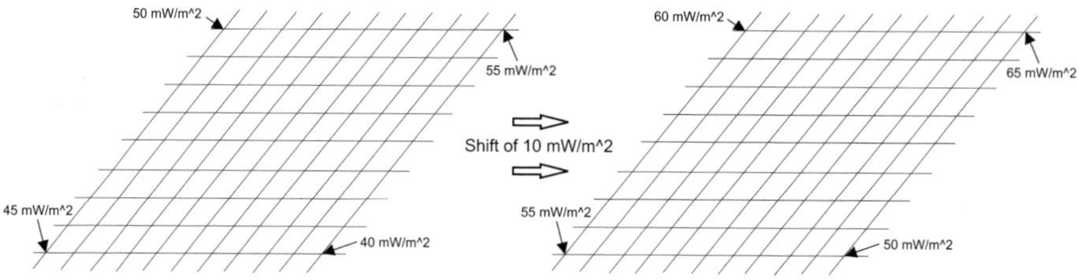

Fig. 7.14. Shift of (gridded) basal heat flow map

Complex structural uncertainties can easily be risked with the prize of restriction to special forms of variation (Fig. 3.36).

7.2.3 Latin Hypercube Sampling (LHC)

Arbitrary random sampling of the uncertainty distributions has some drawbacks. Clusters of drawn numbers can occur (Fig. 7.15) and low probability tails of distributions are often not sampled, although they might contribute significantly to the analysis (e.g. calculation of moments) especially if they have a wide range. The statistics (e.g. estimating a mean with the average over a set of random numbers) becomes increasingly better, as less clusters exist and the smoother the sampling is.

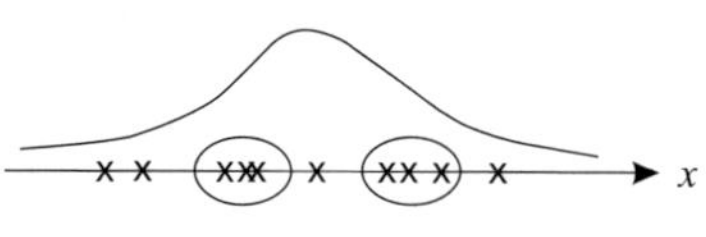

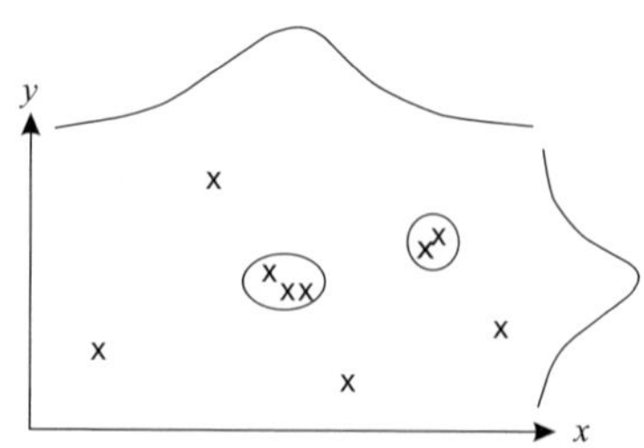

Fig. 7.15. Clustering of random numbers in one and two dimensions. The variables x and y are uncertainty parameters e.g. a heat flow and a SWI temperature shift

Latin hypercube sampling is a technique which helps to avoid clustering and samples low probability tails without affecting basic statistics. It consists of primarily two parts: The first part is an improved drawing algorithm in the series drawn from one distribution. The second part refers to the "hypercube" and deals with multiple drawings in the multi–dimensional uncertainty "hyperspace".

Latin Hypercube Sampling in One Dimension

The interval within which the uncertainty parameter is defined can be divided into intervals of the same cumulative probability which are called "strips" (Fig. 7.16). Drawing a random number is now performed in two steps: First, a strip is selected. Then, a random number is drawn according to the probability distribution in this strip (Fig. 7.17). It is not allowed to use a strip again until all others have been selected for drawing. The best efficiency is obviously achieved if the number of drawings equals the number of strips or is a small integer multiple.

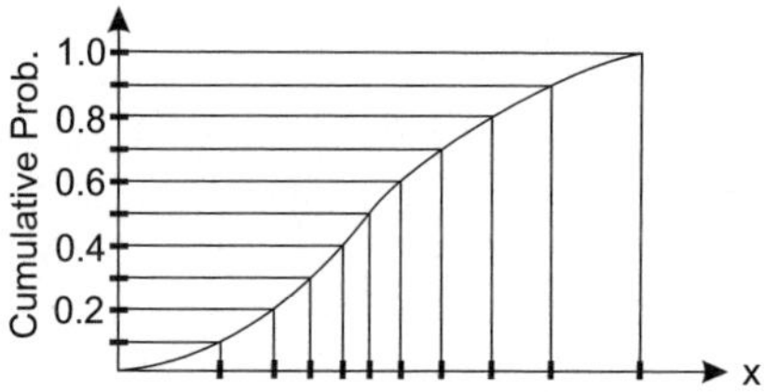

Fig. 7.16. Segmentation into ten equal probable intervals

LHC sampling is three times more efficient for the calculation of basic statistical quantities such as means or confidence intervals (Newendorp and Schuyler, 2000). When considering the huge size of simulation efforts for big basin models, this is a good deal for the price.

On the other hand it is easy to see, that this method does not reproduce auto–correlations between successive drawings. In practical implementations, the additional effort for the calculation of the strips and for the bookkeeping of used and unused strips has also to be taken into account.

Latin Hypercube Sampling in Multiple Dimensions

In more than one dimension, each distribution is segmented into the same number of strips. When drawing the random numbers, it is necessary to avoid correlations between the selection of the strips. Therefore the strips must be selected randomly too.

The final result is a subdivision of uncertainty space into equal probable hypercubes (Fig. 7.17). In the case that the number of drawings equals the number of strips, each cube contains only one drawn number at maximum.

In two dimensions each column or row contains one drawn number and in N dimensions each $N-1$ subspace contains exactly one number.

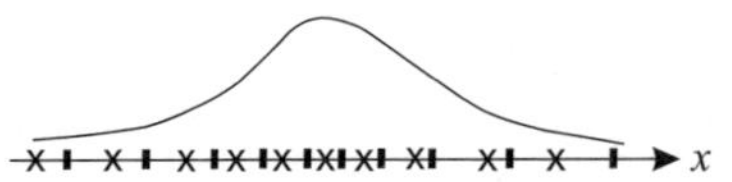

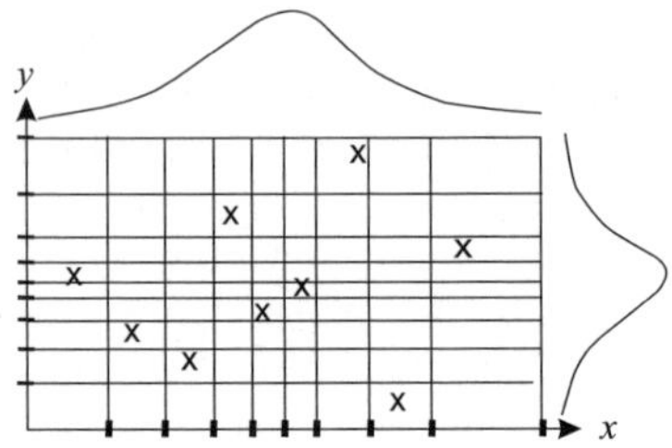

Fig. 7.17. Latin hypercube segmentation with random numbers in one and two dimensions. Compare with Fig. 7.15

Again, bookkeeping of strips has to be performed but the advantages are the same as in the one–dimensional case. LHC is a very efficient method for global sampling of the uncertainty space.

7.2.4 Uncertainty Correlations

Up to now, it was assumed that the uncertainty parameters were independent. The opposite of independence is dependency. This does not need to be further discussed, as a dependent uncertainty parameter can obviously be eliminated from the list of uncertainties and treated like a simulation result. Besides these two extremes, the region of correlation exists where specified combinations of the drawn numbers are favored above others.

An example could be the thermal conductivity of two layers which are known to have similar lithologies but it is not known what they are. So, for heat flow analysis a modeler would prefer to study combinations of similar conductivities.

A complete joint probability distribution, which defines the probability for all combinations of all values of the uncertainty parameters, would be the most thorough description. However, data and theoretical foundations of certain joint probability distribution forms usually do not exist. In practice it is sufficient to deal with correlation coefficients which are used to link marginal distributions. The rest of the joint probability distribution remains unspecified.

Nevertheless, drawing random numbers of correlated distributions is problematic enough. Explicit formulas exist for correlated Gauss distributions. The simplest case are two correlated Gauss distributions which have the following form (Beyer et al., 1999)

$$p(\mathbf{x}) = \frac{1}{2\pi\sqrt{|\Sigma|}} e^{-\frac{1}{2}\mathbf{x}^T \Sigma^{-1} \mathbf{x}} \,. \tag{7.8}$$

with two variables $\mathbf{x}^T = (x_1, x_2)$. The correlation is defined by the covariance matrix

$$\Sigma = \begin{pmatrix} \sigma_1^2 & \rho\sigma_1\sigma_2 \\ \rho\sigma_1\sigma_2 & \sigma_2^2 \end{pmatrix} \tag{7.9}$$

which is symmetric and positive definite (Fahrmeir and Hamerle, 1984). Here $\sigma_i =< x_i^2 >$ are the variances and $\rho =< x_1x_2 > /\sigma_1\sigma_2$ is the correlation coefficient with $-1 \leq \rho \leq 1$. Without loss of generality $\sigma_1 = \sigma_2 = 1$ is further assumed. The correlation matrix can be Cholesky decomposed (Beyer et al., 1999; Press et al., 2002) into

$$\Sigma = A^T A \qquad \text{with} \qquad A = \begin{pmatrix} 1 & \rho \\ 0 & \sqrt{1-\rho^2} \end{pmatrix} . \tag{7.10}$$

Thus $\mathbf{x}^* = A\mathbf{x} = (x_1 + \rho x_2, \sqrt{1-\rho^2}x_2)^T$ is Gauss distributed without correlation. Obviously, this can easily be generalized to higher dimensions.

Correlations of arbitrary marginal distributions can be forced with numerical methods. At least three different algorithms are known to exist (Miller, 1998). In the case of many distributions with many correlations, these algorithms become computationally very expensive. Especially, if one parameter is correlated multiple times with other parameters, these methods are not affordable anymore. Another disadvantage of these algorithms is their incompatibility with latin hypercube sampling. Abdication of LHC sampling reduces the performance significantly.

A new approximative method to get correlated random numbers is now described: All random numbers can be drawn before performing the risk runs if the total number of simulation runs is known at the beginning and correlations are ignored. These random numbers can be sorted afterwards with the following algorithm: An uncertainty parameter is randomly selected and after that two random numbers out of its sequence are randomly selected again. These numbers are swapped, if the resulting covariance matrix approximates the target covariance matrix more closely than before. This procedure can be repeated until a high degree of accuracy is reached. The sum of the squared deviations of all correlation coefficients can be taken as a measure for the total deviation.

It is clear that this permutation procedure certainly does not lead to a sufficient approximation and never to the exact reproduction of all correlation coefficients if the number of simulations is small.[3] But experience has shown that an almost exact numerical agreement can be reached very fast on modern computers in practical relevant examples. Even for about twenty runs with a few correlated parameters a good numerical agreement could be achieved.

[3] An extension of this method with some acceptance/rejection probability of a number swap would transfer it into a "Markov chain Monte Carlo" (MCMC) algorithm. It can be proved that such an algorithm finds the optimal approximation over longer time intervals. Due to rejection the MCMC algorithm shows generally a poorer performance. By experience, the authors found the MCMC algorithm here not necessary. MCMC algorithms in general will be discussed in more detail in Sec. 7.5.

Another big advantage of this procedure is its compatibility with latin hypercube sampling. Hypercubes and therefore higher performance coming from a lower number of necessary risk runs can be conserved.

An example of a correlation matrix constructed with this permutation method is given here: the matrix which links some marginal probability distributions of uniform, triangular, normal, and lognorm form is defined as

$$\begin{pmatrix} 1 & & & & & & & \\ 0 & 1 & & & & & & \\ 0 & 0.2 & 1 & & & & & \\ 0 & 0.3 & 0 & 1 & & & & \\ -0.1 & 0 & 0 & 0 & 1 & & & \\ -0.7 & 0 & -0.1 & 0.1 & 0 & 1 & & \\ 0 & -0.2 & 0 & 0 & 0 & 0 & 1 & \\ -0.6 & 0.2 & 0 & 0 & 0 & 0 & -0.5 & 1 \end{pmatrix}.$$

Because of symmetry, only the lower triangular part of the matrix is shown here. For 20 random numbers, which correspond to 20 simulation runs, the following approximation could be achieved with a maximum deviation of 0.0321 of any correlation value

$$\begin{pmatrix} 1 & & & & & & & \\ -0.0085 & 1 & & & & & & \\ -0.0092 & 0.2088 & 1 & & & & & \\ 0.0131 & 0.3004 & -0.0008 & 1 & & & & \\ -0.0978 & -0.0060 & -0.0005 & 0.0057 & 1 & & & \\ -0.6705 & -0.0063 & -0.0973 & 0.1091 & 0.0006 & 1 & & \\ 0.0178 & -0.1956 & 0.0008 & -0.0033 & -0.0015 & 0.0172 & 1 & \\ -0.5679 & 0.1957 & -0.0102 & -0.0078 & 0.0024 & 0.0310 & -0.4865 & 1 \end{pmatrix}$$

and for 100 runs

$$\begin{pmatrix} 1 & & & & & & & \\ -0.0009 & 1 & & & & & & \\ 0.0004 & 0.2000 & 1 & & & & & \\ 0.0000 & 0.2998 & 0.0002 & 1 & & & & \\ -0.1000 & -0.0002 & 0.0002 & 0.0005 & 1 & & & \\ -0.6938 & 0.0000 & -0.0998 & 0.0999 & -0.0002 & 1 & & \\ 0.0031 & -0.1997 & 0.0001 & 0.0000 & 0.0016 & -0.0023 & 1 & \\ -0.5928 & 0.1992 & 0.0005 & 0.0000 & 0.0011 & 0.0051 & -0.4965 & 1 \end{pmatrix}$$

with a maximum deviation of 0.0072. The maximum relative error of any correlation coefficient with an absolute value bigger than 0.1 is then less than 2%. This error is usually far beyond the accuracy of the knowledge of the correlation coefficients.

7.2.5 Analysis of Results

Many good textbooks are available about probability theory and statistics, e.g. Beyer et al. (1999) or Spiegel and Stephens (1999). Due to some specific problems associated with basin modeling some subjects are reviewed here.

An introduction to statistical analysis has already been given in Sec. 7.2. Histograms and cross plots are used to visualize the data while numbers, such as average values, can be calculated for further analysis. Statements can be quantified, e.g. with percentiles. A histogram is a binned approximation of a probability distribution. The width of a bin should neither be too narrow nor too wide because the visualization would become meaningless or grouping errors would become an issue (Spiegel and Stephens, 1999). Especially the calculation of percentiles for risking directly from histograms yields only "gridded values" (Fig. 7.18). In this case it is often better to use a linear interpolated form of cumulative probability. It is more precise for the calculation (e.g. percentiles) and has a smoother visualization (Fig. 7.18) but the data have to be available in raw form. This corresponds to a higher allocation of resources in a computer implementation.

Nevertheless, the binning width of a well sampled histogram indicates the statistical error of extracted quantities such as mean values or percentiles. Analysis (Fig. 7.18) shows that at least about 100 data values and therefore 100 risk runs are necessary for statistics with an acceptable relative error of around a few percent.

Fig. 7.18. Histogram with cumulative frequency of 100 drawn temperature values on the left and linear interpolated cumulative probability based on the same data on the right

Correlation coefficients are usually calculated for the analysis of possible dependencies. One should not forget that correlation is necessary but not sufficient for dependency. So, finding dependencies is not only part of analysis but also interpretation. Unfortunately, dependencies can have a variety of forms. Standard approaches of statistics test only for special forms.

Most commonly used is the Pearson correlation coefficient. It is a measure for the deviation of a cross plot from a straight line. If its correlation value is 1 then the cross plot fits perfectly to a straight line, which is positively inclined, and if it is -1 to a line, which is negatively inclined. If the correlation is 0 then there is no similarity to a straight line at all. Intermediate values indicate an approach to a straight line which becomes better with increasing absolute values (Fig. 7.19).

A straight line is the most important form of dependency but it is also a strong restriction to a special form. The Spearman rank order correlation coefficient is more general. It models the order of data points and is a measure of the deviation to an arbitrary monotonic increasing or decreasing correlation. Even a little more general but nearly the same is "Kendall's tau". It relies more on relative ordering and less on ranks. Some example cross plots with different correlation coefficients are shown in Fig. 7.19.

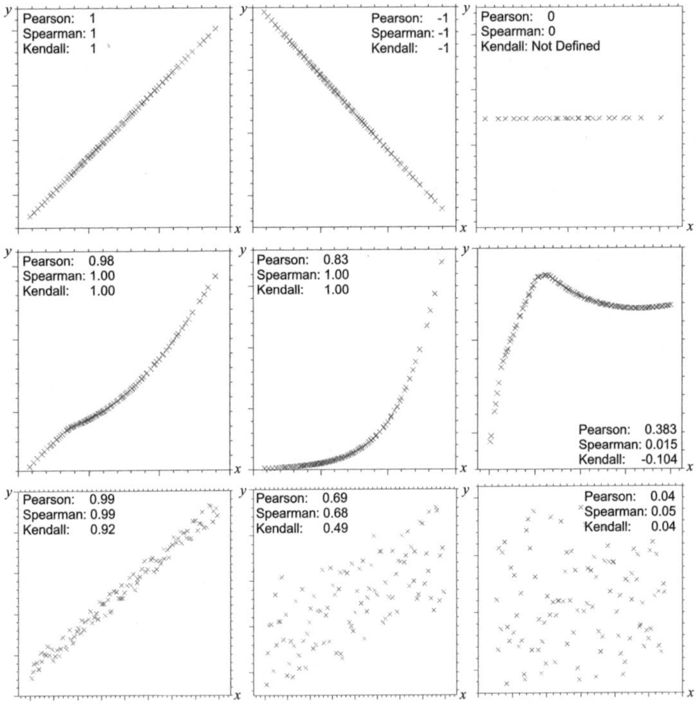

Fig. 7.19. Some examples of cross plots and their correlation coefficients

Spearman's rank order coefficient ranges also from -1 to 1 but it is overall known to be more robust than Pearson's correlation coefficient (Press et al., 2002). Commonly, it is used for tornado diagrams where lists of correlations are ranked and visualized (Fig. 7.9).

The existence of correlation can generally not be determined with a correlation coefficient alone. It only describes the strength of the correlation of a specific data set. For example a small dataset can be randomly correlated. The most extreme case are two points which fall on a straight line every time. But fortunately it is often possible to estimate significance levels for the existence of correlation. For a more detailed discussion see Press et al. (2002).

Nominal distributions must be treated differently. Because of missing order relations a correlation cannot be defined properly anymore. Instead, associations are calculated. Typical association values are Cramer's V or the contingency coefficient C. Their interpretation can be complicated. Alternative measures of association exist and are based on entropy (Press et al., 2002).

The values of association reach from 0 to 1 from no– to full– association. They cannot be negative just like their continuous counterparts, which must be kept in mind when plotted e.g. in tornado diagrams. Similar to continuous correlations, significance levels can also be estimated for associations.

If a uncertain nominal parameter is associated with a result distribution of the corresponding model, the result distribution must obviously be available in discretized, e.g. binned, form.

7.2.6 Model Data

Basin models are usually very large in size and contain a vast amount of data. It is not possible to store all the data of each Monte Carlo run completely. Only selected and restricted amounts of data can be handled and therefore not all statistical methods can be used for analysis every time.

The calculation of the average is exceptional. It can be obtained just by adding the results of each run and finally, after the last run, by dividing through the number of runs. Hence a storage of all results of all runs can be avoided. Variance can be treated in a similar way. More precisely, average and variance of a quantity need only resources of the same size necessary for the storage of the corresponding result values of two simulation runs. It is therefore possible to store them for all quantities of interest even for grid based spatial overlays on huge three dimensional models.

Sophisticated statistical analysis can only be performed if the complete data sets or at least histograms are available. They are usually collected only at some special points or for quantities of special interest. In basin modeling it is common to collect all the results of the different simulation runs at all well locations with logging information because calibration against these measurements might be performed. Additionally, the sizes of petroleum accumulations and column heights as primary targets of petroleum systems modeling are tracked over all risk runs.

It is common to define "risk points" which are special points of interest, where additionally all the data of all runs is collected. These are usually points in source rocks which are of interest for maturation and expulsion timing or points located at faults which can e.g. be important for petroleum migration.

If the spatial density of these points, with full risk data, is high and the intermediate behavior of the fields belonging to the stored values smooth, then it is possible to interpolate the data through space for full reconstruction of each risk run. Especially, in combination with predictive methods for risking, which are discussed later in Sec. 7.5, these methods can be used to estimate full modeling results without performing the accordant expensive risk run. Such forecasted data sets can be generated very fast and used for further more sophisticated statistical analyses.

Besides the risk points it is also common to explicitly track hydrocarbon mass amounts related to layers, facies, faults or individual structures such as introduced in Sec. 6.10. Especially the characteristic HC masses of a petroleum system are subjected to statistical analysis (Sec. 6.10.2). Analyses of individual reservoir structures and accumulations raise similar problems for identification and tracking of a structure in different risk runs as in different events (Sec. 6.10.3). It can be solved in the same manner as in Sec. 6.10.3 just by treating a risk run similarly as a paleo event. Again, the same structure and accumulation tracking problems arise but are assumed to decrease with an increasing number of grouped drainage areas.

7.3 Bayesian Approach

Calibration can be non–unique or numerically unstable dependent on the available data. A bayesian approach for generalized calibration is presented in this section. It can be read almost independently from the rest of this chapter and can also be skipped if calibration topics are not of special interest.

In the following, it is assumed that N calibration data values $\mathbf{d}^T = (d_1, \ldots, d_N)$ are available. They are measured values and have an error, so they can be described by $d_i \pm \Delta_i$. Further, it is assumed that M uncertainties x_k exist. Performing a simulation run with fixed values $\mathbf{x}^T = (x_1, \ldots, x_M)$ yields a model with simulation results $f_i(\mathbf{x})$ as model data, which can be compared to the calibration data.

In arbitrary calibrations it is possible to calculate the probability of how calibration data fit a given model. Under the assumption of small error bars and a statistical independency of measurements belonging to the data points, it is postulated that the measurement values are normally distributed. So, the probability of how well a model fits the calibration data is given by

$$p(\mathbf{d}|\mathbf{x}) \propto \prod_{i=1}^{N} \exp\left[-\frac{1}{2}\left(\frac{d_i - f_i(\mathbf{x})}{\Delta_i}\right)^2\right] . \tag{7.11}$$

Calibration does now imply a search of the values $\mathbf{x}$, which fit the calibration data best. An obviously good criterion for the best fit is looking for the highest probability, which is called the "Maximum Likelihood" method in statistics (Beyer et al., 1999). Caused by the minus sign in the exponent of (7.11) it is equivalent to the search for the minimum of

$$\chi^2 = \sum_{i=1}^{N} \left(\frac{d_i - f_i(\mathbf{x})}{\Delta_i} \right)^2 \tag{7.12}$$

which is the classic chi–square formula for fitting models to data. Its interpretation is easy and can be visualized for simple cases: Equation (7.12) becomes a linear regression of a straight line with the assumption of $M = 2$, a simple "simulator" $f_i = x_1 r_i + x_2$ and data values measured at some locations r namely $d_i = d_{r_i}$ (Fig. 7.20).

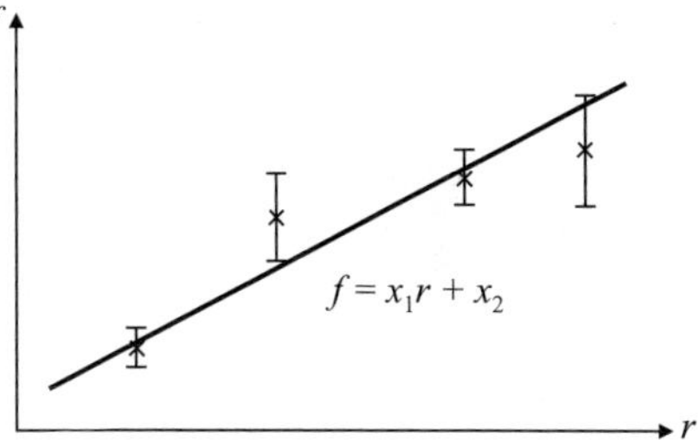

Fig. 7.20. Regression for the fit of a straight line through a "cloud" of measurement data. The inverse of the error bar size determines the weight of each point

Some problems can arise with (7.12) in practice. Often it occurs that a calibration is not unique (Fig. 7.21). In such a case the practitioner would choose a value somewhere out of the middle or at the highest probability of the according uncertainty distribution of the heat flow.

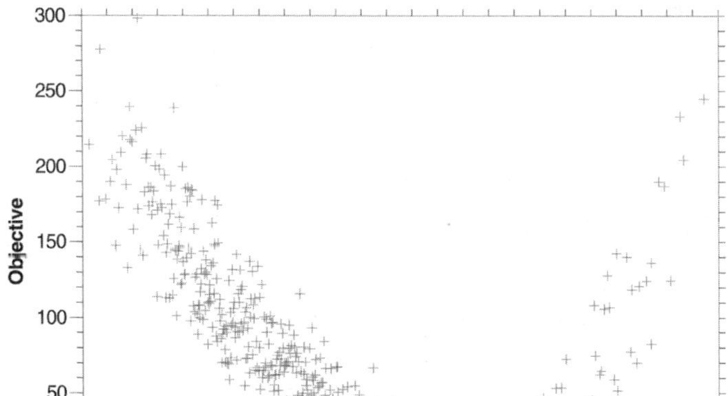

Fig. 7.21. Example with $M = 5$, $N = 19$, and objective χ^2 plotted against heat flow shift, which is known to be the most sensitive parameter. The calibration of heat flow is not unique in the range of $-5 \ldots 3\,\mathrm{mW/m^2}$

Even worse, occasionally a calibration is numerically unstable or yields completely unrealistic results caused by insufficient data points combined with some outliers. Such awkward effects can already be found in mathematically very simple situations: Without loss of generality $\sigma_i = 1$ is assumed for the following discussion. Further a simple "linear simulator" with

$$f_i = \sum_{j=1}^{M} R_{ij} x_j \qquad \text{or} \qquad \mathbf{f} = \mathbf{R} \cdot \mathbf{x} \tag{7.13}$$

is studied. The matrix **R** describes this simple "linear simulator". It is easy to show that minimizing (7.12) directly leads to

$$\mathbf{R}\mathbf{x} = \mathbf{d} \tag{7.14}$$

which is a set of linear algebraic equations with **x** as unknowns. This leads directly to the following statements:

1. The inverse R^{-1} does not exist in general, especially if $M > N$ which denotes that a unique calibration is not possible.
2. If R^{-1} exists it could be numerically unstable (Press et al., 2002).
3. If a solution is found, it is not ensured to be physically or geologically meaningful.

The first statement means that calibration data could be insufficient for calibration. For example present day temperature data alone is never sufficient for paleo–heat flow calibration. The second statement expresses that calibration data might be inconsistent leading to possibly different calibration scenarios and the third states that calibration might be optimal outside of the allowed parameter range of the model, e.g. a negative thermal conductivity.

The problem now, is how to get rid of these possibly awkward calibration behaviors and introduce a method which is at least as good as the workflow of the practitioner.

A possible solution could be a so called "Singular Value Decomposition" of the matrix R (Press et al., 2002). This is a projection on parameters which can be calibrated numerically stable with the available data. The rest of the uncertainty parameters are ignored. This method has two drawbacks: First of all it is only well defined for linear problems such as the "linear simulator". A generalization to non–linear problems would be very complicated if possible at all. Second, there is still a problem with the parameters which cannot be calibrated. Which value should they have?

Regularization is another attempt which can be tried. Instead of minimizing χ^2 it is proposed to minimize

$$\chi^2 + \lambda \left(\mathbf{x}^T \mathbf{x} \right) \tag{7.15}$$

with a number λ which has to be selected properly. It is easy to see that at least the first and second statement are solved with this method because (7.14) changes to

$$\left(\mathbf{R}^T\mathbf{R} + \lambda\mathbf{1}\right)\mathbf{x} = \mathbf{R}^T\mathbf{d}$$

which are regularized normal equations of (7.14).[4] But which value should be taken for λ? A few ideas can be found in (Press et al., 2002) but in general the problem remains unsolved.

All three problems are of principle nature and it is necessary to go back to the basics of the probability definition (7.11). It is written in conditional form stating a probability of calibration data fitting a given model. Instead it is possible to evaluate the probability of models fitting given calibration data following Bayes law

$$p(\mathbf{x}|\mathbf{d}) = p(\mathbf{x})\frac{p(\mathbf{d}|\mathbf{x})}{p(\mathbf{d})} \tag{7.16}$$

which leads down to the roots of probability theory and logic (Jaynes, 2003; Robert, 2001). The term on the left side is called the "posterior", the classical probability (7.11) is situated in the nominator and called the "likelihood" and the first term $p(\mathbf{x})$ on the right side the "prior". The term in the denominator does not play a central role, it is for normalization only.

It is possible to evaluate (7.16) similar to (7.11) under the assumption that all distributions have Gaussian form. One yields a minimization rule for the objective function $\varPhi$ with

$$\varPhi = \sum_{i=1}^{N}\left(\frac{d_i - f_i(\mathbf{x})}{\varDelta_i}\right)^2 + \sum_{i=1}^{M}\left(\frac{x_i - \mu_i}{\sigma_i}\right)^2 \tag{7.17}$$

where μ_i and σ_i are means and standard deviations of the uncertainty distributions. The first term on the right side is the "classical" χ^2 followed by an additional term. It is derived from the uncertainty distributions and implies that the knowledge for the definition of their shape has the same value as the knowledge about error bars of calibration data and should be taken into account with the same weight for calibration. The knowledge entering the definition of the uncertainty distributions is therefore called "prior information" and the distributions often just "priors".

In case of the linear simulator, the objective

$$\varPhi = \sum_{i=1}^{N}\left(\frac{d_i - \sum_{j=1}^{M} R_{ij}x_j}{\varDelta_i}\right)^2 + \sum_{i=1}^{M}\left(\frac{x_i - \mu_i}{\sigma_i}\right)^2$$

must be minimized. This formula has basically the same form as (7.15). In case of $\mu_i = 0$ and $\sigma_i = \sigma$ for all i they are the same with $\lambda = 1/\sigma^2$.[5] It is

[4] This is only half of the truth because it is known that normal equations usually have worse numerical properties than their "non–normal" counterparts. Depending on the numeric value of λ, the stability can still be a problem.

[5] This relation yields some additional hints to how parameters such as λ should be chosen in regularization problems.

easy to see that the first and second statements concerning the existence and numerical stability of R^{-1} vanish with the usage of (7.17).

The term associated with the prior tries to move the calibration in the direction of the μ_i where the center of the distribution is located. The most extreme case would be if there were no given calibration data. Then (7.17) would lead to $x_i = \mu_i$. In general, one can assume that the priors are defined for physically and geologically meaningful parameter ranges. The attraction of the parameters into this region by the prior therefore ensures meaningful solutions and solves for the problem of the third statement. This behavior automates the procedure of the practitioner.

On the other hand if either a huge amount of data or qualitatively very good calibration data is available the prior term can be neglected and calibration approaches the classical χ^2 method. In the intermediate region both terms balance Φ in the same way as different data points balance pure χ^2 calibration.

The discussion is the same for the nonlinear case (7.17) and therefore it is expected that the prior term removes the problems associated with all three statements in almost all cases.

The formula (7.17) provides the very simple interpretation that an uncertainty parameter is used for calibration in exactly the same manner as a calibration data point. For example, if definitions of uncertainty distributions are deduced from measurements, there is no reason why they should not be used for calibration in the same way as calibration data.

A calibration with the classical χ^2 takes only calibration data into account, whereas calibration with the objective Φ calibrates the whole model including parameter uncertainties as well as calibration data uncertainties.

The important point about a Bayesian approach for calibration is the definition of the prior distribution. If it is derived from measurements with error bars, everything is o.k. But very often priors are defined just through the experience of the modeler. So, e.g. the basement heat flow is simply known not to be below $20\,\mathrm{mW/m^2}$ and never to be above $140\,\mathrm{mW/m^2}$. With the definition of a prior such knowledge is taken quantitatively into account and must now withstand critical considerations.

An iterative refinement of uncertainties as feedback of risk results is not allowed in the Bayesian approach because independency of all calibration data values must be ensured. This is completely different to the classic approach where calibration error bars are usually mapped to uncertainty parameter ranges. These ranges are afterwards often taken as "obvious" limits for uncertainty distribution definitions. Distributions constructed in such a way are not allowed to be used as priors in objective functions. Nevertheless, they are often a good choice for Monte Carlo simulations in general.

7.3.1 Prior Information of Derived Parameters

It is sometimes problematic to use the Bayesian approach with uncertain derived parameters. For example the shift of a whole basement heat flow in a huge basin model is taken into account with the same weight as e.g. one measured bottom hole temperature value by (7.17). The weights are only given by the size of the uncertainty but one shift of a basement heat flow, shifts many grid values. Its prior knowledge is not based on experience alone but form a variety of argumentations, consistency arguments, indirect observations, etc. So the prior information which enters the calibration is obviously larger than assumed by (7.17). For that reason it should be possible to increase the weight of such a parameter.

There is no fixed rule for how much the weight should be increased. If the relative uncertainties of the calibration data and the model parameter are of the same size, then the prior term should be multiplied by approximately the number of correlated calibration data points. If it would be much less, the prior term would not affect the calibration and if it would be much bigger, the calibration data would not show significant contributions. In balance, the prior information is believed to be as important as the calibration data itself, which is a reasonable starting point in many cases.

In the example which is shown in Figs. 7.21 and 7.22 the uncertainty in the heat flow shift could be reduced by about two thirds by using the Bayesian approach.

7.3.2 Correlations of Priors

Correlations of priors can be directly taken into account in the Bayesian approach. Instead of using only the diagonal elements of the covariance matrix in (7.17) the whole matrix Σ which is explicitly written down in the two dimensional case in (7.9) must be used:

$$\Phi = (\mathbf{d} - \mathbf{f}(\mathbf{x}))^T \mathbf{C}^{-1} (\mathbf{d} - \mathbf{f}(\mathbf{x})) + (\mathbf{x} - \mu)^T \mathbf{\Sigma}^{-1} (\mathbf{x} - \mu) \; . \tag{7.18}$$

Here a matrix notation with $C_{ij} = \Delta_i \delta_{ik}$ with $\delta_{ik} = 1$ for $i = j$ and $\delta_{ik} = 0$ else was chosen.

7.3.3 Prior Information of Nominal Uncertainties

Nominal distributions (Fig. 7.13) need a special treatment in the Bayesian framework. For example, one continuous uncertainty of the objective is

$$\Phi = \chi^2 + \Phi_{pc} \tag{7.19}$$

with the continuous prior term

$$\Phi_{pc} = \left(\frac{x - \mu}{\sigma} \right)^2 \tag{7.20}$$

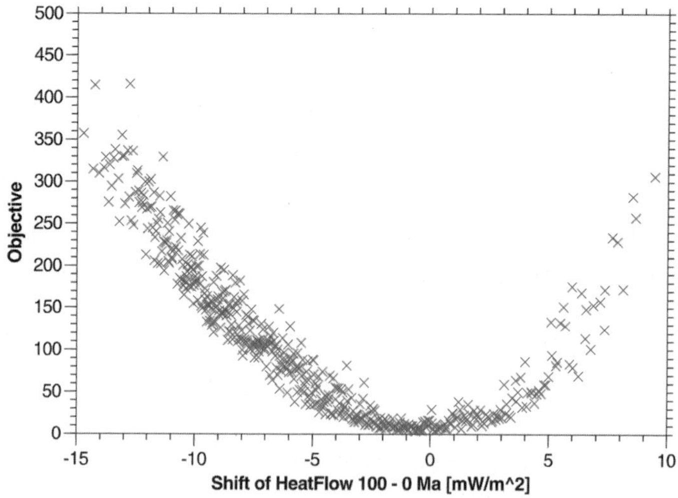

Fig. 7.22. The same example as Fig. 7.21 but with Bayesian objective Φ plotted against the heat flow shift. The calibration is almost unique with a shift in the range of $-2\ldots1\,\mathrm{mW/m^2}$. An extra prior weight of 19 was assumed for the heat flow shift and further rise narrows the range and moves it continuously into the direction of the master run without shift

but for nominal distributions a mean μ or a variance σ does not exist by definition.

In Sec. 7.2.5 association was used instead of correlation. Variance is an "auto–correlation" so it is obvious to try

$$\Phi_{pn} = \sum_{i=1}^{n} \frac{(N_i - n_i)^2}{n_i} \tag{7.21}$$

with n defined as the number of bins of the distribution, $n_i = Np_i$ with N as the number of samples, p_k the probability of the bin k and N_i the number of samples in bin i. It is $\sum_{i=1}^{N} p_i = 1$. Equation (7.21) is known to follow χ^2 statistics as well as its continuous counterpart Φ_{pc} (Press et al., 2002).

The prior must be calculated for one run so $N = 1$ and $N_i = \delta_{ik}$ with k as the bin of the drawn sample. Evaluation of (7.21) yields

$$\Phi_{pn} = \frac{1}{p_k} - 1\,. \tag{7.22}$$

This is a reasonable choice because the objective Φ_{pn} is decreasing with increasing p_k similar as Φ_{pc} with σ^2. With rising uncertainty the prior becomes less important.

In practice, it is possible to add a constant which does not influence the minimization procedure of the objective and use

$$\Phi_{pn} = \frac{1}{p_k} - \frac{1}{p_m} \tag{7.23}$$

with m as the index of the bin of the master run instead of (7.22). This has the advantage that $\Phi_{pn} = 0$ for the master run. If $x = \mu$ is chosen in the continuous case then it is analogously $\Phi_{pc} = 0$ and $\Phi = \chi^2$ for the master run.

7.4 Deterministic Sampling

In the previous section it was shown that the Monte Carlo method is very general. Many topics such as risking, calibration and understanding could be treated simultaneously by just analyzing the results of one Monte Carlo simulation. Occasionally, one is interested only in special questions which are not related to the topic of general risking. In these cases a random and global sampling of the space of uncertainty is often not necessary anymore and it is possible to avoid expensive simulation runs.

The most extreme cases are special algorithms for highly specialized questions, e.g. one is only interested in classical calibration. This can be seen as a minimization problem of one χ^2 function. Special algorithms exist to optimize such a minimization (Press et al., 2002). Expensive simulations are avoided and high numerical accuracy is achieved. However, such algorithms have some serious drawbacks. First, in basin modeling high numerical accuracy is usually not needed because of many uncertainties. Second, these algorithms are so specialized that expensive simulations performed for a minimization of χ^2 are not reusable for a minimization of Φ, which is also often an issue.

Other disadvantages are technical in nature (e.g. bad parallelization properties) because many sophisticated algorithms are of primarily sequential nature, e.g. following a gradient downhill to the minimum.

Thus one is looking for methods, which are more efficient than arbitrary Monte Carlo simulations, for the price of losing generality and which are less special than sophisticated algorithms with high numerical accuracy. Obviously, the targets of interest must be specified exactly before starting to search for appropriate methods.

Risking is the part of Monte Carlo simulations which is most dependent on the random structure of sampling caused by the nature of probabilities. Thus one has to dispense with risking in its general form. On the other hand, one does not need to dispense with "simple" risking such as the calculation of minimum and maximum scenarios. Meaningful targets are hence "simple" risking, calibration, and "simple" understanding as far as understanding can be found without the calculation of statistical quantities.

Other targets, which will be treated more explicitly in Sec. 7.5, are inter– and extrapolation techniques between different simulations for forecasting results. Abdication of risking is not necessary anymore. It can be studied with forecasted models.

7.4.1 Cubical Design

The most simple uncertainty sampling design, which fulfills the conditions of the previous discussion, is simply sampling all combinations of minimum and maximum choices of all uncertainty parameters. In uncertainty space this has the form of a (hyper)cube (Fig. 7.23) and is therefore called cubical design.[6]

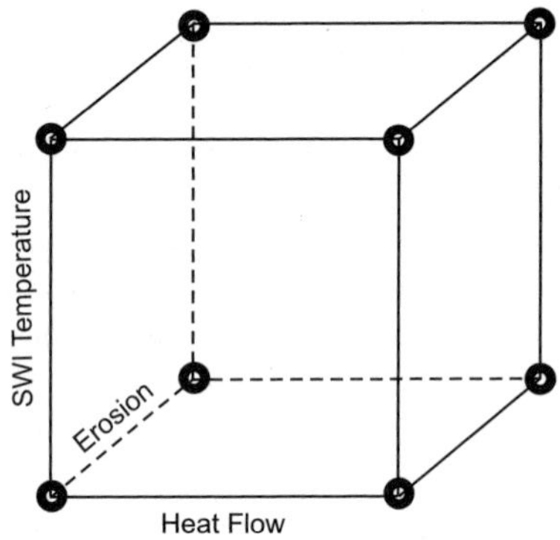

Fig. 7.23. Example of cubical design in uncertainty space with three parameters. The bold circles depict the parameter choices for the simulation runs

Cubical sampling can be used for "simple" risking, calibration, "simple" understanding, and forecasting (Sec. 7.5): the topic of "simple" risking is solved under the assumption of "non–pathological" behavior of the simulator. In such a case, minimum and maximum values of uncertainty parameters would map to minimum and maximum simulation results and thus to result ranges similar to error bars. It is clear that complicated processes such as migration cannot be treated this way.

Calibration is performed analogously. Cubical design samples the uncertainty space regularly and so simulation results are easy to interpolate for good calibration.

Understanding is improved because at least all extreme combinations are simulated. Again under the assumption of smoothness it is possible to calculate interaction effects out of the results (Montgomery, 2001). In general, all results can be inter– and extrapolated, e.g. linearly, which is forecasting. More about this in Sec. 7.5.1.

Nevertheless, cubical design has the serious drawback that the number of simulation runs to be performed increases exponentially with the number of uncertain parameters. For that reason one is often forced to omit certain combinations. There exists a whole theory of "Design of Experiments (DOE)" treating problems such as this (Montgomery, 2001). Keywords are "fractional factorial design" for omitting special combinations of uncertainty parameters or "screening". Screening is important because it tries to find the important

[6] It is exactly cubical if the units of the uncertainties are chosen so that the minimum and maximum values have the same numerical value for all parameters.

and sensitive parameters. The other parameters could be omitted, which drastically reduces the number of combinations. "Blocking" is another important method, which omits sensitive parameters for the better recognition of the effects of the less important parameters. Other designs such as pure cubical designs are proposed, too, e.g. cubical face centered, which obviously can be very valuable. But the theory was created for engineering needs based on real experiments and not on deterministic reproducible computer simulations. Thus only basic ideas such as screening or blocking can be transferred.

Simulation runs for the cubical design can be performed in parallel. The number of parallel runs is restricted to one cube. Multiple cubes themselves are evaluated sequentially if their design is iteratively refined as proposed in Montgomery (2001). Thus simulation runs can to some degree be performed in parallel but not in such a general way as for arbitrary Monte Carlo simulations.

Cubical designs are very valuable for fast uncertainty analyses especially if the number of uncertainties is small.

7.4.2 Other Deterministic Designs

A method that is similar to latin hypercube sampling is Sobol' sequence of "quasi–random" numbers (Press et al., 2002). It guarantees a smoother and more homogeneous sampling than pure "pseudo–random" sampling.[7] The sampling is smoothly refined by increasing numbers in the sequence and it is not influenced by extra parameters such as strip widths.

Sobol' sequence generates quasi–random numbers, which are calculated in any case with a deterministic algorithm, whereas LHC sampling is based on pseudo–random numbers.[8] It can therefore be expected, that LHC sampling generates random numbers with better statistical properties.[9] Additionally, an implementation of Sobol' sequence, such as in Press et al. (2002), does not allow an arbitrary number of independent and different sampling realizations, which can be achieved easily for LHC sampling by different initialization of the random number generator. Independently created samplings can thus usually not be merged to one large with better statistical properties. In practical work this is e.g. a drawback for parallel processing or merging of different risk scenarios. Workarounds, such as sequential precalculation of the random numbers, must be performed (Bücker et al., 2008).

[7] "Pseudo–random" numbers are deterministic numbers which are generated in a way that they pass statistical tests for random numbers. Therefore they are random in practice. "Quasi–random" numbers only appear to be "random".

[8] Obviously, it is even possible to combine LHC sampling with real random numbers.

[9] Better statistical properties are here defined as a larger number of passed statistical tests.

7.5 Metamodels

Interpolation and extrapolation of results between different simulations is forecasting. A method which forecasts all important results is called a "metamodel" or a "surrogate" (Simpson et al., 1997). A metamodel can be used for everything that can be done with the corresponding "real model", e.g. risking.

Metamodels are very important in basin modeling because of the high simulation effort, especially the long simulation times of 3D basin models. In contrast, forecasting is usually very fast, often by a factor of more than a million.

Forecasting does usually not produce the exact results but only approximations. Thus metamodels are often restricted in their applicability. Highly non–linear effects such as hydrocarbon spilling can usually not be forecasted with metamodels.

An overview of common metamodeling methods is given in the following. After this the usage of metamodels for calibration is discussed. Based on the high performance of response surface based metamodels, it is shown that Markov chain Monte Carlo algorithms can be applied.

7.5.1 Response Surfaces

The usage of response surfaces for interpolation and extrapolation is very popular in many fields of science. Very good textbooks are available (Myers and Montgomery, 2002; Box and Draper, 1987; Khuri and Cronell, 1996; Montgomery, 2001). The method is also becoming popular in basin modeling (Wendebourg, 2003).

Response surfaces are low order multivariate polygons which are fitted with least squares regression techniques to the simulation results. Thus, for example a model $f(x_1, x_2)$ with two uncertainty parameters, is typically approximated by

$$f \approx b_0 + b_1 x_1 + b_2 x_2 + b_{11} x_1^2 + b_{22} x_2^2 + b_{12} x_1 x_2 \tag{7.24}$$

with b_i and b_{ik} calculated from a least squares fit (Fig. 7.24).

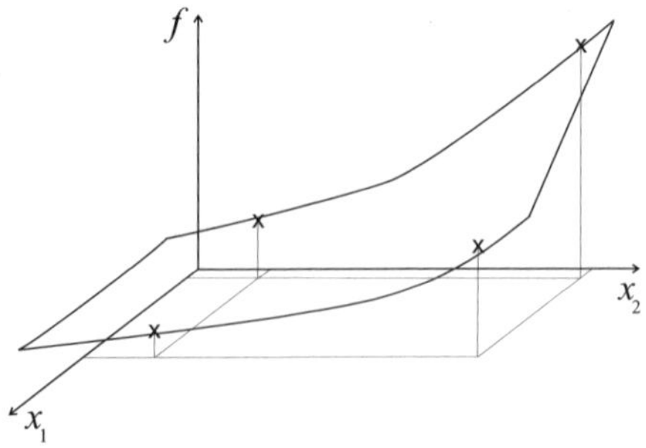

Fig. 7.24. Illustration of a response surface with two uncertainty parameters x_1 and x_2. Crosses indicate simulation results for given x_i e.g. temperatures for given heat flow and bulk conductivity values. Generally, they do not match the response surface exactly

Therefore response surfaces are ideal for approximating smooth and continuous dependencies. Discontinuities and oscillations cannot be reproduced.

It is common to introduce a short hand notation for quadratic terms so that (7.24) becomes

$$f \approx b_0 + b_1 x_1 + b_2 x_2 + b_3 x_3 + b_4 x_4 + b_5 x_5 \quad \text{with} \quad x_3 = x_2 x_2,\ b_3 = b_{22},\ \ldots .$$

In general there are k parameters x_j with $j = 1, \ldots, k$. The number k is determined by the number of uncertainties M and it is $k = M$ for linear response surfaces or $k = M(M+3)/2$ for approximations including quadratic terms.

For the fit, some data values $y_i = f_i(x_{i1}, \ldots, x_{ik})$ of already performed simulations are needed. So finally a vector $\mathbf{b}^T = (b_0, \ldots, b_k)$ for optimization of approximation

$$y_i \approx b_0 + b_1 x_{i1} + b_2 x_{i2} + \ldots + b_k x_{ik}$$

or in vector notation $\mathbf{y} \approx \mathbf{Xb}$ with $X_{ij} = x_{i,j-1}$ for $j > 1$ and $X_{i1} = 1$ is searched. A least squares fit results in minimization of $(\mathbf{y} - \mathbf{Xb})^2$ and evaluation yields [10]

$$\mathbf{b} = (\mathbf{X}^T \mathbf{X})^{-1} \mathbf{X}^T \mathbf{y} . \tag{7.25}$$

A measure of goodness σ_g of this approximation can be evaluated by

$$\sigma_g^2 = \frac{(\mathbf{y} - \mathbf{Xb})^2}{N} . \tag{7.26}$$

This simply denotes the quality of a fit by summing up the quadratic deviations and dividing through the number of points.[11] A safer alternative, which takes outliers into account, can be defined by the maximal deviation

$$\sigma_g = \max_i \left| y_i - \sum_k X_{ik} b_k \right| . \tag{7.27}$$

Design forms of cubical type are very often used as the "natural" sampling procedure for the creation of response surfaces (Myers and Montgomery, 2002; Montgomery, 2001). Under the assumption of smooth behavior of the approximated model, it is obvious that cubical design is an effective sampling

[10] Equation (7.25) is known to be highly unstable and badly conditioned in many practical examples. More robust for the solution of $\mathbf{b}$ is a decomposition of $\mathbf{X}$ into singular values and direct solution of $\mathbf{y} = \mathbf{Xb}$ in appropriate subspaces (Press et al., 2002).

[11] At first glance this seems to be in contradiction to unbiased estimators of variance such as $\sigma^2 = (\mathbf{y} - \mathbf{Xb})^2/(N-k-1)$ and defined in Myers and Montgomery (2002). But this formula is linked tightly to the variation of $\mathbf{b}$ under random variation of $\mathbf{X}$ due to measurement uncertainties. This is a completely different objective.

strategy as all minimum–maximum combinations are studied. Additionally, the number of unknowns to be determined for a quadratic response surface $k+1 = M(M+3)/2+1$ almost matches, in cases of small numbers of uncertainties, M the number of simulations 2^M+1 which have to be performed:[12]

M	1	2	3	4	5	...
$k+1$	3	6	10	15	21	...
2^M+1	3	5	9	17	33	...

Thus only an optimal small number of simulations have to be performed and good matches at the points of simulation itself, leading to small σ_g, are enforced.

In Figs. 7.25 and 7.26 two typical diagrams of response surface models are shown. The formulas describing these isolines can easily be extracted and used for further studies. A high value of the coefficient of a cross term, e.g. x_1x_2, indicates an interaction between the impact of the corresponding uncertainty parameters. This is information which may help to better understand a model.

In Fig. 7.26 negative values for the transformation ratio appear. This is due to the polygonal form of the method. It can therefore not be used in these regions. Generally, it often occurs that simulation results vary faster than can be approximated with simple polygons. In such cases response surface modeling is often performed only in limited regions with adapted sampling of the uncertainty space (Montgomery, 2001).

Another example, which could not be treated in general by response surfaces, are pressure calibrations via variation of permeabilities, especially if permeability is expressed in logarithmic units. Pore pressure is restricted to the lower limit by hydrostatic pressure and to lithostatic pressure at the upper limit but a polynomial fit of response surface type is generally unlimited for infinitely increasing or decreasing uncertainty parameters. However, in a limited region of permeability variations a smooth behavior of pore pressure with good fitting response surfaces can be obtained.

The creation of a metamodel is mainly determined by the solution of a linear set of equations of dimension $(k+1)\times(k+1)$ for each result point which is modeled. In case of calibration purposes this number is given by the number of calibration points N. Thus, the creation of a response surface metamodel is performed within seconds on modern computers because in practice mostly $M < 10$ and $N < 1000$.

The calculation of response surface metamodel results is an evaluation of simple polygons and this is extremely fast. In most computer applications this appears to be almost instantaneous.

7.5.2 Fast Thermal Simulation

Fast thermal simulation is a special method of fast heat flow analysis (Nielsen, 2001). It is based on the approximative linear form of the partial differential

[12] All combinations plus master run.

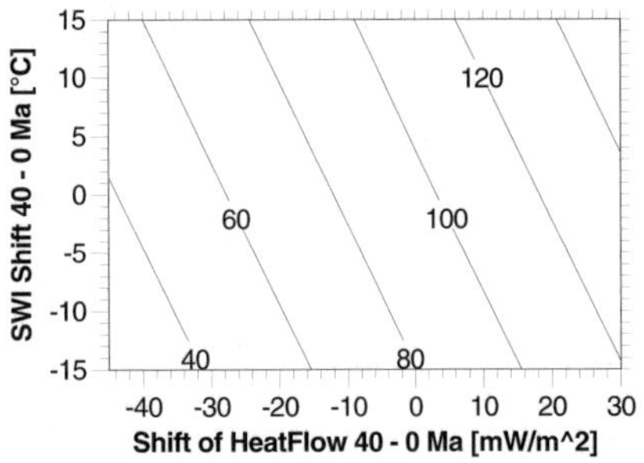

Fig. 7.25. Response surface isolines for temperature in Celsius in a source rock. Dependent on heat flow and SWI temperature variations the isolines are linear as expected

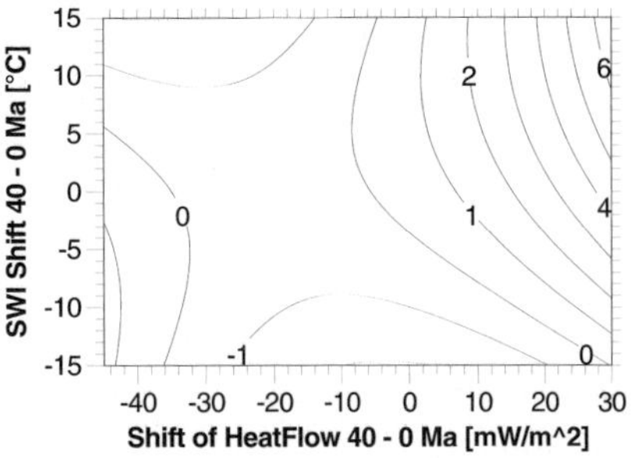

Fig. 7.26. Response surface isolines for transformation ratio in [%] at same point as in Fig. 7.25. Negative values indicate a region where the response surface method cannot be used

equation

$$\rho c \frac{\partial T}{\partial t} - \boldsymbol{\nabla} \cdot (\boldsymbol{\lambda} \cdot \boldsymbol{\nabla} T) = Q \tag{7.28}$$

of heat flow. Here T is the temperature, $\boldsymbol{\lambda}$ are the thermal conductivities, t is the time, ρ the density, c the specific heat capacity, and Q are external heat sources. For a unique compilation of heat flow analysis, boundary and initial conditions must be specified. At the top of the basin usually the temperature is given, at the sides, a condition of prohibited horizontal heat flow is applied and at the bottom, basement heat flows are specified:

$$\begin{aligned} T &= T_{\mathrm{SWI}} && \text{on top,} \\ \boldsymbol{\nabla} T &= 0 && \text{at the sides,} \\ -\boldsymbol{\lambda} \cdot \boldsymbol{\nabla} T &= \mathbf{q} && \text{at the bottom,} \\ T|_{t=\mathrm{initialtime}} &= T_0 && \text{at initial time} \end{aligned} \tag{7.29}$$

with T_{SWI} the "Sediment Water Interface" temperature at top of the basin, $\mathbf{q}$ the basement heat flow, and T_0 the temperature profile at initial time.

If $\boldsymbol{\lambda}$, ρ, and c are assumed to be smooth and weakly temperature–dependent then (7.28) is almost linear. This property can be utilized:

Firstly, one should take a look at the following boundary value problem:

$$\begin{aligned} \rho c \frac{\partial \tilde{T}}{\partial t} - \boldsymbol{\nabla} \cdot (\boldsymbol{\lambda} \cdot \boldsymbol{\nabla} \tilde{T}) &= 0 \text{ with} \\ \tilde{T} &= 0 && \text{on top,} \\ \boldsymbol{\nabla} \tilde{T} &= 0 && \text{at the sides} \\ -\boldsymbol{\lambda} \cdot \boldsymbol{\nabla} \tilde{T} &= \tilde{\mathbf{q}} && \text{at the bottom and} \\ T|_{t=\mathrm{initialtime}} &= 0 && \text{at initial time.} \end{aligned} \tag{7.30}$$

A temperature profile $T + x_q\tilde{T}$ with T as the solution of (7.28) with boundary and initial conditions (7.29) and $\tilde{T}$, a solution of (7.30) is a solution of (7.28) with boundary and initial conditions such as (7.29), which must only be modified at the bottom by $-\boldsymbol{\lambda} \cdot \boldsymbol{\nabla} T = \mathbf{q} + x_q\tilde{\mathbf{q}}$. Herein, x_q is just an arbitrary number, which can be interpreted as a derived uncertainty parameter for a variation of form $\tilde{\mathbf{q}}$.

The important point is that with only two solutions T and $\tilde{T}$ one can construct multiple solutions $T + x_q\tilde{T}$ for heat flow variations $\mathbf{q} + x_q\tilde{\mathbf{q}}$ for any value x_q just by linear combination. The form of $x_q\tilde{\mathbf{q}}$ defines the space of possible heat flow variations. Obviously, multiple variations $x_{q,i}\tilde{\mathbf{q}}_i$ can be combined just by summing up the solutions $x_{q,i}\tilde{T}_i$. Hence it is possible to quickly create flexible variations of the original heat flow and temperature pattern.

It has been proposed by Nielsen (2001) to vary the heat flow below each of the four model corners.[13] Each paleo–heat flow map is calculated by interpolation with two dimensional form functions analogously to (8.22). A corner variation with its shape function states one variation $\tilde{\mathbf{q}}_i$ with $i = 1, \ldots, 4$ for all four corners. The sum of all four corner variations describes tilting and twisting variations of the original heat flow distribution. The method should not be applied in cases when mismatches to the measured data require heat flow shifts very differently in many locations, e.g. from well to well. For such cases methods as described in Sec. 3.9 are more advantageous.

Additionally it must be noted, that $\tilde{\mathbf{q}}$ can vary independently to $\mathbf{q}$ through time. Hence heat flow variations in time can be incorporated. For example, it is possible to shift the heat flow of each of the four corner points linearely with time.

The set T and $\tilde{T}$ can be interpreted as a metamodel for forecasting manifold heat flow histories.

A response surface which is created as an interpolation of two different solutions T_1 and T_2 of (7.28) with the boundary conditions (7.29) once with bottom heat flow $\mathbf{q}$ and the other with $\mathbf{q}+\tilde{\mathbf{q}}$, yields almost the same results as the fast thermal simulation caused by the linearity of the differential equation system. The main difference between both methods comes from the fact, that the differential equation is usually not exactly linear. Parameters such as the thermal conductivity are typically weak but non–linearly temperature dependent. The response surface can then be interpreted rather as a "secant approach", whereas the fast thermal simulation is following more a "tangent" (Fig. 7.27). Response surfaces can incorporate smooth non–linearities to some degree of accuracy with their quadratic terms (7.24). On the other hand, as they are caused by the linear regression, they do not need to match the simulation results from which they were created exactly, whereas fast thermal

[13] It is assumed here that the model has a rectangular base area. Generally, it does not matter if the corner points are not inside of the model.

simulations reproduce, at least in a region of small variation, the original models exactly.

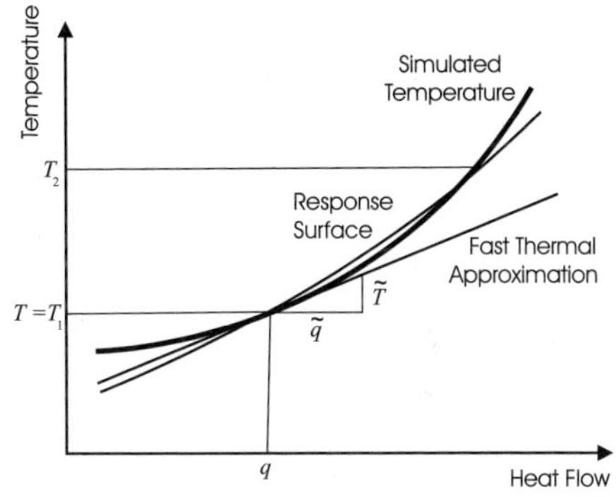

Fig. 7.27. Illustration of the differences between response surface and fast thermal simulations. The curvature of temperature, which comes from the nonlinearities is exaggerated for demonstration. Due to quadratic terms the response surface is able to approximate non–linearities whereas the fast thermal simulation approximation is restricted to a straight "tangential" line. The response surface does not in general need to match the temperature exactly at any point whereas the fast thermal simulation matches at $T = T_1$

The calculation of T and $\tilde{T}$ needs about the same effort. The evaluation of a forecast is just the evaluation of $T + x_q\tilde{T}$ and thus can be performed almost instantaneously. Thus the effort and the needed resources for the creation of the metamodel as well as the evaluation performance of the fast thermal simulation and the response surface are almost the same.

Fast thermal simulations are not limited to heat flow variations only but can also be applied to thermal conductivity variations. This is achieved by introducing a derived uncertainty parameter x_λ for the variation of thermal conductivity according to $\boldsymbol{\lambda} + x_\lambda\hat{\boldsymbol{\lambda}}$ with a $\hat{\boldsymbol{\lambda}}$ describing the form of the variation. The solution $\hat{T}$ of the differential equation

$$\nabla \cdot (\boldsymbol{\lambda} \cdot \nabla \hat{T}) = -\nabla \cdot (\hat{\boldsymbol{\lambda}} \cdot \nabla T) \tag{7.31}$$

with boundary conditions such as (7.29) but with $\mathbf{q} = 0$ can be added to the solution T in the same manner as $\tilde{T}$ to construct valid heat flow histories for thermal conductivity variations. Here, the fast thermal simulation is restricted to "small" variations in conductivity because quadratic terms are neglected in the deviation of (7.31).

7.5.3 Kriging

Kriging is another method for interpolation and extrapolation in multi dimensional spaces. It is based on the minimization of statistical correlations and derived as the best linear unbiased estimator. Originally it was developed for spatial inter– and extrapolation only but it can also be applied to abstract uncertainty spaces. Various different methods of kriging exist. To the authors knowledge it has not been applied up to now in any case as a metamodel in basin modeling and thus we refer only to the literature of geostatistics (Davis, 2002). However, it can be expected that kriging might yield good results in many cases especially when simple functions such as (7.24) are not appropriate at all for the description of the model or process of interest.

7.5.4 Neural Networks

Neural Networks can be interpreted as metamodels.

Neural Networks must be trained. They learn. Three classes of learning are usually distinguished (Zell, 1997):

- supervised learning
- reinforcement learning
- unsupervised learning

Supervised learning is based on comparison with correct results. These results are simulation results in basin modeling. Supervised learning is usually the fastest way of learning. Nevertheless many expensive simulation runs must be performed for this way of learning.

Reinforcement learning is based on reduced feedback. The network is taught only with information about the correctness of its output but not the correct result itself. Therefore reinforced learning networks need even more training than supervised learning networks. Although the amount of feedback data is small it, too, must be available. This means that many expensive simulation runs must be performed for this method.

Unsupervised learning is performed without feedback. The network should learn by classifications in its own right e.g. by "self organization". Caused by the complexity of a typical basin model, it is expected that unsupervised learning neural networks will be improper for result predictions.

The high effort for learning leads to the conjecture that neural networks are not the best alternative for basin metamodeling.

7.5.5 Other Methods for Metamodeling

Methods such as rule based expert systems or decision trees are obviously limited in their applicability for forecasting. Under special circumstances they can be interpreted as metamodels but not in general.

Other special techniques are based on analysis in frequency space. Due to the complex geometry in geology, these methods cannot usually be applied to basin modeling.

7.5.6 Calibration with Markov Chain Monte Carlo Series

The Markov chain Monte Carlo (MCMC) method is designed for the sampling of multi–variate probability distributions (Neal, 1993; Besag, 2000).

Calibration is a search of high probability regions where data fits the model which is not trivial in high dimensional spaces. So MCMC can be misused to find the regions of calibration. Additionally, the sampling allows the error bars of measurement values to be mapped to uncertainty distributions.[14] The whole subject of search and mapping is called "inversion" and thus MCMC is also a method for inversion.

Several different algorithms for MCMC exist, which can be shown to be directly related (Neal, 1993; Besag, 2000). This section is restricted to the classical Metropolis algorithm. It basically works as follows:

According to a distribution, with some special properties which are of no interest here, random jumps are performed in uncertainty space. If the probability density of the distribution, which should be sampled increases, the jump is accepted. If the density decreases it can be rejected or accepted by a special criterion with a random level of acceptance. This ensures that MCMC also samples low probability regions but it focuses primarily on the highly probable regions. In Fig. 7.28 such a MCMC "random walk" is illustrated. Sampling can only be performed on a small fraction of jumps typically every 100th or less to ensure independency of the samples.

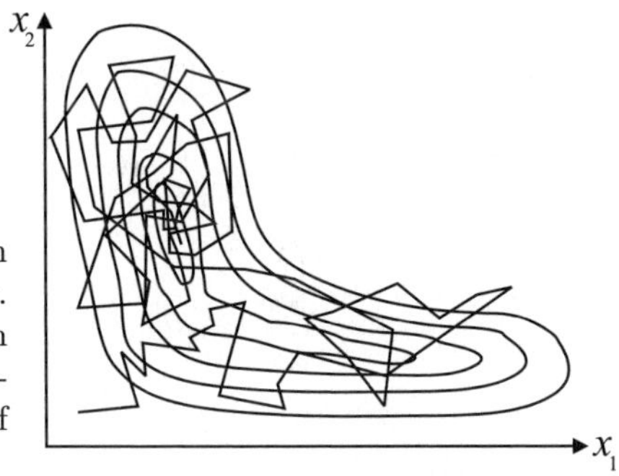

Fig. 7.28. Illustration of Markov chain Monte Carlo sampling with a random walk. Isolines indicate the probability density in the $x_1 - x_2$ uncertainty diagram. The "random walk" is attracted by the region of high probability

Obviously, it is clear that MCMC is not very efficient in sampling because most of the jumps which are model results are ignored for sampling, some of the jumps are rejected and additionally, the random walks can become very long before reaching their high probability calibration targets. Proper MCMC sampling can become a delicate choice of the start point and jump width. Therefore, in basin modeling MCMC is only usable with fast metamodels such as response surfaces or fast thermal simulations. On the other hand, in theory MCMC guarantees to find the regions of interest.

[14] These distributions are not allowed to be used in a Bayesian approach, see Sec. 7.3.

Summary: Models are usually constructed on the basis of uncertain data. These uncertainties cause additional tasks during comprehensive model analysis. Firstly, modeled results must be classified according to their probability. For example, confidence intervals of special output scenarios should be specified or even more concrete a risk of failure must be quantified. Secondly, the behavior of model results with the variation of uncertain parameters should be understood. Which parameter affects which part of the result? Finally, uncertainties should be reduced by comparison with additional calibration data. The three tasks are "risking", "understanding", and "calibration".

Obviously, all three tasks can be studied with multiple simulation runs. Uncertain parameters must therefore be varied according to their range of uncertainty. Monte Carlo simulations are an effective method of treating all three tasks simultaneously. Multiple simulation runs with randomly drawn uncertainty parameters, according to their probability of occurrence, are performed. Another advantage of the approach is the possibility of unrestricted parallel processing. This is especially valuable because simulation runs are often very time consuming and therefore expensive. The method can further be optimized with latin hypercube sampling, which avoids clustering of parameter combinations.

A model can be calibrated in two different ways, with and without consideration of information which describes data uncertainties of the model, e.g. limits or ranges of an uncertain input parameter. This "prior" information is taken into account in the Bayesian approach. Ambiguous and geologically meaningless calibrations can be avoided with this approach.

Simulation runs are very time consuming. Response surfaces are a method for fast interpolation between simulation results. Other methods for rapid result prediction, such as the fast thermal simulation, are also discussed. Particularly with regard to heat flow problems, response surfaces and fast thermal simulations can be used efficiently for calibration. A very robust algorithm concerning inversion is the Markov Chain Monte Carlo (MCMC) sampling, which in principal guarantees the best possible calibration due to random jumps in the uncertainty space.

References

J. Besag. Markov Chain Monte Carlo for Statistical Inference. Working paper, Center for Statistics and the Social Sciences, University of Washington, 2000.

O. Beyer, H. Hackel, V. Pieper, and J. Tiedge. *Wahrscheilichkeitsrechnung und Statistik.* B. G. Teubner Stuttgart Leipzig, 8th edition, 1999.

G. E. P. Box and N. R. Draper. *Empirical Model–Building and Response Surfaces.* John Wiley & Sons, Inc., 1987.

H. M. Bücker, A. I. Kauerauf, and A. Rasch. A smooth transition from serial to parallel processing in the industrial petroleum system modeling package petromod. *Computers & Geosciences*, 34:1473–1479, 2008.

J. C. Davis. *Statistics and Data Analysis in Geology.* John Wiley & Sons, 3rd edition, 2002.

L. Fahrmeir and A. Hamerle, editors. *Multivariate statistische Verfahren.* Walter de Gruyter & Co., 1984.

E. T. Jaynes. *Probability Theory: The Logic of Science.* Cambridge University Press, 2003.

A. I. Khuri and J. A. Cronell. *Response Surfaces: Designs and Analyses.* Marcel Dekker, Inc., second edition, 1996.

I. Lerche. *Geological Risk and Uncertainty in Oil Exploration.* Academic Press, 1997.

J. O. Miller. Bivar: A program for generating correlated random numbers. *Behavior Research Methods, Instruments & Computers*, 30:720–723, 1998.

D. C. Montgomery. *Design and Analysis of Experiments.* John Wiley & Sons, Inc., 5th edition, 2001.

R. H. Myers and D. C. Montgomery. *Response Surface Methodology: Process and Product Optimization Using Designed Experiments.* John Wiley & Sons, Inc., second edition, 2002.

R. M. Neal. Probabilistic Inference Using Markov Chain Monte Carlo Methods. Technical Report CRG–TR–93–1, Department of Computer Science, University of Toronto, 1993.

P. Newendorp and J. Schuyler. *Decision Analysis for Petroleum Exploration.* Planning Press, second edition, 2000.

Søren Nielsen. Århus University, Denmark, *Private communication*, 2001.

W. H. Press, S. A. Teukolsky, W. T. Vetterling, and B. P. Flannery. *Numerical Recipes in C++.* Cambridge University Press, second edition, 2002.

H. Rinne. *Taschenbuch der Statistik.* Verlag Harri Deutsch, second edition, 1997.

C. P. Robert. *The Bayesian Choice.* Springer–Verlag New York, Inc., second edition, 2001.

T. W. Simpson, J. D. Peplinski, P. N. Koch, and J. K. Allen. On the use of statistics in design and the implications for deterministic computer experiments. In *Proceedings of DETC.* ASME Design Engineering Technical Conferences, 1997.

M. R. Spiegel and L. J. Stephens. *Schaum's Outline of Theory and Problems of Statistics.* The McGraw–Hill Companies Inc., New York, 1999.

R. O. Thomsen. Aspects of applied basin modelling: sensitivity analysis and scientific risk. In S. J. Düppenbecker and J. E. Iliffe, editors, *Basin Modelling: Practice and Progress*, number 141 in Special Publication, pages 209–221. Geological Society of London, 1998.

J. Wendebourg. Uncertainty of petroleum generation using methods of experimental design and response surface modeling: Application to the Gippsland

Basin, Australia. In S. Düppenbecker and R. Marzi, editors, *Multidimensional Basin Modeling*, volume 7 of *AAPG/Datapages Discovery Series*, pages 295–307, 2003.

A. Zell. *Simulation neuronaler Netze*. R. Oldenburg Verlag München Wien, 1997.

8

Mathematical Methods

8.1 Introduction

Basin modeling is a framework of adapted geological and physical models. An overall implementation contains a wide range of algorithms and methods each of them appropriate for each "submodel". A detailed discussion of all approaches goes beyond the scope of this volume. Many algorithms are standard in other fields, such as statistics or computer science. In these cases additional information can be found in the cited literature. An excellent general overview over numerical methods is given by Press et al. (2002). Adapted or special algorithms of basin modeling, which are important for understanding and which are not too comprehensive, are outlined together with the fundamental theory in the previous chapters.

From the viewpoint of numerics differential equations constitute the largest class of specific problems in basin modeling. They are fundamental for basin modeling, complex and costly to solve. Therefore they play a central role in all respects. Temperature and pressure fields, which are here denoted with u, are modeled with parabolic diffusion equations in the form of

$$\partial_t u - \boldsymbol{\nabla} \cdot \boldsymbol{\lambda} \cdot \boldsymbol{\nabla} u = q \ . \tag{8.1}$$

Hence this chapter focuses on the methods necessary for the solution of such equations in basin modeling.

Mathematical equations, such as (8.1), are formulated in terms of physical quantities, such as temperature or stress. These quantities have some important basic properties, which are shortly summarized in Sec. 8.2. Very often physical quantities are bulk values. Macroscopic averages between different rock and fluid types must be specified. Typical examples are presented in Sec. 8.3. Multi–dimensional differential equations with complicated geometries are usually solved with the finite element method whereas finite differences are often used in special cases of approximately one–dimensional problems, such as simplified crustal layer models. Finite differences are more elementary

T. Hantschel, A.I. Kauerauf, *Fundamentals of Basin and Petroleum Systems Modeling*, DOI 10.1007/978-3-540-72318-9_8,
© Springer-Verlag Berlin Heidelberg 2009

and therefore discussed first in Sec. 8.4. Afterwards Sec. 8.5 deals with the finite element method. Control volumes must be mentioned because they are sometimes an alternative especially for flow and pressure modeling. They are discussed in Sec. 8.6. All these methods map the differential equations to a system of linear algebraic equations. Necessary for its solution are solvers, which are briefly summarized in Sec. 8.7. High performance modeling can be obtained with parallelization and is discussed in the Sec. 8.8. Finally, some approaches of local grid refinement (LGR) are discussed in Sec. 8.9.

8.2 Physical Quantities

Many physical quantities are functions in space and geological time, e.g. temperature $T(\mathbf{x}, t)$, pressure $p(\mathbf{x}, t)$ or flow velocity $\mathbf{v}(\mathbf{x}, t)$. They are scalars, vectors or tensors (Fig. 8.1). Scalars are undirected, that means they are represented by just one single value, such as temperature or pressure at a given location $\mathbf{x}, t$ in space and time. A vector has both, size and direction. Thus three independent numbers are necessary to describe a vector at a given location, e.g. a vertical and two horizontal components for a water flow velocity

$$\mathbf{v} = \begin{pmatrix} v_x \\ v_y \\ v_z \end{pmatrix} = (v_x, v_y, v_z)^T . \tag{8.2}$$

The index T indicates here a transposition of rows and columns.

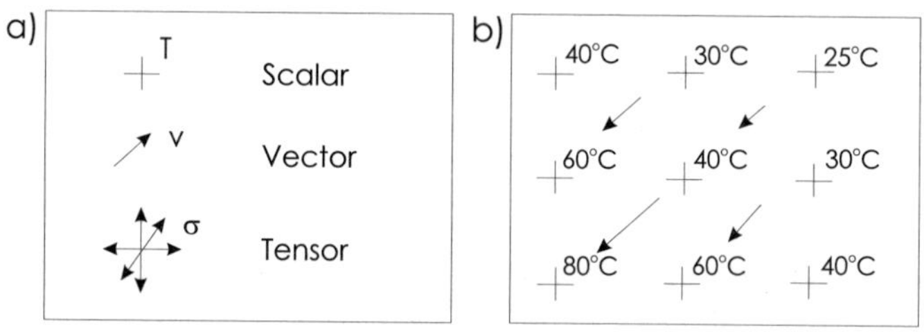

Fig. 8.1. (a) Temperature, velocity and stress are examples for scalars, vectors, and tensors. (b) A field for temperature with temperature gradients

A tensor $\boldsymbol{\sigma}$ has a different size in each direction and is therefore characterized by three vectors. It consists of normal components σ_{ii} and shear components σ_{ij} for $i \neq j$:

$$\boldsymbol{\sigma} = \begin{pmatrix} & \sigma_{ij} & \end{pmatrix} = \begin{pmatrix} \sigma_{xx} & \sigma_{yx} & \sigma_{zx} \\ \sigma_{xy} & \sigma_{yy} & \sigma_{zy} \\ \sigma_{xz} & \sigma_{yz} & \sigma_{zz} \end{pmatrix} . \tag{8.3}$$

Physical quantities, such as the stress tensor, are usually described by symmetrical tensors with $\sigma_{ij} = \sigma_{ji}$. The number of independent values of a symmetrical tensor is six. These six values can be represented by three orthogonal main directions and corresponding sizes. For example, the tensorial character of a physical quantity can be described with normal and tangential vectors acting at the surface of a volume element. There always exist three perpendicular planes, where these vectors are oriented in normal direction of the corresponding surfaces. These surfaces normals are the main directions of the tensor and characterize the tensor similar as the direction characterizes a vector. Corresponding sizes of a tensor are called principal values or eigenvalues λ_k and can be calculated as the solutions of

$$\sigma_{ij} - \lambda_k \delta_{ij} = 0 \tag{8.4}$$

with the unit tensor $\delta_{ij} = 1$ for $i = j$ and $\delta_{ij} = 0$ else. The principal values λ_k are usually sorted according to size and renamed to σ_k with $\sigma_1 > \sigma_2 > \sigma_3$. The main directions of many tensors are parallel and perpendicular to the layer directions in geological models. Principal values in horizontal direction are usually of same size and thus the two different principal values are often named σ_h and σ_v.

For example, thermal conductivities and permeabilities are tensors. In principle they contain direction dependent components, although they are determined in practice by only two values along and across the layering. The terms horizontal and vertical conductivities are used although the values represent layer directions and not fixed space axes. The anisotropy factor is the ratio between the horizontal and the vertical component.

Other characteristics of a tensor are the invariants

$$\begin{aligned}
I_1 &= \sigma_{xx} + \sigma_{yy} + \sigma_{zz} \\
I_2 &= -\sigma_{xx}\sigma_{yy} - \sigma_{xx}\sigma_{zz} - \sigma_{yy}\sigma_{zz} + \sigma_{xy}^2 + \sigma_{xz}^2 + \sigma_{yz}^2 \\
I_3 &= \sigma_{xx}\sigma_{yy}\sigma_{zz} + 2\sigma_{xy}\sigma_{xz}\sigma_{yz} - \sigma_{xx}\sigma_{yz} - \sigma_{yy}\sigma_{xz} - \sigma_{zz}\sigma_{yz} \, ,
\end{aligned} \tag{8.5}$$

which are independent of the choice of the coordinate system. The first invariant I_1 represents the total value of the normal components. The average value of the normal components is defined as $\bar{\sigma} = I_1/3$. Any tensor can be represented as the sum of a pure normal and a deviatoric part s_{ij} according to

$$\sigma_{ij} = \bar{\sigma}\, \delta_{ij} + s_{ij} \; . \tag{8.6}$$

The invariants of the deviatoric tensor are often called J_1, J_2, J_3 with $J_1 = 0$. Deviatoric invariants of the stress tensor are important for the determination of failure criteria, since these are related to distortion.

The value change per distance of a quantity into a specified direction (over an infinitesimal distance) is called its derivative. A gradient is a vector with partial derivatives of a scalar quantity in x, y, and z – direction as components.

It points into the steepest upward direction (Fig. 8.1) and can be calculated as

$$\operatorname{grad} T = \nabla T = \left(\frac{\partial T}{\partial x}, \frac{\partial T}{\partial y}, \frac{\partial T}{\partial z} \right)^T . \tag{8.7}$$

The (scalar) size of a gradient can be calculated as a derivative in steepest upward direction. For example, the main direction of temperature change is approximately vertically downward. A temperature change per distance into downward direction is therefore often called the vertical temperature gradient.

Note, that flow vectors, such as heat or fluid flow, point typically in downward and not in upward direction of the corresponding temperature or pressure field. They are thus proportional to the negative gradient.

8.3 Mixing Rules and Upscaling

Most bulk property values can approximately be derived from single mineral and fluid component values, such as defined in App. A, E, I, by suitable mixing and upscaling rules. Such component based properties are thermal conductivities, heat capacities or densities. The mixed rock matrix and pore fluid values are usually calculated first. The corresponding properties of the minerals, lithologies, organic compounds, and the fluid phase values are taken into account, respectively. Then the bulk value is derived from the two averages of the rock and pore fluid (Fig. 8.2).

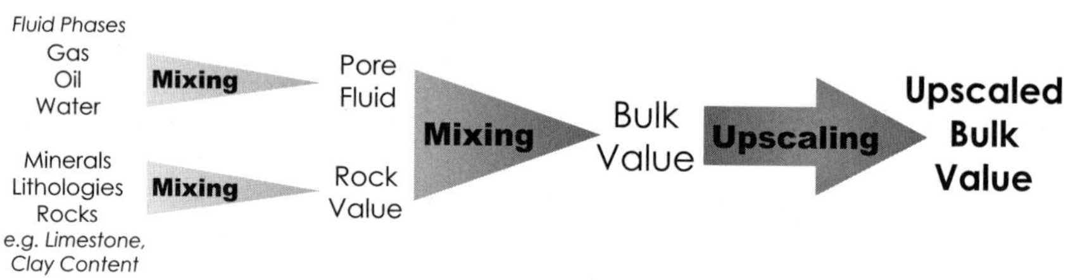

Fig. 8.2. Mixing and Upscaling

Other properties such as permeabilities, compressibilities, and capillary entry pressures are rather defined as porosity dependent functions instead of component based mixtures. Fluid properties need not to be taken into account. However, mixing rules are still needed when the rock matrix consists of various parts of different lithologies.

Generally, mixing and upscaling can become very complex and dependent on details of the specific system and quantities which are mixed. For example, thermal conductivities should be mixed with a sophisticated procedure which incorporates the Bunterbarth formula, the principle of constant mean conductivity, and the variation of anisotropy with porosity (Chap. 3). However, there are three basic methods for mixing and upscaling, namely arithmetic, harmonic, and geometric averaging, which are often used when detailed information is missing. If necessary, the accuracy can often be improved by additional or modified rules for specific parameters or special arrangements of the components.

The arithmetic average of a property value λ is the sum of the component property values λ_i weighted with the mass or volume fraction p_i according to

$$\lambda = \sum_{i=1}^{n} p_i \, \lambda_i \qquad \text{with} \qquad \sum_{i=1}^{n} p_i = 1 \, . \tag{8.8}$$

The geometric mean averages the order of magnitude. Therefore, it is also the arithmetic average of the log–values with

$$\lambda = \prod_{i=1}^{n} (\lambda_i)^{p_i} \qquad \text{or} \qquad \log \lambda = \sum_{i=1}^{n} p_i \log \lambda_i \, . \tag{8.9}$$

The harmonic mean averages the inverse values according to

$$\frac{1}{\lambda} = \sum_{i=1}^{n} \frac{p_i}{\lambda_i} \, . \tag{8.10}$$

A modified rule is the square–root mean which arithmetically averages the square root of the component values

$$\sqrt{\lambda} = \sum_{i=1}^{n} p_i \, \sqrt{\lambda_i} \tag{8.11}$$

and increases the importance of small fractions p_i. This rule is recommended for thermal conductivities (Beardsmore and Cull, 2001).

Generally, it can mathematically be proven that arithmetic averages are larger than geometric averages which are again larger than harmonic averages (Aigner and Ziegler, 2004). The square–root mean is usually found between arithmetic and geometric mean. The difference between the mean values increases with larger variance of the single values (Fig. 8.3). The selection of the appropriate averaging method is less important for property values which are spread over a small interval.

In case of fluid phase mixing the fraction values of the mixing equation are saturations, in case of rock mixing they are mineral fractions, and when pore and rock values are mixed together they are porosity ϕ and rock volume fraction $1 - \phi$.

In table Table 8.1 typical mixing methods for different quantities are listed. The quantities are often distinguished between extensive, conductivity or transport parameters, and critical or threshold properties. Sometimes the list is extended for intensive quantities. Each of them usually requires different methods for mixing and upscaling.

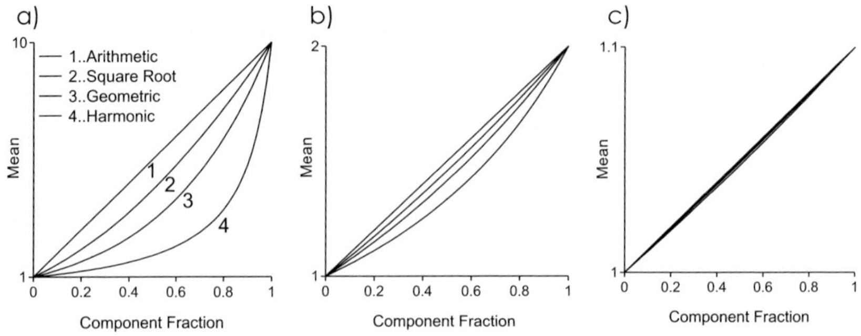

Fig. 8.3. Mean values of two components versus fractions. The first one component value is equal to 1 while the second value is 10, 2, and 1.1 respectively

	Example	Mixing
Extensive Quantity	Density Heat Capacity Porosity Clay Content Radioactive Heat Specific Surface Area	A
Transport Parameter	Thermal Conductivity Permeability	Fluids: G Rocks: GAH, S Bulk: GAH
Threshold Property	Critical Oil Saturation	A
	Critical Gas Saturation	A
	Connate Water Saturation	A
	Capillary Entry Pressure	G or M
	Fracture Limit	A
	Compressibility	A or M

Table 8.1. Arithmetic (A), geometric (G), harmonic (H), and maximum based (M) or special (S) mixing rules, such as square root mixing, are used for different variables and mixture types. GAH indicates that the geometrical average is used for a homogeneous structure and the arithmetic and harmonic average is used for vertical and horizontal components of a layered structure

Extensive properties affect the bulk value simply proportional to their abundance and should obviously be mixed with arithmetic averages. Examples are heat capacity, density, and porosity. Exceptions are for example compressibility values for vertical compaction when the less compactable rock components consist of vertically well connected columns. Then compressibility behaves similar as a threshold property and the minimum compressibility value should be used.

Transport parameters, such as thermal conductivity, are calculated with different mixture and upscaling rules dependent whether and how the components have been layered or arranged (Fig. 8.4). In case of straight layering the mixed parameter is equal to the arithmetic mean for the flow along and to the harmonic mean for the flow perpendicular to the layering direction. For mixed domains the mean is somewhere between arithmetic and harmonic mean. Usually the geometric average is taken. In case of porosity dependent permeability curves each curve point has to be mixed separately. Kozeny–Carman relations are based on special parameters (Sec. 2.2.3).

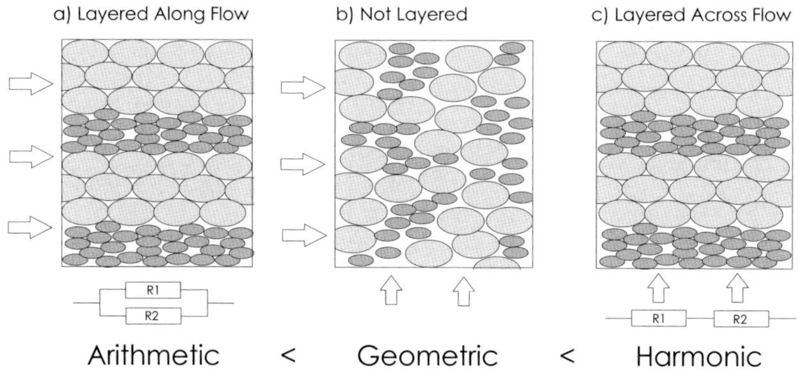

Fig. 8.4. Averaging types for conductivity values. The corresponding resistance R in an electrical circuit is calculated in the case a) from parallel and in the case c) from sequential placed resistances R_1 and R_2

Threshold properties such as the capillary entry pressure often require special considerations for mixing and upscaling based e.g. on the flow and saturation pattern. Mixing and upscaling are discussed in the chapters where the properties are introduced. Due to fractal flow patterns, upscaling must often be treated differently to mixing with special methods.

Without known correlations, fluid properties of mixtures of fluids often do not behave proportional to the abundance of their components. These quantities are often more of intensive nature and are therefore often mixed with the geometric mean. An exception are fluid densities, which are mixed arithmetically. However, phase properties can be derived from the chemical components using the properties of the pure substances. This is especially needed for fluid analysis with the corresponding flash calculations (Sec. 5). Special rules such as Lee–Kesseler mixing (5.6) have a proven track record.

8.4 Finite Differences

The numerical solution of a partial differential equation can only be specified and evaluated for a limited set of discrete points. Hence the first step is the discretization of space and time. For the finite difference method the gridding is usually chosen in such a way that expressions containing derivatives are approximated by

$$\frac{\partial u}{\partial x} \approx \frac{\Delta u}{\Delta x} \tag{8.12}$$

with Δu, Δx determining the difference between u and x at adjacent gridpoints. An appropriate discretization fulfills the conditions of orthogonal grid directions, which ensures the independency of directions and "small" distances Δ in space and time in consistency with approximation (8.12). Sometimes higher order terms in Δ are taken into account by following a Taylor expansion.

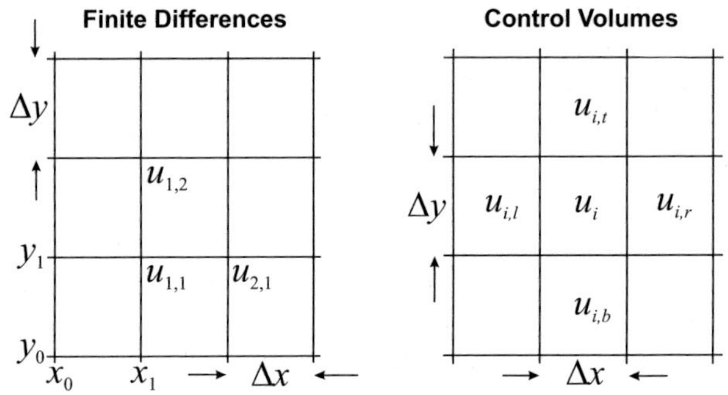

Fig. 8.5. Cutouts of two–dimensional finite difference and control volume grids

A typical two dimensional finite difference grid is depicted in Fig. 8.5. Here, the discretization in space for the diffusion term of (8.1) with constant $\boldsymbol{\lambda}$ works out to be

$$\frac{\partial^2 u}{\partial x^2} + \frac{\partial^2 u}{\partial y^2} \approx \frac{u_{i+1,k} - 2u_{i,k} + u_{i-1,k}}{(\Delta x)^2} + \frac{u_{i,k+1} - 2u_{i,k} + u_{i,k-1}}{(\Delta y)^2} \tag{8.13}$$

with $u_{i,k}$ as the solution at the gridpoint at x_i and y_k.

The discretization in time can be performed independently with the same approach (8.12) and yields

$$\begin{aligned} \frac{u^{n+1} - u^n}{\Delta t} &= \int_{t^n}^{t^{n+1}} dt(\nabla \cdot \boldsymbol{\lambda} \cdot \nabla u + q) \\ &= \left[\eta(\nabla \cdot \boldsymbol{\lambda} \cdot \nabla u^{n+1} + q^{n+1}) + (1-\eta)(\nabla \cdot \boldsymbol{\lambda} \cdot \nabla u^n + q^n)\right] \Delta t \end{aligned} \tag{8.14}$$

with $\Delta t = t^{n+1} - t^n$, discrete time steps t^n, solutions u^n at these time steps, and $0 \leq \eta \leq 1$ according to the midpoint rule of integration. The integral can now be approximated with predefined values of η. The choice $\eta = 0$ is called the "explicit scheme". It has the advantage that u^{n+1} can be evaluated through time from u^n explicitly without the inversion of a system of linear equations. But it can be shown that explicit schemes are only stable for very small Δt (Patankar, 1980; Press et al., 2002). The choice $\eta = 1$ is called the "fully implicit scheme", which can be shown to be unconditionally stable. But it has the disadvantage that a system of linear equations has to be inverted for the calculation of u^{n+1} from u^n. Additionally, it is only accurate to first order in Δt. Alternatively it is possible to choose $\eta = 1/2$, which is called "Crank–Nicolson scheme". This is also stable with higher accuracy. But it is more complicated and it might generate physically unrealistic solutions, which have to be corrected (Patankar, 1980; Press et al., 2002).

The accuracy which can be reached with an adapted Crank–Nicolson scheme goes beyond the common accuracy of basin models. Crank–Nicolson schemes conserve small scale features in the solution of differential equations. Due to upscaling approximations in model building it can be assumed that these features cannot be modeled correctly anyway. Therefore it is the authors' opinion that the fully implicit scheme is sufficient in most cases of basin modeling.

The finite differences method is always used for the discretization in time but rarely in space. It has big disadvantages concerning irregular spatial grids and discontinuities because the fields u, $\boldsymbol{\lambda}$ and q are assumed to vary smoothly from grid point to grid point. For example it is problematic to mimic permeability jumps over orders of magnitude at sharp edges of layer boundaries even if they are following exactly the grid directions. For that reason more sophisticated methods, such as finite elements or control volumes, must be used.

8.5 Finite Element Method

The finite element (FE) method was first developed for mechanical engineering purposes. Many good textbooks are available, e.g. Schwarz (1991) or Zienkiewicz (1984).

Unlike the finite difference method the finite elements method is not restricted to rectangular grid cells only. The grid cells are now called finite elements or just elements and the gridpoints belonging to one element nodes. Finite elements are very flexible for gridding irregular geometries. They can for example easily be used for compaction processes with changing layer thicknesses. In Fig. 8.6 a typical finite element grid is depicted.

In finite element approximation fields, such as pressure or temperature, are approximated by mathematical rather simple but unique and continuous differentiable functions inside the elements. These so called "form" or "shape functions" usually obey some simplified continuity conditions at the element interfaces and it is necessary that they are zero at all nodes except the one which they are related to. In practice low order multi–variate polygons, typically up to quadratic order, are widely used.

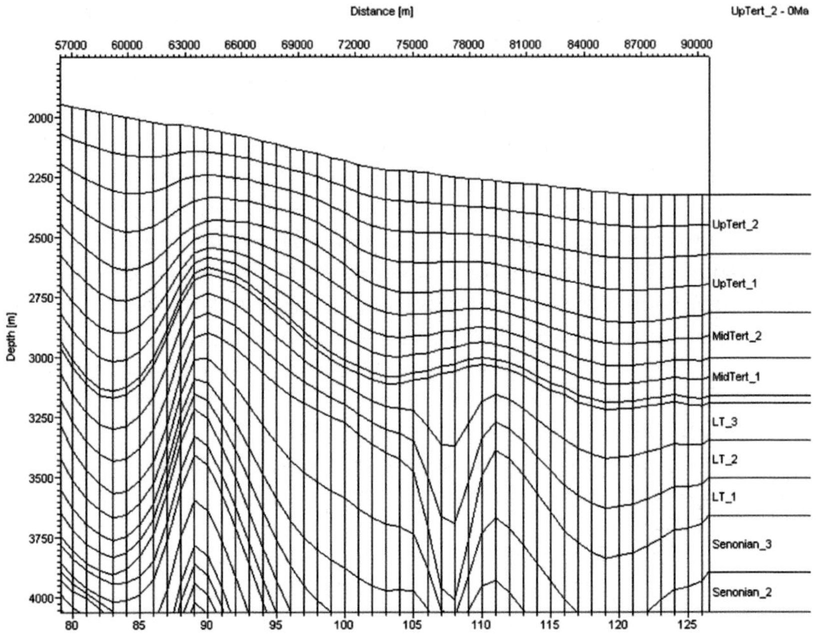

Fig. 8.6. Zoomed cutout of a two dimensional finite element grid. The annotation on the right side shows layer names

In total the finite element approach can be written as

$$u(\mathbf{x}) \approx \sum_{k=1}^{n} u_k N_k(\mathbf{x}) \tag{8.15}$$

with n as the total number of gridpoints and $\mathbf{x} = (x, y, z)^T$.[1] It is $N_k(\mathbf{x}) = 0$ outside of the elements containing the node k. Inside $N_k(\mathbf{x})$ can be written as a sum over these elements

$$N_k(\mathbf{x}) = \sum_{e(k)} N_k^e(\mathbf{x}) . \tag{8.16}$$

The functions N_i^e are the shape functions. Substitution of (8.16) in (8.15) and changing the order of summation yields

$$u(\mathbf{x}) \approx \sum_{e} \sum_{i=1}^{p^e} u_i N_i^e(\mathbf{x}) . \tag{8.17}$$

The first sum is over all elements e, p^e is the number of nodes of element e and i a numbering of nodes inside of each element. The shape functions

[1] The discretization of time is omitted for simplicity of the description. It can be performed the way outlined in the last section.

must obey $N_i^e \neq 0$ only in element e and $N_i^e(x_j, y_j, z_j) = 1$ for $i = j$ and $N_i^e(x_j, y_j, z_j) = 0$ for $i \neq j$. This ensures that for an evaluation of u at a given location $\mathbf{x}$ only the grid values u_i of the element surrounding $\mathbf{x}$ are needed.

The u_k must now be specified so, that (8.15) and (8.17) become numerically good approximations. Galerkin proposed the following method: The approximation (8.15) is used in the original differential Equation (8.1). Then the equation is multiplied from the left side with N_k and integrated over the whole volume Ω of the boundary value problem. The approximation (8.15) is not continuously differentiable. Hence second derivatives of (8.1) cannot be evaluated but this problem can be bypassed. Under consideration of boundary values it is possible to evaluate the integral with the method of partial integration and Gauss' integral formula. The resulting equations contain only first order derivatives (Schwarz, 1991). In total, a linear set of equations for the u_k is generated.

In practice it is common not to evaluate integrals which have been created by multiplication of (8.1) with N_k but with the shape functions N_k^e. Then integrals of type

$$\sum_{i=1}^{p^e} \int_{\Omega} u_i N_k^e \left(\partial_t N_i^e - \boldsymbol{\nabla} \cdot \boldsymbol{\lambda}^e \cdot \boldsymbol{\nabla} N_i^e - q^e N_i^e\right) dV \tag{8.18}$$

must be evaluated. The sum over the elements was already been skipped because the functions N_i^e and N_k^e are zero outside of element e. The conductivity coefficients $\boldsymbol{\lambda}$ and the source term q are assumed to be constant in element e with values $\boldsymbol{\lambda}^e$ and q^e. A partial integration yields now expressions in the form of

$$\begin{aligned}\sum_{i=1}^{p^e} \Bigg(& \partial_t u_i \int_{\Omega} N_k^e N_i^e dV + u_i \int_{\Omega} \boldsymbol{\nabla} N_i^e \cdot \boldsymbol{\lambda}^e \cdot \boldsymbol{\nabla} N_k^e dV \\ & - t^e \int_{\partial\Omega} N_k^e N_i^e dS - q^e \int_{\Omega} N_k^e N_i^e dV \Bigg)\end{aligned} \tag{8.19}$$

Here $\partial\Omega$ is the border of the volume Ω. The value t^e represents a Neumann boundary condition of form

$$\mathbf{n} \cdot \boldsymbol{\lambda} \cdot \boldsymbol{\nabla} u = t \tag{8.20}$$

at the border of element e with normal vector $\mathbf{n}$. If there is no Neumann boundary condition the term can be skipped. Alternatively there might be a Dirichlet boundary condition with given v of form $u = v$ at the border of element e, which can easily be represented by setting $u_i = v_i^e$ at location i. According to (8.15) – (8.17) expressions of form (8.19) can be summed up and set to zero. So it is possible to construct a set of linear equations of form

$$\sum_{i=1}^{n} (A_{ki} u_i + \partial_t u_i) = f_k \, . \tag{8.21}$$

with f_k containing all the terms with t^e, v^e, and q^e and matrix elements A_{ki} containing the terms coming from the integrals over the derivatives of the shape functions.

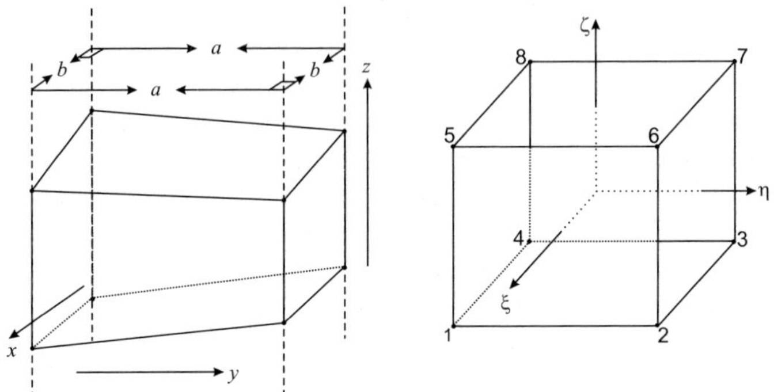

Fig. 8.7. Finite element with rectangular x-y projection on the left and cubical finite element in a ξ–η–ζ coordinate system on the right

The specifications of element types and appropriate shape functions used in basin modeling have not been completed for now. Sedimentary basins usually have a lateral extension which is at least one order of magnitude larger than their thicknesses. Most quantities, such as temperature or pressure, vary strongly in depth but only smoothly in lateral directions. Thus for numerical reasons gridding in depth should not be mixed with gridding in lateral directions.[2] This argumentation implies that gridpoints should be arranged in columns of vertical direction. The finite element method allows variable distances of the gridpoints in depth direction. Hence the most simple choice are cuboid elements of hexahedron type with the nodes at the corners determining layer interfaces Fig. 8.7.

In practice data is often available in form of regularly gridded depth maps. So hexahedrons, which are rectangular in top to bottom view, are not only optimal from a numerical point of view but also for the availability of data.

The shape functions for a cube can explicitly be denoted. It is common to use a notation with "normalized coordinates" $\boldsymbol{\xi} = (\xi, \eta, \zeta)^T$ instead of $\mathbf{x} = (x, y, z)^T$ in space. Hence for a cube with nodes at the corner points at $\xi = \pm 1$, $\eta = \pm 1$, and $\zeta = \pm 1$ a set of simple shape functions is given by

[2] Small variations are mapped in form of small contributions to the matrix elements and large effects in form of large contributions. Small contributions have numerically a more stable impact if they are separated from large contributions. In many cases this can be achieved by separation of directions.

$$\begin{aligned}
N_1^e &= \frac{1}{8}(1+\xi)(1-\eta)(1-\zeta), \; N_2^e = \frac{1}{8}(1+\xi)(1+\eta)(1-\zeta), \\
N_3^e &= \frac{1}{8}(1-\xi)(1+\eta)(1-\zeta), \; N_4^e = \frac{1}{8}(1-\xi)(1-\eta)(1-\zeta), \\
N_5^e &= \frac{1}{8}(1+\xi)(1-\eta)(1+\zeta), \; N_6^e = \frac{1}{8}(1+\xi)(1+\eta)(1+\zeta), \\
N_7^e &= \frac{1}{8}(1-\xi)(1+\eta)(1+\zeta), \; N_8^e = \frac{1}{8}(1-\xi)(1-\eta)(1+\zeta)
\end{aligned} \tag{8.22}$$

with a node numbering according to Fig. 8.7.These shape functions are linear in ξ, η, and ζ at the edges and therefore called "tri–linear". The shape functions of a general hexahedron can only be specified implicitly. Assuming that the corner points are located at the positions $\mathbf{x}_i = (x_i, y_i, z_i)^T$ it is possible to state a coordinate transformation from $\mathbf{x}$ to $\boldsymbol{\xi}$ in such a way, that (8.22) can be used for the evaluation of the shape functions:

$$\mathbf{x} = \sum_{i=1}^{8} N_i^e(\boldsymbol{\xi})\mathbf{x}_i \, . \tag{8.23}$$

Note, that the shape functions are used again for the transformation. It is easy to show that for an element with rectangular projection in the x–y plane and corner points at $x_i = a\xi_i/2$ and $y_i = b\eta_i/2$ the transformation simplifies to

$$x = \frac{a}{2}\xi, \qquad y = \frac{b}{2}\eta, \qquad z = \sum_{i=1}^{8} N_i^e(\xi,\eta,\zeta)z_i \, . \tag{8.24}$$

It is possible now to explicitly invert (8.24). With knowledge of ξ and η from x and y one yields a linear equation for ζ. Hence shape functions can also be explicitly specified but the expressions become lengthy. In a computer program an inversion can be processed almost effortlessly. The calculation of integrals, which are occurring in expressions such as (8.19), can now be drastically simplified. Using the normalized coordinates for integration the Jacobi–matrix $\mathbf{J}$ simplifies to

$$\mathbf{J} = \frac{\partial(x,y,z)}{\partial(\xi,\eta,\zeta)} = \begin{pmatrix} \frac{a}{2} & 0 & \frac{\partial z}{\partial \xi} \\ 0 & \frac{b}{2} & \frac{\partial z}{\partial \eta} \\ 0 & 0 & \frac{\partial z}{\partial \zeta} \end{pmatrix} \tag{8.25}$$

and the determinant of the Jacobi–matrix becomes

$$\det(\mathbf{J}) = \frac{ab}{4}\frac{\partial z}{\partial \zeta} \, . \tag{8.26}$$

The volume of the finite element can for example be calculated analytically and becomes

$$V^e = \frac{ab}{4}(z_5 - z_1 + z_6 - z_2 + z_7 - z_3 + z_8 - z_4) \,. \tag{8.27}$$

The restriction to rectangular projectable elements ensures that it is possible to evaluate integrals of type (8.19) numerically exact, which in general increases the overall numerical stability.

Other types of elements can be used, e.g. hexahedrons with additional nodes (Schwarz, 1991; Zienkiewicz, 1984). More complicated elements give rise to more complicated shape functions with higher order polygons. Higher order does not in general increase accuracy but it ensures higher effort and less numerical stability especially if small elements are located next to larger ones. In basin modeling such effects can occur if e.g. thin layers are modeled. So it is the authors' opinion that hexahedrons with shape functions of type (8.22) are a good choice and high accuracy should be achieved with high grid resolution.

8.6 Control Volumes

In contrast to finite differences or the finite element method where the field u of (8.1) is approximated following a rather technical argumentation control volumes (CV) were developed with the aim to use physical argumentations directly for the construction of the discretized set of equations. Mass or energy conservation with respect to the source term is the basic principle but other points, such as appropriate averages for permeabilities or conductivities, are also taken into account. A very good introduction is given in the textbook of Patankar (1980).

The time discretization is usually performed analog to the finite differences following the argumentation of Sec. 8.4.

The CV formulation starts with a discussion of a small discrete volume around a grid point. Following mass or energy conservation the total flux across the surface of this volume must equal the generated or absorbed amount inside the volume. Hence the gridpoints are usually located inside the cell of discretization but not at its interface such as typically within the finite element method.

For simplicity reasons the method is outlined here only for two dimensions, steady state and isotropic conditions. The generalization up to three dimensions and inclusion of transient as well as non–isotropic effects is straightforward. The integration of (8.1) over a volume according to Fig. 8.5 yields

$$\begin{aligned} \frac{\bar{\lambda}_{i,r}(u_{i,r} - u_i)\Delta y}{\Delta x} &- \frac{\bar{\lambda}_{i,l}(u_i - u_{i,l})\Delta y}{\Delta x} + \\ \frac{\bar{\lambda}_{i,t}(u_{i,t} - u_i)\Delta x}{\Delta y} &- \frac{\bar{\lambda}_{i,b}(u_i - u_{i,b})\Delta x}{\Delta y} + \bar{q}_i \Delta x \Delta y = 0 \,. \end{aligned} \tag{8.28}$$

Here u_i is the approximation of u and $\bar{q}_i$ the average of q inside the control volume i. The quantities referenced by the expressions $\bar{\lambda}_{i,k}$ are the "conductivities" between the control volume i and the neighboring volume in the direction k with k as r, l, t, or b indicating "right", "left", "top" or "bottom". The first four terms describe fluxes across the surfaces of the control volume.

In case of constant conductivity λ_i inside of a control volume but variation across the surfaces it can be shown with a simple physical argumentation (Patankar, 1980) that the $\bar{\lambda}_{i,k}$ must be approximated by a harmonic average such as

$$\frac{1}{\bar{\lambda}_{i,r}} = \frac{1}{2}\left(\frac{1}{\lambda_i} + \frac{1}{\lambda_{i,r}}\right) . \tag{8.29}$$

The terms of (8.28) can finally be rearranged to

$$a_i u_i = \sum_k a_{i,k} u_{i,k} + b_i \tag{8.30}$$

with the sum over all neighbors k. It is $b_i = \bar{q}_i \Delta x \Delta y$ and the coefficients a_i, $a_{i,k}$ can easily be extracted from (8.28) as

$$a_i = (\bar{\lambda}_{i,r} + \bar{\lambda}_{i,l})\frac{\Delta y}{\Delta x} + (\bar{\lambda}_{i,t} + \bar{\lambda}_{i,b})\frac{\Delta x}{\Delta y} \tag{8.31}$$

and

$$\begin{aligned} a_{i,r} &= \bar{\lambda}_{i,r}\frac{\Delta y}{\Delta x}, \qquad & a_{i,l} &= \bar{\lambda}_{i,l}\frac{\Delta y}{\Delta x}, \\ a_{i,t} &= \bar{\lambda}_{i,t}\frac{\Delta x}{\Delta y}, \qquad & a_{i,b} &= \bar{\lambda}_{i,b}\frac{\Delta x}{\Delta y} . \end{aligned} \tag{8.32}$$

They are always positive and due to the linear nature of the differential equation (8.1) it is

$$a_i = \sum_k a_{i,k} . \tag{8.33}$$

The inclusion of transient terms leads to a modification of (8.30), which is similar to (8.21). Care must be taken with generalizations of these scheme, especially if non–linearities are introduced (Patankar, 1980). Typically, this might occur with parameters, such as conductivities, capacities or the source term, if they are dependent on the field u itself.

A straightforward generalization of the scheme to non–regular cells is not possible. Obviously, the fluxes in x and y direction in (8.28) are independent. This is not the case for non–orthogonal directions of arbitrary cells. In Øye (1999) this problem is solved by subdivision of control volumes into finite elements for the calculation of fluxes through the surfaces. Unfortunately, this can only be done with huge effort. Maybe it would be possible to find an easier description if rectangular projectable volumes, such as in the previous section, were used.

A comparison between finite differences, control volumes, and finite elements extends the scope of this chapter. Some detailed arguments can e.g. be found in Gray (1984); Marsal (1976).

8.7 Solver

The finite element and the control volume method both lead to a linear set of equations, namely (8.21) and (8.30), which still must be solved. Many textbooks, such as Press et al. (2002), or software packages, such as LAPACK or LINPACK (www.netlib.org), are available providing methods or programs for the solution of linear sets of equations. Unfortunately the scope of the topic goes beyond the possibilities of this volume. Nevertheless a few facts should be listed. Solutions of linear equation systems are the most time consuming parts of computer simulations in basin modeling.

Both cases, the FE and the CV, yield symmetric and sparse equation systems. The standard approach for the solution are conjugate gradients (Press et al., 2002). A very good description of the method can be found in Shewchuk (1994). Backsubstitution methods are recommended in cases of one dimensional simulations along a well (Press et al., 2002; Patankar, 1980). Because of irregularities in the gridding and wide variations over short distances of parameters such as permeability, it becomes very difficult to use multigrid methods, which are known to have a higher performance in many cases than conjugate gradients.

Iterative methods are an alternative if good estimates for the solution are available. This can be the case if time steps are small or steady state conditions have been reached and the solution of the previous time step is used as the initial step for the actual iteration. Unfortunately, sedimentation, compaction and subsequently changing geometries over geological times prevent this approach in many cases.[3]

Conjugate gradients often work better if appropriate preconditioning is used (Shewchuk, 1994). Even a simple diagonal preconditioning yields a significant improvement of performance. More advanced methods, such as incomplete Cholesky preconditioning, are also common.

8.8 Parallelization

In practice two main computer architectures concerning parallelization must be distinguished. First, separate computers each of them consisting of a processor and memory, which communicate via network. Networks of arbitrary computers are principally not limited to overall computing power and memory and they are relatively cheap if standard PC's are used. The alternative

[3] Strictly spoken the conjugate gradient method can be interpreted as non–iterative (Shewchuk, 1994). Nevertheless, its main calculation steps are called iterations.

architecture is a computer with multiple processors, all of them with access to one shared memory. This has the great advantage that sending and receiving data is not necessary. However, such computers are expensive and do usually have limited capabilities for memory and the number of processors.

Parallelization is complicated from a technical viewpoint The usage of additional software tools is necessary. Parallel processing on networks is often implemented using the Message Passing Interface (MPI), which is the standard in numerics and scientific computing (Gropp et al., 1999; Snir et al., 1998; Gropp et al., 1998). To complete the list, alternatives, such as CORBA (Common Object Request Broker Architecture) or RPC (Remote Procedure Call), are only mentioned with some citations (Linnhoff-Popien, 1998; Redlich, 1996; Fischer and Müller, 1996). Shared memory parallelization is usually implemented with threads. The handling of threads can be improved with tools, such as the OpenMP, which is an advanced software tool for the usage of threads in scientific computing (Chandra et al., 2001).

As has been mentioned in the previous section the most time–intensive operations are solving linear equations. Thus the best starting point is a parallelization of the conjugate gradient method. This is usually done the following way: The model is cut into pieces, each processor computes one piece. Each piece is extended at its boundary by the neighboring and following gridpoints belonging to other pieces. So at each boundary an overlap region is created, which is processed by at least two processors. Data transfer is only necessary for the gridpoints of the overlap region. After each iteration of the conjugate gradient method, the data of the overlap region must be updated between the processors. The amount of data is small and network parallelization can be used if a piece is big compared to its overlap region. In practice the model is cut into slices so that data exchange occurs only between two processors (Fig. 8.9). This again reduces data transfer. Caused by rectangular gridding in lateral direction (Sec. 8.5) the best choice are slices, which are cut vertically in lateral x or y direction. When most of the simulation time is used for the simulation of differential equations, the speedup scales almost linearly with the number of processors, when the number of processors is low (Fig. 8.8).[4] If the number of processors becomes high the speedup converges to a constant value because an overhead of data communication rises. In case of hybrid models there is an important part of time used for reservoir analysis, which cannot be parallelized with the network parallelization so the speedup is low in the start. On the other hand the highly explicit treated Darcy flow part is an efficiently parallelizable part of the program, which still yields speedups where pressure and temperature have already converged. Generally, the speedup for parallelization of migration is unique and can be totally different in different case studies.

[4] Speedup is the rate of processing times. Linear speedup is found when it doubles with the doubling of processors and triples with tripling of processors.

Unfortunately, this approach limits the total number of processors to one fifth of the number of gridpoints perpendicular to the slicing. Theoretically, it would be possible to cut the slices again vertically in the other horizontal direction into sub pieces but this is commonly not done.

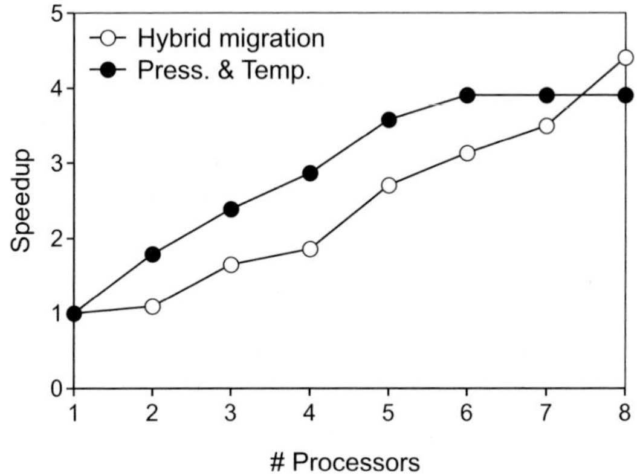

Fig. 8.8. Example speedup of network parallelized model with hybrid migration (Chap. 6.6) or with 3D pressure–temperature calculation only

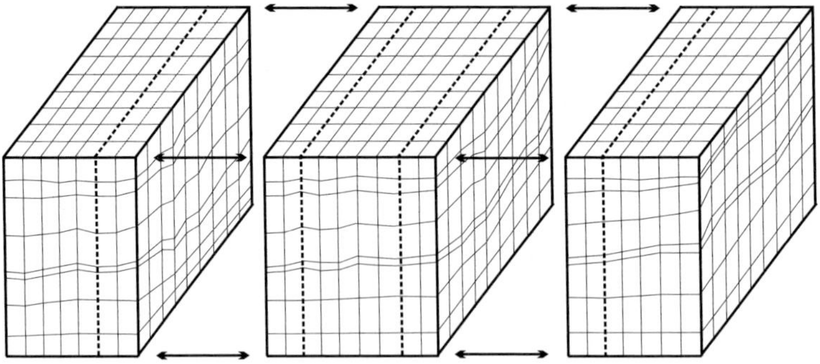

Fig. 8.9. Illustration of model decomposition into slices for parallelization. The overlap regions are clipped by the dashed lines

Caused by the time discretization (8.14) a linear set of equations must be solved for each time step. It must be noted that the performance of parallel computing can be enhanced, if each processor keeps as much data as possible between the time steps to avoid unnecessary communication. With the exception of overlap regions an overall disjunctive data distribution to the processors has the additional benefit of better balance of the computer memory.

This also manifests itself in a better performance. Usually, the effect is negligible but in extreme cases the total speedup can become "superlinear" because different types of computer memory from processor cache to swap space with extremely different performances exist and smaller amounts of data are often stored in faster accessible memory locations. In principle a disjunctive data model is not a major issue, but in practice this needs a lot of detailed work.

Besides the solution of linear sets of equations, reservoir analysis is the second most time consuming part in basin modeling (Chap. 6.5). The hydrocarbons can move large distances almost laterally before they are trapped. Usually, the reservoir analysis is map based. Hence a parallelization founded on vertically cut slices is not possible. Injected amounts can reach the map from almost everywhere in the model and might be redistributed almost everywhere again, especially if conductive faults are taken into account. The amount of transfered data is high if multi–component descriptions of the hydrocarbon phases are necessary. Thus it is not possible to use network parallelization. Additionally, it is impossible to parallel process the reservoirs because hydrocarbons which leave one carrier usually enter another one. So in general, the reservoirs are strongly interacting. They must be processed sequentially from bottom to top, from source to trap. A possible parallelization can only be performed inside each reservoir analysis. Here the most time consuming part is the volumetrics. A thread parallelization of volumetrics in the carriers is possible (Bücker et al., 2008). But speedups as with the solution of differential equations by parallelization of the conjugate gradient algorithm are not to be expected.

8.9 Local Grid Refinement (LGR)

Commonly, two different reasons for LGR are found in basin modeling. Firstly, the overall model resolution is generally too high for processing. In this case, a central area of interest is typically calculated in full resolution whereas the outer rim of the model is sampled. This approach allows for feasible computing times and high resolution results. Secondly, high resolution data is only available for a prospect or field, which is small compared to the extension of a meaningful basin model. Obviously, the high resolution data can be integrated via LGR into the basin model. Principally, processing speed can be optimized if a discretization is adapted rigorously to the availability of data and the areas of interest.

It was argued contrarily in Sec. 8.5 that discretization in basin modeling should be performed on a basis of stacked maps with vertical columns of gridpoints and rectangular base areas in horizontal alignment. The availability and the handling of data in model builders and viewers, which is restricted almost exclusively to rectangular gridded maps, complies very well with this requirement. A slight improvement of common style regular gridding is called the tartan grid and introduced in Sec. 8.9.1. Compromises between slight improvements and a more rigorous local grid refinement are discussed in the following subsections.

8.9.1 Tartan Grid

A tartan grid is shown in Fig. 8.10. It consists of a regular rectangular grid with arbitrary distances in x and y–direction. The central area of interest is here gridded with high resolution. A tartan grid is compatible with the requirements of Sec. 8.5. Such models can be parallelized without any limitations in the same manner as described in Sec. 8.8. Obviously, the number of gridpoints is not optimal. Grid points are lying on grid lines and hence additional gridpoints in an area of interest lead also to additional gridpoints outside of the area of interest. Unnecessary gridpoints reduce the performance.

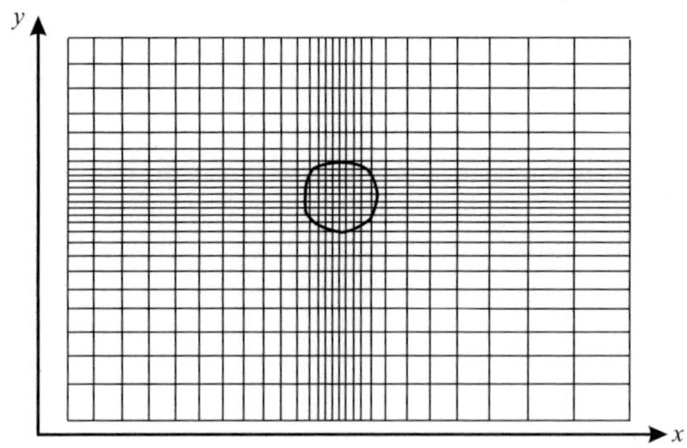

Fig. 8.10. Map view of a tartan grid. The polygon outlines an area of interest which is gridded in higher resolution

8.9.2 Windowing

A low resolution model, which is covering wide areas around a small high resolution model, can be used to calculate boundary conditions (e.g. for heat flow analysis or pressure prediction) at the interface to the small scale high resolution model (Fig. 8.11). Unrealistic "no–flow" Neumann conditions at model sides, such as shown in Fig. 2.5 or Fig. 3.13, are hence shifted away from the area of interest. This approach is often used for heat flow and temperature calculations and sometimes for pressure prediction.

A specification of a Dirichlet boundary condition for an explicitly treated migration model is obviously problematic. Catching of all amounts which enter the area of the small model is possible. These amounts can be injected into the small scale model. However, migration becomes complicated when HCs leave the small model but not the big low resolution area. At a later time they may reenter the area of the small model. A strong coupling in space and time with iterations between both models might be necessary. The whole procedure becomes technically very complicated, when accumulations are located at the boundary between both models.

Alternatively, the big low scale model can be modeled without taking into account any effects of the smaller one. Petroleum amounts which are found in the area of the high resolution model are redistributed according to the high resolution geometry with some final migration steps. Again, both models must be strongly coupled when migration proceeds differently through the high resolution than through the low resolution model. With coupling this approach becomes like a two grid version of a multigrid method.

A big advantage is the possibility of using different migration modeling methods in both models even if they are coupled. For example, a hybrid method with fast long distance migration might be used for the low resolution model and a detailed invasion percolation in the region with the high resolution data. Additionally it must be noted that such an approach is much easier to implement because the model boundary between the low and high resolution area is not of special interest.

Generally, a big advantage of windowing is its compatibility with parallel processing as outlined in Sec. 8.8. Node numbering and slicing remain in each model the same. Especially uncoupled models can be build and calculated with high performance.

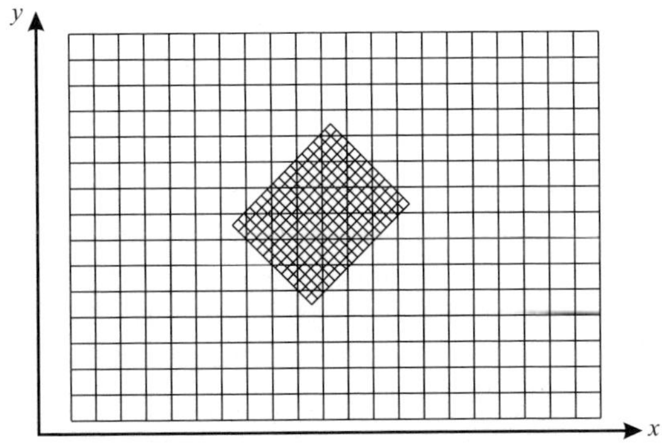

Fig. 8.11. Map view of a high resolution grid, which belongs to a small model, and a low resolution grid of a large model, which covers a wide area around the small model

8.9.3 Coupled Model in Model

In the windowing approach the lowly resolved large area model is simulated before the highly resolved small area model. This might lead to principal problems as outlined in the previous section. Alternatively, both models can be coupled more tightly and simulated consistently so that feedback of each model is taken into account immediately by the other one. For example, both models must be adapted in a way that flow conditions at any location on the interface between the models for all times are consistently fulfilled on both sides. Technically, this can be achieved by special coupling of both models to one large overall model via extra grid cells, extra coupling conditions and/or by iterative modeling where flow amounts at the model interface are stepwise refined until convergence is reached (Fig. 8.12).

Obviously, this approach needs much more computing resources than windowing. Additionally, parallelization becomes more complicated.

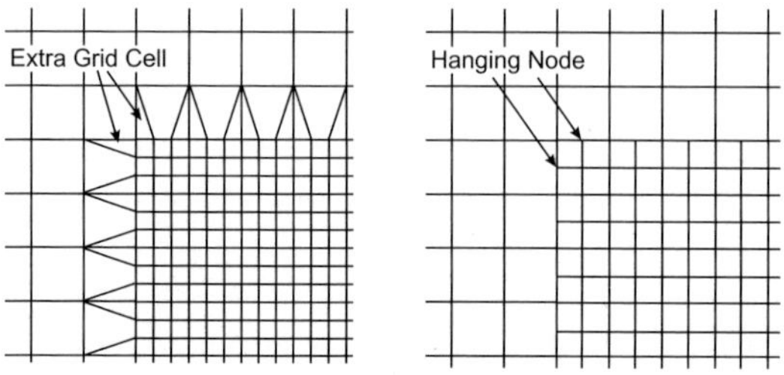

Fig. 8.12. Crossover from a low to a high resolution grid with extra grid cells on the left side and "hanging nodes" without extra grid cells on the right side. Crossovers with hanging nodes are usually treated with extra coupling conditions

8.9.4 Faults

Faults are very thin compared to common grid resolutions. Concerning heat flow or migration they can often be modeled as ideal surfaces (Secs. 6.5.4, 6.6.3, 6.8.6). Contrarily, pressure prediction makes it necessary to take into account volumetric properties of faults, e.g. which determine water flow conductivity (Fig. 2.56). The grid is for that reason often improved locally around faults (Figs. 2.47, 8.13). Practically, this can easily be achieved if the faults are following the surfaces and edges of existing cells of the "volume–grid". A small error in fault location due to gridding is less important than not correctly taking into account water outflow from an overpressured region.

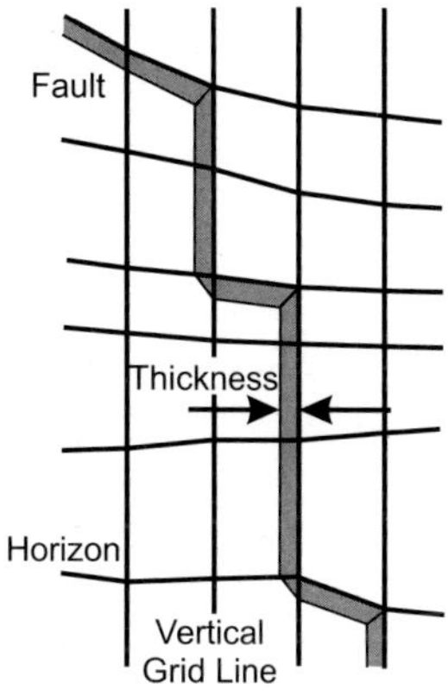

Fig. 8.13. Fault of constant thickness, which is gridded with extra cells (grey). It is following surfaces and edges of the overall "volume–grid"

Summary: Basin modeling makes extensive use of mathematical and numerical methods. The solution of differential equations of diffusion type are the largest group of frequently occurring numerical tasks.

The temporal evolution is often modeled with finite differences. Explicit and implicit schemes are discussed. The spatial behavior of the differential equations is treated with finite elements or control volumes. Both methods are shortly outlined in theory.

Regularly gridded maps are the most common data source for basin modeling. Hexahedrons are for that reason the basic three dimensional spatial building blocks of a model. The finite element method is presented in more detail for hexahedrons with form functions of the most simple tri–linear type. Some typically appearing integrals can be solved analytically for this case.

The numerical solution of the resulting linear equations is summarized. Focus is put on a discussion of computational parallelization. An overall model slicing and separate processing of these slices on different computers is described. This method yields a linear speedup for a low number of processors. However, it is not applicable for flowpath and invasion percolation based migration. Other approaches must be found, e.g. parallelization based on shared memory access.

Local grid refinement (LGR) allows processing of models with high resolution data, which are usually not feasible due to computer performance and memory restrictions. Small scale field or prospect data can be incorporated into basin models. Special areas of interest can be processed in high resolution. LGR is outlined for a continuous crossover from regularly gridded models over the tartan grid and windowing to coupled models in models. Spatially improved fault handling is also described.

Mixing rules for upscaling to bulk values are discussed. Some rules of thumb for mixing and upscaling are given, when more detailed information is missing. The most simple approaches are arithmetic, geometric, and harmonic averages or maximum based upscaling.

References

M. Aigner and G. M. Ziegler. *Das BUCH der Beweise.* Springer, second edition, 2004.

G. R. Beardsmore and J. P. Cull. *Crustal Heat Flow.* Cambridge University Press, 2001.

H. M. Bücker, A. I. Kauerauf, and A. Rasch. A smooth transition from serial to parallel processing in the industrial petroleum system modeling package petromod. *Computers & Geosciences*, 34:1473–1479, 2008.

R. Chandra, L. Dagum, D. Maydan, J. McDonald, and R. Menon. *Parallel Programming in OpenMP.* Morgan Kaufmann Publishers, San Francisco, CA, 2001.

S. Fischer and W. Müller. *Netzwerprogrammierung unter Linux und Unix.* UNIX easy. Carl Hanser Verlag München Wien, 1996.

W. G. Gray. Comparision of finite difference and finite element methods. In J. Bear and M. Y. Corapcioglu, editors, *Fundamentals of Transport Phenomena in Porous Media*, number 82 in NATO ASI Series, E: Applied Science. Martinus Nijhoff, 1984.

W. Gropp, S. Huss-Lederman, A. Lumsdaine, E. Lusk, B. Nitzberg, W. Saphir, and M. Snir. *MPI – The Complete Reference, Volume 2, The MPI Extensions.* Scientific and Engineering Computation. The MIT Press, 1998.

W. Gropp, E. Lusk, and A. Skejellum. *Using MPI – Portable Parallel Programming with the Message–Passing Interface.* Scientific and Engineering Computation. The MIT Press, 2 edition, 1999.

C. Linnhoff-Popien. *CORBA–Kommunikation und Management.* Springer, 1998.

D. Marsal. *Die numerische Lösung partieller Differentialgleichungen in Wissenschaft und Technik.* Bibliographisches Institut, 1976.

G. Å. Øye. *An Object–Oriented Parallel Implementation of Local Grid Refinement and Domain Decomposition in a Simulator for Secondary Oil Migration.* PhD thesis, University of Bergen, 1999.

S. V. Patankar. *Heat Transfer and Fluid Flow.* Hemisphere Publishing Corporation, 1980.

W. H. Press, S. A. Teukolsky, W. T. Vetterling, and B. P. Flannery. *Numerical Recipes in C++.* Cambridge University Press, second edition, 2002.

J.-P. Redlich. *CORBA 2.0.* Addsion–Wesley Publishing Company, 1996.

H. R. Schwarz. *Methode der finiten Elemente.* B. G. Teubner, Stuttgart, 1991.

J. R. Shewchuk. An introduction to the conjugate gradient method without the agonizing pain. Lecture, School of Computer Science, Carnegie Mellon University, Pittsburgh, PA 15213, 1994.

M. Snir, S. Otto, S. Huss-Lederman, D. Walker, and J. Dongarra. *MPI – The Complete Reference, Volume 1, The MPI Core.* Scientific and Engineering Computation. The MIT Press, second edition, 1998.

O. C. Zienkiewicz. *Methode der finiten Elemente.* Carl Hanser, second edition, 1984.

A

Compaction and Flow Parameter

The data for depositional porosities and Athy's compaction parameters vs. depth are worked out by Doug Waples. It is based on a comprehensive literature review and his modeling experience over 20 years. The other compaction parameters are due to IES experience and fit to the curves from Doug Waples. The permeability values are also provided from Doug Waples. The capillary entry pressures are derived from permeabilities with the Hildenbrand equations from Chap. 6.

	Depo. Poro.	Athy k	Athy k	Compress. y Max.	Compress. y Min.	Schneider Factor k_a	Schneider Factor k_b	ϕ
	[%]	$\left[\frac{1}{km}\right]$	$\left[\frac{1}{MPa}\right]$	$\left[\frac{10^{-7}}{kPa}\right]$	$\left[\frac{10^{-7}}{kPa}\right]$	$\left[\frac{1}{MPa}\right]$	$\left[\frac{1}{MPa}\right]$	
Limestone								
ooid grainstone	35.0	0.01	0.001	2.0	1.0	0.001	0.001	0.18
Waulsort. mound	16.0	0.01	0.001	2.0	1.0	0.001	0.001	0.08
micrite	51.0	0.52	0.04766	850.0	19.8	0.03212	0.07894	0.26
shaly	48.0	0.50	0.04493	686.5	18.8	0.03091	0.07090	0.24
org.–rich								
typical	51.0	0.52	0.04939	881.2	20.5	0.03329	0.08185	0.26
1-2% TOC	51.0	0.52	0.04851	865.1	20.1	0.03270	0.08037	0.26
10% TOC	51.0	0.52	0.05365	956.8	22.3	0.03616	0.08890	0.26
chalk, typical	70.0	0.90	0.10546	4613.3	41.6	0.20793	0.05813	0.35
chalk, 95% calcite	70.0	0.90	0.10546	4611.2	41.7	0.20793	0.05813	0.35
chalk, 75% calcite	67.0	0.90	0.10124	3871.1	40.6	0.19000	0.05738	0.34
chalk, 40% calcite	65.0	0.90	0.09870	3453.3	39.9	0.05891	0.18975	0.33
Marl	50.0	0.50	0.04651	778.0	19.8	0.07375	0.03086	0.25
Dolomite								
typical	35.0	0.39	0.03084	252.3	11.0	0.04219	0.02344	0.17
org.–lean, sandy	35.0	0.39	0.03119	255.2	11.2	0.04312	0.02344	0.17
org.–lean, silty	35.0	0.39	0.03138	256.7	11.2	0.04188	0.02406	0.17
org.–rich	35.0	0.39	0.03459	283.0	12.4	0.02656	0.04634	0.17

Table A.1. Compaction Parameter: Carbonate Rocks

	Depo. Poro. [%]	Athy k $\left[\frac{1}{\text{km}}\right]$	Athy k $\left[\frac{1}{\text{MPa}}\right]$	Compress. y Max. $\left[\frac{10^{-7}}{\text{kPa}}\right]$	Compress. y Min. $\left[\frac{10^{-7}}{\text{kPa}}\right]$	Schneider Factor k_a $\left[\frac{1}{\text{MPa}}\right]$	Schneider Factor k_b $\left[\frac{1}{\text{MPa}}\right]$	Schneider Factor ϕ
Biogenic Sediments								
Chalk, typical	70.0	0.90	0.10546	4611.2	41.7	5.00000	0.04500	0.35
Chalk, 95% calcite	70.0	0.90	0.10546	4611.2	41.7	0.20793	0.05813	0.35
Chalk, 75% calcite	67.0	0.90	0.10124	3871.1	40.6	0.19000	0.05738	0.34
Chalk, 40% calcite	65.0	0.90	0.09870	3453.3	39.9	0.05891	0.18975	0.33
Coal, pure	76.0	0.43	1.51396	87407.4	564.6	3.41736	0.74043	0.38
Coal, impure	74.0	0.42	0.22597	11692.7	87.9	0.49695	0.11328	0.37
Coal, silty	68.0	0.40	0.13392	5289.1	53.2	0.07308	0.28035	0.34
Clastic Sediments								
Sandstone								
typical	41.0	0.31	0.02660	274.7	11.5	0.04156	0.01781	0.20
clay–rich	40.0	0.32	0.02661	265.4	11.1	0.04000	0.01812	0.20
clay–poor	42.0	0.30	0.02627	280.5	11.8	0.04047	0.01734	0.21
quartzite								
typical	42.0	0.30	0.02726	291.0	12.2	0.04281	0.01781	0.21
very quartz–rich	42.0	0.27	0.02461	252.6	11.8	0.01615	0.03977	0.21
subarkose								
typical	41.0	0.28	0.02468	244.8	11.4	0.01631	0.03874	0.20
quartz–rich	42.0	0.28	0.02533	263.4	11.9	0.03984	0.01641	0.21
clay–rich	42.0	0.30	0.02643	282.2	11.8	0.04063	0.01750	0.21
clay–poor	42.0	0.30	0.02627	280.5	11.8	0.04047	0.01734	0.21
dolomite–rich	40.0	0.30	0.02572	249.8	11.2	0.04000	0.01688	0.20
arkose								
typical	39.0	0.33	0.02772	267.3	11.2	0.01981	0.04212	0.20
quartz–rich	41.0	0.30	0.02623	267.1	11.6	0.04094	0.01750	0.20
quartz–poor	40.0	0.32	0.02646	263.9	11.1	0.04000	0.01812	0.20
clay–rich	40.0	0.32	0.02661	265.4	11.1	0.04000	0.01812	0.20
clay–poor	39.0	0.32	0.02722	258.9	11.2	0.04063	0.01812	0.20
dolomite–rich	39.0	0.32	0.02658	252.8	11.0	0.03875	0.01812	0.20
wacke	39.0	0.34	0.02773	271.1	10.9	0.01992	0.04192	0.20

Table A.2. Compaction Parameter

	Depo. Poro. [%]	Athy k $\left[\frac{1}{\text{km}}\right]$	Athy k $\left[\frac{1}{\text{MPa}}\right]$	Compress. y Max. $\left[\frac{10^{-7}}{\text{kPa}}\right]$	Min. $\left[\frac{10^{-7}}{\text{kPa}}\right]$	Schneider Factor k_a $\left[\frac{1}{\text{MPa}}\right]$	k_b $\left[\frac{1}{\text{MPa}}\right]$	ϕ
Shale								
typical	70.0	0.83	0.09613	4032.7	40.3	0.19157	0.05270	0.35
org.–lean								
typical	70.0	0.83	0.09613	4032.7	40.3	0.19157	0.05270	0.35
sandy	65.0	0.83	0.08999	3005.2	38.7	0.17479	0.05340	0.33
silty	67.0	0.83	0.09230	3373.4	39.4	0.18394	0.05357	0.34
silic., typical	70.0	0.83	0.09556	4007.5	40.1	0.19043	0.05238	0.35
silic., 95% Opal–CT	80.0	0.83	0.14631	7397.5	37.8	0.33248	0.06983	0.40
black	70.0	0.83	0.10931	4584.3	45.8	0.22348	0.06082	0.35
organic–rich								
typical	70.0	0.83	0.10931	4291.1	42.9	0.18549	0.05265	0.30
3% TOC	70.0	0.83	0.10165	4264.1	42.6	0.20782	0.05656	0.35
8% TOC	70.0	0.83	0.10931	4584.3	45.4	0.22348	0.06082	0.35
20% TOC	70.0	0.83	0.12975	5441.2	54.4	0.26527	0.07219	0.35
Siltstone								
org.–lean	55.0	0.51	0.04907	1036.1	21.1	0.03165	0.08611	0.28
organic–rich								
typical	55.0	0.51	0.04937	1042.2	21.2	0.03185	0.08660	0.28
$>$ 10% TOC	55.0	0.51	0.05459	1152.6	23.5	0.09119	0.03419	0.28
2-3% TOC	55.0	0.51	0.04966	1048.7	21.4	0.03204	0.08714	0.28
Conglomerate								
typical	30.0	0.30	0.02429	142.1	8.8	0.01897	0.03228	0.15
quartzitic	30.0	0.30	0.02429	142.1	8.8	0.01897	0.03228	0.15
Tuff, felsic	60.0	0.35	0.03711	959.1	16.5	0.06961	0.02117	0.3
Tuff, basaltic	60.0	0.35	0.03212	829.3	14.3	0.06000	0.01844	0.3

Table A.3. Compaction Parameter: Clastic Sediments – Part 2

	Aniso-tr.	Porosity [%] at Point 1	2	3	Permeability [log mD] at Point 1	2	3
Biogenic Sediments							
Chalk, typical	1.5	1.00	25	70	−6.75	−3.10	1.00
Chalk, 95% calcite	1.5	1.00	25	70	−6.75	−3.10	1.00
Chalk, 75% calcite	2.0	1.00	25	67	−6.75	−3.10	1.00
Chalk, 40% calcite	3.0	1.00	25	65	−6.75	−3.10	1.00
Coal, pure	4.0	3.02	25	76	−7.50	−2.60	0.00
Coal, impure	4.0	3.17	25	74	−7.50	−2.60	0.00
Coal, silty	4.0	3.39	25	68	−7.50	−2.60	0.00
Diatomite, clay–poor	3.0	1.00	25	80	−2.55	−1.30	3.00
Diatomite, clay–rich	10.0	1.00	25	75	−3.05	−1.80	2.50
Carbonate Rocks							
Limestone							
ooid grainstone	1.1	1.00	15	35	−2.44	2.60	3.00
Waulsortian mound	1.1	1.00	12	16	−2.50	3.00	3.35
micrite	1.1	1.00	25	51	−2.20	1.00	1.52
shaly	2.0	1.00	25	48	−1.99	0.72	1.50
organic–rich, typical	2.0	1.00	25	51	−1.99	0.73	1.00
organic–rich, 1-2% TOC	1.2	1.00	25	51	−1.99	0.73	1.00
organic–rich, 10% TOC	2.0	1.00	25	51	−1.99	0.72	1.00
chalk, typical	1.5	1.00	25	70	−6.75	−3.10	1.00
chalk, 95% calcite	1.5	1.00	25	70	−6.75	−3.10	1.00
chalk, 75% calcite	2.0	1.00	25	67	−6.75	−3.10	0.73
chalk, 40% calcite	3.0	1.00	25	65	−6.75	−3.10	0.54
Marl	1.2	1.00	25	50	−5.05	−2.25	−0.78
Dolomite							
typical	1.1	1.00	25	35	0.11	2.83	3.92
organic–lean, sandy	1.1	1.00	25	35	0.60	2.98	3.93
organic–lean, silty	1.1	1.00	25	35	0.31	2.88	3.91
organic–rich	1.2	1.00	25	35	−0.37	2.69	3.91

Table A.4. Porosity and Permeability Parameter

	An-iso-tr.	Porosity [%] at Point 1	2	3	Permeability [log mD] at Point 1	2	3
Sandstone							
typical	5.0	1	25	41	−1.80	3.00	4.33
clay–rich	10.0	1	25	40	−2.80	2.00	3.62
clay–poor	5.0	1	25	42	−1.80	3.00	4.84
quartzite, typical	3.0	1	25	42	−1.80	3.00	4.84
quartzite, very quartz–rich	2.0	1	25	42	−1.80	3.00	4.84
subarkose, typical	3.0	1	25	41	−1.80	3.00	4.73
subarkose, quartz–rich	3.0	1	25	42	−1.80	3.00	4.84
subarkose, clay–rich	10.0	1	25	42	−2.80	2.00	3.84
subarkose, clay–poor	3.0	1	25	42	−1.80	3.00	4.84
subarkose, dolomite–rich	5.0	1	25	40	−2.30	2.50	4.12
arkose, typical	5.0	1	25	39	−1.80	3.00	4.51
arkose, quartz–rich	4.0	1	25	41	−1.80	3.00	4.73
arkose, quartz–poor	7.0	1	25	40	−2.30	2.50	4.12
arkose, clay–rich	10.0	1	25	40	−2.80	2.00	3.62
arkose, clay–poor	4.0	1	25	39	−1.80	3.00	4.51
arkose, dolomite–rich	5.0	1	25	39	−2.30	2.50	4.01
wacke	7.0	1	25	39	−2.80	2.00	3.51
Shale							
typical	1.2	1	25	70	−8.52	−3.00	−1.00
org.–lean, typical	1.2	1	25	70	−8.52	−3.00	−1.00
org.–lean, sandy	1.5	1	25	65	−8.42	−2.00	1.00
org.–lean, silty	1.3	1	25	67	−8.47	−2.50	0.50
org.–lean, silic., typical)	1.5	1	25	70	−8.47	−2.00	1.30
org.–lean, silic., 95% Opal-CT	1.1	1	25	80	−8.42	−1.50	−1.00
black	5.0	1	25	70	−8.52	−3.00	−1.00
organic–rich, typical	5.0	1	25	70	−8.52	−3.00	−1.00
organic–rich, 3% TOC	2.0	1	25	70	−8.52	−3.00	−1.00
organic–rich, 8% TOC	5.0	1	25	70	−8.52	−3.00	−1.00
organic–rich, 20% TOC	10.0	1	25	70	−8.52	−3.00	−1.00
Siltstone							
organic–lean	10.0	1	25	55	−6.28	−1.00	0.71
organic–rich, typical	10.0	1	25	55	−6.28	−1.00	0.71
organic–rich, > 10% TOC	10.0	1	25	55	−6.28	−1.00	0.71
organic–rich, 2-3% TOC	10.0	1	25	55	−6.28	−1.00	0.71
Conglomerate, typical	1.1	1	25	30	−2.80	2.00	4.00
Conglomerate, quartzitic	1.1	1	25	30	−2.80	2.00	4.00
Tuff, felsic	1.3	1	28	60	−1.87	1.75	4.00
Tuff, basaltic	1.3	1	28	60	−1.87	1.75	4.00

Table A.5. Porosity and Permeability Parameter: Clastic Sediments

	Entry Pr. [MPa] at Porosity		Parameter	
	1%	25%	a [MPa]	b
Biogenic Sediments				
Chalk, typical	15.42	5.78	16.78	−0.037
Chalk, 95% calcite	15.42	5.78	16.78	−0.037
Chalk, 75% calcite	15.42	5.78	16.78	−0.037
Chalk, 40% calcite	15.42	5.78	16.78	−0.037
Coal, pure	23.35	5.36	26.14	−0.049
Coal, impure	23.35	5.36	26.16	−0.050
Coal, silty	23.35	5.35	26.19	−0.050
Diatomite, clay–poor	2.94	1.47	1.56	−0.013
Diatomite, clay–rich	4.30	2.15	2.05	−0.013
Carbonate Rocks				
Limestone				
ooid grainstone	1.42	0.08	1.74	−0.086
Waulsortian mound	1.47	0.06	1.94	−0.120
micrite	1.25	0.26	1.34	−0.032
shaly	1.11	0.32	1.18	−0.027
organic–rich, typical	1.11	0.31	1.18	−0.027
organic–rich, 1-2% TOC	1.11	0.31	1.18	−0.027
organic–rich, 10% TOC	1.11	0.32	1.18	−0.027
chalk, typical	15.42	5.78	16.78	−0.037
chalk, 95% calcite	15.42	5.78	16.78	−0.037
chalk, 75% calcite	15.42	5.78	16.78	−0.037
chalk, 40% calcite	15.42	5.78	16.78	−0.037
Marl	6.03	3.03	6.43	−0.028
Dolomite				
typical	0.35	0.06	0.37	−0.027
organic–lean, sandy	0.27	0.06	0.28	−0.024
organic–lean, silty	0.31	0.06	0.33	−0.026
organic–rich	0.45	0.07	0.49	−0.031

Table A.6. Capillary Entry Pressure

	Entry Pr. [MPa] at Porosity 1%	Entry Pr. [MPa] at Porosity 25%	Parameter a [MPa]	Parameter b
Sandstone				
typical	1.00	0.06	1.12	−0.048
clay–rich	1.74	0.12	1.94	−0.048
clay–poor	1.00	0.06	1.12	−0.048
quartzite, typical	1.00	0.06	1.12	−0.048
quartzite, very quartz–rich	1.00	0.06	1.12	−0.048
subarkose, typical	1.00	0.06	1.12	−0.048
subarkose, quartz–rich	1.00	0.06	1.12	−0.048
subarkose, clay–rich	1.74	0.12	1.94	−0.048
subarkose, clay–poor	1.00	0.06	1.12	−0.048
subarkose, dolomite–rich	1.32	0.08	1.47	−0.048
arkose, typical	1.00	0.06	1.12	−0.048
arkose, quartz–rich	1.00	0.06	1.12	−0.048
arkose, quartz–poor	1.32	0.08	1.47	−0.048
arkose, clay–rich	1.74	0.12	1.94	−0.048
arkose, clay–poor	1.00	0.06	1.12	−0.048
arkose, dolomite–rich	1.32	0.08	1.47	−0.048
wacke	1.74	0.12	1.94	−0.048
Shale				
typical	41.02	5.36	46.58	−0.055
organic–lean, typical	41.02	5.36	46.58	−0.055
organic–lean, sandy	38.82	2.50	45.00	−0.064
organic–lean, silty	39.90	3.66	45.78	−0.060
organic–lean, siliceous, typical	39.90	2.50	46.31	−0.065
organic–lean, siliceous, 95% Opal–CT	38.82	1.71	45.52	−0.069
black	41.02	5.35	46.58	−0.055
organic–rich, typical	41.02	5.35	46.58	−0.055
organic–rich, 3% TOC	41.02	5.35	46.58	−0.055
organic–rich, 8% TOC	41.02	5.35	46.58	−0.055
organic–rich, 20% TOC	41.02	5.35	46.58	−0.055
Siltstone				
organic–lean	11.90	1.17	13.43	−0.053
organic–rich, typical	11.90	1.17	13.43	−0.053
organic–rich, $>$ 10% TOC	11.90	1.17	13.43	−0.053
organic–rich, 2-3% TOC	11.90	1.17	13.43	−0.053
Conglomerate, typical	1.74	0.12	1.94	−0.048
Conglomerate, quartzitic	1.74	0.12	1.94	−0.048

Table A.7. Capillary Entry Pressure: Clastic Sediments

B

Deviation of the Pressure Equation

The deviation of (2.13) follows the appendix in the article from Luo and Vasseur (1992).

The conservation of solid rock material in a fixed coordinate system can be described by a continuity equation of form

$$\frac{\partial(\rho_s(1-\phi))}{\partial t} + \boldsymbol{\nabla} \cdot (\rho_s(1-\phi)\mathbf{v}_s) = 0 \tag{B.1}$$

with the porosity ϕ, the density ρ_s of the solid grains and the velocity $\mathbf{v}_s$ of the compacting and thus moving solid. Note that $\mathbf{v}_s$ is not a bulk velocity. It is the "real" volocity of the moving rock. Under the assumption that the grain density keeps constant it becomes

$$\boldsymbol{\nabla} \cdot \mathbf{v}_s = \frac{1}{1-\phi} \frac{d\phi}{dt} \tag{B.2}$$

with

$$\frac{d}{dt} = \frac{\partial}{\partial t} + \mathbf{v}_s \cdot \boldsymbol{\nabla} \quad . \tag{B.3}$$

The continuity equation for a fluid of density ρ which is moving with velocity $\mathbf{v}$ is

$$\frac{\partial(\rho\phi)}{\partial t} + \boldsymbol{\nabla} \cdot (\rho\phi\mathbf{v}) = 0 \; . \tag{B.4}$$

Again $\mathbf{v}$ is the "real" and not the bulk velocity. Darcy's law states

$$\phi(\mathbf{v} - \mathbf{v}_s) = -\frac{\mathbf{k}}{\mu} \cdot \boldsymbol{\nabla} u \tag{B.5}$$

with the permeability $\mathbf{k}$, mobility μ, and overpressure potential $u = p - p_h$. The term $\mathbf{v} - \mathbf{v}_s$ is the relative velocity of the fluid in the rock and $\phi(\mathbf{v} - \mathbf{v}_s)$ is the relative bulk velocity. Equations (B.5) and (B.2) can be inserted in (B.4). Rearranging yields

$$\frac{\rho}{1-\phi}\frac{d\phi}{dt} = \boldsymbol{\nabla}\cdot\left(\frac{\mathbf{k}}{\mu}\cdot\boldsymbol{\nabla}u\right) - \phi\frac{d\rho}{dt}\,. \tag{B.6}$$

According to the compaction law

$$\frac{d\phi}{dt} = -C\frac{d\sigma'_z}{dt} \tag{B.7}$$

it becomes with $\sigma'_z = u_l - u$

$$\frac{\rho}{1-\phi}\frac{du}{dt} - \boldsymbol{\nabla}\cdot\left(\frac{\mathbf{k}}{\mu}\cdot\boldsymbol{\nabla}u\right) = \frac{\rho}{1-\phi}\frac{du_l}{dt} - \phi\frac{d\rho}{dt}\,. \tag{B.8}$$

A transition to a moving coordinate system, which moves with the solid rock, yields a substitution of d/dt with $\partial/\partial t$ and finally to equation (2.13). The result differs only by some $1/(1-\phi)$ factors from the expected result. A direct usage of (2.10) in the moving coordinate system yields the same result.

X. Luo and G. Vasseur. Contributions of compaction and aquathermal pressuring to geopressure and the influence of environmental conditions. *AAPG Bulletin,* **76**(10):1550–1559, 1992.

C

Analytic Groundwater Flow Solution from Tóth

The steady state groundwater flow for an idealized block as shown in Fig. C.1 can be calculated analytically. The boundary condition on top is given by a hydraulic head of form $h_0(1 - x/w)$. At bottom and at the sides a no–flow condition is assumed. All properties such as permeability k or water density ρ_w within the block are constant and without anisotropy. The pressure equation reduces for the potential u hence to

$$\frac{\partial^2}{\partial x^2} u(x,z) + \frac{\partial^2}{\partial z^2} u(x,z) = 0 \tag{C.1}$$

with the boundary conditions

$$\begin{aligned} u(x,0) &= \rho_w g h_0 (1 - \frac{x}{w}) && \text{for} \quad 0 < x < w \\ \frac{\partial}{\partial z} u(x,z)\Big|_{z=h} &= 0 && \text{for} \quad 0 < x < w \\ \frac{\partial}{\partial x} u(x,z)\Big|_{x=0} &= \frac{\partial}{\partial x} u(x,z)\Big|_{x=w} = 0 && \text{for} \quad 0 < z < h\,. \end{aligned} \tag{C.2}$$

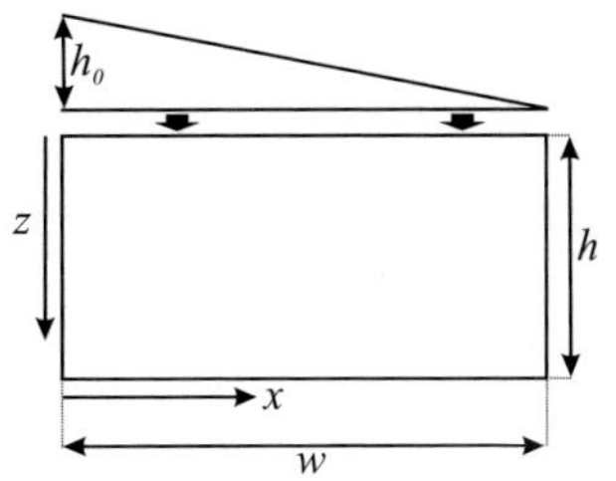

Fig. C.1. Groundwater flow model according to Tóth (1962)

The analytical solution has the form

$$u = \frac{\rho_w g h_0}{2} \left(1 + 8 \sum_{n=0}^{\infty} \frac{\cosh(\mu_n(h-z))\cos(\mu_n x)}{\mu_n^2 w^2 \cosh(\mu_n h)} \right) \tag{C.3}$$

with $\mu_n = (2n+1)\pi/w$. An example is shown in Figs. C.2 and C.3. The water density has been chosen as $\rho_w = 1019\,\mathrm{kg/m^3} = 1000.0/0.980665\,\mathrm{kg/m^3}$ so that $u(0,0) = 10\,\mathrm{MPa}$. The model has also been calculated numerically as a benchmark with finite elements on a regular grid with 100×20 cells. Errors were below 0.01 MPa.

The differential equation and the solution are formally very similar to those of the heat flow example of Sec. F.8. Obviously, a heat flow example with a linearly varying SWI temperature can easily be constructed from (C.1), (C.2) and (C.3).

J. Tóth. A theory of groundwater motion in small drainage basins in central Alberta, Canada. *Journal of Geophysical Research*, 67:4375–4387, 1962

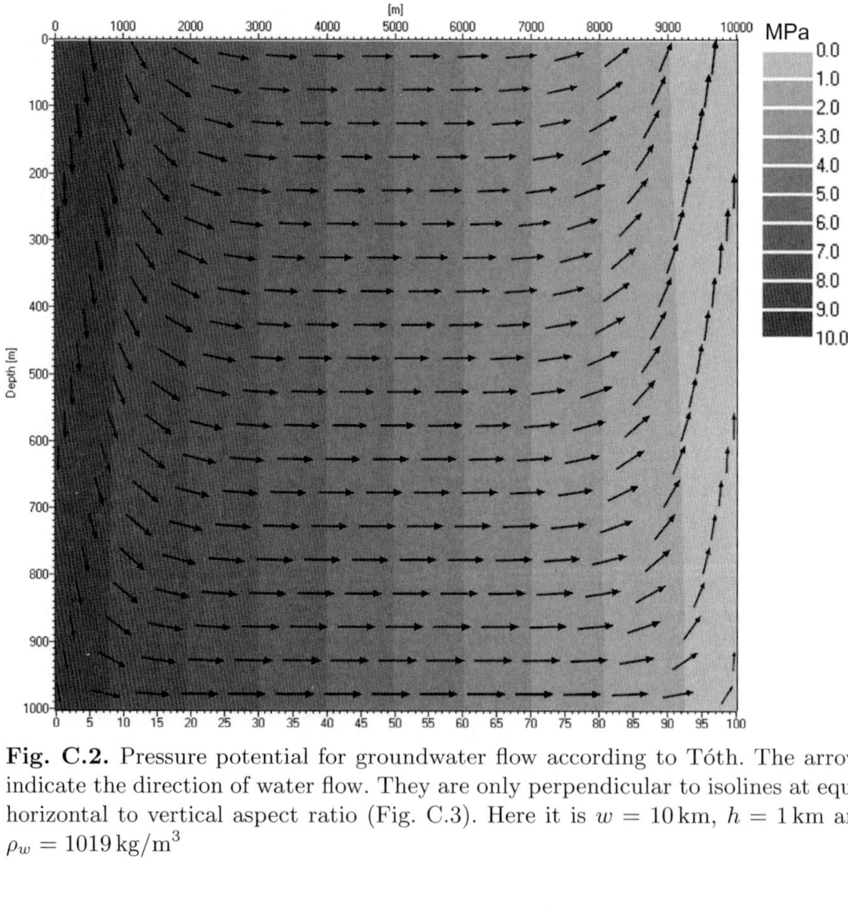

Fig. C.2. Pressure potential for groundwater flow according to Tóth. The arrows indicate the direction of water flow. They are only perpendicular to isolines at equal horizontal to vertical aspect ratio (Fig. C.3). Here it is $w = 10\,\mathrm{km}$, $h = 1\,\mathrm{km}$ and $\rho_w = 1019\,\mathrm{kg/m^3}$

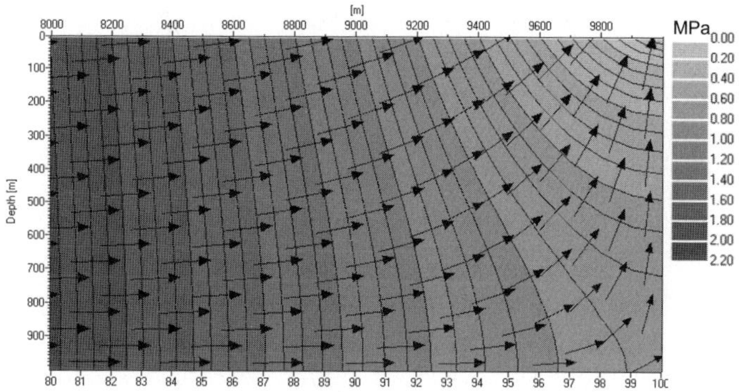

Fig. C.3. Cutout from Fig. C.2 with almost the same horizontal to vertical aspect ratio

D

One Dimensional Consolidation Solution from Gibson

Gibson (1958) found a closed expression for the compaction problem

$$\frac{\mu}{C}\frac{\partial^2 u}{\partial x^2} = \frac{\partial u}{\partial t} - \Delta\rho g \frac{dh}{dt} \tag{D.1}$$

with overpressure u, $x = h(t) - z$, depth z, and the boundary conditions

$$u(h,t) = 0, \qquad \left.\frac{\partial}{\partial x} u(x,t)\right|_{x=0} = 0 \tag{D.2}$$

for all t and height $h(t) = S\,t$ (Fig. D.1). The constant factor S determines the velocity with which the layer increases in thickness. Density contrast $\Delta\rho = \rho_r - \rho_w$, rock density ρ_r, water density ρ_w, mobility μ, and compressibility C are assumed to be constant in space and time. The assumption of constant compressibility is only valid for shallow depths and therefore called consolidation. It has been found as

$$\frac{u}{\Delta\rho g h} = 1 - \frac{4}{\sqrt{\pi K^3}} e^{-x'^2 K/4} \int_0^\infty \xi \tanh\xi \cosh x'\xi \, e^{-\xi^2/K} d\xi \tag{D.3}$$

with $x' = x/h$, $K = kh$, and $k = SC/\mu$. It is drafted in Fig. D.2.

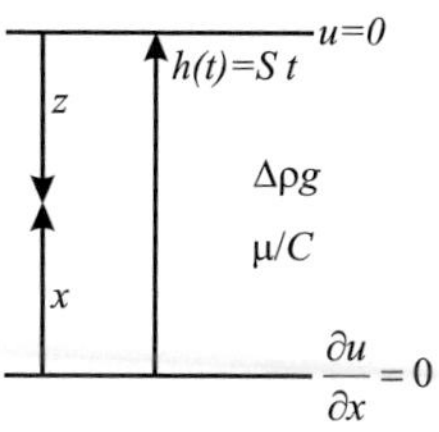

Fig. D.1. Illustration of (D.1) and (D.2)

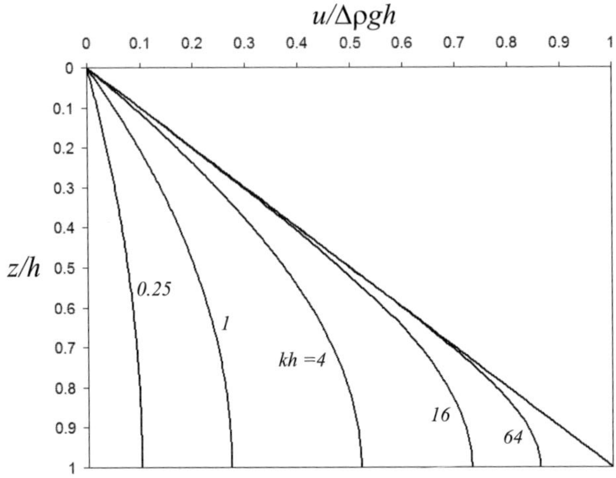

Fig. D.2. Example curves according to (D.3) for depth $z = h - x$

The factor K can roughly be estimated as $K = 0.04$ for a sandstone with $C = 0.05/\text{MPa}$, $\mu = 10^{-16}\text{m}^2/0.5\,\text{mPa}\,\text{s}$, a sedimentation rate of $S = 1\,\text{km/Ma}$ and a height $h = 5\,\text{km}$. Hence a sandstone will never build up pressure on its own. When the permeability decreases about one or two orders of magnitude and K is approaching $K \approx 1$, pressure build up might start.

However, (D.3) can be expanded for small K. The exponential function in the integrand of (D.3) is decaying much faster for small K than other terms rise. Hence the tanh– and the cosh–terms can be Taylor–expanded up to second order. The resulting expression can be integrated analytically and finally yields

$$\frac{u}{\Delta \rho g h} \approx \frac{K}{4}(1 - 2x'^2). \tag{D.4}$$

This shows again that practically no overpressure is existing for $K \to 0$.

Gibson found additional closed expressions of related problems e.g. with drainage boundary condition $u = 0$ at $x = 0$ or $h(t) \sim \sqrt{t}$.

R. E. Gibson. The progress of consolidation in a clay layer increasing in thickness with time. *Géotechnique*, 8:171–182, 1958

E

Thermal Properties

The following data of this section is provided by Doug Waples. It is based on a comprehensive literature review and his modeling experience over 20 years.

	Vert. Cond. $\left[\frac{W}{mK}\right]$	An-iso-tr.	Sp. Heat Cap. $\left[\frac{J}{kgK}\right]$	Den. $\left[\frac{kg}{m^3}\right]$	U [ppm]	Th [ppm]	K [%]	Radio-active Heat $\left[\frac{\mu W}{m^3}\right]$
Amphibolite	2.40	1.25	1130	2960	0.50	1.60	1.00	0.37
Eclogite	3.55	1.10	750	3400	0.20	0.40	0.20	0.12
Gneiss	2.70	1.40	800	2740	5.00	13.00	3.00	2.50
Marble	2.80	1.02	860	2700	1.00	1.00	0.20	0.34
Quartzite	5.40	1.03	770	2650	0.40	2.20	1.10	0.35
Schist	2.90	1.35	920	2740	2.10	9.70	2.20	1.44
Slate	1.80	1.75	860	2750	3.00	12.00	4.00	2.01
Serpentinite	2.60	1.45	785	2900	0.03	0.07	0.005	0.01

Table E.1. Thermal Properties: Metamorphic Rocks

	Vert. Cond. $\left[\frac{W}{mK}\right]$	An-iso-tr.	Sp. Heat Cap. $\left[\frac{J}{kgK}\right]$	Den. $\left[\frac{kg}{m^3}\right]$	U [ppm]	Th [ppm]	K [%]	Radio-active Heat $\left[\frac{\mu W}{m^3}\right]$
				Rock Forming Minerals				
Quartz	7.69	1.00	740	2650	0.70	2.00	0.00	0.31
K–Spar	2.40	1.00	675	2550	1.60	5.00	14.05	1.96
Plagioclase	1.85	1.00	740	2680	2.50	1.50	0.00	0.74
Mica	0.60	6.00	770	2860	12.00	12.00	9.40	5.08
Chlorite	1.55	2.65	865	2800	1.00	4.00	0.65	0.62
Mixed–layer Clays	0.85	2.75	835	2780	1.00	4.00	4.00	0.94
Calcite	3.25	1.00	820	2710	0.00	0.00	0.00	0.00
Dolomite	5.30	1.00	870	2870	0.00	0.00	0.00	0.00
Halite, Mineral	6.50	1.00	865	2160	0.00	0.00	0.00	0.00
Anhydrite, Mineral	6.30	1.00	550	2960	0.00	0.00	0.00	0.00
Gypsum, Mineral	1.30	1.15	1070	2310	0.00	0.00	0.00	0.00
Olivine	4.20	1.00	750	3320	0.01	0.01	0.00	0.00
Orthoclase	2.30	1.00	620	2570	0.00	0.00	14.05	1.26
				Rock Fragments				
Igneous, Granite	2.60	1.15	800	2650	4.00	13.50	3.50	2.25
Volcanic, Basalt	2.10	1.17	790	2870	0.90	2.70	0.80	0.52
Metamorphic, Schist/Slate	2.35	1.55	890	2745	2.55	10.85	3.10	1.73
Opaques, Amphibole	2.90	1.00	750	3100	0.00	0.00	0.20	0.02
				Other Minerals				
Serpentine	2.60	1.00	785	2600	0.00	0.00	0.00	0.00
Siderite	3.00	1.00	700	3940	0.00	0.00	0.00	0.00
Sphalerite	12.70	1.00	460	4090	0.00	0.00	0.00	0.00
Sphene, Titanite	2.35	1.00	840	3520	400.00	350.00	0.00	165.58
Sylvite	6.70	1.00	685	1990	0.00	0.00	52.45	3.63
Talc	5.85	1.00	840	2780	0.00	0.00	0.00	0.00
Topaz	11.25	1.00	790	3200	0.00	0.00	0.00	0.00
Tourmaline	4.25	1.00	540	3100	0.00	0.00	0.00	0.00
Volcanic glass	1.35	1.00	770	2400	5.00	15.00	5.00	2.48
Zeolite	1.80	1.00	1050	2250	0.00	0.00	2.00	0.16
Zircon	5.00	1.00	630	4670	3000	2000	0.00	1573

Table E.2. Thermal Properties: Minerals – Part 1

	Vert. Cond. $\left[\frac{W}{mK}\right]$	An-iso-tr.	Sp. Heat Cap. $\left[\frac{J}{kgK}\right]$	Dens. $\left[\frac{kg}{m^3}\right]$	U [ppm]	Th [ppm]	K [%]	Radio-active Heat $\left[\frac{\mu W}{m^3}\right]$
				Other Minerals				
Kaolinite	1.25	2.67	950	2590	1.00	4.00	0.50	0.56
Smectite	0.85	2.80	808	2610	0.50	2.00	0.50	0.30
Illite	0.85	2.75	822	2660	1.50	5.00	6.03	1.28
TOC (weight%)	0.28	1.10	1500	1100	100	0.00	0.00	10.47
Amphibole	2.90	1.00	750	3100	0.00	0.00	0.2	0.02
Analcime	1.40	1.00	960	2260	0.00	0.00	0.00	0.00
Anorthite	1.70	1.00	740	2760	0.00	0.00	0.00	0.00
Apatite	1.35	1.00	700	3180	30.00	150.00	0.00	21.29
Aragonite	2.25	1.00	800	2930	0.00	0.00	0.00	0.00
Barite	1.35	1.00	460	4480	0.00	0.00	0.00	0.00
Brucite	12.90	1.00	1260	2370	0.00	0.00	0.00	0.00
Chalcopyrite	8.20	1.00	535	4200	0.00	0.00	0.00	0.00
Chert, Mineral	5.00	1.00	740	2600	0.00	0.00	0.00	0.00
Clinoptilolite	1.40	1.00	1140	2100	0.00	0.00	3.30	0.24
Diopside	4.90	1.00	760	3280	0.00	0.00	0.00	0.00
Epidote	2.80	1.00	770	3590	35.00	275.00	0.00	37.24
Fluorite	9.50	1.00	870	3180	0.00	0.00	0.00	0.00
Forsterite	5.10	1.00	820	3210	0.00	0.00	0.00	0.00
Galena	2.70	1.00	210	7600	0.00	0.00	0.00	0.00
Garnet	3.50	1.00	750	3800	0.00	0.00	0.00	0.00
Gibbsite	2.60	1.00	1170	2440	0.00	0.00	0.00	0.00
Glauconite	1.60	1.00	770	2300	0.00	0.00	5.49	0.44
Goethite	2.70	1.00	825	4270	0.00	0.00	0.00	0.00
Hematite	11.30	1.00	625	5280	0.00	0.00	0.00	0.00
Hornblende	2.80	1.00	710	3080	15.00	25.00	1.75	6.56
Hypersthene	4.40	1.00	800	3450	0.00	0.00	0.00	0.00
Ilmenite	2.40	1.00	700	4790	25.00	0.00	0.00	11.40
Magnesite	5.80	1.00	880	3010	0.00	0.00	0.00	0.00
Magnctite	5.10	1.00	620	5200	15.00	10.00	0.00	8.76
Microcline	2.50	1.00	710	2560	0.00	0.00	14.05	1.25
Opal–CT	1.70	1.00	725	2300	0.00	0.00	0.00	0.00
Polyhalite, Min.	1.55	1.00	700	2780	0.00	0.00	12.97	1.25
Pyrite	3.00	1.00	510	5010	0.00	0.00	0.00	0.00
Pyrophyllite	3.75	1.00	805	2820	0.00	0.00	0.00	0.00
Pyroxene	4.20	1.00	770	3500	20.00	13.00	0.00	7.83
Rutile	5.10	1.00	700	4250	0.00	0.00	0.00	0.00

Table E.3. Thermal Properties: Minerals, Part 2

	Vert. Cond. $\left[\frac{W}{mK}\right]$	An-iso-tr.	Sp. Heat Cap. $\left[\frac{J}{kgK}\right]$	Den. $\left[\frac{kg}{m^3}\right]$	U [ppm]	Th [ppm]	K [%]	Radio-active Heat $\left[\frac{\mu W}{m^3}\right]$
Albitite	2.00	1.00	785	2600	0.02	0.02	0.40	0.04
Andesite	2.70	1.00	820	2650	1.35	2.50	1.40	0.64
Anorthosite	1.70	1.00	750	2800	1.00	3.50	0.45	0.56
Basalt, normal	1.80	1.17	800	2870	0.90	2.70	0.80	0.52
Basalt, weathered	2.10	1.17	790	2870	0.90	2.70	0.80	0.52
Bronzitite	3.80	1.00	760	3450	0.02	0.02	0.05	0.01
Diabase	2.60	1.00	800	2800	0.25	0.90	0.45	0.18
Diorite	2.70	1.00	1140	2900	2.00	8.50	1.10	1.29
Dolerite	2.30	1.00	900	2930	0.40	1.60	0.70	0.30
Dunite	3.80	1.30	720	3310	0.004	0.004	0.01	0.00
Gabbro	2.60	1.00	800	2870	0.25	0.80	0.50	0.18
Granite, 150 My old	2.60	1.15	800	2650	6.50	17.00	5.70	3.32
Granite, 500 My old	2.60	1.15	800	2650	4.50	15.00	4.50	2.57
Granite, > 1000 My old	2.60	1.15	800	2650	4.00	13.50	3.50	2.25
Granodiorite	2.60	1.00	730	2720	2.30	9.30	2.80	1.51
Harzburgite	6.90	1.00	760	3300	0.02	0.02	0.05	0.01
Hypersthenite	4.10	1.00	760	3450	0.02	0.02	0.05	0.01
Lamprophyre	2.50	1.00	760	3000	1.20	5.50	4.10	1.19
Lherzolite	3.30	1.00	780	3150	0.015	0.05	0.0006	0.01
Monzonite	2.70	1.00	740	2600	2.70	8.50	4.00	1.60
Monzonite, quartz	2.80	1.00	880	2700	6.50	27.00	3.40	3.86
Nepheline syenite	2.40	1.00	750	2650	9.10	27.00	4.00	4.50
Norite	2.20	1.00	670	2860	0.15	0.50	0.30	0.11
Olivinite	3.20	1.00	800	3450	0.01	0.08	0.05	0.02
Peridotite, typical	4.00	1.00	800	3200	0.025	0.025	0.055	0.02
Peridotite, serpentinized	2.20	1.00	700	3100	0.03	0.03	0.01	0.01
Pyroxenite	3.80	1.00	1000	3220	0.20	0.55	0.75	0.19
Quartz monzonite	2.80	1.00	880	2700	6.50	27.00	3.40	3.86
Rhyolite	2.60	1.20	800	2500	5.80	13.00	3.70	2.53
Syenite, typical	2.60	1.00	800	2760	2.70	15.00	4.70	2.22
Syenite, nepheline	2.40	1.00	750	2650	9.10	27.00	4.00	4.50
Tonalite	2.55	1.00	800	2800	0.00	12.00	2.50	1.10
Ultramafics	3.80	1.00	900	3310	0.001	0.004	0.003	0.00

Table E.4. Thermal Properties: Igneous Rocks

	Vert. Cond. $\left[\frac{W}{mK}\right]$	An-iso-tr.	Sp. Heat Cap. $\left[\frac{J}{kgK}\right]$	Th. Sort. Fac. f	Den. $\left[\frac{kg}{m^3}\right]$	U [ppm]	Th [ppm]	K [%]	Radio-active Heat $\left[\frac{\mu W}{m^3}\right]$
Sandstone									
typical	3.95	1.15	855	1.00	2720	1.30	3.50	1.30	0.70
clay–rich	3.35	1.20	860	1.00	2760	1.50	5.10	3.60	1.10
clay–poor	5.95	1.05	820	1.00	2700	0.70	2.30	0.60	0.40
quartzite, typical	6.15	1.08	890	1.00	2640	0.60	1.80	0.90	0.36
quart., very quartz–rich	6.45	1.06	890	1.00	2640	0.50	1.70	0.70	0.30
subarkose, typical	4.55	1.14	870	1.00	2680	1.00	3.30	1.30	0.60
subarkose, quartz–rich	5.05	1.13	880	1.00	2650	0.90	2.70	0.90	0.49
subarkose, clay–rich	3.40	1.40	870	1.00	2690	1.20	3.90	2.30	0.79
subarkose, clay–poor	4.80	1.09	870	1.00	2700	0.70	2.30	0.60	0.40
subarkose, dolomite–rich	4.10	1.07	840	1.00	2710	0.90	2.70	0.90	0.50
arkose, typical	3.20	1.25	845	1.00	2730	3.00	8.00	2.50	1.58
arkose, quartz–rich	4.05	1.20	860	1.00	2690	2.00	6.00	0.90	1.01
arkose, quartz–poor	2.00	1.50	835	1.00	2770	3.00	8.00	2.50	1.60
arkose, clay–rich	2.30	1.45	865	1.00	2760	2.00	7.00	3.60	1.37
arkose, clay–poor	4.00	1.12	835	1.00	2710	2.00	7.00	1.00	1.10
arkose, dolomite–rich	4.35	1.05	850	1.00	2750	1.50	5.10	3.60	1.10
wacke	2.60	1.40	850	1.00	2780	2.50	8.00	2.50	1.47
Shale									
typical	1.64	1.60	860	1.38	2700	3.70	12.0	2.70	2.03
organic–lean, typical	1.70	1.55	860	1.35	2700	3.70	12.0	2.70	2.03
organic–lean, sandy	1.84	2.00	860	1.64	2700	2.80	11.0	2.50	1.71
organic–lean, silty	1.77	1.80	860	1.51	2700	3.00	11.0	2.60	1.78
organic–lean									
siliceous, typical	1.90	1.17	860	1.00	2710	2.00	4.5	2.00	1.02
siliceous, 95% Opal–CT	1.75	1.16	860	1.00	2330	1.00	1.3	1.00	0.38
black	0.90	3.27	940	2.40	2500	19.00	11.0	2.50	5.44
organic–rich, typical	1.25	2.19	900	1.76	2600	5.00	12.0	2.80	2.29
organic–rich, 3% TOC	1.45	1.84	880	1.54	2610	5.00	12.0	2.80	2.30
organic–rich, 8% TOC	1.20	2.20	900	1.74	2500	10.00	11.0	2.90	3.34
organic–rich, 20% TOC	0.80	3.09	980	2.22	2270	20.00	11.0	2.60	5.17
Siltstone									
organic–lean	2.05	1.50	910	1.00	2720	2.00	5.00	1.00	0.96
org.–rich, typical	2.01	1.71	940	1.47	2710	2.00	5.00	1.00	0.96
org.–rich, > 10% TOC	1.82	2.64	960	2.03	2550	15.00	6.00	2.00	4.21
org.–rich, 2-3% TOC	2.00	1.76	930	1.52	2700	2.50	6.50	2.00	1.28
Conglomerate, typical	2.30	1.05	820	1.00	2700	1.50	4.00	2.00	0.85
Conglomerate, quartzitic	6.10	1.05	780	1.00	2700	1.00	4.00	1.20	0.65
Tuff, felsic	2.60	1.17	830	1.00	2650	3.00	6.50	3.90	1.56
Tuff, basaltic	1.90	1.17	830	1.00	2900	0.65	2.20	0.85	0.43

Table E.5. Thermal Properties: Sedimentary Rocks, Part 1 – Clastic Sediments

	Vert. Cond. $\left[\frac{W}{mK}\right]$	An-iso-tr.	Sp. Heat Cap. $\left[\frac{J}{kgK}\right]$	Th. Sort. Fac. f	Den. $\left[\frac{kg}{m^3}\right]$	U [ppm]	Th [ppm]	K [%]	Radio-active Heat $\left[\frac{\mu W}{m^3}\right]$
Biogenic Sediments									
Chalk, typical	2.90	1.07	850	1.00	2680	1.90	1.40	0.25	0.60
Chalk, 95% calcite	3.00	1.04	830	1.00	2680	1.90	1.40	0.25	0.60
Chalk, 75% calcite	2.65	1.15	840	1.00	2680	1.90	1.40	0.25	0.60
Chalk, 40% calcite	3.20	1.10	860	1.00	2680	1.90	1.40	0.25	0.60
Coal, pure	0.30	1.20	1300	1.03	1600	1.50	3.00	0.55	0.38
Coal, impure	1.00	1.15	1200	1.00	1600	2.00	3.00	0.70	0.47
Coal, silty	1.60	1.30	1100	1.00	1600	2.00	3.00	0.70	0.47
Diatomite, clay–poor	1.60	1.00	790	1.00	2100	1.30	1.00	1.00	0.39
Diatomite, clay–rich	1.50	1.10	800	1.00	2300	3.00	2.60	2.50	1.01
Kerogen	1.00	1.00	6.3	1.00	1100	100.00	0.00	0.00	10.47
Carbonate Rocks									
Limestone									
ooid grainstone	3.00	1.19	835	1.11	2740	1.00	1.00	0.20	0.35
Waulsortian mound	3.00	1.19	835	1.11	2740	1.00	1.00	0.20	0.35
micrite	3.00	1.19	835	1.11	2740	1.00	1.00	0.20	0.35
shaly	2.30	1.70	850	1.38	2730	2.00	4.00	1.00	0.89
org.–rich, typical	2.00	1.95	845	1.47	2680	5.00	1.50	0.26	1.40
org.–rich, 1-2% TOC	2.63	1.40	840	1.16	2710	2.50	1.70	0.27	0.79
org.–rich, 10% TOC	1.45	2.68	850	1.84	2550	10.00	1.40	0.25	2.54
chalk, typical	2.90	1.07	850	1.00	2680	1.90	1.40	0.25	0.60
chalk, 95% calcite	3.00	1.04	830	1.00	2680	1.90	1.40	0.25	0.60
chalk, 75% calcite	2.65	1.15	840	1.00	2680	1.90	1.40	0.25	0.60
chalk, 40% calcite	3.20	1.10	860	1.00	2680	1.90	1.40	0.25	0.60
Marl	2.00	1.45	850	1.00	2700	2.50	5.00	2.00	1.18
Dolomite									
typical	4.20	1.06	860	1.00	2790	0.80	0.60	0.40	0.29
organic–lean, sandy	4.25	1.06	860	1.00	2770	0.90	0.90	0.70	0.37
organic–lean, silty	3.00	1.10	860	1.00	2760	0.90	0.90	0.80	0.38
organic–rich	2.15	2.44	910	1.76	2600	10.00	1.40	0.60	2.62

Table E.6. Thermal Properties: Sedimentary Rocks, Part 2

	Vert. Cond. $\left[\frac{W}{mK}\right]$	Aniso-tr.	Sp. Heat Cap. $\left[\frac{J}{kgK}\right]$	Th. Sort. Fac. f	Den. $\left[\frac{kg}{m^3}\right]$	U [ppm]	Th [ppm]	K [%]	Radio-active Heat $\left[\frac{\mu W}{m^3}\right]$
Chemical Sediments									
Polyhalite	1.00	1.00	700	1.00	2780	0.02	0.01	12.90	1.25
Salt	6.50	1.01	860	1.00	2740	0.02	0.01	0.10	0.02
Sylvinite	1.00	1.02	685	1.00	2100	0.02	0.01	20.60	1.51
Anhydrite	6.30	1.05	750	1.00	2970	0.10	0.30	0.40	0.09
Chert	4.80	1.00	890	1.00	2650	1.00	1.00	0.70	0.38
Gypsum	1.50	1.15	1100	1.00	2320	0.08	0.20	0.30	0.05
Halite	6.50	1.01	860	1.00	2200	0.02	0.01	0.10	0.01

Table E.7. Thermal Properties: Sedimentary Rocks, Part 3

F

Analytic Solutions to Selected Heat Flow Problems

Heat flow and temperature distributions in sedimentary basins are usually formulated in terms of differential equations of diffusion type with boundary conditions $T = T_{\text{swi}}$ at the sediment water interface, $\nabla T = 0$ at the sides and $-\lambda \nabla T = \mathbf{q}$ below basement. The first boundary condition is obvious. The second describes a condition which does not allow lateral heat flow at the sides. This approximation can be justified if the lateral extension of the model is large compared to its thickness. The last condition is not obvious from scratch. The basal heat flow is not well known and it does not seem natural to choose the heat flow as the important parameter describing the boundary condition. Temperature seems to be more obvious but is usually not known either. However, it is often assumed that the temperature distribution is in or near steady state and that the heat flow boundary condition is independent of the thickness of the basin. It is therefore robust under variations of the stratigraphic interpretation, whereas the temperature in the basement varies strongly with its thickness and structure. Following an argumentation which includes crust and mantle in the heat flow analysis, such as the McKenzie stretching models, it is possible to estimate the heat flow by

$$q = \frac{T_a - T_s}{h} \lambda \tag{F.1}$$

with T_a as the temperature of the asthenosphere, T_s the temperature below the sediments, h the thickness of the basin, and λ its average thermal conductivity (Sec. 3.8). An approach with inclusion of the sediments into (F.1) yields only minor corrections to the overall temperature difference in the numerator, thickness in the denominator and overall thermal conductivity. Thus numerous uncertain parameters such as the thermal conductivity of crust and mantle and their thicknesses are mapped into one key quantity, namely the basal heat flow.

A few heat flow examples that can be solved analytically are presented in the following. These examples can be used for basic understanding and for validating numerical algorithms.

F.1 Influence of Radiogenic Heat Production on a Steady State Temperature Profile

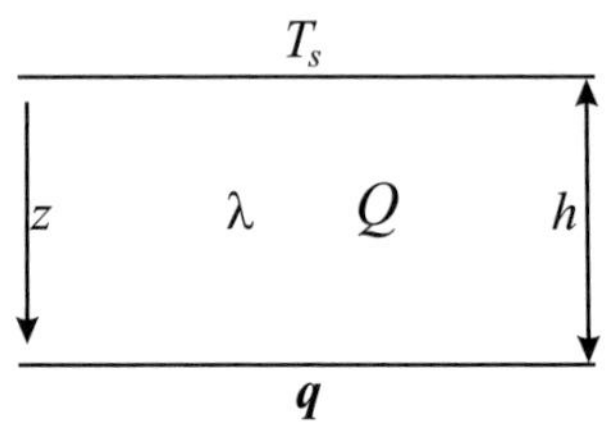

Fig. F.1. Model with radioactive heat production

Figure F.1 exhibits a simple one layer model. The major difference from the steady state models of Sec. 3.2.1 is the inclusion of radioactive heat production. The differential equation becomes

$$\lambda \frac{d^2}{dz^2} T = -Q \tag{F.2}$$

with constant thermal conductivity λ and radiogenic heat production Q. The boundary conditions are $T = T_s$ at top surface and $\lambda T'(h) = q$ with q as the basal heat flow. The solution can be easily evaluated to

$$T = T_s + \frac{q}{\lambda} z - \frac{Q}{\lambda} \frac{z(z-2h)}{2} \ . \tag{F.3}$$

Some example curves are shown in Fig. F.2 Thus, in steady state situations, a constant heat flow q is related to a linear temperature rise. A constant heat production Q introduces a parabolic dependence of temperature in depth. The heat flow is linearly depth dependent and becomes $\lambda T' = q - Q\,(z-h)$. Note that the heat flow is defined here by $\mathbf{q} = -q\mathbf{e}_z$ with $\mathbf{e}_z$ pointing downward and q positive for flow upwards.

F.2 Steady State Temperature Profile with a Lateral Basal Heat Flow Jump

A model with a lateral basal heat flow jump from q_1 to q_2 is studied. It is exhibited in Fig. F.3, does not contain radioactive heat production and can be formulated as

$$\frac{\partial^2}{\partial x^2} T(x,z) + \frac{\partial^2}{\partial z^2} T(x,z) = 0 \tag{F.4}$$

with the boundary conditions

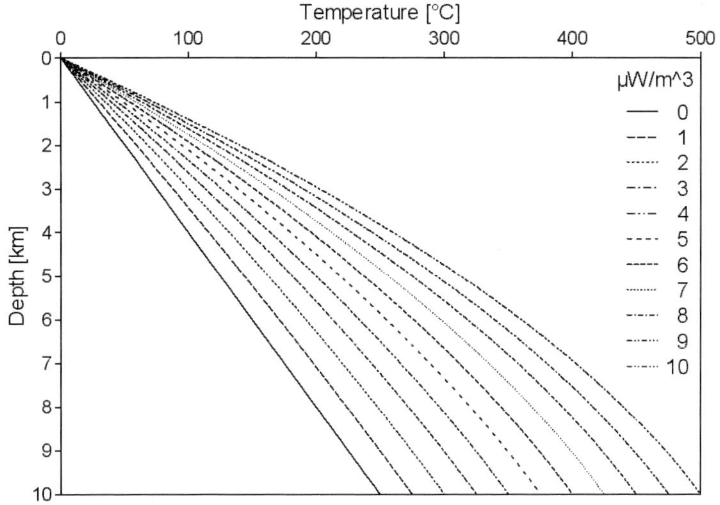

Fig. F.2. Some temperature profiles with varying radiogenic heat production. Here $T_s = 0°\text{C}$, $\lambda = 2\,\text{W/m/K}$ and $q = 50\,\text{mW/m}^2$

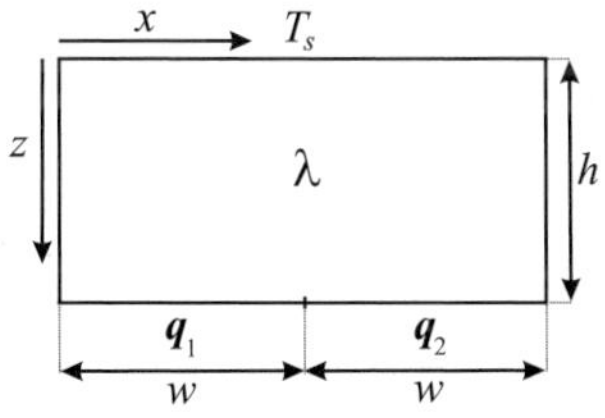

Fig. F.3. Model with lateral heat flow jump

$$
\begin{aligned}
T(x,0) &= T_s && \text{for} \quad 0 < x < 2w \\
\frac{\partial}{\partial z} T(x,z)\Big|_{z=h} &= \frac{q_1}{\lambda} && \text{for} \quad 0 < x < w \\
\frac{\partial}{\partial z} T(x,z)\Big|_{z=h} &= \frac{q_2}{\lambda} && \text{for} \quad w < x < 2w \\
\frac{\partial}{\partial x} T(x,z)\Big|_{x=0} &= \frac{\partial}{\partial x} T(x,z)\Big|_{x=2w} = 0 && \text{for} \quad 0 < z < h\,.
\end{aligned}
\tag{F.5}
$$

The analytical solution can be obtained via a separation technique. It is

$$
T(x,z) = T_s + \frac{q_1 + q_2}{2\lambda} z + \frac{q_2 - q_1}{\lambda} \sum_{n=1}^{\infty} \frac{(-1)^n \sinh(\mu_n z) \cos(\mu_n x)}{w \mu_n^2 \cosh(\mu_n h)} \tag{F.6}
$$

with $\mu_n = (n - 1/2)\pi/w$. An example is shown in Fig. F.4.

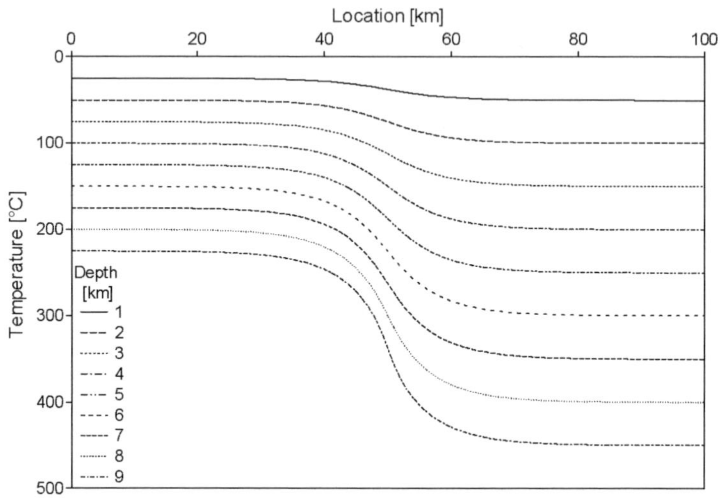

Fig. F.4. Temperatures at given depth according to (F.6) with $w = 50\,\mathrm{km}$, $h = 10\,\mathrm{km}$, $T_s = 0°\mathrm{C}$, $\lambda = 2\,\mathrm{W/m/K}$, $q_1 = 50\,\mathrm{mW/m^2}$ and $q_2 = 100\,\mathrm{mW/m^2}$

F.3 Steady State Temperature Profile with SWI Temperature Jump

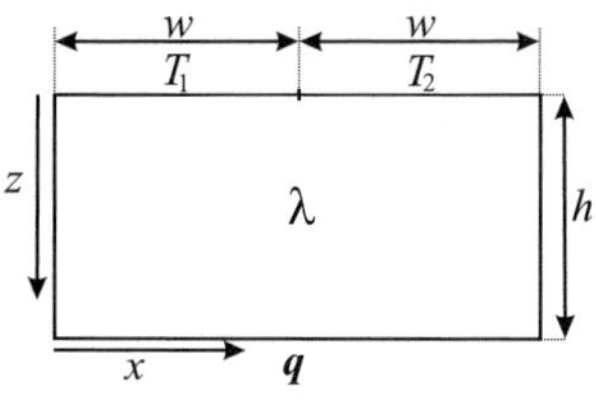

Fig. F.5. Model with SWI temperature jump

This model is similar to Fig. F.3. It differs in that there is a jump in SWI temperature instead of basal heat flow as shown in Fig. F.5. Thus only the first three boundary conditions of (F.5) must be modified to

$$
\begin{aligned}
T(x,0) &= T_1 \qquad \text{for} \qquad 0 < x < w \\
T(x,0) &= T_2 \qquad \text{for} \qquad w < x < 2w \\
\frac{\partial}{\partial z} T(x,z)\bigg|_{z=h} &= \frac{q}{\lambda} \qquad \text{for} \qquad 0 < x < 2w\,.
\end{aligned}
\tag{F.7}
$$

The solution is

$$T(x,z) = \frac{q}{\lambda}z + \frac{T_1 + T_2}{2} + (T_2 - T_1)\sum_{n=1}^{\infty}(-1)^n \frac{\cosh(\mu_n(z-h))\cos(\mu_n x)}{\mu_n w \cosh(\mu_n h)} \tag{F.8}$$

again with $\mu_n = (n - 1/2)\pi/w$. An example is shown in Fig. F.6.

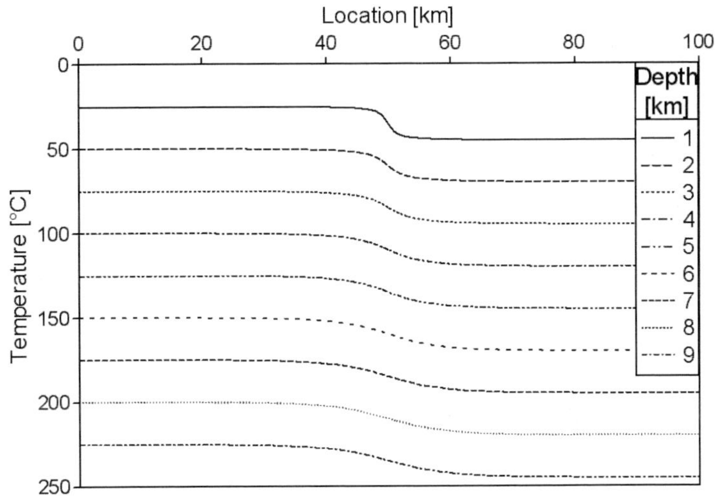

Fig. F.6. Temperatures at given depth according to (F.8) with $w = 50\,\mathrm{km}$, $h = 10\,\mathrm{km}$, $T_1 = 0°\mathrm{C}$, $T_2 = 20°\mathrm{C}$, $\lambda = 2\,\mathrm{W/m/K}$ and $q = 50\,\mathrm{mW/m^2}$

F.4 Steady State Temperature Profile for a Two Block Model

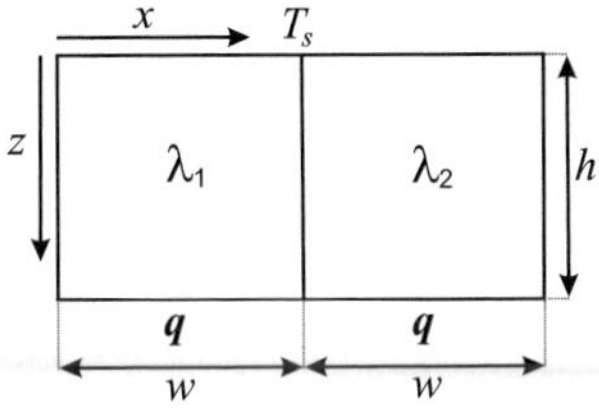

Fig. F.7. Two block model with jump in thermal conductivity λ

The model consists of two equal sized rectangular blocks with different thermal conductivities. It is depicted in Fig. F.7 and its mathematical formulation is given by

$$\begin{aligned}
T(x,0) &= T_s && \text{for} \quad -w < x < w \\
\frac{\partial}{\partial z}T(x,z)\bigg|_{z=h} &= \frac{q}{\lambda_1} && \text{for} \quad 0 < x < w \\
\frac{\partial}{\partial z}T(x,z)\bigg|_{z=h} &= \frac{q}{\lambda_2} && \text{for} \quad -w < x < 0 \\
\frac{\partial}{\partial x}T(x,z)\bigg|_{x=-w} &= \frac{\partial}{\partial x}T(x,z)\bigg|_{x=w} = 0 && \text{for} \quad 0 < z < h \\
T(0+,z) &= T(0-,z) && \text{for} \quad 0 < z < h \\
\lambda_1 \frac{\partial}{\partial x}T(x,z)\bigg|_{x=0+} &= \lambda_1 \frac{\partial}{\partial x}T(x,z)\bigg|_{x=0-} && \text{for} \quad 0 < z < h\,.
\end{aligned} \tag{F.9}$$

Explicitly one obtains

$$\begin{aligned}
T(x,z) = T_s + \frac{q}{\lambda_1}z + \frac{q}{\lambda_2}\frac{\lambda_2-\lambda_1}{\lambda_1+\lambda_2} \times \\
\sum_{n=-\infty}^{\infty} \frac{(-1)^n}{h\mu_n^2} \sin(\mu_n z)\left(\cosh(\mu_n x) - \tanh(\mu_n w)\sinh(\mu_n x)\right)
\end{aligned} \tag{F.10}$$

for $0 < x < w$ and

$$\begin{aligned}
T(x,z) = T_s + \frac{q}{\lambda_2}z - \frac{q}{\lambda_1}\frac{\lambda_2-\lambda_1}{\lambda_1+\lambda_2} \times \\
\sum_{n=-\infty}^{\infty} \frac{(-1)^n}{h\mu_n^2} \sin(\mu_n z)\left(\cosh(\mu_n x) + \tanh(\mu_n w)\sinh(\mu_n x)\right)
\end{aligned} \tag{F.11}$$

for $-w < x < 0$ with $\mu_n = (n + 1/2)\pi/h$. An example is shown in Fig. F.8.

F.5 Non Steady State Model with Heat Flow Jump

The model consists of a layer, which is extended to infinity in the horizontal direction. In the vertical direction the temperature is set to T_s at the upper side. In addition the layer is exposed to a constant heat flow at its bottom side. At time $t = 0$ the heat flow is switched to another value. The heat flow equation becomes

$$\rho c \frac{\partial}{\partial t}T(z,t) = \lambda \frac{\partial^2}{\partial z^2}T(z,t)\,. \tag{F.12}$$

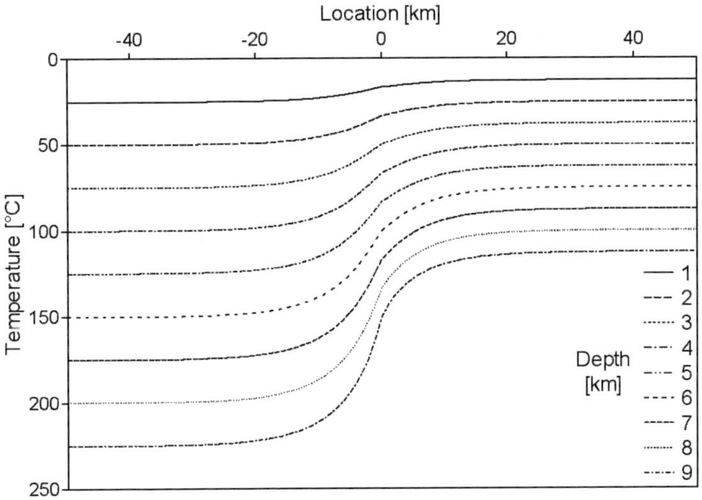

Fig. F.8. Temperatures at given depth according to (F.10) and (F.11) with $w = 50\,\mathrm{km}$, $h = 10\,\mathrm{km}$, $T_S = 0°\mathrm{C}$, $\lambda_1 = 2\,\mathrm{W/m/K}$, $\lambda_2 = 4\,\mathrm{W/m/K}$, and $q = 50\,\mathrm{mW/m^2}$

The density ρ, the specific heat capacity c, and the thermal conductivity are assumed constant. The boundary conditions are

$$\begin{aligned} T(0,t) &= T_s && \text{for} && \text{all} \quad t \\ \left.\frac{\partial}{\partial z}T(z,t)\right|_{z=h} &= \frac{q_1}{\lambda} && \text{for} && t < 0 \\ \left.\frac{\partial}{\partial z}T(z,t)\right|_{z=h} &= \frac{q_2}{\lambda} && \text{for} && t > 0\,. \end{aligned} \tag{F.13}$$

The temperature should be in steady state for $t < 0$:

$$T(z,t) = T_s + \frac{q_1}{\lambda} z\,. \tag{F.14}$$

For $t > 0$ it becomes in explicit form

$$T(z,t) = T_s + \frac{q_2}{\lambda} z + \frac{q_1 - q_2}{\lambda} \sum_{n=-\infty}^{\infty} \frac{(-1)^n}{h\mu_n^2} \sin(\mu_n z) \exp(-\mu_n^2 \lambda t/\rho c) \tag{F.15}$$

with $\mu_n = (n + 1/2)\pi/h$. An example is shown in Fig. F.9.

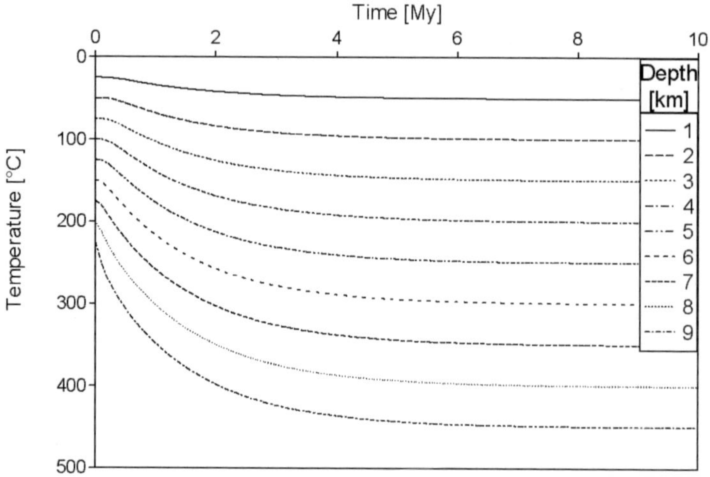

Fig. F.9. Temperatures at given depth according to (F.15) with $h = 10\,\mathrm{km}$, $T_s = 0°\mathrm{C}$, $q_1 = 50\,\mathrm{mW/m^2}$, $q_2 = 100\,\mathrm{mW/m^2}$, $\lambda = 2\,\mathrm{W/m/K}$, $\rho = 2700\,\mathrm{kg/m^3}$, and $c = 860\,\mathrm{J/kg/K}$

F.6 Non Steady State Model with SWI Temperature Jump

The model is almost the same as in Sec. F.5. Only the boundary conditions (F.13) are different:

$$\begin{aligned} \frac{\partial}{\partial z} T(z,t)\bigg|_{z=h} &= \frac{q}{\lambda} \qquad &\text{for} \qquad &\text{all} \quad t \\ T(0,t) &= T_1 \qquad &\text{for} \qquad &t < 0 \\ T(0,t) &= T_2 \qquad &\text{for} \qquad &t > 0\,. \end{aligned} \tag{F.16}$$

Again temperature is in steady state for $t < 0$:

$$T(z,t) = T_1 + \frac{q}{\lambda} z\,. \tag{F.17}$$

For $t > 0$ it is

$$T(z,t) = T_2 + \frac{q}{\lambda} z + (T_2 - T_1) \sum_{n=-\infty}^{\infty} \frac{(-1)^n}{\mu_n h} \sin(\mu_n (z-h)) \exp(-\mu_n^2 \lambda t/\rho c) \tag{F.18}$$

with $\mu_n = (n + 1/2)\pi/h$.

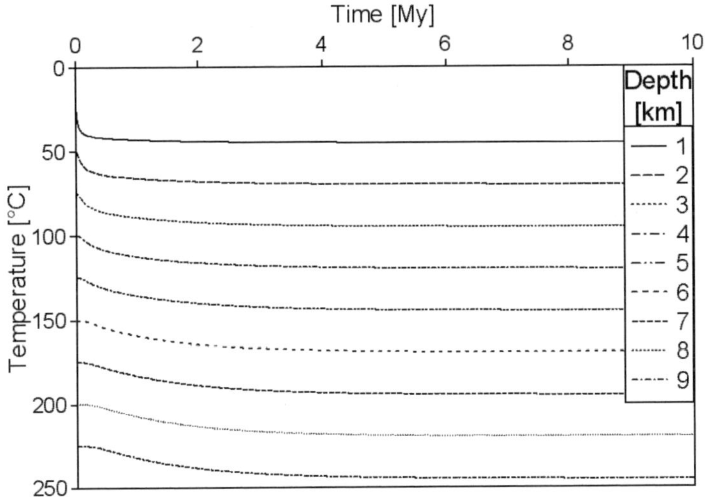

Fig. F.10. Temperatures at given depth according to (F.18) with $h = 10\,\mathrm{km}$, $T_1 = 0°\mathrm{C}$, $T_2 = 20°\mathrm{C}$, $q = 50\,\mathrm{mW/m^3}$, $\lambda = 2\,\mathrm{W/m/K}$, $\rho = 2700\,\mathrm{kg/m^3}$, and $c = 860\,\mathrm{J/kg/K}$

F.7 An Estimate for the Impact of Continuous Deposition on Heat Flow

An example of continuous deposition, a hiatus, and a following erosion with its impact on heat flow is shown in Fig. 3.4 and discussed in Sec. 3.2.2. A derivation of (3.8) is given below.

For a formulation of the heat flow problem with continuous deposition as shown in Fig. 3.4 it must be considered that the bulk rock including its captured heat subsides with velocity $v = S$ which is here equal to the sedimentation rate S. Equation (3.7) must therefore be enhanced by an additional term proportional to $\rho\, c\, v \partial T/\partial z$ similar as in (3.29) or (3.42). It becomes

$$\frac{\partial T}{\partial t} + S\frac{\partial T}{\partial z} - \frac{\lambda}{\rho\, c}\frac{\partial^2 T}{\partial z^2} = 0 \tag{F.19}$$

with boundary conditions

$$T(0,t) = 0, \qquad \lambda \frac{\partial}{\partial z} T(z,t)\bigg|_{z-h} = q \tag{F.20}$$

for all t and $h(t) = S\, t$ (Fig. F.11).

Transient heat flow effects are often negligible in geological systems. Tentatively, the $\partial T/\partial t$ term is discarded here. The boundary value problem can now easily be solved and the temperature becomes

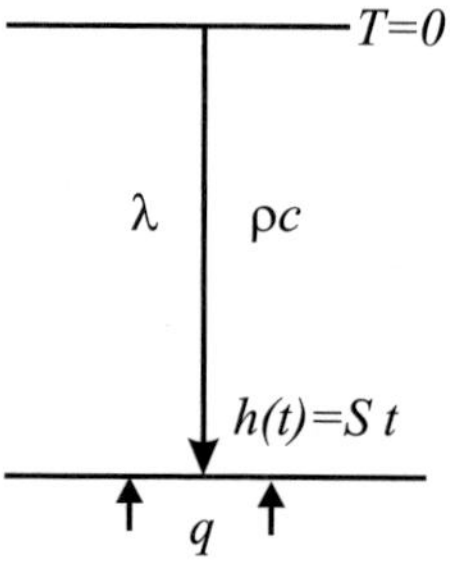

Fig. F.11. Illustration of (F.19) and (F.20)

$$T = \frac{q}{\lambda k} e^{-kSt} \left(e^{kz} - 1 \right) \tag{F.21}$$

with $k = S\rho c/\lambda$. The factor k is independent of time t and therefore it follows directly

$$\frac{\partial T}{\partial t} = -kST \, . \tag{F.22}$$

An example value of k can be calculated for a typical shale with 10 % porosity, a heat capacity of 4000 J/kg/K for water and a sedimentation rate of $S = 1\,\mathrm{km/Ma}$ as

$$\begin{aligned} k &= \frac{S\rho c}{\lambda} \\ &= \frac{1000\,\mathrm{m}\;2700\,\mathrm{kg/m^3}}{3.1536 \times 10^{13}\mathrm{s}} (0.9 \times 860 + 0.1 \times 4000) \frac{\mathrm{J}}{\mathrm{kg\,K}} \frac{1}{1.64\,\mathrm{J/s/m/K}} \\ &= 0.0613/\mathrm{km} \, . \end{aligned} \tag{F.23}$$

This is a rather small value and the factor kS is here with $kS = 0.0613/\mathrm{Ma}$ also very small. Finally it can be assumed that the transient term $\partial T/\partial t$ is commonly small compared with other terms in (F.19).[1] Formula (F.21) is thus a good approximation to a solution of (F.19) if the sedimentation rate S is not too large.[2]

It is further possible to expand the approximation (F.21) for small $K = kh$ which yields for the heat flow

$$\lambda \frac{\partial T}{\partial z} = q - \Delta q \quad \text{with} \quad \Delta q = qkx = \frac{qS\rho c}{\lambda} x \tag{F.24}$$

[1] More evidence can be achieved with following argument: a small geological temperature gradient is 30 °C/km. The first term of (F.19) can be estimated with kST and the second with $S\,30\,°\mathrm{C/km}$. The first term can hence in example Fig. 3.4 be neglected if $T \ll 30/0.0613 \approx 500°\mathrm{C}$.

[2] The pre-factor kS in (F.22) is proportional to S^2 whereas the spatial variation is proportional to S for small S according to (F.21) or (F.24). Hence the transient term decays much faster than the spatial terms for a small decreasing S.

and $x = h - z$ which is the height above the basement. For a typical shale it is hence $\Delta q = 0.0613\, q\bar{S}\bar{x}$ with $\bar{S}$ and $\bar{x}$ as dimensionless numbers. It is $\bar{S}$ in km/Ma and $\bar{x}$ in km. It follows with $q = 60\,\mathrm{mW/m^2}$ for an example such as in Fig. 3.4 $\Delta q = 3.68\,\bar{S}\bar{x}\,\mathrm{mW/m^2}$. This approximation is only valid for $K = 0.0613\,\bar{S}\bar{h} \ll 1$ again with $\bar{h}$ as h in units of km. This implies roughly $\bar{S}\bar{h} < 10$ or $Sh < 10\,\mathrm{km^2/Ma}$.

However, a closed expression for a solution without approximation can also be achieved by first transforming the differential equation from depth z to the height $x = h - z$ over the basement and introduction of the function $u = T - q(x-h)/\lambda$. It can easily be shown, that the differential equation and the boundary conditions for $u(x,t)$ are identical to the consolidation problem from Gibson in App. D. Obviously, the parameter $\Delta\rho g$ must be substituted by $-q/\lambda$, the other parameter names are already chosen identical. Additionally, the solution can be expanded for small $K = kh$ as already demonstrated in (D.4) which yields for $K \to 0$ the same result as (F.24).

A few example curves of the full solution for (F.19) and (F.20) are plotted in Fig. F.12 and the approximation (F.21) in Fig. F.13. The error is below 0.3% for the case $kh = 0.001$ and below 8% for $kh = 0.1$. Approximation (F.21) can be improved for larger kh if k is replaced by k' according to Fig. F.14. The value of k' is calculated from the exact solution at $z = h$.

It must finally be noted that (F.21) does not incorporate transient effects which arise due to improper initial conditions. Such effects might need a few million years to decay, which can be seen very clearly in Fig. 3.4 for the case of erosion.

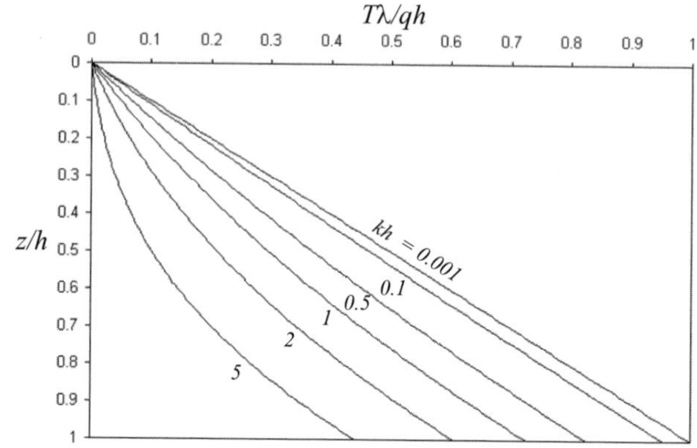

Fig. F.12. Example curves of the solution for (F.19) and (F.20)

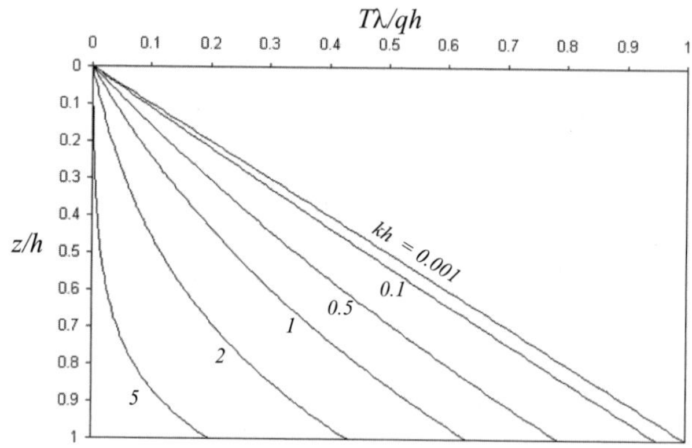

Fig. F.13. Approximation according to (F.21)

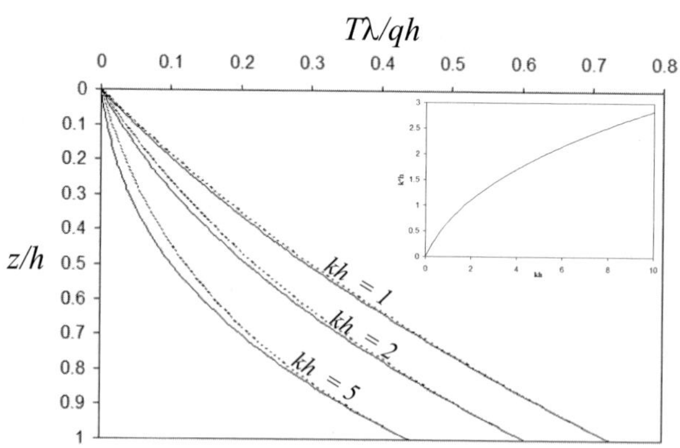

Fig. F.14. Approximation (dashed) according to (F.21) with a newly adapted k'. Here it is found $k'h = 0.678$ for $kh = 1$, $k'h = 1.103$ for $kh = 2$, and $k'h = 1.941$ for $kh = 5$

G

Petroleum Kinetics

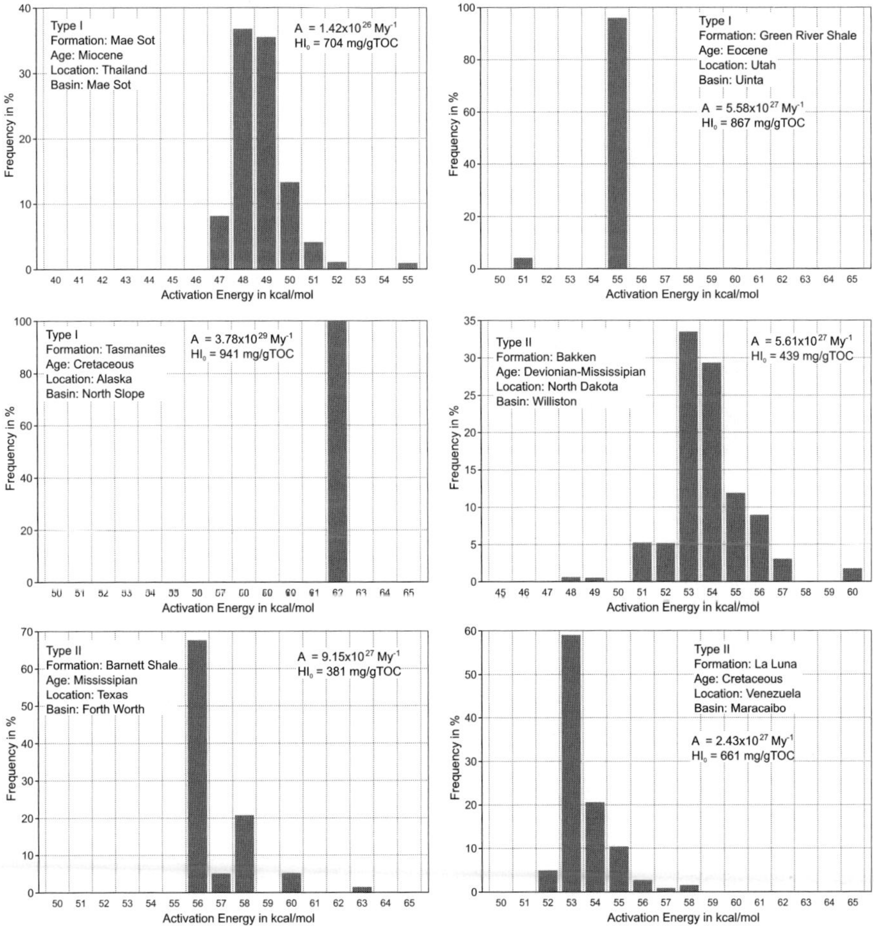

Fig. G.1. Bulk kinetics after Tegelaar and Noble (1994)

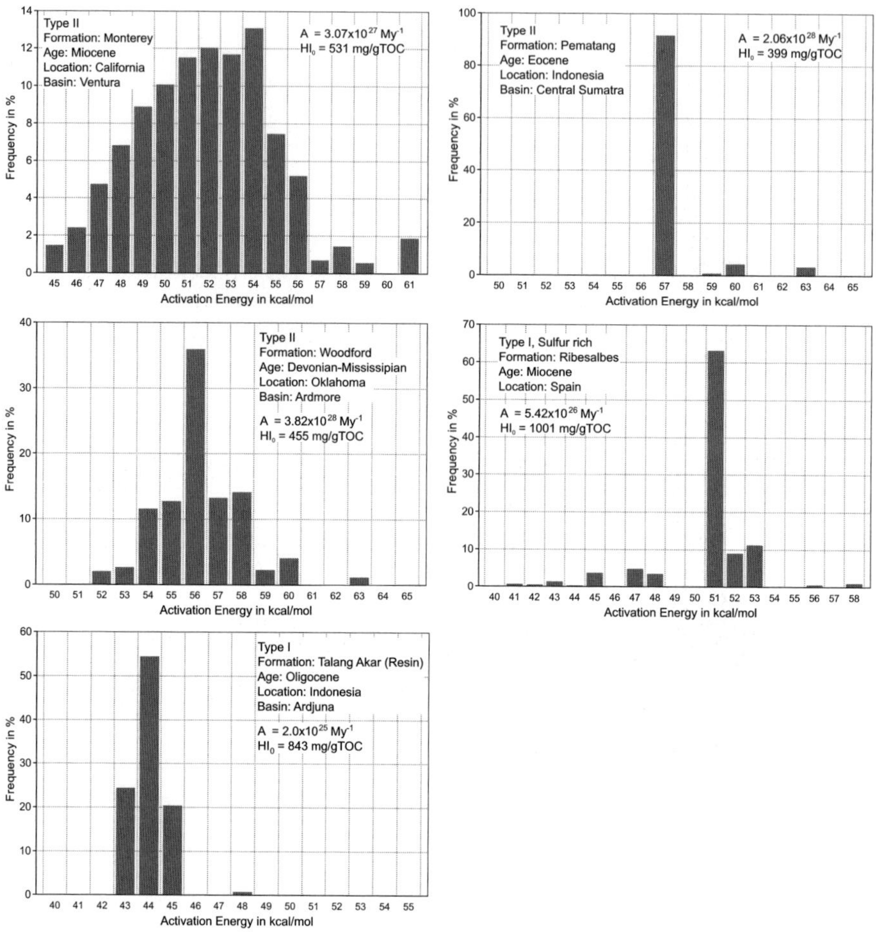

Fig. G.2. Bulk kinetics after Tegelaar and Noble (1994) continued

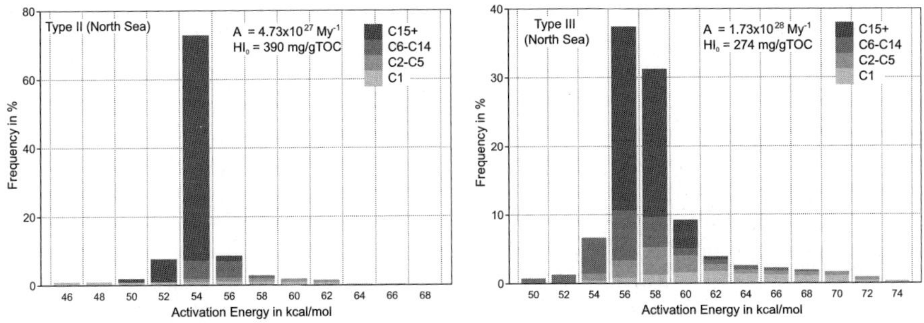

Fig. G.3. Multicomponent kinetics after Ungerer (1990)

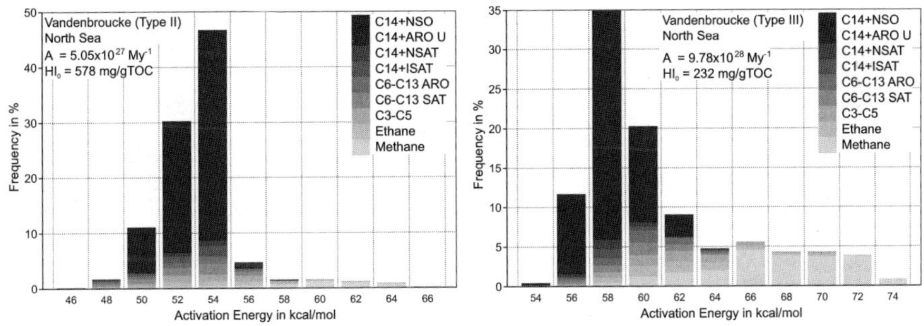

Fig. G.4. Multicomponent kinetics after Vandenbroucke et al. (1999)

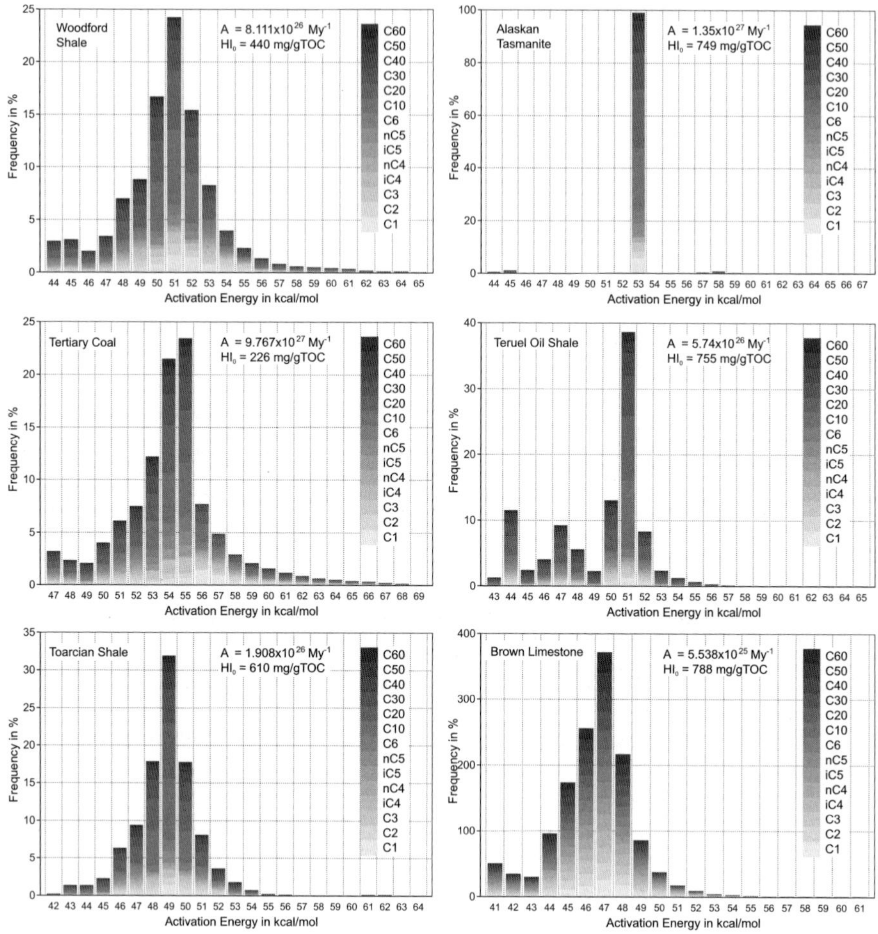

Fig. G.5. 14 component kinetics after di Primio and Horsfield (2006)

H

Biomarker

With much help from K. E. Peters the following tables are compiled from K. E. Peters, C. C. Walters, and J. M. Moldowan. *The Biomarker Guide*, volume 1 and 2. Cambridge University Press, second edition, 2005.

Compound	Biological Origin	Environment
2-Methyldocosane	Bacteria?	Hypersaline
Mid-chain monomethylalkanes	Cyanobacteria	Hot springs, marine
Pristane/phytane (low)	Phototrophs, archaea	Anoxic, high salinity
PMI (PME)*	Archaea, methanogens and methanotrophs	Hypersaline, anoxic
Crocetane	Archaea, methanotrophs?	Methane seeps?
C_{25} HBI**	Diatoms	Marine and lacustrine
Squalane	Archaea	Hypersaline?
Botryococcane, polymethylsqualanes	Green algae, *Botryococcus*	Lacustrine-brackish-saline

Table H.1. Acyclic biomarkers as indicators of biological input or depositional environment (assumes high concentration of component)

* PMI: 2,6,10,15,19-pentamethylicosane (current IUPAC nomenclature), previously spelled pentamethyleicosane (PME).
** C_{25} HBI: 2,6,10,14-tetramethyl-7-(3-methylpentyl)-pentadecane.

Compound	Biological Origin	Environment
	Saturates	
$C_{25} - C_{34}$ macrocyclic alkanes	Green algae, *Botryococcus*	Lacustrine-brackish
β-carotane	Cyanobacteria, algae	Arid, hypersaline
Phyllocladanes	Conifers	Terrigenous
$C_{19} - C_{30}$ tricyclic terpanes	*Tasmanites?*	Marine, high latitude
C_{30} 24-n-propylcholestanes (4-desmethyl)	Chrysophyte algae	Marine
Pregnane, homopregnane	Unknown	Hypersaline
Diasteranes	Algae/higher plants	Clay-rich rocks
Dinosteranes	Dinoflagellates	Marine, Triassic or younger
28,30-bisnorhopane	Bacteria	Anoxic marine, upwelling?
Bicadinanes	Higher plants	Terrigenous
23,28-bisnorlupanes	Higher plants	Terrigenous
Gammacerane	Tetrahymanol in ciliates feeding on bacteria	Stratified water, sulfate-reducing, hypersaline (low sterols)
18α-oleanane	Cretaceous or younger, higher plants	Paralic
Hexahydro-benzohopanes	Bacteria	Anoxic carbonate-anhydrite
	Aromatics	
Benzothiophenes, alkyldibenzo-thiophenes	Unknown	Carbonate/evaporite
Methyl n-pristanyl, methy i-butyl maleimides, isorenieratane	*Chlorobiaceae*, anaerobic green sulfur bacteria	Photic zone anoxia
Trimethyl chromans*	Phytoplankton	Saline photic zone?

Table H.2. Cyclic biomarkers as indicators of biological input or depositional environment (assumes high concentration of component)

* Trimethyl chromans: 2-methyl-2-(4,8,12-trimethyl-tridecyl)-chromans.

Characteristics	Shales	Carbonates
Non-biomarker Parameters		
API, Gravity	Medium-High	Low-Medium
Sulfur, wt.%	Variable	High (marine)
Thiophenic sulfur	Low	High
Biomarker Parameters		
Pristane/phytane	High (≥ 1)	Low (≤ 1)
Steranes/17α-hopanes	High	Low
Diasteranes/steranes	High	Low
C_{24} Tetra-/C_{26} tricyclic diterpanes	Low-Medium	Medium-High
C_{29}/C_{30} Hopane	Low	High (> 1)
C_{35} Homohopane index	Low	High
Hexahydrobenzohopanes and benzohopanes	Low	High
Dia/(Reg + Dia) MA-steroids*	High	Low
Ts/(Ts + Tm)**	High	Low
C_{29} MA-steroids	Low	High

Table H.3. Some characteristics of petroleum from carbonate versus shale source rocks

* Monoaromatic-steroid ratio

** Tm: C_{27} 17α-trisnorhopane, Ts: C_{27} 18α-trisnorhopane II

Property	Marine	Terrigenous	Lacustrine
Sulfur (wt.%)	High (anoxic)	Low	Low
Pristane/phytane	< 2	> 3	$\sim 1 - 3$
$C_{27} - C_{29}$ steranes	High C_{28}	High C_{29}	High C_{27}
C_{30} 24-n-propylcholestane	Low	Low or absent	Absent
Steranes/hopanes	High	Low	Low
Bicyclic sesquiterpanes	Low	High	Low
Tricyclic diterpanes	Low	High	High
Tetracyclic diterpanes	Low	High	Low
Lupanes, bisnorlupanes	Low	High	Low
28,30-bisnorhopane	High (anoxic)	Low	Low
Oleananes	Low or absent	High	Low
β-carotane	Absent	Absent	High (arid)
Botryococcane	Absent	Absent	High (brackish)
V/(V + Ni)	High (anoxic)	Low or absent	Low or absent

Table H.4. Generalized geochemical properties* differ between nonbiodegraded crude oils from marine, terrigenous, or lacustrine source-rock organic matter (modified from Peters and Moldowan, 1993)

* Quoted properties encompass most samples, but exceptions occur. For example, many nearshore oxic marine environments resulted in source rocks that generated oils with low sulfur and some very high sulfur oils originated from source rocks deposited in hypersaline lacustrine settings. The terms marine, terrigenous, and lacustrine can be misleading. “Marine oil” might refer to: (1) oil produced from marine reservoir rock, (2) oil generated from source rock deposited under marine conditions, or (3) oil derived from marine organic matter in the source rock. The table refers to provenance of the organic matter (3).

I

Component Properties

The following tables with component properties are compiled from:

A. Danesh. *PVT and Phase Behaviour of Petroleum Reservoir Fluids.* Number 47 in Developments in petroleum science. Elsevier, 1998,

R. C. Reid, J. M. Prausnitz, and B. E. Poling. *The Properties of Gases and Liquids.* McGraw–Hill Book Company, 4th edition, 1987,

IES PetroMod®, Petroleum Systems Modeling Software, Release 10.0, 2007, URL `ies.de`.

Cells are blanked if there was no data available. The definition of the listed quantities can be found in Chapter 5. Note that some values, such as the molar weight of C7+ or C15+, are obviously not representative for all fluids. This are example values which must be adapted.

Name	MW $\left[\frac{g}{mol}\right]$	T_c $[K]$	P_c [MPa]	v_c $\left[\frac{m^3}{kmol}\right]$	Acentric Fac. ω_c	Rackett Z_{Ra}	Shift Fac. S_E	ρ $\left[\frac{kg}{m^3}\right]$
Oil: Alkanes								
n–Hexane	86.177	507.6	3.025	0.371	0.3013	0.26355	−0.01478	663.8
n–Heptane	100.204	540.2	2.740	0.428	0.3495	0.26074	0.02509	688.2
n–Octane	114.231	568.7	2.490	0.486	0.3996	0.25678	0.04808	707.0
n–Nonane	128.258	594.6	2.290	0.544	0.4435	0.25456	0.06799	721.9
n–Decane	142.285	617.7	2.110	0.600	0.4923	0.25074	0.08546	734.2
n–Undecane	156.312	639.0	1.949	0.659	0.5303	0.24990	0.10100	744.5
n–Dodecane	170.338	658.0	1.820	0.716	0.5764	0.24692	0.11500	752.7
n–Tridecane	184.365	675.0	1.680	0.775	0.6174	0.24698	0.12764	761.7
n–Tetradecane	198.392	693.0	1.570	0.830	0.6430	0.24322	0.13923	763.3
n–Pentadecane	212.419	708.0	1.480	0.889	0.6863	0.23030	0.14989	772.2
n–Hexadecane	226.446	723.0	1.400	0.944	0.7174	0.22760	0.15974	777.2
n–Heptadecane	240.473	736.0	1.340	1.000	0.7697	0.23431	0.16890	779.7
n–Octadecane	254.500	747.0	1.270	1.060	0.8114	0.22917	0.17744	782.0
n–Nonadecane	268.527	758.0	1.210	1.120	0.8522	0.21580	0.18545	786.9
n–Eicosane	282.553	768.0	1.160	1.170	0.9069	0.22811	0.19297	792.7
n–Heneicosane	296.580	781.7	1.147	1.198	0.9220	0.20970	0.20007	795.4
n–Docosane	310.610	791.8	1.101	1.253	0.9550	0.20680	0.20678	798.1
n–Tricosane	324.630	801.3	1.059	1.307	0.9890	0.20380	0.21314	800.4
n–Tetracosane	338.680	810.4	1.019	1.362	1.0190	0.20110	0.21919	802.5
Oil: Alkenes								
1–Hexene	84.161	504.03	3.140	0.3540	0.2800	0.26600		676.9
1–Heptene	98.188	537.29	2.830	0.4130	0.3310	0.26150		701.5
Oil: Aromatics								
Benzene	78.114	562.16	4.898	0.2589	0.2108	0.26967	−0.04870	882.9
Toluene	92.141	591.79	4.109	0.3158	0.2641	0.26390	−0.01450	874.3
Ethylbenzene	106.167	617.17	3.609	0.3738	0.3036	0.26186	0.01397	874.4
o–Xylene	106.167	630.37	3.734	0.3692	0.3127	0.26200	0.01397	884.9
m–Xylene	106.167	617.05	3.541	0.3758	0.3260	0.26200	0.01397	869.4
p–Xylene	106.167	616.26	3.511	0.3791	0.3259	0.28700	0.01397	866.6
Oil: Cycloalkanes								
Cyclopentane	70.134	511.76	4.502	0.2583	0.1943	0.26824	−0.11868	760.3
Methylcyclopent.	84.161	532.79	3.784	0.3189	0.2302	0.27040	−0.07227	754.0
Cyclohexane	84.161	553.54	4.075	0.3079	0.2118	0.27286	−0.07227	783.5
Methylcyclohex.	98.188	572.19	3.471	0.3680	0.2350	0.26986	−0.03454	774.8
Ethylcyclopent.	98.188	569.52	3.397	0.3745	0.2715	0.26670	−0.03454	771.2
Ethylcyclohex.	112.215	609.15	3.040	0.4500	0.2455	0.26900	−0.00292	792.1
Oil: Methyl–Alkanes								
2 2-Dimethylprop	72.150	433.78	3.199	0.3036	0.1964	0.27570	−0.04350	597.4
2–Methylpentane	86.177	497.50	3.010	0.3664	0.2781	0.26620	−0.01478	657.8

Table I.1. Component parameters – Part 1

Name	MW $\left[\frac{g}{mol}\right]$	T_c $[K]$	P_c [MPa]	v_c $\left[\frac{m^3}{kmol}\right]$	Acentric Fac. ω_c	Rackett Z_{Ra}	Shift Fac. S_E	ρ $\left[\frac{kg}{m^3}\right]$
Oil: Single Carbon Number (SCN)								
C6	84	510	3.271	0.348	0.251	0.269	−0.01478	690
C7	96	547	3.071	0.392	0.280	0.266	0.01745	727
C8	107	574	2.877	0.433	0.312	0.263	0.03669	749
C9	121	603	2.665	0.485	0.352	0.260	0.05804	768
C10	134	627	2.481	0.532	0.389	0.256	0.07540	782
C11	147	649	2.310	0.584	0.429	0.253	0.09088	793
C12	161	670	2.165	0.635	0.467	0.250	0.10583	804
C13	175	689	2.054	0.681	0.501	0.247	0.11932	815
C14	190	708	1.953	0.727	0.536	0.244	0.13243	826
C15	206	727	1.853	0.777	0.571	0.240	0.14512	836
C16	222	743	1.752	0.830	0.610	0.237	0.15670	843
C17	237	758	1.679	0.874	0.643	0.234	0.16669	851
C18	251	770	1.614	0.915	0.672	0.232	0.17536	856
C19	263	781	1.559	0.951	0.698	0.229	0.18235	861
C20	275	793	1.495	0.997	0.732	0.226	0.18900	866
C21	291	804	1.446	1.034	0.759	0.224	0.19729	871
C22	300	815	1.393	1.077	0.789	0.221	0.20174	876
C23	312	825	1.356	1.110	0.815	0.219	0.20743	881
C24	324	834	1.314	1.147	0.841	0.217	0.21286	885
C25	337	844	1.263	1.193	0.874	0.214	0.21849	888
C26	349	853	1.230	1.226	0.897	0.212	0.22346	892
C27	360	862	1.200	1.259	0.944	0.200	0.22784	896
C28	372	870	1.164	1.296	0.968	0.198	0.23244	899
C29	382	877	1.140	1.323	0.985	0.196	0.23614	902
C30	394	885	1.107	1.361	1.008	0.194	0.24044	905
C31	404	893	1.085	1.389	1.026	0.193	0.24390	909
C32	415	901	1.060	1.421	1.046	0.191	0.24759	912
C33	426	907	1.039	1.448	1.063	0.189	0.25117	915
C34	437	914	1.013	1.480	1.082	0.188	0.25464	917
C35	445	920	0.998	1.502	1.095	0.187	0.25710	920
C36	456	926	0.974	1.534	1.114	0.185	0.26040	922
C37	464	932	0.964	1.550	1.124	0.184	0.26275	925
C38	475	938	0.941	1.583	1.142	0.182	0.26589	927
C39	484	943	0.927	1.604	1.154	0.181	0.26840	929
C40	495	950	0.905	1.636	1.172	0.180	0.27139	931
C41	502	954	0.896	1.652	1.181	0.179	0.27325	933
C42	512	959	0.877	1.680	1.195	0.178	0.27586	934
C43	521	964	0.864	1.701	1.207	0.177	0.27815	936
C44	531	970	0.844	1.733	1.224	0.175	0.28065	938
C45	539	974	0.835	1.749	1.232	0.174	0.28261	940

Table I.2. Component parameters – Part 2

Name	MW $\left[\frac{g}{mol}\right]$	T_c $[K]$	P_c [MPa]	v_c $\left[\frac{m^3}{kmol}\right]$	Acentric Fac. ω_c	Rackett Z_{Ra}
Oil: Boiling Range Groups						
Oil–range Mahakam(waxy crude)	133.440	617.95	2.600	0.5103	0.3795	0.25725
Oil–range Tuscaloosa(Live Oil)	159.140	657.60	2.334	0.5920	0.4451	0.25137
Oil–range Smackover(S–rich crude)	184.440	693.37	2.105	0.6773	0.5083	0.24593
Oil BO	196.032	705.54	2.026	0.7084	0.5392	0.24244
Oil–range Tualag(waxy crude)	201.620	713.24	1.998	0.7237	0.5485	0.24155
Oil: Compound Classes						
C6–C14ARO	89.616	588.92	4.321	0.3047	0.2546	0.26788
C6–C14SAT	121.422	587.96	2.518	0.4980	0.4003	0.25812
C5–C14	132.314	595.03	2.305	0.5594	0.4546	0.25367
C6–C14	137.378	605.58	2.215	0.5801	0.4717	0.25242
C15+	311.950	824.32	1.392	1.0926	0.8079	0.21739
C6–C13ARO	104.484	618.13	3.703	0.3669	0.3077	0.26604
C6–C13SAT	137.936	609.86	2.191	0.5792	0.4730	0.25313
C7+	209.610	724.56	1.932	0.7551	0.5686	0.23983
C14+ISAT	299.180	813.82	1.442	1.0524	0.7797	0.22017
C14+NSAT	299.180	813.82	1.442	1.0524	0.7797	0.22017
C14+ARO U	299.180	813.82	1.442	1.0524	0.7797	0.22017
C15+ARO	282.553	768.00	1.165	1.1700	0.9069	0.22811
C15+SAT	255.342	747.73	1.309	1.0531	0.8085	0.22382
C14+NSO	402.700	891.20	1.109	1.3697	1.0050	0.19590
NSO	338.680	810.40	1.009	1.3620	1.0190	0.20110
Standard Oils						
Volatile Oil	53.135	367.79	4.373	0.1956	0.1267	0.27919
Light Oil	48.439	357.38	4.389	0.1895	0.1238	0.27968
Black Oil	90.072	491.62	3.535	0.3121	0.2423	0.26892
Medium Oil	101.141	511.92	3.459	0.3302	0.2606	0.26722
Heavy Oil	127.492	568.42	3.041	0.4080	0.3290	0.26099
Tuscaloosa T2 Live Oil	57.930	400.71	4.112	0.2241	0.1634	0.27631
Smackover T2–S Live Oil	142.418	604.26	2.669	0.4839	0.3930	0.25608
Tualag T1 High Waxy	167.860	650.49	2.298	0.5886	0.4750	0.24916
Mahakam T3 High Waxy+Aro	102.266	529.78	3.082	0.3768	0.3165	0.26429

Table I.3. Component parameters – Part 3

Name	MW $\left[\frac{g}{mol}\right]$	T_c $[K]$	P_c [MPa]	v_c $\left[\frac{m^3}{kmol}\right]$	Acentric Fac. ω_c	Rackett Z_{Ra}	Shift Fac. S_E	ρ $\left[\frac{kg}{m^3}\right]$
Gas: Alkanes								
Methane	16.043	190.56	4.599	0.0986	0.0115	0.28941	−0.15400	300.0
Methane–C13	17.043	190.56	4.599	0.0986	0.0115	0.28941	−0.15400	318.7
Ethane	30.070	305.32	4.872	0.1455	0.0995	0.28128	−0.10020	356.2
Propane	44.096	369.83	4.248	0.2000	0.1523	0.27664	−0.08501	507.0
n–Butane	58.123	425.12	3.796	0.2550	0.2002	0.27331	−0.06413	584.0
n–Pentane	72.150	469.70	3.370	0.3130	0.2515	0.26853	−0.04183	631.1
Gas: Alkenes								
Ethylene	28.054	282.36	5.032	0.1291	0.0852	0.28054		500.0
Propadiene	40.065	393.15	5.470	0.1620	0.1596	0.27283		599.7
Propylene	42.081	364.76	4.612	0.1810	0.1424	0.27821		521.0
1–Butene	56.107	419.58	4.020	0.2399	0.1867	0.27351		600.5
cis–2–Butene	56.107	435.58	4.206	0.2340	0.2030	0.27044		628.6
trans–2–Butene	56.107	428.63	4.103	0.2382	0.2182	0.27212		611.2
1 2–Butadiene	54.092	444.00	4.500	0.2190	0.2509	0.26850		657.6
1 3–Butadiene	54.092	425.37	4.330	0.2208	0.1932	0.27130		627.3
1–Pentene	70.134	464.78	3.529	0.2960	0.2329	0.27035		645.8
cis–2–Pentene	70.134	475.93	3.654	0.3021	0.2406	0.26940		659.8
trans–2–Pentene	70.134	475.37	3.654	0.3021	0.2373	0.26970		652.4
Gas: Methyl–Alkanes								
i–Butane	58.123	408.14	3.648	0.2627	0.1770	0.27569	−0.07935	562.9
i–Pentane	72.150	460.43	3.381	0.3058	0.2275	0.27060	−0.04350	624.7
Gas: Methyl–Alkenes								
2–Methyl–1–Butene	70.134	465.00	3.400	0.2920	0.2287	0.2705		656.3
3–Methyl–1–Butene	70.134	450.37	3.516	0.3021	0.2286	0.2705		632.2
2–Methyl–2–Butene	70.134	471.00	3.400	0.2920	0.2767	0.2663		668.3

Table I.4. Component parameters – Part 4

Name	MW $\left[\frac{g}{mol}\right]$	T_c $[K]$	P_c [MPa]	v_c $\left[\frac{m^3}{kmol}\right]$	Acentric Fac. ω_c	Rackett Z_{Ra}	ρ $\left[\frac{kg}{m^3}\right]$
Gas: Compound Classes							
C2–C4	47.130	379.57	4.192	0.2120	0.1615	0.27623	
C2–C5	53.070	400.97	3.996	0.2360	0.1828	0.27440	
C3–C5	62.752	430.72	3.599	0.2721	0.2008	0.27340	
Standard Gases							
Hydrogen	2.016	33.18	1.313	0.0642	−0.2150	0.31997	
Carbon Monoxide	28.010	132.92	3.499	0.0931	0.0663	0.28966	
Nitrogen	28.014	126.10	3.394	0.0901	0.0403	0.28971	809.4
Oxygen	31.999	154.58	5.043	0.0734	0.0218	0.28962	1142.1
Hydrogen Sulphide	34.082	373.53	8.963	0.0985	0.0827	0.28476	801.4
Carbon Dioxide	44.010	304.19	7.382	0.0940	0.2276	0.27275	818.0
Sulfur Dioxide	64.065	430.75	7.884	0.1220	0.2451	0.26729	1394.6
Dry Gas	17.943	197.42	4.850	0.0977	0.0221	0.27992	
Wet Gas	30.186	272.40	4.801	0.1345	0.0624	0.28493	
Water	18.015	647.13	22.055	0.0560	0.3449		1000.0

Table I.5. Component parameters – Part 5

	CO_2	H_2S	N_2	CO	CO_2	H_2S	N_2	CO
H_2S	0.097				0.099			
N_2	−0.017	0.000			−0.032	0.000		
Methane	0.092	0.000	0.031	0.030	0.093	0.000	0.028	0.032
Ethylene	0.055	0.083	0.086	0.000	0.053	0.085	0.080	0.000
Ethane	0.132	0.000	0.052	−0.023	0.136	0.000	0.041	−0.028
Propylene	0.093	0.000	0.090	0.000	0.094	0.000	0.090	0.000
Propane	0.124	0.088	0.085	0.026	0.129	0.088	0.076	0.016
Isobutane	0.120	0.047	0.103	0.000	0.128	0.051	0.094	0.000
n–Butane	0.133	0.000	0.080	0.000	0.143	0.000	0.070	0.000
Isopentane	0.122	0.000	0.092	0.000	0.131	0.000	0.087	0.000
n–Pentane	0.122	0.063	0.100	0.000	0.131	0.069	0.088	0.000
n–Hexane	0.110	0.000	0.150	0.000	0.118	0.000	0.150	0.000
n–Heptane	0.100	0.000	0.144	0.000	0.110	0.000	0.142	0.000
n–Decane	0.114	0.000	0.000	0.000	0.130	0.000	0.000	0.000
Cyclohexane	0.105	0.000	0.000	0.000	0.129	0.000	0.000	0.000
Benzene	0.077	0.000	0.164	0.000	0.077	0.000	0.153	0.000
Toluene	0.106	0.000	0.000	0.000	0.113	0.000	0.000	0.000

Table I.6. Binary Interaction Parameter (BIP) for the Peng–Robinson (left) and the Soave–Redlich–Kwong (right) equation of state according to Reid et. al. (1987)

	N_2	CO_2	C1	Ethyl.	C2	Propyl.	C3	iC4	nC4
N_2	0.0000								
CO_2	0.0000	0.0000							
C1	0.0311	0.1070	0.0000						
Ethylene	0.0500	0.1200	0.0215	0.0000					
C2	0.0515	0.1322	0.0026	0.0089	0.0000				
Propylene	0.0600	0.1300	0.0330	0.0000	0.0890	0.0000			
C3	0.0852	0.1241	0.0140	0.0100	0.0011	0.0100	0.0000		
iC4	0.1000	0.1400	0.0256	0.0200	−0.0067	0.0080	−0.0078	0.0000	
nC4	0.0711	0.1333	0.0133	0.0200	0.0096	0.0080	0.0033	0.0000	0.0000
iC5	0.1000	0.1400	−0.0056	0.0250	0.0080	0.0080	0.0111	−0.0040	0.0170
Neopent.	0.1000	0.1400	−0.0056	0.0250	0.0080	0.0080	0.0111	−0.0040	0.0170
nC5	0.1000	0.1400	0.0236	0.0250	0.0078	0.0100	0.0120	0.0020	0.0170
nC6	0.1496	0.1450	0.0422	0.0300	0.0140	0.0110	0.0267	0.0240	0.0174
Methyl-cyc.pent.	0.1500	0.1450	0.0450	0.0310	0.0141	0.0120	0.0270	0.0242	0.0180
Cyc.hex.	0.1500	0.1450	0.0450	0.0310	0.0141	0.0120	0.0270	0.0242	0.0180
nC7	0.1441	0.1450	0.0352	0.0300	0.0150	0.0140	0.0560	0.0250	0.0190
Methyl-cyclohex.	0.1500	0.1450	0.0450	0.0300	0.0160	0.0150	0.0580	0.0250	0.0200
Toluene	0.1700	0.1800	0.0600	0.0400	0.0200	0.0210	0.0600	0.0300	0.0110
o–Xylene	0.1500	0.1400	0.0470	0.0300	0.0160	0.0150	0.0590	0.0260	0.0120
nC8	0.1500	0.1400	0.0470	0.0300	0.0160	0.0150	0.0590	0.0260	0.0120
nC9	0.1550	0.0145	0.0474	0.0400	0.0190	0.0200	0.0070	0.0060	0.0100
nC10–nC14	0.1550	0.0145	0.0500	0.0450	0.0300	0.0250	0.0200	0.0100	0.0010
nC15–nC19	0.1550	0.0145	0.0600	0.0500	0.0400	0.0300	0.0250	0.0150	0.0010
nC20–nC24	0.1550	0.0145	0.0700	0.0600	0.0500	0.0350	0.0300	0.0200	0.0015

Table I.7. Binary Interaction Parameter (BIP) for the Peng–Robinson equation of state according to Danesh (1998)

	N_2	CO_2	C1	Ethyl.	C2	Propyl.	C3	iC4	nC4
N_2	0.0000								
CO_2	0.0000	0.0000							
C1	0.0278	0.1107	0.0000						
Ethyl.	0.0300	0.1000	0.0189	0.0000					
C2	0.0407	0.1363	−0.0078	0.0026	0.0000				
Propyl.	0.0800	0.1000	0.0289	0.0000	0.0200	0.0000			
C3	0.0763	0.1000	0.0080	0.0080	−0.0220	0.0033	0.0000		
iC4	0.0944	0.1000	0.0241	0.0900	−0.0010	−0.0144	−0.0100	0.0000	
nC4	0.0700	0.1000	0.0056	0.1000	0.0067	0.0000	0.0000	0.0000	0.0000
iC5	0.0867	0.1000	−0.0078	0.0120	0.0050	0.0000	0.0078	0.0000	0.0000
Neopent.	0.0870	0.1000	−0.0078	0.0120	0.0050	0.0000	0.0078	0.0000	0.0000
nC5	0.0878	0.1000	0.0019	0.0120	0.0056	0.0050	0.0230	−0.030	0.0204
nC6	0.1400	0.1000	0.0374	0.0140	−0.0156	0.0050	−0.0022	0.0000	−0.0111
Methyl-cyc.pent.	0.1400	0.1000	0.0400	0.0140	0.0330	0.0050	0.0030	0.0000	0.0000
Cyc.hex.	0.1400	0.1000	0.0333	0.0150	0.0230	0.0050	0.0030	0.0005	0.0000
nC7	0.1422	0.1000	0.0307	0.0144	0.0411	0.0100	0.0044	0.0005	0.0000
Methyl-cyc.hex.	0.1450	0.1000	0.0500	0.0150	0.0230	0.0100	0.0050	0.0005	0.0000
Toluene	0.1500	0.1000	0.0978	0.0300	0.0900	0.0300	0.0300	0.0200	0.0100
o–Xyl.	0.1500	0.1000	0.1000	0.0250	0.0500	0.0300	0.0300	0.0200	0.0100
nC8	0.1500	0.1000	0.0448	0.0200	0.0170	0.0100	0.0040	0.0015	0.0000
nC9	0.1500	0.1000	0.0448	0.0200	0.0170	0.0100	0.0040	0.0015	0.0000
nC10–nC14	0.1500	0.1000	0.0550	0.0300	0.0200	0.0150	0.0040	0.0020	0.0010
nC15–nC19	0.1500	0.1000	0.0600	0.0400	0.0350	0.0250	0.0005	0.0025	0.0010
nC20–nC24	0.1500	0.1000	0.0650	0.0450	0.0400	0.0300	0.0010	0.0050	0.0015

Table I.8. Binary Interaction Parameter (BIP) for the Soave–Redlich–Kwong equation of state according to Danesh (1998)

J

Methane Density

The Modified Benedict–Webb–Rubin (MBWR) EOS has the form

$$
\begin{aligned}
P = \quad & \rho RT + \rho^2(N_1 T + N_2 T^{1/2} + N_3 + N_4/T + N_5/T^2) \\
& + \rho^3(N_6 T + N_7 + N_8/T + N_9/T^2) + \rho^4(N_{10} T + N_{11} + N_{12}/T) \\
& + \rho^5 N_{13} + \rho^6(N_{14}/T + N_{15}/T^2) + \rho^7 N_{16}/T \\
& + \rho^8(N_{17}/T + N_{18}/T^2) + \rho^9 N_{19}/T^2 \\
& + \rho^3(N_{20}/T^2 + N_{21}/T^3)e^{-\gamma\rho^2} + \rho^5(N_{22}/T^2 + N_{23}/T^4)e^{-\gamma\rho^2} \\
& + \rho^7(N_{24}/T^2 + N_{25}/T^3)e^{-\gamma\rho^2} + \rho^9(N_{26}/T^2 + N_{27}/T^4)e^{-\gamma\rho^2} \\
& + \rho^{11}(N_{28}/T^2 + N_{29}/T^3)e^{-\gamma\rho^2} \\
& + \rho^{13}(N_{30}/T^2 + N_{31}/T^3 + N_{32}/T^4)e^{-\gamma\rho^2}
\end{aligned} \tag{J.1}
$$

with $\gamma = 0.0096$, $R = 0.08205616\,\text{atm/mol}/K$, temperature in K, pressure in atm, density ρ in mol/l and

$$
\begin{aligned}
N_1 &= -1.8439486666 \times 10^{-2} & N_2 &= 1.0510162064 \\
N_3 &= -1.6057820303 \times 10 & N_4 &= 8.4844027562 \times 10^{2} \\
N_5 &= -4.2738409106 \times 10^{4} & N_6 &= 7.6565285254 \times 10^{-4} \\
N_7 &= -4.8360724197 \times 10^{-1} & N_8 &= 8.5195473835 \times 10 \\
N_9 &= -1.6607434721 \times 10^{4} & N_{10} &= -3.7521074532 \times 10^{-5} \\
N_{11} &= 2.8616309259 \times 10^{-2} & N_{12} &= -2.8685285973 \\
N_{13} &= 1.1906973942 \times 10^{-4} & N_{14} &= -8.5315715699 \times 10^{-3} \\
N_{15} &= 3.8365063841 & N_{16} &= 2.4986828379 \times 10^{-5} \\
N_{17} &= 5.7974531455 \times 10^{-6} & N_{18} &= -7.1648329297 \times 10^{-3} \\
N_{19} &= 1.2577853784 \times 10^{-4} & N_{20} &= 2.2240102466 \times 10^{4} \\
N_{21} &= -1.4800512328 \times 10^{6} & N_{22} &= 5.0498054887 \times 10 \\
N_{23} &= 1.6428375992 \times 10^{6} & N_{24} &= 2.1325387196 \times 10^{-1} \\
N_{25} &= 3.7791273422 \times 10 & N_{26} &= -1.1857016815 \times 10^{-5} \\
N_{27} &= -3.1630780767 \times 10 & N_{28} &= -4.1006782941 \times 10^{-6} \\
N_{29} &= 1.4870043284 \times 10^{-3} & N_{30} &= 3.1512261532 \times 10^{-9} \\
N_{31} &= -2.1670774745 \times 10^{-6} & N_{32} &= 2.4000551079 \times 10^{-5} \,.
\end{aligned}
\tag{J.2}
$$

For the selection of the appropriate root it is often very helpful to solve a SRK or PR EOS firstly. This can be performed analytically and the solution can then be used as a start value for the numerical solution of the MBWR EOS.

K

Compositions and Components for Fig. 5.14

	Dry Gas	Wet Gas	Gas Condensate	Volatile Oil	Black Oil
Methane	75.0	84.0	79.0	54.0	47.0
Ethane	7.0	8.0	10.0	9.0	6.0
Propane	6.0	3.0	3.0	8.0	4.0
n–Butane	5.0	1.0	2.0	8.0	3.0
n–Pentane	4.0	1.0	1.5	7.0	2.0
n–Hexane	3.0	1.0			
C_{6-14}		2.0	1.7	6.0	10.0
C_{15}			1.8	4.0	9.0
C_{25}			0.7	2.0	8.0
C_{35}			0.2	1.5	6.0
C_{45}			0.1	0.5	5.0

Table K.1. Compositions of the examples in Fig. 5.14. All entries are in molar %

	M [g/mol]	T_c [°C]	P_c [MPa]	v_c [m^3/kmol]	Acentric Factor	Volume Shift [m^3/kmol]
Methane	16.043	82.50	4.599	0.0986	0.0115	0.0007
Ethane	30.070	32.17	4.872	0.1455	0.0995	0.0028
Propane	44.096	96.68	4.248	0.2000	0.1523	0.0052
n–Butane	58.123	151.97	3.796	0.2550	0.2002	0.0080
n–Pentane	72.150	196.55	3.370	0.3130	0.2515	0.0122
n–Hexane	86.177	234.45	3.025	0.3710	0.3013	0.0176
C_{6-14}	133.680	292.11	1.999	0.5801	0.3111	0.0389
C_{15}	183.516	456.45	1.723	0.7770	0.5882	0.0724
C_{25}	323.821	573.739	1.128	1.1930	0.8361	0.1821
C_{35}	442.868	638.133	0.886	1.5020	1.1327	0.3356
C_{45}	537.418	699.490	0.737	1.7490	1.1940	0.4761

Table K.2. Component properties of the examples in Fig. 5.14

L

An Analytic Solution for the Diffusion of Methane Through a Cap Rock

A homogeneous cap rock of thickness h is sketched in Fig. L.1. The concentration at bottom is determined by an methane accumulation with a rather high concentration value c_b. At top the methane might quickly migrate away, which can be described by a rather small concentration c_t. The initial concentration within the cap rock is given by c_0. The mathematical formulation is

$$\frac{\partial}{\partial t} c(z,t) = D \frac{\partial^2}{\partial z^2} c(z,t) \tag{L.1}$$

with boundary conditions

$$c(z,0) = c_0 \quad \text{and} \quad c(0,t) = c_t \ , \quad c(h,t) = c_b \quad \text{for} \quad t > 0 \ . \tag{L.2}$$

The problem is similar to App. F.5 and App. F.6. The solution becomes for $t > 0$

$$\begin{aligned} c(z,t) &= c_t + (c_b - c_t)\frac{z}{h} \\ &+ \sum_{n=1}^{\infty} 2\left[c_0 - c_t + (-1)^n (c_b - c_0)\right] \frac{\sin \mu_n z}{\mu_n h} \exp(-\mu_n^2 D t) \end{aligned} \tag{L.3}$$

with $\mu_n = n\pi/h$.

An example solution is shown in Fig. L.1.More than 1 My are needed for the establishment of a static concentration gradient in the cap rock. Note that the time dependency scales quadratical with height h due to the form of exponent $\mu_n^2 Dt = n^2\pi^2 Dt/h^2$. Hence a cap rock with doubled thickness of 200 m needs four times as long and with 10 times thickness of 1 km about 100 times as long for the convergence to the static solution. These are here times of 4 My and 100 My. Additionally, the time dependency scales linear inversely with the diffusion coefficient. Realistic values can become $D = 10^{-11}\mathrm{m}^2/\mathrm{s}$, which is one order of magnitude smaller than in example Fig. L.1 or even yet smaller (Sec. 6.4).Transient effects decelerate respectively.

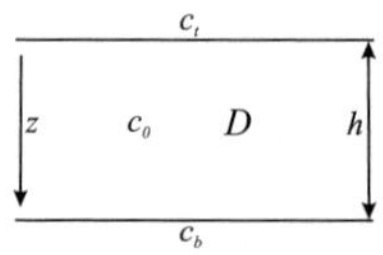

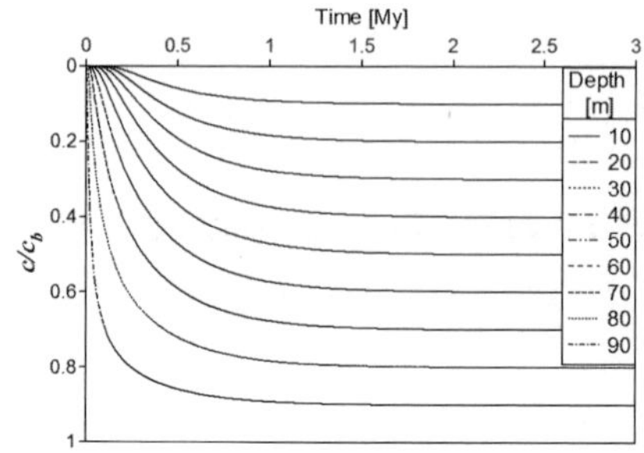

Fig. L.1. Model for diffusion through a cap rock of thickness h and example curves of solution (L.3) for $c_t = 0$, $c_0 = 0$, $h = 100\,\text{m}$, and a rather big diffusion coefficient $D = 10^{-10}\text{m}^2/\text{s}$

The cumulative volume Q which is flown through a horizontal plane of unit size at depth z is given by

$$Q = \int_0^t D\frac{\partial}{\partial z}c(z,t')dt' \tag{L.4}$$

and can be calculated at top $z = 0$ or bottom $z = h$, here again for $c_t = 0$, $c_0 = 0$, as

$$\begin{aligned} \frac{Q_t}{hc_b} &= \frac{Dt}{h^2} - \frac{1}{6} - 2\sum_{n=1}^{\infty}\frac{(-1)^n}{\mu_n^2 h^2}\exp(-\mu_n^2 Dt) \\ \frac{Q_b}{hc_b} &= \frac{Dt}{h^2} + \frac{1}{3} - 2\sum_{n=1}^{\infty}\frac{1}{\mu_n^2 h^2}\exp(-\mu_n^2 Dt)\ . \end{aligned} \tag{L.5}$$

The figure which is based on the same values as Fig. L.1 is shown in Fig. L.2.The vertical difference $\Delta Q = Q_b - Q_t$ between the volume which passed the bottom and the top boundary defines the methane volume in the cap rock. In the static limit for $t \to \infty$ it approaches the expected value $\Delta Q = hc_b/2$.

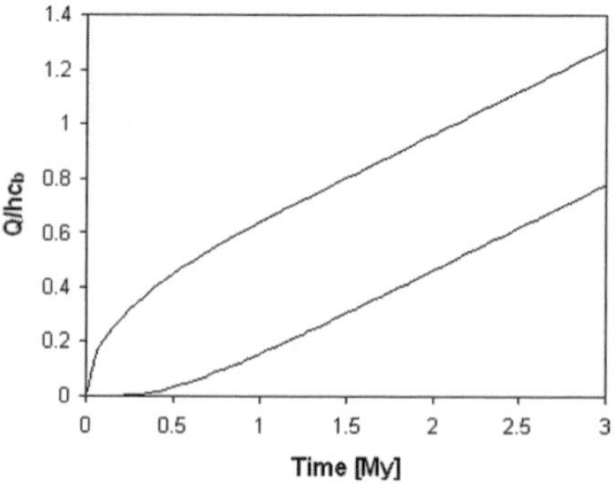

Fig. L.2. Example curves for diffusion amounts according to (L.5). The curve on top quantifies the diffusion amounts on bottom and vice versa

M

Flowpath Bending

The angle ψ which indicates the direction of petroleum flow in a reservoir below a planar seal with dipping angle α in the y–direction and lateral water flow with angle β to the x–axis is sketched in Fig. M.1. The normal vector on the seal is given by $\mathbf{n}_s = -\sin\alpha\,\mathbf{e}_y + \cos\alpha\,\mathbf{e}_z$. The vectors $\mathbf{e}_{x,y,z}$ are unit vectors in the $x, y,$, and z–directions. The petroleum potential u_p according to (6.18) is

$$u_p = u_w - \Delta\rho g z \tag{M.1}$$

with $\Delta\rho = \rho_w - \rho_p$ and $\rho_{w,p}$ the densities of water and petroleum.

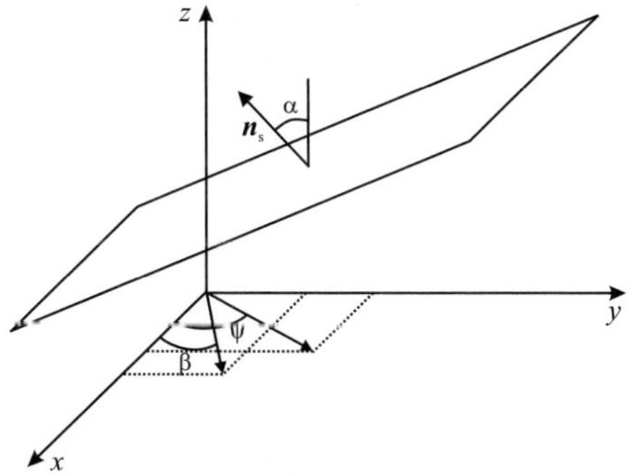

Fig. M.1. Seal with dipping angle α

Without a seal the flow direction of the petroleum is $\mathbf{v}_p \propto -\boldsymbol{\nabla} u_p$. Directly below the seal the petroleum cannot move in the direction of $\mathbf{n}_s$. The part of the flow which points in the direction of $\mathbf{n}_s$ must be subtracted and the condition $\mathbf{v}_p \cdot \mathbf{n}_s = 0$ must be ensured. Hence the final direction of flow is

$$\mathbf{v}_p \propto -\boldsymbol{\nabla} u_p + (\boldsymbol{\nabla} u_p \cdot \mathbf{n}_s)\mathbf{n}_s \,. \tag{M.2}$$

A lateral water flow in direction

$$\mathbf{v}_w \propto \cos\beta\,\mathbf{e}_x + \sin\beta\,\mathbf{e}_y \tag{M.3}$$

exists. The hydraulic head has a dipping angle of γ which leads to $u_w = -\rho_w g \tan\gamma(\cos\beta\,x + \sin\beta\,y)$. Insertion into (M.1) and calculation of the gradient yields

$$-\boldsymbol{\nabla} u_p \propto \rho_w \tan\gamma(\cos\beta\,\mathbf{e}_x + \sin\beta\,\mathbf{e}_y) + \Delta\rho\mathbf{e}_z\;. \tag{M.4}$$

It must be noted that the lateral pressure gradient can become so large that the petroleum flow points away from the seal into the reservoir. Such a flow exists when according to Fig. M.2

$$\tan\alpha > \frac{\Delta\rho}{\rho_w \tan\gamma \sin\beta} \tag{M.5}$$

and need not to be considered further. Equation (M.4) can now be combined with (M.2) to calculate $\mathbf{v}_p$ and afterwards

$$\tan\psi = \frac{v_{py}}{v_{px}} = \tan\beta\cos^2\alpha + \frac{\Delta\rho}{\rho_w}\frac{\cos\alpha\sin\alpha}{\tan\gamma\cos\beta}\;. \tag{M.6}$$

A lateral overpressure which is orthogonal to the dipping direction of the seal is equivalent to the special case $\beta = 0$ which leads to (6.44).

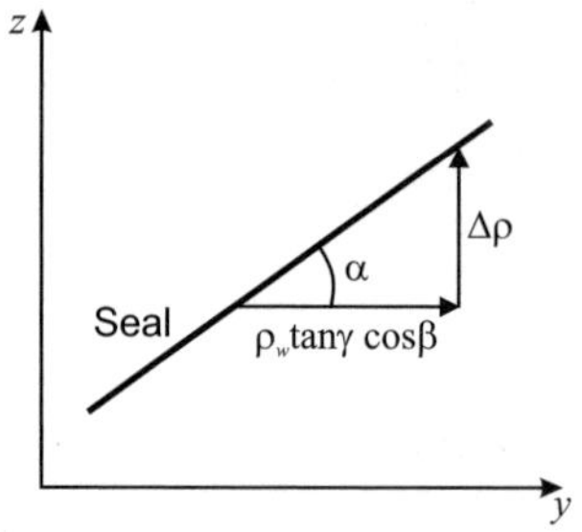

Fig. M.2. Lateral waterflow which has to be overcome to release petroleum from the seal

The result (M.6) can be checked for some special cases. For $\gamma \to 0$ it is $v_{py}/v_{px} \to \infty$ which correctly yields $\psi \to 90°$, the "non–bending" case. For very small α (M.6) can be developed into a Taylor–series and it becomes

$$\tan\psi \to \tan\beta + \frac{\Delta\rho}{\rho_w}\frac{\alpha}{\tan\gamma\cos\beta} \to \tan\beta \quad \text{for} \quad \alpha \to 0\;. \tag{M.7}$$

For a flat untilted surface, the flow follows the lateral overpressure direction. The impact of overpressure on flowpath bending is maximized.

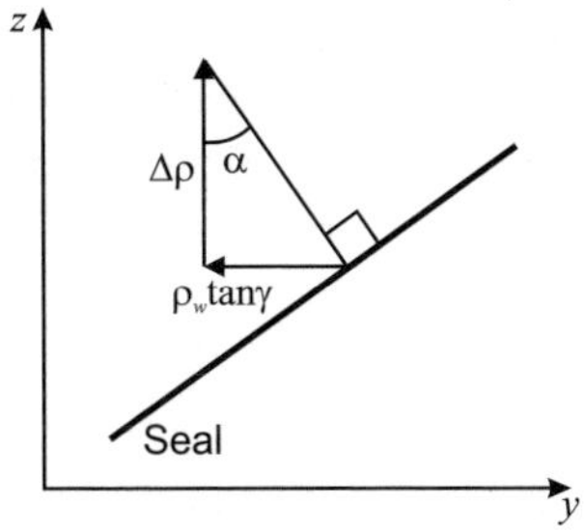

Fig. M.3. Lateral waterflow which has to be overcome to reverse the lateral y–flow direction

A case which is not covered by (M.6) is $\beta = -90°$ which can easily be calculated with Fig. M.3. The lateral flow reverses its direction into the negative y–axis if $\tan\gamma > \Delta\rho \tan\alpha/\rho_w$.

A problem rises from (M.6) for $\Delta\rho = 0$ which results in $\tan\psi = \tan\beta\cos^2\alpha$. This is at a first glance unexpected. Without buoyancy the petroleum flow should follow the lateral overpressure with $\psi = \beta$. However, water flow in the lateral direction passes through the seal Fig. M.4. This is not allowed for the petroleum. The flow has to follow the barrier.

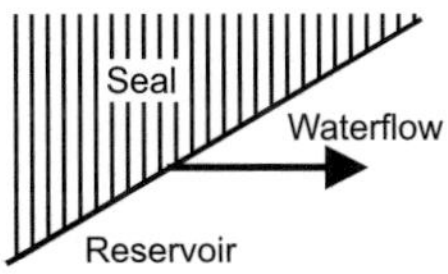

Fig. M.4. Water flow at seal according to (M.3)

A slightly more elegant formula for flowpath bending can be obtained if the water flow is also assumed to follow the dipping of the seal. This behavior can be obtained by the constrains

$$\mathbf{v}_w \cdot \mathbf{n}_s = 0\,, \quad \mathbf{v}_w \cdot \mathbf{e}_x \propto \cos\beta\,, \quad \mathbf{v}_w \cdot \mathbf{e}_y \propto \sin\beta\,. \tag{M.8}$$

The water potential becomes

$$u_w = -\frac{\rho_w g \tan\gamma}{\sqrt{1+\sin^2\beta\tan^2\alpha}}(\cos\beta\, x + \sin\beta\, y + \sin\beta\tan\alpha\, z) \tag{M.9}$$

when these conditions are fulfilled. The same straight forward calculation as for (M.6) yields now

$$\tan\psi = \tan\beta + \frac{\Delta\rho}{\rho_w}\sqrt{1+\sin^2\beta\tan^2\alpha}\,\frac{\cos\alpha\sin\alpha}{\tan\gamma\cos\beta}\,. \tag{M.10}$$

This result is very similar to (M.6). All special cases are identical except for $\Delta\rho = 0$ which now yields the more intuitive result $\psi = \beta$. Orthogonal water flow with $\beta = 0$ is exactly the same (Fig. 6.29). A petroleum release as shown in Fig. M.2 can obviously not occur with a water flow parallel to the barrier. The inversion of the flow direction analogously as in Fig. M.3 for $\beta = -90°$ yields now the condition $\tan\gamma > \Delta\rho \sin\alpha/\rho_w$.

N

Unit Conversions and Constants

Quantity	Unit	Conversion	Unit
Distance/Depth	foot (ft)	0.3048	meter (m)
	yard (yd)	0.9144	
	mile (mi)	1609.344	
Time	million years (My,Ma)	3.1536×10^{13}	second (s)
Pressure	psi	6.89475×10^{3}	Pascal (Pa)
	bar	10^{5}	
	atmosphere (atm)	1.01325×10^{5}	
Temperature	Celsius (°C)	°C = K − 273.15	Kelvin (K)
	Fahrenheit (°F)	°F = 1.8 °C + 32	
Heatflow Unit	HFU	0.041868	W/m^2
Density	g/cm^3	1000	kg/m^3
Mudweight	pounds per gallon (ppg)	119.83	
	°API	$kg/m^3 = 141500/(131.5 + °API)$	
Mass	ton (t)	1000	kilogram (kg)
	pound (lb)	0.45359237	
Force	pound force (lbf)	4.448221615	Newton (N)
	dyne (dyn)	10^{-5}	
Volume	barrel (bbl,bo)	0.158987	m^3
	cubic foot (cf)	0.02831685	
	gallon US (gal)	0.003785412	
	acre–foot (af)	1233.482	
	liter (l)	0.001	
	barrel of oil equivalent gas (boe) *	1068.647751	
Energy	calorie (cal)	4.1868	Joule (J)
Permeability	Darcy (D)	0.98692×10^{-12}	m^2
Viscosity	Poise (P)	0.1	Pa s
Gas Oil Ratio	GOR (SCF/STB)	0.178137	m^3/m^3

* Approximately 6000 cubic feet of natural gas are equivalent to one barrel of crude oil (chevron.com/investor/annual/2006/glossary.asp, moneyterms.co.uk/boe).

Common Unit Prefixes:

tera, trillion (T,t):	10^{12}	giga, billion (G,B):	10^9	mega, million (M):	10^6
kilo (k):	10^3	hecto (h):	10^2	deci (d):	10^{-1}
centi (c):	10^{-2}	milli (m):	10^{-3}	micro (μ):	10^{-6}

Standard acceleration due to gravity: $g = 9.80665\,\mathrm{m/s^2}$
Universal gas constant: $R = 8.31447\,\mathrm{Pa\,m^3/K/mol}$

Index

Printing: Krips bv, Meppel, The Netherlands
Binding: Stürtz, Würzburg, Germany